Also in the Variorum Collected Studies Series:

PETER MEREDITH, edited by JOHN MARSHALL
The Practicalities of Early English Performance: Manuscripts, Records, and Staging
Shifting Paradigms in Early English Drama Studies

MEG TWYCROSS, edited by SARAH CARPENTER and PAMELA KING
The Materials of Early Theatre: Sources, Images, and Performance
Shifting Paradigms in Early English Drama Studies

SEYMOUR DRESCHER
Pathways from Slavery
British and Colonial Mobilizations in Global Perspective

DAVID JACOBY
Medieval Trade in the Eastern Mediterranean and Beyond

GILES CONSTABLE
Medieval Thought and Historiography

GILES CONSTABLE
Medieval Monasticism

MICHAEL J.B. ALLEN
Studies in the Platonism of Marsilio Ficino and Giovanni Pico

ALEXANDRA F. JOHNSTON, edited by DAVID N. KLAUSNER
The City and the Parish: Drama in York and Beyond

BENJAMIN Z. KEDAR
Crusaders and Franks
Studies in the History of the Crusaders and the Frankish Levant

DAVID MILLS, edited by PHILIP BUTTERWORTH
To Chester and Beyond: Meaning, Text and Context in Early English Drama
Shifting Paradigms in Early English Drama Studies

NELSON H. MINNICH
The Decrees of the Fifth Lateran Council (1512–17)
Their Legitimacy, Origins, Contents, and Implementation
www.routledge.com/history/series/VARIORUMCS

VARIORUM COLLECTED STUDIES SERIES

The Christian Epigraphy of Egypt and Nubia

Figure 0 Jacques van der Vliet and Pope Tawadros II, the Coptic Patriarch of Alexandria, at the St. Bishoy Monastery, Wadi al-Natrun, 15 February 2017 (photo: Ibrahim Saweros)

Jacques van der Vliet
edited by Renate Dekker

The Christian Epigraphy of Egypt and Nubia

LONDON AND NEW YORK

First published 2018
by Routledge
2 Park Square, Milton Park, Abingdon, Oxon OX14 4RN

and by Routledge
711 Third Avenue, New York, NY 10017

Routledge is an imprint of the Taylor & Francis Group, an informa business

© 2018 Jacques van der Vliet

The right of Jacques van der Vliet to be identified as author of this work has
been asserted by him in accordance with sections 77 and 78 of the Copyright,
Designs and Patents Act 1988.

All rights reserved. No part of this book may be reprinted or reproduced or
utilised in any form or by any electronic, mechanical, or other means, now
known or hereafter invented, including photocopying and recording, or in any
information storage or retrieval system, without permission in writing from
the publishers.

Trademark notice: Product or corporate names may be trademarks or
registered trademarks, and are used only for identification and explanation
without intent to infringe.

British Library Cataloguing-in-Publication Data
A catalogue record for this book is available from the British Library

Library of Congress Cataloging-in-Publication Data
A catalog record for this book has been requested

ISBN: 978-0-8153-5429-1 (hbk)
ISBN: 978-1-351-13347-0 (ebk)

Typeset in Times New Roman
by Apex CoVantage, LLC

VARIORUM COLLECTED STUDIES SERIES CS1070

CONTENTS

List of figures   xi
List of contributors   xv
Preface   xix
List of abbreviations   xxiii

Part 1: A general introduction

1   The Christian epigraphy of Egypt and Nubia: State of
research and perspectives   3
First published as "L'épigraphie chrétienne de l'Égypte et de Nubie:
bilan et perspectives", in: A. Boud'hors and D. Vaillancourt (eds.),
Huitième congrès international d'études coptes (Paris 2004), I, Paris:
De Boccard 2006, 303–20

Part 2: Egypt

2   "In a robe of gold": Status, magic and politics on inscribed
Christian textiles from Egypt   27
First published in: C. Fluck and G. Helmecke (eds.), *Textile messages:
Inscribed fabrics from Roman to Abbasid Egypt*, Leiden/Boston:
Brill 2006, 23–67

3   *Christus imperat*: An ignored Coptic dating formula   63
First published in: Y.N. Youssef and S. Moawad (eds.), *From Old
Cairo to the New World: Coptic studies presented to Gawdat Gabra
on the occasion of this sixty-fifth birthday*, Leuven/Paris/Walpole, MA:
Peeters 2013, 173–84

4   Perennial Hellenism! László Török and the al-Mu'allaqa lintel
(Coptic Museum inv. no. 753)   75
First published in: *ECA 4* (2007), 77–80

CONTENTS

5 History through inscriptions: Coptic epigraphy in the
Wadi al-Natrun   83
First published in: *Coptica* 3 (2004), 187–207; reprinted in: M.S.A. Mikhail and
M. Moussa (eds.), *Christianity and monasticism in the Wadi al-Natrun*, Cairo/
New York: The American University in Cairo Press 2009, 329–49

6 Reconstructing the landscape: Epigraphic sources for the
Christian Fayoum   99
First published in: Gawdat Gabra (ed.), *Christianity and Monasticism in the
Fayoum Oasis*, Cairo/New York: American University in Cairo Press 2005, 79–89

7 Monumenta fayumica   111
First published in: *Enchoria* 28 (2002–2003), 137–46 (in French, but with
the same title)

8 Monuments of Christian Sinnuris (Fayoum, Egypt)   123
Peter Grossmann, Tomasz Derda and Jacques van der Vliet
First published as "Monuments of Christian Sinnuris
(Fayyum, Egypt)" in: *ECA* 8 (2011), 29–48

9 Four Christian funerary inscriptions from the Fayoum
(I. Dayr al-'Azab 1–4)   151
Tomasz Derda and Jacques van der Vliet
First published as "Four Christian funerary inscriptions
from the Fayum (I. Deir el-cAzab 1–4)" in: *JJP* 36 (2006), 21–33

10 A lintel from the Fayoum in the British Museum   161
Jacques van der Vliet and Adeline Jeudy
First published as "Un linteau du Fayoum au British Museum" in: *ECA* 3 (2006), 73–80

11 A Naqlun monk brought home: On the provenance
of Louvre inv. E 26798–9   173
First published in: *BSAC* 39 (2000), 239–44

12 *I. Varsovie*: Graeco-Coptica   179
First published in: *JJP* 34 (2004), 121–25 (in French).

13 A Coptic funerary stela in the Montreal Museum of Fine Arts   185
Jitse Dijkstra and Jacques van der Vliet
First published as "Une stèle funéraire copte au Musée des
Beaux-Arts de Montréal" in: *CdÉ* 87 (2012), 189–96

14 Snippets from the past: Two ancient sites in the Asyut region:
Dayr al-Gabrawi and Dayr al-'Izam   193
First published in: G. Gabra and H.N. Takla (eds.), *Christianity and monasticism
in Middle Egypt: Al-Minya and Asyut*, Cairo/New York: American University in
Cairo Press 2015, 161–68

viii

CONTENTS

15 Monks and scholars in the Panopolite nome: The epigraphic
evidence   203
Sofia Schaten and Jacques van der Vliet
First published in: G. Gabra and H.N. Takla (eds.), *Christianity and Monasticism in Upper Egypt*, vol. 1: *Akhmim and Sohag*, Cairo/New York: American University in Cairo Press 2008, 131–42

16 *Parerga*: Notes on Christian inscriptions from Egypt and Nubia   215
First published in: *ZPE* 164 (2008), 153–58

17 Epigraphy and history in the Theban region   225
First published in: G. Gabra and H.N. Takla (eds.), *Christianity and Monasticism in Upper Egypt*, vol. 2: *Nag Hammadi – Esna*, Cairo/New York: American University in Cairo Press 2010, 147–55

18 From Naqada to Esna: A late Coptic inscription at Dayr Mari
Girgis (Naqada)   235
Renate Dekker and Jacques van der Vliet
First published as "From Naqada to Esna: A late Coptic inscription at Deir Mari Girgis (Naqada)" in: *ECA* 5 (2008), 37–42

19 "In year one of King Zachari": Evidence of a new Nubian king
from the Monastery of St. Simeon at Aswan   243
Jitse H.F. Dijkstra and Jacques van der Vliet
First published in: *BSF* 8 (2003), 31–39

20 Contested frontiers: Southern Egypt and Northern Nubia,
A.D. 300–1500. The evidence of the inscriptions   253
First published in: G. Gabra and H.N. Takla (eds.), *Christianity and monasticism in Aswan and Nubia*, Cairo/New York: American University in Cairo Press 2013, 63–77

Part 3: Nubia

21 Coptic as a Nubian literary language: Four theses for discussion   269
First published in: W. Godlewski and A. Łajtar (eds.), *Between the Cataracts: Proceedings of the 11th Conference for Nubian Studies, Warsaw University, 27 August – 2 September 2006, II: Session papers*, Warsaw: Warsaw University Press 2010, 765–71.

22 Gleanings from Christian Northern Nubia   279
First published in: *JJP* 32 (2002), 175–94

23 Four north-Nubian funerary stelae from the Bankes collection   295
Jacques van der Vliet and Klaas A. Worp
First published in: A. Łajtar, G. Ochała and J. van der Vliet (eds.), *Nubian Voices II: New texts and studies on Christian Nubia*, Warsaw: Raphael Taubenschlag Foundation 2015, 27–43

CONTENTS

24 Churches in Lower Nubia, old and "new"   309
First published in: *BSAC* 38 (1999), 135–42

25 Two Coptic epitaphs from Qasr Ibrim   317
First published in: *JEA* 92 (2006), 217–23

26 The Church of the Twelve Apostles: The earliest Cathedral of Faras   327
First published as "The Church of the Twelve Apostles: The earliest Cathedral of Faras?" in: *Orientalia* 68 (1999), 84–97

27 Exit Tamer, bishop of Faras (*SB* V 8728)   341
First published in: *JJP* 37 (2007), 185–91

28 Rich ladies of Meinarti and their churches. With an appended list of sources from Christian Nubia containing the expression "having the Church of so-and-so"   347
Adam Łajtar and Jacques van der Vliet
First published in: *JJP* 28 (1998), 35–53

29 From Aswan to Dongola: The epitaph of Bishop Joseph (died A.D. 668)   365
Stefan Jakobielski and Jacques van der Vliet
First published in: A. Łajtar and J. van der Vliet (eds.), *Nubian Voices: Studies in Christian Nubian Culture*, Warsaw: Raphael Taubenschlag Foundation 2011, 15–35

30 Rome – Meroe – Berlin: The southernmost Latin inscription rediscovered (*CIL* III 83)   381
Adam Łajtar and Jacques van der Vliet
First published in: *ZPE* 157 (2006), 193–98

31 "What is man?" The Nubian tradition of Coptic funerary inscriptions   389
First published in: A. Łajtar and J. van der Vliet (eds.), *Nubian Voices: Studies in Christian Nubian Culture*, Warsaw: Raphael Taubenschlag Foundation 2011, 171–224

Index   427

FIGURES

The captions of the images may deviate from those in the original publications.

0	Jacques van der Vliet and Pope Tawadros II, the Coptic Patriarch of Alexandria, at the St. Bishoy Monastery, Wadi al-Natrun, 15 February 2017 (photo: Ibrahim Saweros)	iv
2.1	Antwerp, collection Katoen Natie, inv. 711: upper line of Coptic text (photo: © The Phoebus Foundation)	45
2.2	Antwerp, collection Katoen Natie, inv. 711: lower line of Coptic text (photo: © The Phoebus Foundation)	46
2.3	Paris, Musée du Louvre, inv. E 26792: upper line of Coptic text (photo: © Claire Beugnot, 2016)	49
2.4	Paris, Musée du Louvre, inv. E 26792: lower line of Coptic text (photo: © Claire Beugnot, 2016)	49
5.1	The Church of the Virgin, Dayr al-Suryan: plan, after P. Grossmann, showing the location of some of the recently discovered inscriptions	87
7.1	Stela Allard Pierson Museum inv. no. 7804 (photo: © Allard Pierson Museum)	112
7.2	Stela Dayr Sinnuris (after the original publication)	114
8.1	Interior of the church of St. Michael in Sinnuris in 1977 (photo: Peter Grossmann)	124
8.2	Reconstruction of the medieval church of Sinnuris (drawing: Peter Grossmann)	125
8.3	Reconstruction of the vaulting system of the medieval church (drawing: Peter Grossmann)	128
8.4	Plan of the church with its extensions in 1977 (drawing: Jacek Kościuk)	129
8.5	The stone relief of Phib and Phoibammon in 1977 (photo: Peter Grossmann)	131
8.6	The stone relief of Phib and Phoibammon in 2003 (photo: Jacques van der Vliet)	132

FIGURES

8.7 The inscriptions on the altar screen: middle (photo:
J. van der Vliet) 140

8.8 The inscriptions on the altar screen: right (photo:
J. van der Vliet) 140

8.9 The inscriptions on the altar screen: left (photo:
J. van der Vliet) 141

9.1 Dayr al-'Azab 1 (photo: Jacques van der Vliet) 152

9.2 Dayr al-'Azab 2 (photo: Jacques van der Vliet) 155

9.3 Dayr al-'Azab 3 (photo: Jacques van der Vliet) 156

9.4 Dayr al-'Azab 4 (photo: Jacques van der Vliet) 157

10.1 Lintel British Museum EA 54040: overview (photo: © Courtesy
of the Trustees of the British Museum) 162

12.1 *I. Varsovie*, Appendix I, A8 (drawing: de Ricci; adapted
by Renate Dekker) 181

13.1 Coptic stela, Montreal Museum of Fine Arts, front side
(photo: J.H.F. Dijkstra) 186

13.2 Coptic stela, Montreal Museum of Fine Arts, upper side
(photo: J.H.F. Dijkstra) 187

16.1 Inscription of Heliodoros in the White Monastery (Sohag)
(photo: Jacques van der Vliet) 216

16.2 Inscription of Heliodoros from Akhmim (after Lefebvre's
facsimile) 217

18.1 Stone inscription at Dayr Mari Girgis (Naqada) (photo:
R. Dekker) 236

18.2 *Dipinto* in the church of Dayr al-Fakhuri (Esna) (photo:
R. Dekker) 238

19.1 Monastery of St. Simeon, Aswan: general view of the wall
with the Coptic *dipinto* of Peter in the center (photo:
Kathryn E. Piquette 2017) 245

19.2 Monastery of St. Simeon, Aswan: the *dipinto* of Peter
(photo: Kathryn E. Piquette 2016) 246

22.1 A tombstone for Paulos from Kalabsha (?), *meizoteros*
(*SB Kopt.* I 432) (photo: © Courtesy of the Trustees
of the British Museum) 285

25.1 Coptic stela of Martha, Qasr Ibrim (photo: © National
Museum of Antiquities, Leiden) 318

29.1 The stela of Bishop Joseph at its find-spot (photo: Wojciech
Chmiel) 366

29.2 Plan of the eastern part of the Church showing the place of
the discovery of the stela and its suggested original position
(drawing by Marek Puszkarski; updated by Marta Momot) 367

29.3 The stela of Bishop Joseph after its preliminary conservation
on site (photo: Wojciech Chmiel) 368

xii

FIGURES

29.4 The stela of Bishop Joseph (tracing by Stefan Jakobielski)   369
30.1 Latin visitor's inscription, Musawwarat al-Sufra (Nubia),
SMBK inv. no. 9675 (photo: © Skulpturensammlung und
Museum für Byzantinische Kunst)   382

CONTRIBUTORS

Author

Jacques van der Vliet, Egyptologist and Copticist, is lecturer at Leiden University and Jozef M.A. Janssen Professor of Religions of Ancient Egypt at the Radboud University, Nijmegen. His scholarly interests include the rich Coptic literature from late antiquity, including magical, gnostic and hagiographic texts, as well as documentary texts and inscriptions, which shed light on the world of the monks in late antiquity. He has a special interest in the epigraphic sources from Egypt and Nubia, and he participates in several fieldwork projects as papyrologist and epigrapher.

Editors

Renate Dekker, co-author of Study 18 and main editor of this volume, is a former Ph.D student of van der Vliet at Leiden University. In November 2017, she defended her Ph.D dissertation, *Episcopal Networks and Authority in Late Antique Egypt: Bishops of the Theban Region at Work*, which will be published under the same title (Leuven: Peeters, 2018).

Clara ten Hacken is a former Ph.D student of van der Vliet. She defended her Ph.D dissertation, *The Legend of Saint Aūr and the monastery of Naqlūn: the Copto-Arabic texts* (to be published in Leuven by Peeters), in December 2015, and wrote articles on the Archangel Michael. She assisted during the edition of this volume.

Joost Hagen is Ph.D student of van der Vliet, working on the Coptic manuscript collection of the Ägyptisches Museum und Papyrussammlung, Staatliche Museen zu Berlin. He is writing a Ph.D dissertation about the Coptic literary texts that were found during the excavations at Qasr Ibrim, Nubia in the 1960–1970s. He assisted during the edition of this volume.

Maher Eissa is a former Ph.D student of van der Vliet and Associate Professor of Egyptology/Coptology at Fayoum University. He is currently the Head of the

xv

CONTRIBUTORS

Egyptology Department at Fayoum University and visiting scholar at Leiden University. Many of his articles focus on the Coptic language and on Coptic texts (ostraca, stelae) that are kept in the different museums of Cairo. He assisted during the edition of this volume.

Co-authors

Tomasz Derda, co-author of Studies 8 and 9. Trained as an ancient historian and Greek papyrologist, he is Associate Professor and head of the department of Papyrology at the Institute of Archaeology of the University of Warsaw and one of the chief editors of *The Journal of Juristic Papyrology*.

Jitse H.F. Dijkstra, co-author of Studies 13 and 19. He is Associate Professor of Classics at the University of Ottawa and a member of the joint Swiss-Egyptian archaeological mission at Aswan since 2001. He has published widely on Late Antique Egypt and is the author of *Philae and the End of Ancient Egyptian Religion* (Leuven: Peeters, 2008).

Peter Grossmann, co-author of Study 8. He is an architect and archaeologist specialized in the Christian architecture of late antique and medieval Egypt. Between 1969 and 2005 he directed excavations at the sites of Abu Mina, the Monastery of Apa Jeremiah in Saqqara, Makhura, Luxor, Pharan, Faw Qibli and Hermopolis Magna.

Stefan Jakobielski, co-author of Study 29. He is a Nubiologist, historian, archaeologist, philologist and epigraphist formerly appointed at the Polish Centre of Mediterranean Archeology and at the Cardinal Stefan Wyszynski University in Warsaw. Between 1966 and 2006 he directed the excavations at the Nubian site of Dongola.

Adeline Jeudy, co-author of Study 10. She specialized in Eastern Christian art at Leiden University and worked on her Ph.D dissertation at the l'Institut français d'archéologie orientale in Cairo. Presently, she runs a gallery for urban art in Paris.

Adam Łajtar, co-author of Studies 28 and 30. Being trained as a historian, archaeologist, (Greek) philologist and epigraphist, and Nubiologist, he is professor at the department of papyrology at the Institute of Archaeology of the University of Warsaw and one of the chief editors of *The Journal of Juristic Papyrology*.

Sofia Schaten, co-author of Study 15. She studied Coptology at the University of Münster. She is a member of the German/Swiss archaeological mission in Elephantine and published several studies on Coptic funerary stelae.

xvi

CONTRIBUTORS

Klaas A. Worp, co-author of Study 23. He is a Greek papyrologist formerly employed at the University of Amsterdam and Professor by special appointment at Leiden University at the Leiden Papyrological Institute. He wrote a large number of articles and books on Greek papyri, chronological issues, Greek shorthand, papyri from Kellis and many other topics.

PREFACE

Prof. Dr. Jacques van der Vliet, lecturer in Egyptology and Coptic studies at Leiden University and endowed professor at Radboud University, Nijmegen, is the (co-)author of the thirty-one studies collected in this volume on the Christian epigraphy of Egypt and Nubia. We – four of his Ph.D-students – edited the book as a present for him on the occasion of his retirement in 2018. Now that these studies have been reprinted by Routledge in the *Variorum Collected Studies Series*, we hope that his research is more accessible for scholars worldwide.

In the studies collected here, written in the period 1998–2015, Jacques does not simply edit (mostly Coptic and Greek) inscriptions from Egypt or Nubia for the sake of factual data, but he also examines them as texts in their own right, with particular aims. He demonstrates that they may convey views inspired by literary or liturgical traditions (e.g. Studies 29, 31); reflect the wish of Christians under Muslim rule to have a Christian sovereign (Studies 3 and 19); or belonged to a ritual and monumental setting that involved praying and singing together, which enhanced social cohesion (Study 31). In addition, Jacques redefines the domain of "Coptic epigraphy" and proposes a new approach to the Christian inscriptions from the Nile Valley that will do justice to their diversity at three levels (Study 1):

- at the level of *languages*, by ending the artificial linguistic segregation between Greek and Coptic, which are usually studied separately, and by also including Arabic, Syriac, Old Nubian and Latin, since some of these languages occasionally appear in the same texts or in the same space as Coptic and Greek texts;
- in terms of *sources*, by including a variety of inscriptions, such as epitaphs on funerary stelae; graffiti, *dipinti* and legends that clarify wall paintings; dedicatory texts on tools, vessels, furniture and buildings; magical texts on amulets; and prayers or Psalms on "Christian *tirâz*", or non-liturgical luxury garments (Study 2).
- and finally, in terms of *sites*, by mapping regional traditions, and by combining data from epigraphy, papyrology, archaeology and historiography, in order to reconstruct local history (especially Studies 5–6, 14–15, 17, 20).

The inscriptions discussed in the book range in date from the third/fourth to the fourteenth centuries (see Studies 30 and 18 respectively).

In the initial stage of this project, Jacques himself selected thirty studies (no. 28 was added later at the suggestion of a helpful colleague), divided them in three parts (a general introduction – Egypt – Nubia, in which sites are arranged from north to south), and checked the translations of studies that originally appeared in French (nos. 1, 7, 10, 12–13). The editorial work was done by Renate Dekker, with the assistance of Clara ten Hacken, Joost Hagen and Maher Eissa. Jitse H.F. Dijkstra, the co-author of Studies 13 and 19, edited these two articles and adapted the photograph published as Figure 19.1, whereas Adam Łajtar edited Study 28.

The selected studies were composed during a period of seventeen years and their original texts display different editorial styles. For the sake of cohesion, the editors adopted the most common spelling of personal names and toponyms (e.g. Dayr, Fayoum, Naqlun without circumflex), which resulted in slight adaptations of (*inter alia*) a few study titles.

Bibliographical references were standardized but the original reference systems – author/date and author/title – were left intact, since neither system was fully satisfactory for the entire book. We preferred to keep the articles in their original form as much as possible.

The Greek texts were retyped in IFAO-Grec Unicode and the Coptic and Old Nubian were typeset using Antinoou. At Jacques' request, abbreviations were resolved, with two exceptions: when it was the intention of the author(s) to reproduce a scholar's transcription, or when the adaptations would result in a considerably different text (e.g. Study 28). In such cases, abbreviations are best explained in the translation or commentary, as in the original publications. The editors added updates and cross-references between square brackets, made some corrections, when necessary, and removed Greek words from translations in accordance current editorial practices (unless Greek words highlight the choice for particular formulae, as in Study 31).

A number of relevant articles could not be included, on account of their very recent date, or the limited size of this volume, particularly:

- J. van der Vliet, "Literature, liturgy, magic: a dynamic continuum", in: P. Buzi and A. Camplani (eds.), *Christianity in Egypt: literary production and intellectual trends. Studies in honor of Tito Orlandi*, Roma: Institutum Patristicum Augustinianum, 2011, 555–74.
- "Epigraphy: Coptic", in: P.C. Finney (ed.), *The Eerdmans encyclopedia of early Christian art and archaeology* 1, Grand Rapids, Michigan: Wm. B. Eerdmans Publishing Company, 2017, 481–84.
- "A fifth Nubian funerary stela from the Bankes collection: an addendum to *CIEN* 3, 26–9" with Klaas A. Worp (a complement to Study 23), which will appear in *JJP*.

PREFACE

- "The wisdom of the wall: innovation in monastic epigraphy", in: M. Choat and M. Giorda (eds.), *Writing and communication in early Egyptian monasticism*, Leiden/Boston: Brill, 2017, 151–64.

This volume took shape thanks to the support and permissions given by a large number of people, including Prof. Dr. Olaf E. Kaper, professor of Egyptology at Leiden University; the co-authors Jitse H.F. Dijkstra, Adam Łajtar, Tomasz Derda, Peter Grossmann, Adeline Jeudy, Sofia Schaten, Stefan Jakobielski and Klaas A. Worp; the editors of the original publications, Anne Boud'hors, Cäcilia Fluck (also for sharing a pdf of Study 2), Gisela Helmecke, Youhanna N. Youssef, S. Moawad, Maged S.A. Mikhail, Mark Moussa, Gawdat Gabra, Hany N. Takla and Włodimierz Godlewski, the editors of journals, Michael Zach (*BSF*), Wassif Boutros Ghali, director of the Society of Coptic Archaeology (*BSAC*), Alain Delattre, on behalf of the Association Égyptologique Reine Elisabeth, Brussels (*CdÉ*), Bas ter Haar Romeny, Karel Innemée and Mat Immerzeel (*ECA*), Jan Geishbusch (*JEA*), Jakub Urbanik and Grzegorz Ochała (*JJP*, together with Łajtar and Derda), Iwona Zych of the Polish Centre of Mediterranean Archaeology, Warsaw (*PAM*), Werner Eck (*ZPE*); Nigel Fletcher-Jones and Doug Wallace of the American University in Cairo Press; Laura Westbrook of Koninklijke Brill NV; Daniela Talarico of the Gregorian and Biblical Press, Rome (publisher of *Orientalia*); Andrea Johari of Harrassowitz Verlag, Wiesbaden (publisher of *Enchoria*); Paul Peeters of Peeters Publishers, Leuven; Judith Gonis, director of The Phoebus Foundation, which presently owns the Katoen Natie (Figures 2.1–2.2); Florence Calament, curator at the Musée du Louvre (Figures 2.3–2.4); Willem van Haarlem, curator at the Allard Pierson Museum, Amsterdam (Figure 7.1); Kathryn E. Piquette and Lena Krastel, members of the Deir Anba Hadra project (Figures 19.1–19.2); Elisabeth O'Connell and the Trustees of the British Museum (Figures 10.1, 22.1); Robbert Jan Looman, photographer at the National Museum of Antiquities, Leiden (Figure 25); and Lucas Bosman, who checked the English translation of Study 13.

Together with Professor Kaper, we present this volume to Jacques on behalf of the entire department of Egyptology at Leiden University, his colleagues worldwide and his students, with our best wishes.

RENATE DEKKER, CLARA TEN HACKEN, JOOST HAGEN AND MAHER EISSA
Leiden University

ABBREVIATIONS

AfP	*Archiv für Papyrusforschung*
ArtB	*The Art Bulletin*
ASAE	*Annales du Service des antiquités de l'Égypte*
BASP	*Bulletin of the American Society of Papyrologists*
BCH	*Bulletin de correspondance hellénique*
BIFAO	*Bulletin de l'Institut français d'archéologie orientale*
BiOr	*Bibliotheca orientalis*
BSAC	*Bulletin de la Société d'archéologie copte*
BSF	*Beiträge zur Sudanforschung*
CAVT	J.-C. Haelewyck (ed.), *Clavis apocryphorum Veteris Testamenti*, Turnhout: Brepols 1998
CANT	M. Geerard (ed.), *Clavis apocryphorum Novi Testamenti*, Turnhout: Brepols 1992
CArch	*Cahiers archéologiques*
CCO	Collectanea Christiana Orientalia
CdÉ	*Chronique d'Égypte*
CIG	A. Boekh (ed.), *Corpus inscriptionum graecarum*, 4 vols., Berlin: G. Reimerum 1828–1877
CIL	T. Mommsen et al. (eds.), *Corpus inscriptionum latinarum*, 17 vols. (to be continued), Berlin: G. Reimerum 1853–
CGC	Catalogue général du Musée du Caire
CoptEnc	A.S. Atiya (ed.), *The Coptic encyclopedia*, 8 vols., New York: MacMillan 1991
CPG	M. Geerard (ed.), *Clavis patrum graecorum*, Turnhout: Brepols 1974–1989
CRIPEL	*Cahiers de recherches de l'Institut de papyrologie et d'égyptologie de Lille*
DACL	F. Cabrol and H. Leclercq (eds.), *Dictionnaire d'archéologie chrétienne et de liturgie*, 30 vols., Paris: Letouzey et Ané 1907–1953
DOP	*Dumbarton Oaks Papers*
ECA	*Eastern Christian Art*

ABBREVIATIONS

ÉtudTrav	*Études et travaux*
FHN III	T. Eide, T. Hägg, R.H. Pierce, and L. Török (eds.), *Fontes historiae Nubiorum* III, Bergen: University of Bergen 1998
GM	*Göttinger Miszellen*
IGBulg.	G. Mihailov (ed.), *Inscriptiones graecae in Bulgaria repertae*, Sofia: Academia Litteratum Bulgarica 1956 (1970²)– 1966
I. Khartoum Copt.	J. van der Vliet, *Catalogue of the Coptic inscriptions in the Sudan National Museum at Khartoum (I. Khartoum Copt.)*, Louvain, etc.: Peeters 2003
I. Khartoum Greek	A. Łajtar, *Catalogue of the Greek inscriptions in the Sudan National Museum at Khartoum (I. Khartoum Greek)*, Louvain, etc.: Peeters 2003
I. Qasr Ibrim	A. Łajtar and J. van der Vliet, *Qasr Ibrim: the Greek and Coptic inscriptions*, Warsaw: University of Warsaw/Taubenschlag Foundation 2011
I. Varsovie	A. Łajtar and A. Twardecki (eds.), *Catalogue des inscriptions grecques du Musée National de Varsovie*, Warsaw: University of Warsaw/Taubenschlag Foundation 2003
JARCE	*Journal of the American Research Center in Egypt*
JbAC	*Jahrbuch für Antike und Christentum*
JCoptS	*Journal of Coptic Studies*
JDAI	*Jahrbuch des Archäologischen Instituts*
JEA	*Journal of Egyptian Archaeology*
JJP	*The Journal of Juristic Papyrology*
JRS	*Journal of Roman Studies*
LAAA	*Liverpool Annals of Archaeology and Anthropology*
MDAIK	*Mitteilungen des Deutschen Instituts für ägyptische Altertumskunde in Kairo*
PAM	*Polish Archaeology in the Mediterranean*
PBA	*Proceedings of the British Academy*
P.KRU	W.E. Crum and G. Steindorff, 1912. *Koptische Rechtsurkunden des achten Jahrhunderts aus Djême (Theben)*, Leipzig: J.C. Hinrichs
P.Lond.Copt. I	W.E. Crum, *Catalogue of the Coptic manuscripts in the British Museum*, London: British Museum 1905
P.Lond.Copt. II	B. Layton, *Catalogue of the Coptic literary manuscripts in the British Library acquired since the year 1906*, London: British Library 1987
OCP	*Orientalia christiana periodica*
RAC	*Reallexikon für Antike und Christentum*, 26 vols., Stuttgart 1950–2015
RdTr	*Recueil de travaux relatifs à la philologie et à l'archéologie égyptiennes et assyriennes*

ABBREVIATIONS

REG	*Revue d'égyptologie*
RevÉtByz	*Revue des études byzantines*
RivArchCrist	*Rivista di archeologia cristiana*
SB	*Sammelbuch griechischer Urkunden aus Ägypten*, Strasbourg: K.J. Trübner (presently, Wiesbaden: Harrassowitz), 1915–
SB Kopt. I–III	M.R.M. Hasitzka (ed.), *Koptisches Sammelbuch*, Vienna: Hollinek 1993–2004; Munich/Leipzig: K.G. Saur 2006
SEG	J.E. Hondius et al. (eds), *Supplementum epigraphicum graecum*, Leiden: A.W. Sijthoff (continued in Amsterdam: J.C. Gieben, then Leiden: Brill), 1923–
SNR	*Sudan Notes and Records*
SOC	*Studia orientalia christiana*
Studia Pontica III	J.G.C. Anderson, F. Cumont, and H. Grégoire, *Studia Pontica* III: *recueil des inscriptions grecques et latines du Pont et de l'Arménie*, Brussels: H. Lamertin 1910
WB	F. Preisigke, *Wörterbuch der griechischen Papyrusurkunden*, 3 vols., Berlin: privately published 1925–1931
WZKM	*Wiener Zeitschrift zur Kunde des Morgenlandes*
ZÄS	*Zeitschrift für ägyptische Sprache*
ZPE	*Zeitschrift für Papyrologie und Epigraphik*

Part 1

A general introduction

1

THE CHRISTIAN EPIGRAPHY
OF EGYPT AND NUBIA
State of research and perspectives[1]

Jacques van der Vliet

Introduction

The present keynote paper is – as far as I am aware – the first one in the history of the congresses of the International Association for Coptic Studies to be exclusively dedicated to epigraphy, the study of inscriptions. For this reason, I will not limit myself to a simple enumeration of publications that have appeared in recent years and the discussion of their merits or possible shortcomings, but will rather present some general and methodological reflections. These reflections will focus first on the specificity of the field of research: what constitutes the domain of Coptic epigraphy and what are the main historical factors that determined its shape? Next, I will briefly attempt to outline the discipline and its developments in the past century and to judge its achievements until present. Finally, I will indicate which direction the study of the Christian inscriptions from Egypt and Nubia could and – in my opinion – should take.

I

Although this paper represents an innovation in the context of the Coptic congresses, the epigraphy of Christian Egypt has nonetheless been the subject of synthetic essays at several times in the past decades. In 1986, Kent Brown published what he called "a brief review" of Coptic and Greek inscriptions from Christian Egypt in an American volume on the origins of Egyptian Christianity.[2] Then, in 1991, in the fourth volume of the *Coptic encyclopedia*, Martin Krause devoted an important article to the same subject.[3] Most recently, during the 11th International Congress of Greek and Latin Epigraphy, held in Rome in 1996, Jean Bingen presented a paper on "the Greek epigraphy of post-Constantinian Egypt", which addresses several major problems of our discipline.[4] In the following pages I will take a stand vis-à-vis these and other authors, yet without mentioning them by name each time.

The titles that I just quoted already indicate one of the main problems of our sources, that is, their linguistic diversity. Coptic epigraphy is "Coptic" only to a certain extent: after the extinction of the hieroglyphic writing tradition, which was linked to the traditional religion, the public space of Christian Egypt was almost exclusively Greek, until a time that is difficult to specify with certainty.[5] In the fourth and fifth centuries, it is unlikely that the Coptic language was ever used for monumental inscriptions. When Coptic joins Greek in this domain, probably from the sixth century onwards, the epigraphic dossier from Christian Egypt remains deeply marked by the coexistence of the two languages, until the time when the epigraphic tradition inherited from antiquity itself petered out, around the eleventh-twelfth centuries. In Nubia, Greek still remained in use for some time after the disappearance of Sahidic Coptic, which occurred towards the end of the eleventh century. For both countries, the epigraphic dossier is therefore essentially bilingual.

This bilingualism is a well-known fact. By way of example, one has only to mention the funerary stelae from the Fayoum, which use two different languages, Greek and Fayoumic Coptic, while sharing the same artistic idiom that is strongly marked by Hellenism, or the epitaphs from southern Egypt and Nubia, where the language switches within a single text, sometimes several times. Even in the monasteries, which are considered to be the vital centers of the Coptic language, bilingualism is the rule, as is shown by the *dipinti* and graffiti discovered on the walls in Kellia.

More rarely, during the Middle Ages, other languages than Coptic and Greek made their appearance in the Christian epigraphy of Egypt. Among them, Arabic comes first to mind, of course, but the other languages of the Christian East have also left their traces. For instance, work in recent years at Dayr al-Suryan brought to light a significant group of wall inscriptions in Syriac.[6] In this case, the presence of Syriac is all but incidental. The coappearance of Syriac and Bohairic Coptic texts (and more rarely Greek ones) in the single space of the Church of the Virgin in this monastery testifies to a living multilingualism, where the selection of the language was dictated by the communicative needs of the monastic community.[7]

If it is therefore difficult to define so-called Coptic epigraphy from a linguistic point of view, it is also difficult to define it from the point of view of genre. The epigraphic culture of classical antiquity was dominated by monumental inscriptions in stone, which were directly linked to the public life of urban communities. This epigraphic tradition has always been weak in Egypt, both before and after Constantine.[8] As a result of the Arab conquest, public epigraphy became, with some exceptions, linguistically Arabic and religiously Muslim. The principal domain where epigraphy in stone survived is that of funerary inscriptions. On the other hand, over the centuries Christian epigraphic culture had expanded and diversified both with regard to the form and the function of inscriptions. In Byzantine Egypt, as elsewhere in the Christian Empire, the walls and floors of churches and monasteries became the bearers of various kinds of inscriptions that were directly related to the function and decoration of Christian sacred spaces. I only need to recall the legends and dedication inscriptions that accompany paintings or mosaics, or the graffiti and *dipinti* left by pilgrims. Likewise, the tools, vessels and furniture used in churches and monasteries, and not just those intented for the liturgy, often bore inscriptions. Many of these "Christian epigraphies" are intimately linked to ecclesiastical circles, whereas others have a more secular character, for example in the case of amulets and dress. Following the Arab conquest, the Muslim fashion of the so-called *tirâz*, long inscriptions woven into luxury garments, was adopted and imitated also by the Christians of Egypt. That is why we have a small corpus of "Christian *tirâz*", costly non-liturgical garments, the decoration of which includes texts in Greek, Coptic or Arabic.[9]

To sum up, the epigraphy of Christian Egypt was diversified in several respects: with regard to language use as well as to bearers and function. We can add a third source of diversity. This is the well-known fact of regional or even local variation in the domain of funerary epigraphy. In late antiquity, the main regions of Egypt were distinguished by a marked preference for certain textual formulas, in Greek or Coptic, and for certain types of decoration. Thus, there are particular types of funerary stelae that are characteristic for the Fayoum, Middle-Egypt, Esna, Nubia or even certain monastic centers, such as the monasteries of Anba Hadra in Aswan or Apa Jeremiah in Saqqara. An often very rich documentation shows how local schools and traditions survived for many generations after the Arab conquest.

The triple diversity that I have just sketched is, so to speak, interior: it is inherent in the epigraphic production itself. Another distinctive feature of the Christian epigraphy of Egypt rather concerns its documentary context. It is part of a much broader documentation that mainly belongs to the domain of papyrology, but also of archaeology, art history and literature. There is hardly need here to emphasize the natural links in the Nile Valley that connect epigraphy and papyrology on the one hand, and epigraphy and archaeology on the other. In no other country in the Mediterranean is it possible to reconstruct the economic, social and religious life of a monastery simultaneously on the basis of original documents on papyrus, important architectural and artistic remains, and inscriptions, as is the case for example in Bawit. In the First Cataract region, to mention another example, the situation may be different because of the nature and distribution of the historical sources, yet their variety is no less great. The epigraphy of Christian Egypt is marked at the same time by diversity of form and contents, and abundance of contextual documentation.

II

The study of the Christian epigraphy of Egypt has its roots in the early nineteenth century, but fully developed only in the twentieth century. As regards its present state, we may deplore – as others did before me – the lack of handbooks as well as bibliographical and critical tools, since the *Supplementum epigraphicum graecum*, for instance, is limited to Greek texts and the pre-Islamic period. Given this state of affairs, access to the sources is not easy, which is particularly true for the Coptic material, due to the great dispersion of the publications. The well-known *Recueil* by Gustave Lefebvre is still useful, but dates back to a century ago and, worse still, is limited to Greek texts.[10] On the other hand, the *Koptisches Sammelbuch* deals with Coptic material only and publishes – so it seems – what happens to fall into the hands of the editor, without justifying the principles that guided the selection and edition of the texts, if there were any.[11] Nobody can dispute the usefulness of this collection, however heterogeneous it may be, but one must regret the decision to adopt mechanically the very nineteenth-century concept of the Greek *Sammelbuch*, instead of critically rethinking the objectives and methods of such an undertaking. Another negative factor concerns the great collections in Egypt. As is well known, the collections of the principal Egyptian museums, especially those of Cairo and Alexandria, remain poorly accessible and poorly published.[12]

If this picture appears rather dark at first glance, there is no lack of light. Already at an early stage, the great local variety in the field of funerary epigraphy, which gave rise to specific forms of texts and decoration, attracted the interests of researchers. This resulted in synthetic studies that, although they date from the first half of the last century, have often not been replaced. They include publications by Lefebvre, particularly the seminal introduction to his *Recueil*,[13] entries by Henri Leclercq et Alexis Mallon in the *Dictionnaire d'archéologie chrétienne et de liturgie*, groundbreaking at the time,[14] and the well-known study by Dora Zuntz,[15] followed by Maria Cramer in several of her works. For the interbellum, the best publication in this genre undoubtedly remains Hermann Junker's study on the funerary inscriptions from Nubia.[16] He goes beyond establishing local dossiers, to offer a thorough analysis of the Greek and Coptic textual formulas, their mutual relations and their position with regard to the Egyptian sources. This study has rightly remained a classic. A second publication that still deserves to be quoted with distinction after so many years is Maria Cramer's study of a group of Coptic epitaphs from the region of Antinoopolis.[17] These are characterized by a specific literary form, after which they retained the name of "funerary laments" ("Totenklage"). Other important regional ensembles, each containing hundreds of funerary stelae, were published by Henri Munier (Monastery of Anba Hadra, Aswan) and Togo Mina (necropolis of Toshka West in Sakinya, northern Nubia). These groups allow us to follow the development of local textual formulas for several centuries.[18]

Particular mention must be made here of those publications that situate the epigraphic material in a wider context, for example by integrating it into the ensemble of archaeological or textual documents relating to a particular monastic complex. Thus, at the beginning of the twentieth century, excavations at the monastic complexes of Bawit and Saqqara yielded important collections of inscriptions.[19] Although, in the cases mentioned, the excavators limited themselves to the more or less felicitous publication of the texts, these sources have the advantage of being found *in situ* and being part of a wider documentation. Owing to these circumstances, the inscriptions can be restored to the history of a particular site or a particular community. Thus, epigraphy plays a modest but not unimportant role in Hugh Evelyn White's masterful synthetic study of the monasteries in the Wadi al-Natrun.[20]

In the second half of the twentieth century, it was precisely by collecting and studying local and regional dossiers that the most notable progress was made. In this regard, mention should be made of the regional and thematic dossiers assembled by André and Étienne Bernand, whose concept can serve as a model, even though Christian texts occupy only a marginal place in most of their publications. Among these, however, the second volume of inscriptions from Philae, published in 1969 by Étienne Bernand, deserves to be singled out as a major exception.[21] One of the great successes of Nubian epigraphy in the second half of the twentieth century likewise concerns a particular site. In the third volume of the series devoted to the Polish excavations at Faras, Stefan Jakobielski did not merely collect and publish the newly discovered Coptic inscriptions, but in scrupulously sorting the Coptic and Greek epigraphic documentation as well as archaeological reports, he succeeded in bringing this ancient capital of Nobadia to life.[22] The harvest is richer for the many monastic sites of the Nile Valley than for urban sites. As early as the 1950s, Serge Sauneron, later assisted by René-Georges Coquin, initiated an ambitious project aimed at assembling a dossier as complete as possible on the history of Esna. Despite the absence of a final synthesis and the bibliographic dispersion of materials, Esna is now one of the best-documented Christian centers in Egypt.[23] Even though the excavations at the Monastery of Saint Jeremiah in Saqqara, carried out at the beginning of the last century, had made a significant number of inscriptions available, we had to wait until the 1990s for someone trying to make use of them. The book by Cäcilia Wietheger (now Cäcilia Fluck), published in 1992, has the great merit of integrating the epigraphic sources into the full documentation concerning this monastic center.[24] The more recent excavations at Kellia, another major monastic site, have revealed large numbers of wall inscriptions.[25] Since their discovery, they have proved to be a rich source not only for the history of monasticism in Lower Egypt, but for the study of the Bohairic language and the history of local piety as well. Thanks to the efforts of a team of scholars, a wide and varied series has already been made available. At present, the systematic assembling of these inscriptions is a priority research task.[26] The projects and publications that I have just mentioned, to which others could be added, highlight the richness of local and regional epigraphic ensembles. This richness is not primarily due to the quality of the epigraphic sources, but rather to their capacity of integrating a wider documentation, which allows the study of major social, political and religious developments from close-by, in the intimacy of the towns and monasteries of the Egyptian countryside. Thus, the inscriptions of Philae and its neighborhood occupy a central place in the study of the Christianization of southern Upper Egypt and Nubia.[27] Similarly, a more rigorous analysis of the abundant monastic documentation will bring out its importance for the history of the institutions and mentalities of Egyptian monasticism.[28]

III

From the previous remarks, on the specificity of the Christian epigraphy of Egypt and on the nature of the discipline and its achievements in the last century, the core of a program already emerges. The final paragraphs of this paper venture to sketch the outlines of what – in my opinion – this program should more or less look like. As a model for future research, I propose an approach to the Christian epigraphy of the Nile Valley that is aimed at integration at three different levels: integration in terms of *languages*, integration in terms of *sources*, and finally, integration in terms of *sites*.

Let us return to the question of languages first. Given the multilingualism of Egyptian society, dossiers should no longer be organized according to the language of the sources, but preferably according to other criteria, such as topographical or thematic ones. Instead of perpetuating segregation, the integration of linguistic dossiers is in order. In this respect, it is surprising to still hear calls urging for a remake of the *Recueil des inscriptions grecques-chrétiennes* by Gustave Lefebvre, published in 1907. Such an undertaking seems hardly desirable. Particularly in the case of funerary material, but also for other kinds of inscriptions, the segregation of Greek and Coptic is completely artificial. Even if each obviously has his or her own particular competence, specialists of Greek and Coptic should no longer be allowed to confine themselves to their respective linguistic domains. Here I allow myself a word of criticism about one of my own publications. It is precisely the absence of the integration proposed here that I regret in the two volumes of the inscriptions from Khartoum, published by Adam Łajtar and myself.[29] Instead of integrating the Greek and the Coptic inscriptions, often originating from the same site and the same social milieu, practical and academic considerations have led the authors to split up what essentially constitutes a single bilingual dossier.

The second form of integration that I propose here concerns the sources themselves. Instead of constituting a homogeneous corpus, they are diversified in several respects. They show great variation, not only in language, but also – as we pointed out above – in genre and bearers. In Christian Egypt, monumental inscriptions on stone are no longer the only expression of epigraphic culture. Here again, the authors of future manuals should accept that some distinctions are artificial. We find, for example, the same textual formulas on amulets and clothes as in the monumental inscriptions that mark the construction of a house of the dedication of a wall painting.[30] To emphasize the diversity of the bearers would obscure the unity of the human habitat, where each of these inscriptions occupied its proper place. From this point of view, the entire corpus of inscriptions is, so to speak, an open corpus. To give one example, eloquent in its simplicity: the Church of the Virgin at Bulgusuq, a small town in the southeast of the Fayoum, is only attested by three textual sources. One of them is an inscription, not on stone, but on a liturgical vessel,[31] whereas the two other ones are non-epigraphic. The second source is the colophon of a literary manuscript on parchment from the ninth or tenth century;[32] the third one a letter on paper that may date from the tenth or eleventh century.[33] In order to situate the Church of the Virgin at Bulgusuq in the Christian history of the Fayoum, no one would be satisfied with the single epigraphic attestation to the exclusion of the other two non-epigraphic ones. Here again, distinctions prove to be artificial. Only by integrating the epigraphic data into the entire body of available documentation they can become meaningful.[34] In

Christian Egypt, the epigraphist who confines himself to the domain of inscriptions or, even worse, inscriptions on stone, handicaps himself.

The third level at which integration is in order is that of sites. The inscription differs from the literary text in the special relations that connect it, in most cases, to a particular spatial setting. The fact of being localized gives the inscription its full historical interest. It follows that, on the level of the analysis, epigraphic sources should be restored to their original setting as much as possible, instead of being treated as isolated artifacts. This original setting may be a church, a cemetery or a monastic center, a particular city or even a wider region. The work of assembling local and regional dossiers, which started a hundred years ago, should be continued and broadened in the next century, but preferably in a more synthetic way. It does not suffice to assign a particular inscription to a particular provenance. Simultaneously, the epigraphic artifact should be integrated into the entire body of historical sources that constitute the local dossier. Synthetic studies organized around local and regional ensembles that envisage the dual restoration (topographically and in terms of documentation) of the artifact to its original context will enable us eventually to reconstruct the different landscapes that made up Christian Egypt in the different periods of its history.[35]

In this respect, I would like to point out an endeavor that, in my view, can serve as a model. The biannual conferences organized *in situ* by Coptic scholarly societies, initiated by the Saint Mark Foundation for Coptic History, have set themselves the task of writing the history of the centers of Egyptian monasticism. Two of these conferences have already taken place, the first in the Wadi al-Natrun and the second one in this very year, 2004, in the Fayoum.[36] Their objective is to assemble multidisciplinary dossiers around each of these monastic centers, bringing together an extremely varied documentation. Fortunately, an important place in this endeavor is given to the epigraphic sources.

If I had to make the *Sammelbuch* that I dream of, it would rather consist of collections that bring together epigraphic material, of course, in several languages, but also papyrological, archaeological, literary and artistic material that relates to a specific region or city, to a particular village or monastery. Each inscription originates in a landscape and needs to be restored to this landscape, which is neither a piece of land nor the corner of a street, but a moral landscape in which man has created himself a world.

Notes

1   I would like to thank Harry Blonk and Ewa D. Zakrzewska, my wife, for their valuable advice. In the following, the footnotes are reduced to a minimum; for recent literature, see the bibliography for the years 2000–2004 in the appendix. Renate Dekker kindly undertook the uneasy task of translating my original French text.

2   S.K. Brown, "Coptic and Greek inscriptions from Christian Egypt: a brief review", in: B.A. Pearson and J.E. Goehring (eds.), *The roots of Egyptian Christianity*, Philadelphia: Fortress Press 1986, 26–41.

3   M. Krause, "Inscriptions", in: *CoptEnc* 4, 1290–99.

4   J. Bingen, "L'épigraphie grecque de l'Égypte post-constantinienne", in: *XI Congresso internazionale di epigrafia greca e latina, Roma, 18–24 settembre 1997. Atti* II, Rome: Edizioni Quasar 1999, 613–24.
5   Cf. Bingen, "L'épigraphie grecque", 613–14.
6   See the bibliography below, under 5.1.
7   For Dayr al-Suryan, see the observations made by L. van Rompay, especially in "Deir al-Surian: Miscellaneous reflections", and J. van der Vliet, "History through inscriptions: Coptic epigraphy in the Wadi al-Natrun", *Coptica* 3 (2004), 195 [Study 5] (both are listed in the bibliography, under 5.1). For Old Nubian, see A. Łajtar, "Wall inscriptions in the Banganarti churches: a general note after three seasons of work", *JJP* 33 (2003) (in the bibliography, under 5.2).
8   Cf. Bingen, "L'épigraphie grecque", 615.
9   The renewed study of inscribed "Coptic" fabrics is a recent phenomenon; see the exhibition catalogues *Égyptes . . . L'égyptien et le copte*, ed. N. Bosson and S.H. Aufrère, Lattes: Musée archéologique Henri Prades 1999 (in particular the contribution by D. Bénazeth, M. Durand, and M.-H. Rutschowscaya, "Le rapport de l'objet copte à l'écrit: des textiles aux vêtements liturgiques", 145–54), and *Égypte, la trame de l'histoire* (see the bibliography, under 3). A conference dedicated to this topic ("Textiles and inscriptions from Egypt of the first millennium A.D.") took place in Berlin in January 2003. Its proceedings are published in C. Fluck and G. Helmecke (eds.), *Textile messages: inscribed fabrics from Roman to Abbasid Egypt*, Leiden/Boston: Brill 2006.
10 G. Lefebvre, *Recueil des inscriptions grecques-chrétiennes d'Égypte*, Cairo: IFAO 1907; reprinted Chicago: Ares 1978.
11 M.R.M. Hasitzka, *Koptisches Sammelbuch* I, Vienna: Hollinek 1993 (*SB Kopt.* I); a second volume appeared shortly after the Paris congress.
12 For the Coptic Museum in Cairo, the results of the work of our colleague Sofia Schaten are awaited.
13 Lefebvre, *Recueil*, XXVI–XXVIII.
14 A. Mallon, art. "Copte (Épigraphie)", in *DACL* 3/2 (1914), col. 2819–86; H. Leclercq, art. "Égypte", in *DACL* 4/2 (1921), col. 2401–2571 at 2486–2521 ("Épigraphie").
15 D. Zuntz, "Koptische Grabstelen: Ihre zeitliche und örtliche Einordnung", *MDAIK* 2 (1932), 22–38.
16 H. Junker, "Die christlichen Grabsteine Nubiens", *ZÄS* 60 (1925), 111–48.
17 M. Cramer, *Die Totenklage bei den Kopten*, Vienna/Leipzig: Hölder-Pichler-Tempsky 1941.
18 H. Munier, "Les stèles coptes du Monastère de Saint-Siméon à Assouan", *Aegyptus* 11 (1930–1931), 257–300, 433–84; T. Mina, *Inscriptions coptes et grecques de Nubie*, Cairo: Société d'archéologie copte 1942; cf. M. Krause, "Die Formulare der christlichen Grabsteine Nubiens", in: K. Michałowski (ed.), *Nubia: récentes recherches*, Warsaw: Musée national 1975, 76–82.
19 For the respective bibliographical records, see J. Clédat, *Le monastère et la nécropole de Baouit*, ed. D. Bénazeth et al., Cairo: IFAO 1999; C. Wietheger, *Das Jeremias-Kloster zu Saqqara unter besonderer Berücksichtigung der Inschriften*, Altenberge: Oros 1992.
20 H.G. Evelyn White, *The monasteries of the Wâdi 'n Natrûn*, New York: Metropolitan Museum of Art 1926–1933. Similarly in the synthetic study of a monastic site in the Theban region: H.E. Winlock, and W.E. Crum, *The monastery of Epiphanius at Thebes*, New York: Metropolitan Museum of Art 1926. For the same region, see also the more recent monographs on two monasteries named after Saint Phoibammon: Ch. Bachatly et al., *Le monastère de Phoebammon dans la Thébaïde*, Cairo: Societe d'archéologie copte 1961–1981; W. Godlewski, *Le monastère de St. Phoibammon*, Warsaw: PWN–Éditions scientifiques de Pologne 1986.

THE CHRISTIAN EPIGRAPHY OF EGYPT AND NUBIA

21 E. Bernand, *Les inscriptions grecques et latines de Philae* II: *Haut et Bas Empire*, Paris: Centre National de la Recherche Scientifique 1969.

22 S. Jakobielski, *A history of the bishopric of Pachoras on the basis of Coptic inscriptions*, Warsaw: PWN–Éditions scientifiques de Pologne 1972.

23 See the various publications by S. Sauneron, J. Leroy and R-G. Coquin, for which it is impossible to give the bibliography here; see also M. Krause, "Verlorene Inschriften und Beischriften der Eremitensiedlungen 3 und 4 bei Esna", in: B. Schmitz and A. Eggebrecht (eds.), *Festschrift Jürgen von Beckerath*, Hildesheim: Gerstenberg 1990, 147–70, which lists the main references. Coquin also edited the wall inscriptions of the monasteries of Saint Antony and Saint Paul, in P.P.V. van Moorsel, *Les peintures du monastère de Saint-Antoine près de la mer Rouge*, Cairo: IFAO 1995–1997; and idem, *Les peintures du monastère de Saint-Paul*, see the bibliography, under 5.1).

24 Wietheger, *Jeremias-Kloster*.

25 For the bibliographic record, see especially the references listed in N. Bosson, "Les inscriptions", in P. Ballet, N. Bosson, and M. Rassart-Debergh (eds.), *Kellia* II: *l'ermitage copte QR 195*, Cairo: IFAO 2003, 209–326, 214, n. 6 (in the bibliography, under 5.1).

26 The project of a corpus of Kellia inscriptions is being undertaken by our colleague Nathalie Bosson.

27 See Bingen, "L'épigraphie grecque", 623–24, and most recently, S.G. Richter, *Studien zur Christianisierung Nubiens* (in the bibliography, under 5.2).

28 See the remarks by E. Wipszycka, "Les recherches sur le monachisme égyptien, 1997–2000", M. Immerzeel and J. van der Vliet (eds.), *Coptic studies on the threshold of a new millennium: proceedings of the Seventh International Congress of Coptic Studies, Leiden, 27 August-2 September 2000*, II, Louvain/Paris/Dudley, MA: Peeters 2004, 831–55 at 835–36.

29 See the bibliography, under 2.

30 Particularly the invocations that include the verb βοηθέω; cf. my remarks in " 'In a robe of gold': status, magic and politics on inscribed Christian textiles from Egypt", in: C. Fluck and G. Helmecke (eds.), *Textile messages: inscribed fabrics from Roman to Abbasid Egypt*, Leiden/Boston 2006 (see above, n. 9) [Study 2].

31 See A. Boud'hors in the exhibition catalogue *Égyptes . . .* (listed above, n. 9), 297–98, no. 127.

32 L. Depuydt, *Catalogue of Coptic manuscripts in the Pierpont Morgan Library*, Louvain: Peeters 1993, I, 367–68, no. 182; II, pl. 404.

33 See Yassâ 'Abd al-Masîh, "Letter from a bishop of Al-Fayyûm", *BSAC* 7 (1941), 15–18.

34 Among recent publications, the book by A. Papaconstantinou, *Le culte des saints* (see the bibliography, under 1), provides an excellent example.

35 In recent years, two much different monographies offered excellent examples of such a local synthesis, both according an important place to epigraphy: J-Y. Carrez-Maratray, *Péluse et l'angle oriental du delta égyptien aux époques grecque, romaine et byzantine*, Cairo: IFAO 1999 (urban site); and E.S. Bolman (ed.), *Monastic visions* (monastic site; see the bibliography, under 5.1). For other sites, see the bibliography, under 5.1.

36 The acts of the first conference appeared in the journal *Coptica* 2 (2003) and 3 (2004), reprinted in M.S.A. Mikhail and M. Moussa (eds.), *Christianity and monasticism in the Wadi al Natrun: essays from the 2002 international symposium*, Cairo/New York: American University in Cairo Press 2009. Those of the second one are published in Gawdat Gabra (ed.), *Christianity and monasticism in the Fayoum Oasis*, Cairo/New York 2005.

37 Other general introductions to ancient epigraphy fail to take note of the Christian material from Egypt; e.g. J. Bodel (ed.), *Epigraphic evidence: ancient history from inscriptions*, London/New York: Routledge 2001; B.H. McLean, *An introduction to Greek epigraphy of the Hellenistic and Roman periods from Alexander the Great down to the reign of Constantine (323 B.C.–A.D. 337)*, Ann Arbor: University of Michigan Press 2002.

JACQUES VAN DER VLIET

Bibliography for the years 2000–2004

1. Critical and bibliographical tools; reference works

Bagnall, R.S., and K.A. Worp, *Chronological systems of Byzantine Egypt: second edition*, Leiden: Brill 2004: a completely revised edition.

Bérard, F. et al., *Guide de l'épigraphiste: bibliographie choisie des épigraphies antiques et médiévales*, third edition, Paris: Éditions Rue d'Ulm/Presses de l'École normale supérieure 2000, 80–3: Egypt/Nubia; 325–28: the Coptic domain.[37]

Bernand, E., "État du corpus des IG d'Égypte et de Nubie", *ZPE* 139 (2002), 119–26: bibliographic overview.

Bingen, J., "Bulletin épigraphique: Égypte et Nubie", *REG* 113 (2000), 572–80 (nos. 687–731); 114 (2001), 589–96 (nos. 512–58); 115 (2002), 744–52 (nos. 502–35); 116 (2003), 675–81 (nos. 613–44).

Förster, H., *Wörterbuch der griechischen Wörter in den koptischen dokumentarischen Texten*, Berlin/New York: de Gruyter 2002: "Inschriften wurden . . . nur am Rande zugezogen und nicht systematisch exzerpiert" (p. x).

Hasitzka, M.R.M., *Koptisches Sammelbuch* II, Vienna: Verlag Brüder Hollinek 2004: nos. 1055–1252 are Coptic inscriptions from Egypt and Nubia.

Krause, M., "Referat der koptischen Urkunden von 1998 und 1999", *AfP* 47 (2001), 229–44: especially 237–39 ("III: Inschriften und Graffiti").

Morss, C., "Byzantine letters in stone", *Byzantion* 73 (2003), 488–509: paleographic observations; very little about Egypt (497).

Papaconstantinou, A., *Le culte des saints en Égypte des Byzantins aux Abbasides: l'apport des inscriptions et des papyrus grecs et coptes*, Paris: Éditions du CNRS 2001.

Richter, S.G., and G. Wurst, "Referat über die edition koptischer literarischer Texte und Urkunden", *AfP* 49 (2003), 127–62 at 158–59, nos. 8.1–8.8 ("Inschriften und Graffiti").

SEG XLVII (1997), Amsterdam 2000: Egypt, nos. 2087–2156; Nubia, nos. 2157–61; XLVIII (1998), Amsterdam 2001: Egypt, nos. 1951–2042; Nubia, nos. 2043–44; XLIX (1999), Amsterdam 2002: Egypt, nos. 2106–2334; Nubia, nos. 2335–48; L (2000), Amsterdam 2003: Egypt, nos. 1546–1624; Nubia, nos. 1625–27.

Stillman, Y.K., and P. Sanders, "Tirâz", in: *Encyclopaedia of Islam* 10, new edition, Leiden: Brill 2000, 534–38.

Various authors, "Bibliographische Notizen und Mitteilungen", annually in the *Byzantinische Zeitschrift*: especially "9. Epigraphik" (the entries on Egypt are mainly due to P. Grossmann).

2. Collections, museography

Berlin, flasks

Witt, J., *Menasampullen*, Wiesbaden: Reichert Verlag 2000 (*non vidi*; see "Bulletin épigraphique", *REG* 115 (2002), no. 512, D. Feissel).

Berlin, textiles

Fluck, C., P. Lindscheid, and S. Merz, *Textilien aus dem Vorbesitz von Theodor Graf, Carl Schmidt und dem Ägyptischen Museum Berlin*, Wiesbaden: Reichert Verlag 2000.

Khartoum

Łajtar, A., *Catalogue of the Greek inscriptions in the Sudan National Museum at Khartoum (I. Khartoum Greek)*, Louvain/Paris/Dudley, MA: Peeters 2003.

Van der Vliet, J., *Catalogue of the Coptic inscriptions in the Sudan National Museum at Khartoum (I. Khartoum Copt.)*, Louvain/Paris/Dudley, MA: Peeters 2003: the inscriptions on stone in two volumes.

Lyon

Galliano, G. (ed.), *Recueil des inscriptions lapidaires grecques, latines et coptes du musée des Beaux-Arts et du muséum d'Histoire naturelle de Lyon*, Lyon: Association des Amis du Musée des Beaux-Arts 2001: several Christian epitaphs from Egypt, in Coptic (nos. 1–5, by M. Pezin and J. van der Vliet) and in Greek (nos. 36–40, by P-L. Gatier); Greek no. 41 (= *I. Recueil* 807) is in Coptic; see A. Łajtar, "Notationes legentis", *JJP* 33 (2003), 181–87 at 184–85.

Old Cairo

Bénazeth, D., *Catalogue général du Musée copte du Caire* I: *Objets en métal*, Cairo: IFAO 2001: various inscribed objects; for no. 270, see G. Nachtergael, "Un verset du Psaume 140 sur un encensoir du Musée Copte du Caire", *ZPE* 141 (2002), 148: Ps. 140:2 (in Greek).

Warsaw

Łajtar, A., and A. Twardecki, *Catalogue des inscriptions grecques du Musée national de Varsovie*, Warsaw: University of Warsaw/Raphael Taubenschlag Foundation 2003: Christian inscriptions from Egypt (nos. 94–100) and Nubia (nos. 101–14) by A. Łajtar (nos. 118 and A.8 are in Coptic; cf. J. van der Vliet, "*I. Varsovie*: Græco-coptica", *JJP* 34 (2004), 121–25 [Study 12]).

Worms

Renner-Volbach, D., *Die sogenannten koptischen Textilien im Museum Andreasstift der Stadt Worms: Bestandskatalog*, Wiesbaden: Reichert Verlag 2002: some legends.

3. Exhibition catalogues, books on art

Égypte, la trame de l'histoire: textiles pharaoniques, coptes et islamiques, eds. M. Durand and F. Saragoza, exhibition in Rouen, Roanne, Paris 2002–2004, Paris: Somogy 2002: inscribed textiles, especially from the Fayoum (see below).

Gabra, G., *Art in the Coptic Museum*, Cairo: Egyptian International Pub. Co./Longman 2000: for a wide audience; reproduces several inscriptions at the Coptic Museum (Greek, Coptic, Arabic).

L'art copte en Égypte: 2000 ans de christianisme, exhibition in Paris/Cap d'Agde 2000–2001, Paris: Gallimard 2000: many objects inscribed in Greek, Coptic and Arabic; several funerary stelae (nos. 8, 11, 28, 102–17); a wall inscription (no. 89).

4. Monographic studies

Davis, S. J., *The cult of Saint Thecla: a tradition of women's piety in late antiquity*, Oxford: Oxford University Press 2000: makes use of inscriptions from Egypt.

Fossile, E., "Le indizioni nell'epigrafia cristiana", *RivArchCrist* 76 (2000), 589–99.

Papaconstantinou, A., "Les sanctuaires de la Vierge dans l'Égypte byzantine et ommeyade", *JJP* 30 (2000), 81–94: partly after epigraphic sources.

Pearson, B.A., "Enoch in Egypt", in: R.A. Argall, B. Bow, and R.A. Werline (eds.), *For a later generation: the transformation of tradition in Israel, early Judaism, and early Christianity* (*Fs. G. Nickelsburg*), Harrisburg, Pa.: Trinity Press International 2000, 216–31: on Enoch in inscriptions, especially at Saqqara (228–30); reprinted in: B.A. Pearson, *Gnosticism and Christianity in Roman and Coptic Egypt*, New York/London: T & T Clark International 2004, 132–52.

Scheidel, W., *Death on the Nile: disease and the demography of Roman Egypt*, Leiden/Boston/Cologne: Brill 2001: uses data from Greek and Coptic funerary epigraphy.

Thomas, Th.K., *Late antique Egyptian funerary sculpture: images for this world and the next*, Princeton: Princeton University Press 2000: on the symbolism of decorated funerary monuments.

Tidda, F., "Terminologia della luce e battesimo nelle iscrizioni greche cristiane", *Vetera Christianorum* 38 (2001), 103–24: mentions some inscriptions from Egypt.

5. Sites and sanctuaries, arranged from north to south

5.1 Egypt

Miscellaneous or unknown provenances

Nachtergael, G., "La stèle de Tsôphia", *CdÉ* 75 (2000), 387–98: dossier of (Greek-Coptic) funerary stelae with the formula "Peace (be) with this holy mountain" (see also *I. Lyon*, Coptic nos. 2–3, listed above).

Nachtergael, G., "Sceaux et timbres de bois d'Égypte II: les sceaux de grand format", *CdÉ* 76(2001), 231–57; "Sceaux et timbres de bois d'Égypte III: la Collection Froehner (suite et fin)", *CdÉ* 78 (2003), 277–93: inscriptions in Greek and Coptic.

Papaconstantinou, A., "Antioche ou l'Égypte? Quelques considérations sur l'origine du 'Danielstoff'", *CArch* 48 (2000), 5–10 (cf. M. Stein, "Die Inschriften", listed below).

Papaconstantinou, A., "La manne de saint Jean: à propos d'un ensemble de cuillers inscrites", *RevÉtByz* 59 (2001), 239–46: Greek inscriptions (*SB* I 5977).

Pillinger, R., "Drei Amulettarmbänder mit Psalmzitaten", in: U. Horak (ed.), *Realia coptica: Festgabe zum 60. Geburtstag von Hermann Harrauer*, Vienna: Holzhausen 2001, 75–80.

Pillinger, R., "Elf Stofffragmente mit Inschriften und christlichen Darstellungen im Museum für angewandte Kunst in Wien", *Mitteilungen zur christlichen Archäologie* 7 (2001), 35–42: texts in Greek (Ps. 8: 6–7) and in Sahidic Coptic (prayer); seventh/eighth century?

Rutschowscaya, M-H., "Le peigne d'Helladia", in: N. Bosson (ed.), *Études coptes VII: Neuvième journée d'études, Montpellier 3–4 juin 1999*, Louvain/Paris: Peeters 2000, 235–44: comb with an interesting inscription (= *I. Louvre Dain* 217).

THE CHRISTIAN EPIGRAPHY OF EGYPT AND NUBIA

Stein, M., "Die Inschriften auf dem Daniel- und dem Petrusstoff in Berlin", in: M. Hutter, W. Klein, and U. Vollmer (eds.), *Hairesis. Fs. K. Hoheisel*, Münster: Aschendorff 2002, 84–98 (cf. A. Papaconstantinou, "Antioche ou l'Égypte?", listed above).

von Falck, M., and C. Fluck, *Die ägyptische Sammlung des Gustav-Lübcke-Museums Hamm*, Bönen: Kettler 2004: no. 203 (Coptic funerary stela, seventh/eighth century?, Edfu?).

Wiese, A. (with S. Winterthaler and A. Brodbeck), *Antikenmuseum Basel und Sammlung Ludwig: die ägyptische Abteilung*, Mainz: Philipp von Zabern 2001: no. 161 (funerary stela with architectural decoration; southern Upper Egypt?).

Alexandria

Boussac, M.F., and J-Y. Empereur, "Les inscriptions", in: J-Y. Empereur and M-D. Nenna (eds.), *Nécropolis* 1, Cairo: IFAO 2001, 225–42: four Christian inscriptions from the necropolis of Gabbari (nos. 19–22).

Heinen, H., "Dipinti aus der West-Nekropole Alexandrias: Gabbari", in: J-Y. Empereur and M-D. Nenna (eds.), *Nécropolis* 2, Cairo: IFAO 2003, 639–52: a small, but interesting group, in Greek; see in the same volume: J. Gascou, "Nécropolis byzantine: IVe-VIIe siècles", 653–8.

Tell al-Makhzan, near Pelusium

Carrez-Maratray, J-Y., in Ch. Bonnet and Mohamed Abd el-Samie, "Les églises de Tell el-Makhzan: les campagnes de fouille de 1998 et 1999", *CRIPEL* 21 (2000), 95, pl. 14: Greek inscription, sixth century?

Kellia

Bosson, N., "Les inscriptions", in P. Ballet, N. Bosson, and M. Rassart-Debergh (eds.), *Kellia* II: *l'ermitage copte QR 195*, Cairo: IFAO 2003, 209–326: group of 155 wall inscriptions in Greek and Bohairic Coptic.

Bridel, P., and D. Sierro, *Explorations aux Qouçoûr Hégeila et 'Éreima lors des campagnes 1987, 1988 et 1989*, Louvain: Peeters 2003, 143: wall inscriptions in Greek and Bohairic Coptic, edited by N. Bosson, R. Kasser and J. Partyka.

Partyka, J.S., "Nowe ustalenia w topografii Cel (Kellia) na podstawie inskrypcji", in: B. Iwaszkiewicz Wronikowska and D. Próchniak (eds.), *Topografia świata wczesnochrześcijańskiego między starożytnością i średniowieczem*, Lublin: Towarzystwo Naukowe Katolickiego Uniwersytetu Lubelskiego 2001, 235–40: "New findings on the topography of Kellia on the basis of the inscriptions" (*non vidi*).

Wadi al-Natrun

Van der Vliet, J., "History through inscriptions: Coptic epigraphy in the Wadi al-Natrun", *Coptica* 3 (2004), 187–207: overview; bibliography [Study 5].

Dayr al-Suryan

Innemée, K.C., and L. van Rompay, "Deir al-Surian (Egypt): new discoveries of January 2000", *Hugoye: Journal of Syriac Studies* 3/2 (July 2000): wall inscriptions, Bohairic Coptic and Syriac (http://syrcom.cua.edu/hugoye).

Innemée, K.C., and L. van Rompay, "Deir al-Surian (Egypt): new discoveries of 2001–2002", *Hugoye: Journal of Syriac Studies* 5/2 (July 2002): Bohairic Coptic and Syriac legends; Syriac graffiti.

Martin, M.J., "A Syriac inscription from Deir al-Surian", *Hugoye: Journal of Syriac Studies* 5/2 (July 2002): dedicatory inscription.

van Moorsel, P.P.V, "La grande annonciation de Deir es-Sourian", in *Called to Egypt: collected studies on painting in Christian Egypt*, Leiden: NINO 2000, 203–24: corrected reading of the Greek and Bohairic legends (at 213 L. van Rompay, "Deir al-Surian: miscellaneous reflections", *Essays on Christian art and culture in the Middle East* 3 (2000), 80–7: discussion of the Bohairic Coptic and Syriac wall inscriptions.

van Rompay, L., "The Syriac texts of the flabellum", in B. Snelders and M. Immerzeel (eds.), "The thirteenth-century flabellum from Deir al-Surian in the Musée royal de Mariemont (Morlanwelz, Belgium)", *ECA* 1 (2004), 113–39 at 134–37.

van Rompay, L., and A.B. Schmidt, "A new Syriac inscription in Deir al-Surian (Egypt)", *Hugoye: Journal of Syriac Studies* 4/1(January 2001): wood, A.D. 1285/1286.

van Rompay, L., and A.B. Schmidt, "Takritans in the Egyptian desert: the monastery of the Syrians in the ninth century", *Journal of the Canadian Society for Syriac Studies* 1 (2001), 51–9 at 49–51: discussion of the Syriac wall inscriptions.

Old Cairo

Girgis Daoud Girgis, "A new interpretation of the wooden lintel no. 753 of the Coptic Museum in Old Cairo, Cairo", *BSAC* 40 (2001), 187–90 (in Arabic): on the al-Mu'allaqa lintel (Greek dedicatory inscription from the Church of the Virgin; see also Study 4).

Skálová, Z., and S. Davis, "A medieval icon with scenes from the life of Christ and the Virgin in the Church of Abu Seifein, Cairo: an interdisciplinary approach", *BSAC* 39 (2000), 211–38: the Greek and Arabic legends are only partly provided.

Fayoum

Boud'hors, A., and F. Calament, "Un ensemble de stèles", Appendix (see "Fayoum, Tutun" below).

Van der Vliet, J., "Monumenta fayumica", *Enchoria* 28 (2002–2003), 137–46 [Study 7]: a Coptic epitaph (provenance unknown) and a Greek one (Sinnuris); Coptic lintel (re-edition; from the area of Tutun?).

Fayoum, Naqlun

Godlewski, W., "Les peintures de l'église de l'archange Gabriel à Naqlun", *BSAC* 39 (2000), 89–101: refers to wall inscriptions (Fayoumi-Sahidic).

Godlewski, W., "Les textiles issus des fouilles récentes de Naqlûn", in cat. *Égypte, la trame de l'histoire*, 100–04 (see above, under 3).

Godlewski, W., "Naqlun: excavations, 2000", *PAM* 12: *Reports 2000* (2001), 149–61: Greek funerary stela; textile inscribed with Arabic.

Van der Vliet, J., "A Naqlûn monk brought home: on the provenance of Louvre inv. E 26798–9", *BSAC* 39 (2000), 239–44 [Study 11]: Coptic inscription (Fayoumi-Sahidic) of the *tirâz* type, on a tunic; tenth-eleventh centuries?

THE CHRISTIAN EPIGRAPHY OF EGYPT AND NUBIA

Fayoum, Tutun

Boud'hors, A., and F. Calament, "Un ensemble de stèles fayoumiques inédites: à propos de la stèle funéraire de Pantoleos de Toutôn", in: M. Immerzeel and J. van der Vliet (eds.), *Coptic studies on the threshold of a new millennium: proceedings of the Seventh International Congress of Coptic Studies, Leiden, 27 August-2 September 2000*, Louvain/Paris/Dudley, MA: Peeters 2004, I, 447–75: three Coptic stelae (Fayoumi-Sahidic), tenth century; in the appendix, 461–9, inventory of the Coptic epigraphy from the Fayoum.

Durand, M., and S. Rettig, "Un atelier sous contrôle califal identifié dans le Fayoum: le *tirâz* privé de Tutûn", in cat. *Égypte, la trame de l'histoire*, 167–70: group of textiles inscribed with Coptic (Sahidic with Fayoumic tendency) and Arabic.

Middle Egypt

Doresse, J., *Les anciens monastères coptes de Moyenne Égypte, du Gebel-et-Teir à Kôm-Ishgaou, d'après l'archéologie et l'hagiographie*, Ph.D dissertation, Paris 1971; printed in Yverdon-les-Bains: Institut d'archéologie yverdonnoise 2000, 503–51: a still useful discussion of the epigraphic dossier, with an appendix containing a bibliography by Ph. Luisier.

Abusir

Oerter, W.B., "Koptische Funde aus Abusir", in: M. Bárta and J. Krejčí (eds.), *Abusir and Saqqara in the year 2000*, Prague: Academy of Sciences of the Czech Republic, Oriental Institute 2000, 55–66: some inscriptions on stone and pottery (at 57–8, cf. 62).

Saqqara

Ghaly, Holeil, "Koptische Grabsteine aus dem Gebiet des Jeremias-Kloster zu Saqqara", *BSAC* 40 (2001), 117–23: two Coptic stelae.

Raven, M.J., *The tomb of Maya and Meryt* II, Leiden/London: Rijksmuseum van Oudheden/Egypt Exploration Society 2001, 39–41, pl. 35–6: fragments of inscribed pottery (in Coptic and Greek).

Heracleopolite nome

Froschauer, H., "Tradition im koptischen Bestattungswesen: Ein christliches Mumientäfelchen aus den Beständen Tamerit in der Papyrussammlung der Österreichischen Nationalbibliothek", *Eirene* 40 (2004), 91–100: painted mummy label, seventh-eighth centuries?

Behnasa

Subías Pascual, E. (ed.), *La corona immarcescible: pintures de l'antiguitat tardana de la necròpolis alta d'Oxirrinc (Mínia, Egipte)*, Tarragona: Institut Català d'Arqueologia Clàssica 2003: at 37–42, discussion of the Greek wall inscriptions by C. Piedrafita.

Antinoopolis

Attia, Sobhi Shenuda, Youhanna Nessim Youssef, and Z. Skálová, "Coptic Archangel Michael, the Archon of the hosts of Heaven in the Coptic Museum at Cairo, painted on canvas and dated to 1328", *BSAC* 39 (2000), 43–8: dedicatory inscriptions in Sahidic Coptic; on the provenance, see L. Langener, "Verschollen und wieder aufgetaucht: das Grabtuch mit dem Erzengel Michael Koptisches Museum Kairo, Inv.-Nr. 8452", *ibid.*, 157–59.

Manfredi, M., "Iscrizione funeraria da Antinoe", in: H. Harrauer and R. Pintaudi (eds.), *Gedenkschrift Ulrike Horak*, Florence: Gonnelli 2004, 143–47: Greek stela from the sixth century.

Dayr Abu Hennis

van Loon, G.J.M., and A. Delattre, "La frise des saints de l'église rupestre de Deir Abou Hennis", *ECA* 1 (2004), 89–112: includes edition of the legends by A. Delattre.

Hermopolis Magna

Grossmann, P., and D. Bailey, "Report on the excavation in the South Church at Hermopolis-Ashmunâyn (Winter 1991)", *JCoptS* 3 (2001), 45–61: Greek epitaph that was re-used in the baptistery of the church (53, pl. 8.b); for a better text, see A. Łajtar, "Bemerkungen zu drei kürzlich veröffentlichten griechischen Inschriften aus der Spätantike", *ZPE* 114 (1996), 141–46 (at 143–46); reprinted in *SEG* XLIV 1455.

Reiramun, near Mallawi

de Martino, M., "Sul significato del termine στήλη in un'iscrizione onoraria di età teodosiana", *RivArchCrist* 77 (2001), 407–17: on *I. Métriques* 123.

Dayr al-Muharraq

Youssef, Youhanna Nessim, "A Coptic inscription from the Monastery of the Virgin Mary known as al-Moharraq Monastery", *GM* 195 (2003), 109–11: fragment of a commemorative inscription.

Manqabad

Abdal-Tawab, Abdal-Rahman, Adel Hassan al Mansub Ibrahim, and S. Schaten, "Coptic stelae from Manqabâd", *BSAC* 40 (2001), 53–8: five Coptic stelae found in 1976.

Panopolite

Criscuolo, L., "A textual survey of Greek inscriptions from Panopolis and the Panopolite", in: A. Egberts, B.P. Muhs, and J. van der Vliet (eds.), *Perspectives on Panopolis*, Leiden/Boston/Cologne: Brill 2002, 55–69: overview, including Christian material.

Sohag

Bénazeth, D., and A. Boud'hors, "Les clés de Sohag: somptueux emblèmes d'une austère réclusion", in: Chr. Cannuyer (ed.), *Études coptes VIII: Dixième journée d'études, Lille 14–16 juin 2001*, Lille/Paris: Association francophone de coptologie 2003, 19–36.

THE CHRISTIAN EPIGRAPHY OF EGYPT AND NUBIA

Orlandi, T., "The library of the Monastery of Saint Shenute at Atripe", in: A. Egberts, B.P. Muhs, and J. van der Vliet (eds.), *Perspectives on Panopolis*, Leiden/Boston/Cologne: Brill 2002, 211–31: the inscriptions in the library, after W.E. Crum (213–15).

Abydos

Delattre, A., "Les graffitis coptes d'Abydos et la crue du Nil", in: Chr. Cannuyer (ed.), *Études coptes VIII: Dixième journée d'études, Lille 14–16 juin 2001*, Lille/Paris: Association francophone de coptologie 2003, 133–46: provides the texts in an appendix, 139–45, after U. Bouriant and M.A. Murray/W.E. Crum.

Koptos

Cat. *Coptos: l'Egypte antique aux portes du désert*, Lyon: Musée des beaux-arts 2000: two Sahidic funerary stelae with an uncertain provenance (nos. 176, 177 = *I. Lyon*, see above, Coptic nos. 1 and 3).

Western Thebes

Delattre, A., "Graffitis de la montagne thébaine. I", *CdÉ* 76 (2001), 333–39: nos. 3122, 2706, 3631 (Greek and Coptic); "Graffitis de la montagne thébaine. II", *CdÉ* 78 (2003), 371–80: two dossiers of Coptic inscriptions.

Delattre, A., "Un 'Notre-Père' en copte à Médinet Abou", *BASP* 39 (2002), 13–16.

Gonis, N., "Some Egyptian datings from the reign of Phocas", in: T. Gagos and R.S. Bagnall (eds.), *Essays and texts in honor of J. David Thomas*, Oakville, CT: American Society of Papyrologists 2001, 255–62 at 261–62: on *P.Mon.Epiph.* I, 11–12 (Coptic inscription; cf. Bagnall and Worp, *Chronological systems*, 52, 266).

Heurtel, Ch., *Les inscriptions coptes et grecques du temple d'Hathor à Deir al-Médîna*, Cairo: IFAO 2004: an important group of graffiti from the church of a martyr saint Isidore.

Łajtar, A., "Ein übersehener inschriftlicher Beleg für den Praefectus Praetorio Orientis Hephaistos (Mitte des VI. Jh.s)", *ZPE* 137 (2001), 188: *I. Syringes* 788; re-edition of the text (Greek); cf. J.L. Fournet, "Reddite quae sont Horionis Horioni", *ZPE* 139 (2002), 138: on the identity of the individual.

Lecuyot, G., "Le Ramesseum à l'époque copte: à propos des traces chrétiennes au Ramesseum", in: A. Boud'hors (ed.), *Études coptes VI: Huitième journée d'études, Colmar 29–31 mai 1997*, Louvain/Paris: Peeters 2000, 121–34: graffiti.

Rzepka, S., "Graffiti survey", *PAM* 12: *reports 2000* (2001), 229–34 ("Deir el-Bahari cliff mission"): mentions Greek and Coptic graffiti; no texts given.

Aswan

Dijkstra, J.H.F., "Late antique inscriptions from the First Cataract area discovered and rediscovered", *JJP* 33 (2003), 55–66: Greek inscriptions from Philae (east church) and Elephantine (quay wall).

Dijkstra, J.H.F., and J. van der Vliet, " 'In year one of King Zakhari': evidence of a new Nubian king from the Monastery of St. Simeon at Aswan", *BSF* 8 (2003), 31–9 [Study 19]: Sahidic Coptic *dipinto*, A.D. 962.

Richter, S.G., *Studien zur Christianisierung Nubiens*, Wiesbaden: Reichert Verlag 2002: examines various inscriptions from the region, especially *SB Kopt*. I 302 (Philae, 28–135).

Eastern Desert

Bolman, E.S. (ed.), *Monastic visions: wall paintings in the Monastery of St. Antony at the Red Sea*, Cairo: ARCE/ New Haven: Yale University Press 2002: chapters on the graffiti (various languages; S.H. Griffith at 185–93) and the Coptic wall inscriptions (re-edited by B.A. Pearson at 217–39).

Cuvigny, H., A. Bülow-Jacobsen, and N. Bosson, "Le paneion d'Al-Buwayb revisité", *BIFAO* 100 (2000), 243–66 at 251–52: *I. Ko.Ko.* 176 (Coptic graffito); three monograms (Byzantine period).

Mahmoud Abd el-Raziq et al., *Les inscriptions d'Ayn Soukhna*, Cairo: IFAO 2002, 72–80, nos. 45–58: Coptic inscriptions (graffiti).

Meyer, C. et al., *Bir Umm Fawakhir survey project 1993: a Byzantine gold-mining town in Egypt*, Chicago: The Oriental Institute 2000, 25–6: Greek *dipinti* on amphorae (by T. Wilfong).

Rohl, D. (ed.), *The followers of Horus: Eastern Desert survey report* I, Basingstoke, Hants.: Institute for the Study of Interdisciplinary Sciences 2000: some "Coptic" inscriptions are mentioned.

Sidebotham, S.E., R.E. Zitterkopf, and C.C. Helms "Survey of the Via Hadriana: the 1998 season", *JARCE* 37 (2000), 115–26 at 124–26: Greek epitaph of a native of Pharan, sixth century (cf. *SEG* 50, no. 1573).

van Moorsel, P., *Les peintures du monastère de Saint-Paul près de la mer Rouge*, Cairo: IFAO 2002: edition of the Bohairic and Arabic legends by R.-G. Coquin.

Western Desert

Khargah Oasis, al-Bagawat
Boyaval, B., "Jeu de mots dans un distique chrétien d'Égypte?" *CRIPEL* 23 (2003), 95–6: on *I. Métriques* 173.

Khargah Oasis, Dayr Mustafa Kashef
Bagnall, R.S., "A Coptic graffito from the Valley building at Deir Mustafa Kashef", in: T. Gagos and R.S. Bagnall (eds.), *Essays and texts in honor of J. David Thomas*, Oakville, CT: American Society of Papyrologists 2001, 263: witness of a local cult of St. Michael; cf. in the same volume at 3–9.

Theban region
Darnell, J.C., *Theban desert road survey in the Egyptian Western Desert* I: *Gebel Tjauti rock inscriptions 1–45 and Wadi el-Hôl inscriptions 1–45*, Chicago: The Oriental Institute 2002: several inscriptions in Sahidic Coptic.5.2 Nubia

General

See "2. Collections, museography" above: Khartoum, Warsaw.

Satzinger, H., "Some peculiarities of Greek and Coptic epigraphy from Nubia", in: M. Immerzeel and J. van der Vliet (eds.), *Coptic studies on the threshold of a new millennium: proceedings of the Seventh International Congress of Coptic Studies, Leiden, 27 August-2 September 2000*, Louvain/Paris/Dudley, MA: Peeters 2004, I, 529–35: influence of Old Nubian.

Tsakos, A., "Linguistic notes on two funerary steles with the 'Euchologion Mega' type of prayer for the dead from Christian Nubia", *CCO* 1 (2004), 287–92: general observations; no texts given.

Northern Nubia

Richter, S.G., *Studien zur Christianisierung Nubiens*, Wiesbaden: Reichert Verlag 2002: discusses various inscriptions from Philae, especially *SBKopt* I 302, and northern Nubia.

Schenke, G., "Ein koptischer Grabstein aus Nubien", *ZPE* 132 (2000), 176–78: provenance unknown; eighth century?

Van der Vliet, J., "Gleanings from Christian Northern Nubia", *JJP* 32 (2002), 175–94 [Study 22]: Tafa, Kalabsha, Ikhmindi.

Tafa

Cat. *Bogowie starożytnego Egiptu/The gods of Ancient Egypt*, eds. K. Babraj and H. Szymańska, Krakow: Muzeum Archeologiczne w Krakowie 2000: nos. 137–39 (three Greek funerary stelae, from the necropolis of Ginari = Firth, "Grave 302"; "Grave 95"; "Grave 122").

Cat. *Coptic texts and artifacts hidden in Amsterdam*, ed. M. Kuhn, Leiden: Faculty of Arts /Amsterdam: Allard Pierson Museum 2000: no. 29 (Greek funerary stela, from the necropolis of Ginari = Firth, "Grave 839"; no text given).

Felle, A.E., "Epigrafia e Sacra Scrittura: un'iscrizione nubiana scomparsa (CIG, IV 8888)", *Vetera Christianorum* 40 (2003), 71–91: biblical inscription (Odes) in Greek, probably from the southern temple (Lefebvre, *Recueil*, no. 614).

Kalabsha

Łajtar, A., "Varia Nubica VI-VII", *ZPE* 137 (2001), 183–86 ("VI. Eine wenig bekannte Grabinschrift aus Kalabscha", 183–84): re-edition of a Greek epitaph.

Faras

Cat. *Faras: Die Kathedrale aus dem Wüstensand*, ed. W. Seipl, Vienna/Milan: Kunsthistorisches Museum 2002: only previously known pieces, from the National Museum at Warsaw: several inscriptions linked to wall paintings; nos. 30–4 and 53 are inscriptions on stone; nos. 60–2, inscribed bricks.

Argin, Nag' al-Arab

Torallas Tovar, S., "A Coptic epitaph from Nubia at the Museo Arqueológico Nacional, Madrid", in: S. Bay (ed.), *Studia palaeophilologica Professoris G.M. Browne in honorem oblata*, Champaign, Ill.: Stipes 2004, 19–22: double epitaph, Sahidic, eighth-tenth centuries.

Argin, ad-Donga

Torallas Tovar, S., and K.A. Worp, "A Greek epitaph from Nubia rediscovered", *JJP* 32 (2002), 169–74: stela discovered in the 1960s, A.D. 1093.

Meinarti

Adams, W.Y., *Meinarti* I-V, Oxford: Archaeopress 2000–2003: complete publication, in four volumes, of the excavations of the 1960s, including the inscriptions.

Saï

Grimal, N., and E. Adly, "Fouilles et travaux en Égypte et au Soudan, 2002–2003", *Orientalia* 73 (2004), 1–149: Coptic funerary stela (130–31, fig. 41, no text given).

Łajtar, A., "Varia Nubica V: SB I 3897: Ein Nilhochwasser auf der Insel Saï unter dem König Stephanos", *ZPE* 136 (2001), 62–4: bilingual (Greek/Old Nubian) graffito, eleventh century; cf. G.M. Browne, "Ad Varia Nubica V", *ZPE* 139 (2002), 194: corrected reading.

Reinold, J. et al., *Archéologie au Soudan: les civilisations de Nubie*, Paris: Editions Errance 2000, 126: photo of a fragmentary "Coptic" stela = *I. Khartoum Greek* 14.

Dongola

Godlewski, W., "Old Dongola: Kom A, 2001", *PAM* 13: *reports 2001* (2002), 203–16: some inscribed amphorae (at 208–10).

Jakobielski, S., "Old Dongola: season 2000", *PAM* 12: *reports 2000* (2001), 265–79: inscribed amphora fragments; wall inscriptions (Greek/Old Nubian; no texts given); "Old Dongola: Fieldwork in 2002", *PAM* 14: *reports 2002* (2003), 211–29: several minor inscriptions.

Łajtar, A., "Georgios, Archbishop of Dongola (+ 1113) and his epitaph", in *Εὐεργεσίας χάριν: studies presented to B. Bravo and E. Wipszycka*, Warsaw: University of Warsaw/ Raphael Taubenschlag Foundation 2002, 159–92: study of the Greek/Old Nubian dossier of a prominent personality.

Łajtar, A., "Heb 5. 4 in a graffito in the Western Annex of the Monastery on Kom H at Old Dongola", *EtudTrav* 19 (2001), 210–15: Greek/Old Nubian, twelfth century?

Łajtar, A., "Terracotta funerary stele of the monk Iohannes from Old Dongola", in: S. Jakobielski and P.O. Scholz (eds.), *Dongola-Studien: 35 Jahre polnischer Forschungen im Zentrum des makuritischen Reiches*, Warsaw: ZAŚ PAN 2001, 327–34: Greek stela, eighth-ninth centuries? (see *OrChr* 81 [1997], 123, no. 8).

Łajtar, A., and K. Pluskota, "Inscribed vessels from the Monastery of the Holy Trinity at Old Dongola", in: Jakobielski and Scholz, *Dongola-Studien*, 335–55: major collection with names of monastic dignitaries, ca. twelfth-thirteenth centuries.

Müller, C.D.G., "Schutzinschriften einer Grablege in Alt-Dongola: Zu nubischen Geheim wissenschaften", in: Jakobielski and Scholz, *Dongola-Studien*, 321–26: on the wall inscriptions in the tomb of Archbishop Georgios (Greek, Sahidic Coptic).

Banganarti

Łajtar, A., "Three Greek epitaphs from Banganarti", *JJP* 33 (2003), 161–75.

Łajtar, A., "Wall inscriptions in the Banganarti churches: a general note after three seasons of work", *JJP* 33 (2003), 137–59: important ensemble of late inscriptions (Greek/Old Nubian), mostly graffiti.

Żurawski, B., "Dongola Reach: the Southern Dongola Reach Survey, 2001", *PAM* 13 (2002), 217–26 at 221–26; "Dongola Reach: the Southern Dongola Reach Survey, 2002",

PAM 14 (2003), 237–52 at 241–52: preliminary reports on the epigraphic discoveries in the Church of St. Raphaël (see the previous references).

Bakhit

Żurawski, B., "Dongola Reach: the Southern Dongola Reach Survey. Report on fieldwork in 2000", *PAM* 12 (2001), 281–90: inscribed funerary cross (at 290).

Ghazali

Łajtar, A., "Varia Nubica VI-VII", *ZPE* 137 (2001), 183–86 ("VII. Kein Bischof namens Ane in Ghazali", 184–86): re-edition of a Greek epitaph.

Al-Fereikha

Mohamed Ahmed, Salah el-Din, and J. Anderson, "Prospections archéologiques et fouilles de sauvetage dans le voisinage du site de Dangeil (1997 et 1999)", *CRIPEL* 21 (2000), 17–37: inscribed brick (20–21). ·

Part 2

Egypt

2

"IN A ROBE OF GOLD"
Status, magic and politics on inscribed Christian textiles from Egypt[1]

Jacques van der Vliet

Introduction

The present paper addresses a subject that until recently has received scant attention, the texts found on textiles from Christian Egypt in the period between about A.D. 300 and 1200. The relative novelty of the subject may explain the composite structure of the following pages. First, in order to provide a framework for the interpretation, some observations are made on the phenomenon of textiles with writing in general and in late antique and early medieval Egypt in particular. Then, the argument focuses on the characteristics of the specifically Christian *tirâz* style that developed in the wake of the Arab conquest in 641. Finally, a hitherto neglected class of early medieval garments inscribed with verses from the Greek and Coptic Psalter requires a first presentation and a more extensive discussion. In addition to the main text, there are two appendices. The first appendix contains a list of the textiles with Psalm verses that are known so far and the second gives a first edition of the texts on two of them (a shawl from the collections of the Katoen Natie in Antwerp and another from the Louvre in Paris). The reflections and interpretations that follow here are meant to stimulate further discussion. Pending the publication of new material and more thorough research into the subject, they cannot claim to be final or exhaustive.

Inscriptions fall into the domain of epigraphy, the study of texts written on other materials than the usual media, such as papyrus, parchment or paper. In fact, great variation can be observed in the nature of the material that serves for inscriptions. Monumental inscriptions on stone, both funerary (commemorating the dead) and dedicatory (commemorating the foundation or the restoration of e.g. a building or a statue), are probably the best-known objects of epigraphy. Less familiar, but common in Egypt, are inscriptions on wood, metal, household objects and textiles.[2] Epigraphy studies inscriptions from the point of view of their *form*, *contents* and *function*, and the following observations on inscriptions on textiles reflect this epigraphic approach. This paper will therefore deal primarily with the texts, rather than with the textiles themselves. Moreover, it will focus on the Christian material from Egypt, irrespective of the language used (Greek, Coptic or Arabic).

The corpus of Christian inscriptions on textiles that we have at our disposal is regrettably small and lacks uniformity. Although the corpus is not a well-defined unity, two major classes of inscribed textiles can be distinguished. The first belongs to the large number of late antique–type textiles, in which pictorial decoration is predominant and in which inscriptions are allotted only a subordinate role. In many cases, these even take the form of legends, a form of inscription that is entirely subservient to the pictorial image. The second class consists of textiles where the roles of text and decoration are reversed. Here, the texts are either of equal importance to the other decorative elements, including pictorial design, or even constitute the major form of decoration. These texts, unlike legends, have an independent status and represent an innovation that postdates the Arab conquest of Egypt.[3] Formally, they often depart from traditional late antique models in order to follow an originally Islamic fashion.

Quantitatively, compared to the astonishing wealth of decorated late antique and early medieval textiles from Egypt found in so many private and public collections all over the world, the amount of inscribed pieces is rather small. The proportions may be different for material with Arabic inscriptions that are primarily Islamic in nature, but pieces inscribed in Greek or Coptic are percentually a small minority of the vast mass of surviving Egyptian textiles. Apparently, even for many centuries after the Arab conquest, pictorial decoration was considered more important than written text. Access to this small corpus of inscribed Christian textiles is not easy. The pieces are widely dispersed both geographically and bibliographically. The publication of the inscriptions themselves is rarely done to the current standards in Greek or Coptic epigraphy. Apart from a few earlier contributions, among which an article by Stephen Gaselee deserves to be mentioned,[4] only very recently has more serious attention been paid to the subject.[5] Our corpus of inscribed Christian textiles from Egypt is therefore neither very big nor particularly well accessible. Nevertheless, it does deserve more attention than it has hitherto received.

Text as token

Writing, as a systematic way of representing language, is a visible means of communication and this applies to writing on textiles too. Inscriptions on textiles, as with any kind of inscription, establish contacts between persons or groups of persons on the level of written language through the mediation of a particular material object. Inscriptions are, as a rule, meant to be viewed more or less publicly. They are more visible than other kinds of written texts and this accounts for the frequent use of special types and sizes of letters in order to create particular expressive or aesthetic effects. The mediating object determines, to a large extent, the nature of the "message" communicated to the observer. A text on an inscribed lintel over the gate of a house will be different from that on a tombstone. Among textiles, hangings serve representative purposes within a walled and roofed space,[6] whereas elements of dress are both highly exposed and highly personal, being both in close contact with the body and visible wherever the owner wears them.

Communication by inscriptions on textile can, in late antique and early medieval Egypt, serve different purposes and take different forms. Only very rarely is it inspired by practical motives, and laundry marks and care labels are unknown. Isolated names on textiles, however, may have served as owner's marks. This appears to be the case with a fragment, now in the Allard Pierson Museum, Amsterdam, where the name of a Deacon (?) Peter was secondarily embroidered into the textile by an unpracticed hand.[7] It is likely, however, that owner's names on luxury textiles more often served purposes other than that of mere identification. As with the inscriptions on the silver tableware of Petronius' Trimalchio,[8] they are primarily meant to show off. This motivation may be suspected behind, for example, a late antique group of colorful pieces probably from Panopolis/Akhmim that combine a decoration of *ankh*-crosses with names woven into the textile.[9] Such inscriptions with names may be dedicatory in character, as is shown by an extensive text on a fragment from Oxyrhynchus, now in the Victoria and Albert Museum.[10] The much damaged Greek inscription apparently records the donation of the garment in question by a man named ". . . osis . . . at his expenses and of his wife's and children's".[11] The donor's name, although partly lost, occupies a central place in the text, but the text serves as a monument to his wealth and piety rather than as a way of identifying him.

The desire to enhance the owner's or donor's social status is a prominent feature not only of textiles with names but of virtually all inscribed objects regardless of the contents of the text. In "a literate society containing many illiterates",[12] the socially distinctive function of written text can hardly be overestimated. This is particularly obvious in the case of public texts, which include inscriptions on garments that are worn ostensibly or hangings that are exposed in a house or a church. Through the action of being read, inscriptions establish a link between a literate beholder and a literate owner or donor. In addition to a process of intellectual interaction, the written text initiates a process of social interaction.[13] In other words, apart from the message conveyed by their contents, inscriptions articulate status. Thus the well-known phenomenon of pseudo-text on textiles can be explained.[14] Pseudo-texts, at least when viewed from a distance (or by the illiterate), represent a cheap and easy method of obtaining the same primary effect as real text: status.

Beyond general notions such as status and prestige, the precise social and cultural roles of inscriptions can hardly ever be fully understood on the basis of their form, language and contents alone. This is well illustrated by the study of present-day inscribed textiles, such as the East African *leso* or *kanga*, which bear highly stylised proverb-like texts in Swahili. They are in use primarily as gifts among women and convey more or less subtle messages that can mostly not be decoded without information on the occasion of the gift and the relationships between the parties involved.[15] Intimate knowledge of the social context in which these inscriptions function is indispensable in order to understand the significance of the artifacts.[16] Only in a few instances are inscriptions in themselves straight-forward enough to express a clearly defined function or purpose. As far as Egyptian textiles are concerned, this is the case for two classes of inscriptions that may be styled "didactic" and "amuletic" respectively.

The didactic intent of inscriptions on textiles is intimately connected with the status-conveying qualities of literacy and erudition in general. It is most clearly visible in the category of texts known as legends. In late antique Egypt, where pictorial design was so much valued, the role and place of inscriptions on textiles were hardly ever central.[17] In a vast majority of cases, the inscriptions are legends, brief texts commenting on the scenes depicted and only providing the names of the actors (animate or inanimate) involved in those scenes. A representative sample of textiles with legends can be found in Marie-Hélène Rutschowscaya's beautiful book *Tissus coptes*,[18] and particular attention has been paid to the decipherment of legends in recent publications by Maximilien Durand.[19]

The nature of these legends could be considered as "informative" and that is how they tend to be seen: as useful information for both modern scholars and ancient beholders and as tools that assist them in interpreting otherwise ambiguous or unclear scenes. It would not have been possible to establish the identity of the two bearded men on a late antique textile in the Berlin Museum für Byzantinische Kunst, for example, had no inscriptions been added that label them as the Apostles Peter and Paul.[20] The examination of legends in this utilitarian way, however, represents a modern point of view inspired by our own desire for precise information. Instead of merely filling in gaps in the knowledge of the beholder, the legends accompanying scenes on textiles, mosaics or wall paintings actually invite the literate beholder to activate knowledge that he already possesses, e.g. in the field of mythology, sacred history or hagiography. The image of a mounted horseman, for instance, may be taken for granted as a commonplace apotropaic motif.[21] When this figure is identified by a legend as Alexander the Great[22] or Saint Phoibammon,[23] however, the beholder is encouraged to remember a specific saint or hero, his cult and story, as well as his *raison d'être* in this specific setting. The legend is not a stray piece of information, but rather a verbal token that consciously appeals to the beholder's store of knowledge. In other words, legends have a didactic purpose, in addition to defining social roles in terms of literacy and erudition. Other inscriptions have a primarily amuletic function; that is, they have been chosen, wholly or partly, on account of their potential to attract good fortune and/or to ward off bad fortune. This twofold amuletic function, both attractive and protective, was convincingly argued for in the pictorial design of late antique textiles in Henry Maguire's classic study, "Garments pleasing to God".[24] He illustrates his thesis with all sorts of texts, including inscriptions. One of these may be quoted here because it was actually part of the decoration of a textile. A fragment from the Boston Museum of Fine Arts, decorated with an ornate vegetal design, bears the brief Greek inscription: εγϕορι (from εὐφορέω), "flourish!".[25] This is a typical wish for prosperity, close in spirit to the wishes for good luck found on many pieces of personal adornment from the late antique and early Byzantine Mediterranean.[26] Prosperity is conveyed on the Boston textile simultaneously by its abundant floral decoration and by the corresponding inscription.

Apotropaic inscriptions are also amuletic, but are rather more defensive in character, because they have to ward off evil. Their function is perhaps best illustrated by a passage from a Coptic literary text. *The Homily on Saint Michael the Archangel*, attributed to Timothy, Archbishop of Alexandria, belongs to the popular genre of half-narrative, half-paranetic sermons on the angels put into the mouth of famous bishops and preachers.[27] After praising the Archangel for his successful intercession with God in favor of mankind, the unknown author proceeds: "when you wish that the name of Michael shields you in every temptation, then you must write the name of Michael on the four corners of your house, at the inside as well as on the outside, and *also upon the edge of your garment* in order that it will protect you (ⲅⲁⲣⲉⲅ ⲉⲣⲟⲕ) against every evil that may rise up against you, either hidden or manifest".[28]

Shortly before in the text, where the author quotes from a secret revelation communicated by an angel to Saint John the Evangelist, this "name of Michael" itself had been given. This is not the name as it is familiar both to us and to the Coptic audience. Rather, it takes the form of a series of twenty-six numerical cryptograms, written in large red capitals, that protect against the attacks of demons as well as against the schemes of evil men.[29] This is not mere theory, because in 1964 a similar series of numerical cryptograms, now twenty-four in number, were found in Nubia at Qasr Ibrim, on a eleventh-twelfth-century paper phylactery in Greek, that was made to protect (φυλάσσω) a certain Theodore, son of Staura (?).[30] The enigmatic inscribed squares on a textile in the Berlin Museum für Byzantinische Kunst may have served similar apotropaic purposes.[31]

Christian texts and the *tirâz* tradition

The functional scope of inscriptions on textiles was widened considerably with the spread of the Islamic fashion of *tirâz* inscriptions in the centuries following the Arab conquest of Egypt in 641.[32] Not only did the Islamic textiles give a far more prominent place to texts as a decorative element, but also these texts assumed a more active part in assigning specific social, political and religious roles to the textiles themselves. In addition to articulating the general notions of status and "good luck", the often very elaborate classic *tirâz* texts expressed adherence to Islam as well as to the system of Islamic political rule. Moreover, the textiles themselves served as a means of assuring such adherence. At least in theory, they were used as economically valuable and prestigious gifts that were meant to strenghten the system of loyalties supporting Islamic rule, playing a major role in what has been called "une politique des présents".[33] The texts bear out this role and, as a result, contain very specific information that situates the textiles along the parameters of politics, patronage and economic and status value: names of caliphs, names of donating high officials, names of workshops and dates. On the Christian side, the *tirâz* fashion, in spite of its Islamic background, was followed rather than rejected by the population. It even gave rise to new developments within the domain of Christian dress, for it can hardly be doubted that "Islamic" habits of textile decoration, in particular the fashion of *tirâz*-style inscriptions, were strongly influential in shaping an entire class of unmistakenly Christian textiles inscribed in Greek and Coptic in addition to Arabic.[34] Overtly Christian textiles, however, by their very nature are not likely to assume a role within the Islamic system of political rule and patronage. Even apart from religious contents, there is no reason why they should provide the kind of precise politically and socially significant information found on their Islamic counterparts. As a consequence, Christian inscribed textiles provide fewer clues to interpret their function as a means of communication within a given social context. We may possess the artifacts, but knowledge about their possible social roles has to be largely inferred.

Even if the circulation of typically Christian *tirâz*-style inscriptions must by necessitiy have been limited to certain socially and religiously defined circles, there can be no doubt about the popularity and early diffusion of the *tirâz* style among the Christian population as a whole.[35] How profoundly this fashion reshaped the conception of status dress among the Christians of Egypt, even among monks, is illustrated by the thirteenth-century mural paintings in the ancient church of the Monastery of Saint Antony near the Red Sea.[36] The holy bishops on the walls are represented in dresses adorned with bands of decorative Arabic writing that are both expressive of their high social status and of their allegiance to the Savior, *al-fâdi*, i.e. Christ.[37] Even the Three Patriarchs, in the traditional scene where they receive the souls of the deceased in their laps, have bands on their sleeves with inscriptions that identify them, in Coptic, as Abraham, Isaac and Jacob respectively.[38] An earlier example, from the tenth century, is provided by the line of Arabic script on the dress of two equestrian saints in the now lost wall paintings from Tebtynis in the Fayoum.[39]

While in the case of the Three Patriarchs, some significance may perhaps be attached to the choice of language (Coptic), it is clear from the other paintings that medieval Christians did not object to the use of Arabic as a means of adorning dress.[40] This undoubtedly reflects the situation in actual life where by the year 1000 Arabic had largely superseded Greek and Coptic as the colloquial language, also among Christians. Significantly, an Egyptian turban cloth with a single-line Arabic inscription that enjoys a certain fame as "the earliest dated Islamic textile" was made for a Christian, Samuel, son of Mark.[41] The burials excavated by the Polish mission in the monastery at Naqlun also show that Christians could choose to be interred in garments with extensive Arabic or pseudo-Arabic inscriptions, some of them more or less religiously neutral, while others were undeniably Islamic in character.[42] Fashion clearly transcended religious dividing lines. Simultaneously with large-scale adherence to Islamic models, however, a distinctly Christian *tirâz* style was created, characterized by longer lines of texts usually in Greek and/or Coptic, often combined with Christian symbols, usually crosses of various shapes.

In their layout, these Christian *tirâz*-style texts show great variety and describing their formal characteristics is beyond the scope of the present paper. In addition, the fragmentary state of many texts does often not allow precise statements about their relation to the supporting textiles. There is no reason, however, to doubt that most of them adorned elements of dress.[43] Typically, single (or sometimes double) lines of text follow the horizontal or vertical outlines of a shawl, turban or tunic. In the latter case, the texts may be found near the neck[44] or near the end of the sleeves.[45] A tunic recently discovered in Naqlun bears four groups of double parallel lines of text descending vertically over the length of the body (Appendix I, no. 7), and lines of texts may be doubled or framed by decorative patterns. In a number of cases, such doubling can take the shape of a line of real or fake Arabic script, in a "Kufic" or "Kufesque" style.[46] Otherwise, non-script-related decorative elements on these textiles tend to be modest, although they are rarely entirely absent.[47] Obviously, this style of design accords an important place to text and by its nature permits lengthier inscriptions. Such variation in style and layout that can be observed quite plausibly reflects chronological developments as well as regional and local peculiarities, but little is presently known about either. One quite distinct group, partly bilingual (Coptic and Arabic), can be assigned to the Fayoum and even with some probability to the same workshop, the private *tirâz* workshop at Tutun.[48] Two tunics from nearby Naqlun, both inscribed in Coptic, resemble each other but are quite distinct from the items assigned to Tutun.[49] These two tunics may have been produced in some other textile center in the Fayoum.

The language of the texts is usually either Greek or Coptic; both can be combined on one garment as can Coptic and Arabic.[50] The Coptic is sometimes a pure Sahidic or Fayoumic, but more often the Fayoumi-Sahidic that was used in the Fayoum region around the end of the first millennium.[51] In the case of bilingual inscriptions it appears that functional distinctions can be observed: the different languages correspond to different domains or registers of discourse. This is most evident when Arabic is used. In the rare Coptic-Arabic bilingual textiles, the Arabic contains (or was meant to contain) a reference to the workshop where the textile was produced. In the corresponding Coptic text, however, such a reference is lacking. This example pertains to a more general phenomenon. Traditionally, the classic type of Arabic *tirâz* inscriptions provides various types of factual information relevant for the religio-political and economic valuation of the textile itself: names of rulers and donors, dates and specifications concerning the workshops. Greek and Coptic *tirâz*-style inscriptions that are of a Christian nature usually do not, for the reasons explained above. The factual information that these provide centers on the owner or (intended) bearer of the textile: names and more rarely titles or a filiation.[52] Toponyms too, when they are found at all, concern the owner, rather than the textile itself. Thus, Louvre tunic inv. E 26798 declares the owner's affiliation with a local monastic community, "the Monastery Nekloni (Naqlun)".[53] The Coptic-Arabic bilingual shawl Louvre inv. E 25405, made for a priest Raphael, remains the best example of these distinct styles of discourse.[54] Here the Arabic inscription contains an impersonal and religiously neutral wish for prosperity that serves as an introduction to information about the workshop, whereas the Coptic text takes the form of a personalized prayer that contains no factual information other than the name and title of the owner and his father's name.

How Greek relates to Coptic in bilingual inscriptions is less easy to define. Textiles with Psalms, which will be discussed below, may combine a Psalm quote in Greek with a personalized prayer for the owner in Coptic. A significant example is provided by a textile from the Museum für angewandte Kunst in Vienna (Appendix I, no. 1). In addition to figurative decoration, this fragmentary tunic was adorned with a prayer in Coptic for the owner of the garment and with a Greek line of text taken from Psalm 8. Precisely the same division of languages and genres is found in a piece from Akhmim, first published by Forrer (Appendix I, no. 5). At first sight, the division appears to be a logical one. The division can be found in other inscriptions too, for example in the funerary stela of Bishop Kerikos (Kyriakos) of Sai in Nubia (ninth century?), now in the Fitzwilliam Museum, Cambridge, which has next to a personalized epitaph in Coptic in the center of the stone, Psalm verses in Greek (Psalm 19:2–4, LXX) on its rim.[55] Due to the fact that Greek retained a prominent status within the Egyptian liturgy,[56] it may be argued that it would be the first choice for a quote from the Psalms, whereas the prayer for the owner, which includes a certain amount of factual and personal information, would be in the language of everyday communication. Attractive as this explanation may appear, however, a shawl in the St. Petersburg Ermitage reverses the roles (Appendix 1, no. 8). It bears an impeccable text of Psalm 71:10–11 in Sahidic Coptic as well as two badly damaged lines of an extensive personalized text in Greek.[57] The strategies that determined the choice of either language in a specific case are therefore not always obvious.

Blessing from aloft

Two main types of texts can be distinguished within the group of Christian *tirâz* inscriptions. The first one represents the amuletic functions of inscriptions on textiles that were discussed above. The texts are of a predominantly apotropaic nature in order to ward off evil and ask for divine favor and protection. Their effect can be enhanced by the simultaneous use of crosses and of the amuletic pictorial motifs that were inherited from late antiquity.[58] When they are written in the form of prayers, these texts usually contain personalized information on the owner. The second type consists of biblical quotations, in particular from the Greek and Coptic Psalter. Inscriptions of both types can occur together on the same garment, thus in a sense supplementing each other.

Apotropaic inscriptions on Christian textiles of the *tirâz* type may take several different forms. Long lines that repeat divine names and appellations, most often the Greek or Coptic monograms for Jesus or Jesus Christ, are a secure means of protection against all kinds of evil.[59] Inscriptions beginning with the Coptic counterpart of the Arabic *bismillah*: ϩⲙ ⲡⲣⲁⲛ ⲙⲡⲛⲟⲩⲧⲉ . . ., "in the name of God . . .", undoubtedly have the same function.[60] However, both more common and explicit are the short conventional invocations that ask for the Lord's blessing and protection for the owner and bearer of the garment. Similar formulaic texts, more or less directly influenced by famous biblical verses such as Num. 6:24, were also used on the contemporary Jewish dress of Egypt.[61] A characteristic Christian example reads: "O Lord Jesus-Christ, our true God, bless (ⲥⲙⲟⲩ) and help (ⲃⲱⲓⲉⲓ for ⲃⲟⲏⲑⲉⲓ) and protect (ⲅⲁⲣ{ⲓ}ⲉⲅ) the life of Your servant N., the son of N.".[62] Such prayers occur not only on textiles but in all kinds of documents, magical as well as non-magical. The following example is from the opening of a Coptic magical amulet against illness: "Jesus-Christ, guard (ⲣⲟⲉⲓⲥ), help (ⲃⲟⲛⲉⲓⲁ for ⲃⲟⲏⲑⲉⲓ), protect (ⲅⲁⲣⲉⲅ), guide (ⲉⲣⲅⲏⲙⲉ = ⲣⲅⲙⲙⲉ) N., the daughter of N.".[63] The typical verbs used both here and on garments are ⲥⲙⲟⲩ, "to bless", ⲣⲟⲉⲓⲥ, "to guard", ⲅⲁⲣⲉⲅ, "to protect", and in particular ⲃⲟⲏⲑⲉⲓ, a form of the Greek verb βοηθέω, "to come to the aid of, to help".[64] They provide these brief prayers a markedly defensive character. This is true in particular for the popular ⲃⲟⲏⲑⲉⲓ invocation that is well known from a wide variety of contexts.[65] Here, a Coptic literary text may be cited, in order to show how universally appreciated its amuletic or, more precisely, apotropaic properties were among the literate.

In the Sahidic-Coptic *Acts of Andrew and Paul*,[66] a conversation is reported between Jesus and Judas that took place after Christ's descent into hell. Judas explains to Jesus how he went to perdition because he yielded to the Devil when the latter appeared to him in the form of a dragon with his mouth wide opened to devour him. Judas got frightened and out of fear worshiped the Devil. Upon which Jesus comments: "O wretch, if – at the moment he (scil. the Devil) approached you – you would have said: "Jesus, help me (ⲓ(ⲛⲥⲟⲩ)ⲥ ⲃⲟ[ⲛⲉ]ⲉⲓ ⲉⲣⲟⲓ)", you would have been saved!"[67] Apparently, the invocation, which here we find recommended by the Savior himself, protects against even the most terrifying attacks of the Devil in person. Since clothes are worn close on the body, they are the perfect medium for these and similar invocations.

However, in addition to divine protection, also more specific worldly boons may be conferred by inscribed textiles. A characteristic late antique example of a wish for prosperity ("flourish!") was already cited above. Islamic *tirâz* inscriptions with a similar meaning are quite frequent. They may wish the owner, in addition to God's blessing, happiness, grace, joy, prosperity, etc.[68] Such Arabic phrases may also be used on Christian pieces, as in the case of the bilingual Louvre shawl of the priest Raphael, with its standard wish of "perfect grace (*ni'ma*) for its owner".[69] Their counterparts in Coptic exhibit a rare but peculiar benedictory formula that could follow the invocations discussed above. As a typical and reasonably preserved example, the Berlin fragment, Museum für Byzantinische Kunst, inv. 4659, may be quoted, where I propose to interpret the much corrupted text as: "O Lord, Jesus-Christ, help (ⲃⲱⲓⲉⲓ for ⲃⲟⲏⲑⲉⲓ) and p<rotect> (ⲅ<ⲁⲣⲉⲅ>) the life of N., the son of N., that he may wear this garment (ⲅⲉⲧ ⲡⲓⲅⲁⲓϯ) in gladness (ⲗⲉⲱⲓ) and joy (ⲧⲉⲗⲏⲗ? – very uncertain) and happiness (ⲟⲩⲛⲁⲃ)".[70] The doubtful word "ⲧⲉⲗⲏⲗ: joy" is largely in a lacuna,[71] but can be plausibly supplemented from another Berlin garment with the same formula.[72] In one case, it appears that "соφιᴀ: wisdom" is added following "happiness" (in no. 6 of our Appendix I).

The identification and interpretation of this Coptic formula as proposed here was inspired by its occurrence in a number of ninth-century book colophons from the Fayoum that have never been satisfactorily interpreted. These colophons use the identical verb ειτε ("to use, to wear off") in connection with similar expressions for happiness.[73] Thus we find in the manuscript of a Sahidic *Life of Pachomius*: "use it (ϩετϥ, scil. the book in question) in gladness (ρᴀϣε) and joy (τελнλ) and cheerfulness (εὐφροσύνη) and righteousness (δικαιοσύνη) and happiness (ογнοϥ), o my holy fathers . . .", then followed with the names of the heads of the monastic community to which the book was dedicated.[74] In the colophons, the phrase appears to be a typical formula of donation, politely addressing the beneficients of the gift.[75] Its appearance on garments convincingly suggests that these also were gifts and that the pious wishes that it articulated were meant to accompany the act of donation. Unfortunately, all examples presently known are difficult to decipher and require further study.[76]

At this point, another general characteristic of this class of Christian *tirâz* inscriptions deserves to be mentioned. Although many of them were conceived as prayers, none show any sign as being conceived of as a funerary prayer. Most of the inscribed textiles from this period that have come down to us are grave goods and bear the visible traces of their use as shrouds.[77] However, typical funerary formulae such as: "give him/her rest in the Kingdom of Heavens", "have mercy upon his/her soul", etc., frequent on tombstones and in other commemorative inscriptions, are conspicuously absent.[78] Instead, in the Arabic and Coptic formulae just discussed, quite worldly preoccupations are apparent.[79]

In so far as the texts of this type are personalized prayers, mentioning names and sometimes titles, they may inform us about the clientele for whom such inscribed textiles were produced. Regrettably, the facts derived from them are hardly sociologically useful and can be summed up in a few words. Inscribed textiles could apparently be destined for both male and female persons.[80] Prayers for "my beloved son" (on the sleeve of a tunic)[81] or "daughter" (on a shawl or veil)[82] show that at least some of them must have been gifts from parents to their children, perhaps on the occasion of their marriage.[83] In fact, two of the examples quoted preserve remains of the Coptic benedictory formula discussed above. The testimony of the textiles themselves is confirmed by documents from the Cairo Geniza that show well-to-do fathers ordering expensive garments for their sons, inscribed with their names.[84] Textiles that mention filiations up to the grandfather are equally suggestive of a family setting.[85] There is no reason, however, to assume that their role in gift-giving was limited to situations of (real or, perhaps also, spiritual) filiation: Christian *tirâz*-style textiles, like their Islamic (and Jewish) counterparts, may have served as gifts in various circumstances. The much ruined Greek inscriptions on a shawl from the St. Petersburg Ermitage appear to suggest that the piece in question was offered as a present to a high-ranking cleric, one Abba Chael (Appendix I, no. 8). Indeed, whereas professions are rarely mentioned, clerical titles do occur. The tunic from the Vienna Museum für angewandte Kunst bearing verses from Psalm 8 was ordered for the "great master (ⲥⲁ̅ⲡ̅), the Deacon Giorgios" (Appendix I, no. 1). Several other garments (shawls and tunics) belonged to priests who bore the honorific title *papa*,[86] including *papa* Kolthi, who was an inmate of the monastery of Naqlun,[87] and the *papas* Biktor (Victor) and Thoter, both of whom may have also lived in the Fayoum.[88] For different reasons, it can be assumed that the Louvre shawl inv. E 26792 also belonged to a cleric, probably a monk (see Appendix II, no. 2). Someone designated as "my brother Biktor (Victor: ⲡⲁⲥⲟⲛ ⲃⲓ[ⲕ]ⲧⲱⲣ)", on a fragment from the collections of the Katoen Natie in Antwerp, must have been a monk as well, rather than a natural brother.[89] The general absence of lay titles and professions and the prominent presence of clerical ones, does not, however, imply that any of the garments discussed here were, in a technical sense, "ecclesiastical" or "liturgical" vestments,[90] even if some were part of a clerical and, in particular, monastic wardrobe.[91] However meager our information about ownership may be, there is definitely more reason to interpret the phenomenon of Christian *tirâz*-style textiles, which must have been an expensive luxury product under all circumstances, in terms of social interaction (gift-giving, status conveyance) than primarily in terms of purpose (liturgical use, funerary use).

Gifts for the king divine

The second and definitely most remarkable type of Christian *tirâz*-style textiles are those that exhibit biblical quotations. The habit to adorn garments in this way was not an exclusively Christian one. Medieval documents from the Cairo Geniza show that Jewish textiles with biblical quotations must have also existed.[92] Only Christian examples survive, however, and these, as far as they are now known, appear to bear exclusively quotes from the Book of Psalms. They have been assembled in Appendix I, below, in a list that presently numbers twelve textiles. This small but distinct group of inscriptions has never been described as such and even most of the texts themselves remained unidentified. This is hardly surprising since they are usually fragmentary and difficult to decipher. Moreover, although it is theoretically possible to consider them as witnesses for the Greek or Coptic texts of the Psalter,[93] it is clear that sometimes the medieval textile workshops took considerable liberties with both the orthography and the textual form of the verses they quote.[94]

As has been mentioned before, the language of these Psalm verses may be either Greek (five times out of twelve) or Coptic (Sahidic or Fayoumi-Sahidic) and they may occur on the same garment next to the personalized prayers discussed above. Examples of this combination are, in our Appendix I, no. 1 (Coptic prayer, Greek Psalm), no. 3 (Coptic prayer and Psalm), no. 5 (Coptic prayer, Greek Psalm), no. 6 (benedictory formula and Psalm, both Coptic), no. 8 (Greek prayer or dedicatory text, Coptic Psalm), and possibly also no. 10. On the twelve textiles of the list, ten different Psalms are represented. Although some of these (viz. Psalms 44, 71 and

109) occur twice on different pieces, and even in different languages, there is no textual overlap. The length of the quotes ranges from a brief three-word prayer (Appendix I, no. 4) to two full verses (nos. 7 and 8). The closely related shawls nos. 2 and 12 both give the same Psalm verses twice (for the texts, see Appendix II). A tunic from Naqlun (Appendix I, no. 7) repeats the verses Psalms 46:2–3 (in Sahidic Coptic) eightfold, arranged in four groups of two parallel lines. One textile combines parts of two succeeding verses from one Psalm, but in inverse order (no. 6), whereas another item presents verses from two different, though closely related Psalms, 109 and 2, in that order (no. 10). No. 2, the Antwerp shawl, joins related Psalm-like verses to a literal quotation from Psalms 17.

The nature of the small corpus of Psalm quotes that occur on early medieval textiles from Egypt is both unexpected and highly intriguing. The following remarks merely represent a first attempt at their characterization. The list betrays a clear preference for well-known Psalm verses, but these seem to have been carefully selected. In spite of considerable variety (not a single verse is found on two different garments), an undeniable homogeneity can be observed, apparent from a high degree of thematic coherence. A vast majority of the quotes appear to offer variations on the theme of God's generosity towards humanity. He bestows power and dominion (Appendix I, nos. 1, 2, 10 and 11), honor and glory (nos. 1, 5, 6, 8 and 9), a spotless way (no. 2), help (no. 4), or simply everything man needs (no. 3). Moreover, a significant number of them specifically evoke the bestowal of power and honor upon the Davidic king. The Psalms from which these verses were taken have traditionally received a messianic interpretation and, in the Christian tradition were therefore applied to Christ.[95] This is particularly the case with the closely related Psalms 2 and 109, and 44 and 71. Verses from these four Psalms occur on six out of the twelve textiles in our small corpus, whereas parts of three of them are even quoted twice. Apart from the messianic Psalms, kingship, in particular, God's universal kingship, is also markedly present in Psalms 46:2–3, quoted in no. 7. Therefore, two general themes appear to predominate: God's generosity towards mankind and divine kingship. A significant number of quotes combine these themes in depicting the bestowal of God's favor upon the messianic king, i. e. Christ.

Due to the amuletic qualities widely ascribed to Psalm verses,[96] one might be inclined to consider the texts from our corpus as primarily apotropaic or phylacteric. The actual selection of texts, however, hardly leaves room for an interpretation exclusively along these lines. Only one of the Psalm quotes in question has a clear apotropaic ring, with a noun deriving from the verb βοηθέω (Appendix I, no. 4: Psalms 26:9b-beginning, in Greek). The inscription is distinct from the other texts, however, in both its contents and form. Instead of making up one or more long lines of text, this brief invocation is contained within a *tabula*-like frame that is reminiscent of late antique design rather than of the *tirâz* fashion. Another theme in the texts is power over enemies. This is particularly so in Psalms 109:1, quoted on no. 11, and in the second verse on the Antwerp shawl (Appendix II, no. 1).[97] It does not seem, however, to be the predominant concern of the corpus as a whole. In fact, apart from the examples just mentioned, most of the standard formulae invoking God's help or protection, and that often derive from the Psalter, do not occur in our list. In particular, quotations from Psalms 90, the apotropaic Psalm *par excellence*,[98] are conspicuously absent from this group of textiles.

It is less easy to be categorical about the eventual liturgical background of the quotations. The Psalms are the backbone of the Christian liturgy and the monastic one in particular. In Egypt, the liturgical use of Psalms and Psalm verses is even more developed than elsewhere in the Christian East.[99] In principle, therefore, every quote from the Psalms qualifies as liturgical. The Psalm verses represented by our corpus are without exception well known owing to their use in the liturgy. Psalms 44:10b (quoted in Appendix I, no. 6), for instance, is a favorite text in celebrations of the Virgin.[100] No less than five of the quotations from the corpus also occur in the indices of Greek Psalm verses that can probably be connected with the "Hymns" of the medieval Coptic liturgical directories.[101] This partly explains the choice of these particular Psalms: they were well known and, for that reason, were widely appealing and easily identifiable. It does not, however, imply that the motivation behind the selection was specifically liturgical and only one piece expresses an overtly liturgical interest. This piece is the Louvre shawl that bears a double recension of Psalms 133:1 (edited in Appendix II, no. 2). The brief Psalms 133 is an appeal to join the prayer in God's house that, in both Coptic and Latin monasticism, has its typical place as an "invitatorium" in the nocturnal hours.[102] On account of the outspoken character of the text, there can be little doubt that this shawl was indeed made for one of those "servants of the Lord, who stand in the house of the Lord" mentioned in the text, presumably a monk.[103] In the rest of the corpus, such specific references or more general allusions to the Eucharist, for example, are lacking. In spite of its obvious liturgical inspiration, it would be therefore difficult to maintain that this particular selection of verses from the Psalter had its *raison d'être* in liturgical practice.

Another noteworthy feature of the corpus is the double occurrence, on two unrelated textiles and in two different languages, of Psalms 44. Both quotations are well known and both occur in the liturgical indices just mentioned.[104] They are, however, remarkable thematically. In both verses garments themselves play a conspicious role: the king's garments exhale the smell of precious spices (in Psalms 44:9a, see Appendix I, no. 5), and at his right the queen is standing "in a robe of gold, adorned with many colours" (Psalms 44:10b, quoted on no. 6). The occurrence of garments in texts on garments may simply reflect the functional link between text and text-bearer that is characteristic of epigraphic material in general.[105] The link here is not merely functional, however, but also thematical as the text comments on its own material support. This feature again calls to mind the Islamic *tirâz* inscriptions. In the Islamic tradition of a "politique des présents", the garments, as the carriers of the honor and the prestige bestowed upon their owners, are themselves the main theme of the inscriptions. It is precisely for this reason that the texts take such pains to record the names and titles of patrons and donors as well as the workshop, place and date of the textile's production. The quotations from Psalms 44 show a similar concern with the textile, but reflect a different perspective since they connect the garments with the vestiary splendor of the court of the messianic king.[106]

Although further discoveries may change the picture, it would appear that the small corpus of Psalm verses inscribed on Christian *tirâz*-style textiles is not a random collection of pious quotes from Holy Scripture. It betrays a certain common inspiration that clearly shows a preference for the so-called messianic Psalms and the evocation of God's generosity towards mankind in general and the messianic king in particular. Apotropaic and liturgical interests can be observed occasionally, but do not seem sufficiently predominant to explain the peculiarities of the collection as a whole. Instead, both the purely formal characteristics of this particular class of Christian *tirâz* inscriptions and certain thematic similarities (such as the formulae of blessing that may accompany the Psalms, and the interest in dress itself as apparent in the quotations from Psalm 44) suggest that they may have to be considered first of all as Christian variations on an Islamic model. Accordingly, the Islamic fashion of *tirâz* inscriptions would offer not only the best interpretational framework for Christian *tirâz*-style textiles in general, but for those with Psalm quotations as well.

A major theme common to both the classical Islamic *tirâz* and their Christian counterparts is the transfer of divine blessing and honor. The Islamic *tirâz* inscriptions invariably promise God's blessing (*baraka min Allah*) to the garment's owner and these promises may be underscored by appropriate Koran verses.[107] The Christian texts also usually ask for divine assistance and blessing for their owners. These texts, on both sides of the confessional dividing line, represent the amuletic use of inscriptions on textiles. In the Islamic *tirâz*, however, the owner is usually not the first recipient of these promises: in the classical Abbasid and Fatimid *tirâz* inscriptions, a much more prominent place is taken by wishes for the Prophet Muhammed and his family, and in particular by blessings for the reigning caliph who appears with his full titles and dignities as the representative of Islamic authority.[108] The bestowal of honor and prestige in the Arabic *tirâz* inscriptions is tied up with Islamic political rule from which the Christian *tirâz* inscriptions are further removed. Not only are Psalms quoted instead of Koran verses, but the character of most of these Psalm verses is quite peculiar. They do not request divine gifts for the bearer of the garment in a straightforward manner, but evoke the bestowal of divine honor and glory on the messianic king of the Psalms, Christ himself. The theme of the texts is the transfer of divine blessing, as has been observed in the Islamic models, but the religio-political setting of this transfer is different.

Seen against the background of their Islamic models, the Christian *tirâz* inscriptions discussed here appear not only to offer variations on identical religious (and, indeed, "amuletic") themes, but also to reinterpret the political setting of these themes from a distinctively Christian point of view. An ideal image of Christian theocracy, expressed in the poetic language of the Bible, is opposed to the ideal image of Islamic theocracy. This does not apply exclusively to the quotes from messianic Psalms, but also to several others: God is the shepherd who takes care of everything (Appendix I, no. 3); He is an awe-inspiring king who reigns over the whole of the earth and to whom all nations should pay hommage (no. 7). If we are correct that at least a certain number of the garments inscribed with Psalms were made as gifts to be presented to members of the clergy, the analogy would be perfect. In Islamic *tirâz*-style dress, the rulers confer status on their subjects. Likewise, the servants of Christ, prototypically represented by the clergy, derive their prestige and their power not from their own merits but, indirectly, from the master whom they serve. The blessing, prestige and authority promised or ascribed to their Master in the Psalms are not only borne out by exhibiting these texts inscribed on luxury dress, but also reflect on the servants themselves who bear the texts on their bodies and who are already a βασίλειον ἱεράτευμα [royal priesthood] by their very nature (1 Peter 2:9). This interpretation is, in fact, strongly supported by the deliberate change of pronouns in our no. 8, where every "him" of the Psalm was turned into a "you", and by the fusion between royal dress and real dress accomplished by the quotations from Psalms 44 (nos. 5 and 6).

The socio-political interpretation of the Christian *tirâz* inscriptions with Psalms proposed here would not only reflect the character of their Islamic models, but it would also accord well with contemporary trends in Coptic political thought. Other sources indicate that the kingship of Christ was a theme for Egyptian Christians during the Abbasid and Fatimid eras. This is perhaps most apparent in the so-called apocalyptic writings from the same general period, with their message of future political change leading up to the final establishment of Christ's rule.[109] An even more telling example is provided by the dating formula "Christ being king over us" that occurs in Coptic book colophons, inscriptions and documents.[110] It does not simply avoid stating the regnal year of some (temporal) Muslim ruler, but also positively situates the donation of the manuscript in question and thereby the life of the community concerned in the (extra-temporal) reign and under the (ideal) suzerainty of Christ. Examples like these show that, in certain milieus at least, the reflection on Christianity as a polity primarily involved biblical and transcendental models rather than political strategies.[111] The quotations from the Psalms here discussed would represent the same mode of discourse.

Conclusions

In the centuries following the Arab conquest in 641, Egyptian Christians accepted without reservation the Islamic fashion of adorning elements of dress with so-called *tirâz* inscriptions. At least by the tenth century, there was no objection against depicting saints on church walls wearing Arabic-inscribed garments. Christians could be buried in the status dress of the period even if these bore overtly Islamic inscriptions. Nonetheless, a distinctly Christian *tirâz* style developed, using Greek and Coptic texts in addition to Arabic ones. The texts exhibited by these Christian *tirâz* are of an outspoken Christian nature, reinforced by the use of Christian symbols. The inscriptions themselves are either conventional prayers and wishes to ensure the welfare and well-being of the bearers of the dress and/or quotes from the Psalter. A number of these garments were made for members of the clergy, but the habit was certainly not exclusively clerical. Moreover, several inscriptions suggest that they were presented as gifts, as were the Islamic *tirâz*. In spite of some variation, the group of inscriptions taken from the Psalms shows a clear preference for the double theme of bestowal of divine gifts and biblical Christian kingship. Pending further discoveries, this group of inscriptions may tentatively be interpreted as an intentional response to the similar themes borne out by the Islamic *tirâz*: the transmission of prestige by gift-giving within the context of Islamic religio-political rule. Although it is certainly unwarranted to read the inscriptions discussed here as "subversive" texts, it would seem that, while reflecting some of both the formal features and the general themes of contemporary Muslim examples, they do so in a specifically Christian way, exhibiting, through a deliberate choice of Psalm verses, a clear feeling of Christian self-awareness and social identity.

Appendix I

Provisional list of textiles with Psalm verses from early medieval Egypt

The order of the items follows that of the Psalms (but note Psalm 2 under no. 10). The numbering of the Psalms is according to the LXX; the English translations are freely adapted from Thomson 1954; descriptions, notes and bibliographical information are kept very brief. Not included are the dubious piece described in Shier 1952, that according to report bears a Sahidic text of Psalms 22:1–6,[112] and a fragment from the Vienna Papyrussammlung that may contain an echo of Psalms 32:1 (?).[113]

1   *Vienna, Museum für angewandte Kunst, inv. T 10 049: Psalms 8:6b–7, Greek. Probably fragment of a tunic; Pillinger 2001a.*
    (Psalms 8:6b) With glory and honour You have crowned him, (7) and set [him over the work of Your hands].
    Note: Some unidentified remains of text precede.

2   *Antwerp, collection Katoen Natie, inv. 711: Psalms 17:33–34a plus two unidentified verses that look like a paraphrase of Psalms 17:40–41, Fayoumi-Sahidic. Shawl; see below Appendix II, no. 1.*
    (Psalms 17:33) It is God who girds me with strength; He made my way spotless, (34a) (He) who straightens my feet like those of hinds.
    (?) You made me superior over all my adversaries, and humiliated all my enemies behind (?) my (*sic desinit*).

"IN A ROBE OF GOLD"

3 *Athens, Benaki Museum, inv. 15343: Psalms 22:1(-2a?), Fayoumi-Sahidic. Shawl; Cat. Rouen 2002: 199, no. 166.*
(Psalms 22:1) It is the Lord who herds me, and He makes me want nothing; (2a?) and He (*sic desinit*).
Note: Uncertain whether end represents beginning of next verse.

4 *Vienna, Papyrussammlung, collection Tamerit, inv. T 5: Psalms 26:9b, Greek. Fragment of a hood (?); Cat. Vienna 1999: 107–08, no. 100 (U. Horak); Fro-schauer 2006: 138–40, no. 5 and fig. 27.*
(Psalms 26:9b-beginning) Be Thou my helper.

5 *Present whereabouts unknown; according to report from Akhmim: Psalms 44:9, Greek. "Leinentuch"; Forrer 1891: pl. 8, fig. 29; 1893: pl. 8, fig. 12.*
(Psalms 44:9a) Myrrh and stacte and casia (exhale) from your garments.

6 *Vienna, Museum für angewandte Kunst, inv. T 580, according to report from Saqqara: Psalms 44:11 (preceded by an unidentified text) plus 10b, Fayoumi-Sahidic. Part of a tunic (?); Riegl 1889: 52, pl. 9 (identified by J. Krall).*[114]
(unidentified) . . . happiness and wisdom.
(Psalms 44:11) Incline (your) ear, forget your people and [the house of your father].
(Psalms 44:10b) [The queen stood up to the] right of you, o king, in a robe of gold, adorned [with many colours].
Notes: The unidentified text is probably the end of a benedictory formula as discussed above.
The word in v. 10b which is rendered as a vocative, "o king", has no corre-spondance in the Psalm text and must be considered a deliberate insertion rather than a copyist's error.

7 *Naqlun, inv. Nd.00.083; from the medieval cemetery behind Dayr al-Malak: Psalms 46:2–3, Sahidic. Tunic; Czaja-Szewzak 2004.*[115]
(Psalms 46:2) All ye nations, clap your hands; shout to God with a jubilant voice. (3) For the Lord is exalted, He is awe-inspiring; He is a great king over all the earth.
Note. Uncertain continuation.

8 *St. Petersburg, Ermitage, inv. 11186: Psalms 71:10–11, Sahidic. Shawl; unpublished?*[116]
(Psalms 71:10) The kings of Tharsis and the isles will bring gifts for you; the kings of the Arabs and Saba will offer presents to you. (11) And all the kings of the earth will pay you hommage; the nations will serve [you].
Note. Throughout both verses, there has been a deliberate change of pronouns ("you" instead of "him").

9 London, Victoria and Albert Museum, inv. 1176–1900, according to report from
   Dayr al-'Izam, near Asyut: Psalms 71:15, Greek. End of a shawl (?); Kendrick
   1921: 19, no. 330, pl. 11 (identified by J. Strzygowski); cf. Gaselee 1924: 79, no. 9.
   (Psalms 71:15a) [And] he shall live, and to him shall be given of the gold of Arabia.

10 Berlin, Museum für Byzantinische Kunst, inv. 6826a: Psalms 109:1 plus
   Psalms 2:7b-8, Greek. Fragment; Fluck 1996: 166–67, pl. 11, fig. 7; Fluck
   2006: 161–62; Mälck 2006, col. fig. 34.
   (Psalms 109:1) The Lord said to my lord: Sit at my right hand till I make your
   enemies a footstool [for your feet].
   (Psalms 2:7b) The Lord said to me: You [are my son], today I have begotten
   [you. (8) Ask me, and I will give you nations for] your inheritance, and for
   [your] possession [the extremes of the earth].
   Note: Remains of four lines of text, two of which can be read from the published photo (one fragment is placed wrongly).

11 Berlin, Museum für Byzantinische Kunst, inv. 6826d: Psalms 109:3a, Sahidic.
   Fragment; Fluck 1996: 163–64, pl. 9, fig. 2; Fluck 2006: 161 and col. fig. 21.
   (Psalms 109:3a) Your dominion is with you on the day of your power, in the
   splendours of the holy [ones].

12 Paris, Musée du Louvre, inv. E 26792, according to report from the Fayoum:
   Psalms 133:1, Fayoumi-Sahidic. Shawl; see below Appendix II, no. 2.
   (Psalms 133:1) Bless the Lord, all (you) servants of the Lord, who stand in
   the house of the Lord, in the courts of the house (of our God).

Appendix II

New Psalm texts from early medieval Egyptian textiles

In this appendix, the Coptic Psalm texts from two textiles discussed in Cäcilia
Fluck and Gisela Helmecke (eds.), *Textile messages: inscribed fabrics from
Roman to Abbasid Egypt* are edited for the first time. I thank the keepers of both
collections concerned for the permission to present the texts here and for the
excellent photos and additional information supplied by them. The following edition was made after the photos, but the text on the Antwerp piece was collated
with the original. Since only the texts are the concern here, the technical information provided about the textiles themselves is kept brief. Otherwise, the edition
adheres to the procedures current in Coptic epigraphy.

1. Antwerp, collection Katoen Natie, inv. 711 (Figs. 2.1–2.2)

Two horizontal lines of text, yellow on a dark-blue background, near the short
upper and lower ends of a woollen shawl, 220.5 × 105 cm. Each line closely

follows a decorative line of pseudo-Arabic ("Kufesque"), situated immediately beneath (l. 1) and above (l. 2) the Coptic. Contrary to the "Kufesque", the two Coptic lines are not mirrored over the axis of the shawl, but are written in the same direction. The lines almost occupy the full width of the shawl, leaving left- and righthand margins of about 20–25 mm; length of the lines, 98 (l. 1) and 92 (l. 2) cm. Each line is opened and closed by an elaborate cross-like device, while a third cross in the middle divides them into two symmetrical halves. Near the upper and lower edges of the shawl, colorful borders of decorative design can be found. The shawl is complete except for minor damage; l. 2 has some lacunae near the middle of the text.

Script: square and upright uncials, about 8 mm high (ⲥ, 10–11 mm); carefully and well drawn, though a bit stiff; well ruled and aligned. The upper curl of the oddly shaped ("square") ⲥ projects above the line;[117] ⲙ is low and wide; symmetric high-stemmed ⲩ. Only one letter (probably a ⲉ), near the end of l. 2, was badly drawn.

Both lines contain an essentially identical text, although l. 2 continues for a few words more than l. 1. The language is a variety of Fayoumi-Sahidic (S[f]) as can be frequently found in late sources (about tenth-eleventh century) from the Fayoum (see below). Apart from this, neither text nor script provides clues for a more precise dating.

Provenance: unknown. Linguistic and stylistic evidence would be in favor of an origin from the Fayoum.

Bibliography: unpublished, but cf. Boud'hors and Calament 2004: 469, Annexe, no. 38; De Moor, Verhecken-Lammens and van Strydonck 2006: 224–25 and col. fig. 28.

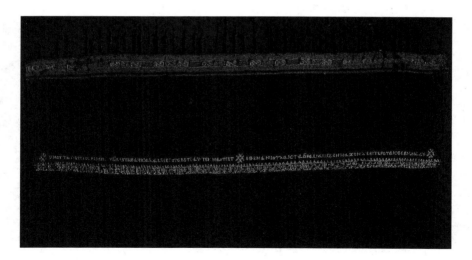

Figure 2.1 Antwerp, collection Katoen Natie, inv. 711: upper line of Coptic text (photo: © The Phoebus Foundation)

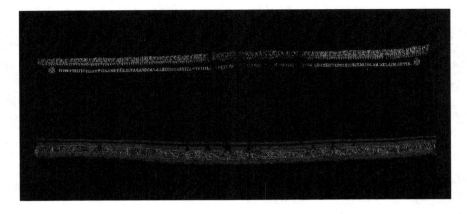

Figure 2.2 Antwerp, collection Katoen Natie, inv. 711: lower line of Coptic text (photo: © The Phoebus Foundation)

(1) ※ ⲡⲛⲟⲩⲧⲉⲓ ⲡⲉⲧⲙⲟⲩⲣ ⲙⲁⲓ ⲛⲟⲩϭⲁⲙ
(2) ※ ⲡⲛⲟⲩⲧⲉⲓ ⲡⲉⲧⲙⲟⲩⲣ ⲙⲁⲓ ⲛⲟⲩϭⲁⲙ

(1) ⲧⲁⲃ ⲁϥⲕⲁ ⲇⲁϩ<ⲓ>ⲉ ⲉⲥⲥⲟⲩⲁⲃ
(2) ⲧⲁⲃ ⲁϥⲕⲁ ⲗⲁϩⲓⲉ ⲉⲥⲥⲟⲩⲁⲃ

(1) ⲉⲧⲥⲁⲩⲧⲉⲛ ⲛⲁⲟⲩⲣⲉⲧ-※ ⲉ ⲑⲉ ⲛⲁ ⲛⲓⲟⲩⲗ
(2) ⲉⲧⲥⲁⲩⲧⲉⲛ ⲛⲁ[ⲟ]ⲩⲣⲉⲧⲉ ⲑⲉ [※] ⲛ̣ⲁ ⲛⲓⲟⲩ[ⲗ

(1) ⲁⲕⲧⲁϭⲣⲁ ⲙⲁⲓ ϩⲓϫⲉⲛ ⲛⲁϫⲉⲛϫⲉⲉⲓ ⲧⲉⲣⲟⲩ
(2) ⲁⲕⲧ]ⲁϭⲣⲁ ⲙⲁⲓ [ϩⲓϫⲉ]ⲛ ⲛⲁϫⲉⲛϫⲉⲉⲓ ⲧⲉⲣⲟⲩ

(1) ⲁⲕⲑⲉⲃⲓⲁ ⲛⲁϫⲁ- ※
(2) ⲁⲕⲑⲉⲃⲓⲁ ⲛⲁϫⲁϫⲉ ϩⲁ ⲡⲉϩⲏⲩ ⲛⲁ- ※

ⲗⲁϩⲓⲉ (l. 2): writing error for ⲇⲁϩⲓⲉ (thus l. 1), both for ⲧⲁϩⲓⲉ; ⲉⲥⲥⲟⲩⲁⲃ: l. ⲉⲥⲟⲩⲁⲃ (or -ⲟⲩⲁⲁⲃ); ⲉⲧⲥⲁⲩⲧⲉⲛ ⲛⲁ-: thus rather than ⲉⲧⲥⲁⲩⲧⲉ (for -ⲥⲁⲃⲧⲉ) ⲛⲛⲁ- (see below); ϩⲁ ⲡⲉϩⲏⲩ (l. 2 only): second ϩ oddly shaped, continuation uncertain (see below).

(Psalms 17:33a) It is God who girds me with strength;
(33b) He made my way spotless,
(34a) (He) who straightens my feet like those of hinds.
(-) You made me superior over all my adversaries;
(-) And humiliated all my enemies behind (?) my (*sic desinit*).

The text comprises a Coptic version of Psalms 17:33ab-34a followed by two other verses in the same vein (including the characteristic *parallelismus membrorum*) that cannot be identified with certainty. Due to the preceding text and a clear similarity in spirit, it seems best to connect these with Psalm 17 as well. As was suggested already by Boud'hors and Calament 2004: 469, they may well represent a paraphrase or abbreviated rendering of Psalms 17:40b-41a.

Since no Fayoumic version of the corresponding verses of Psalm 17 (David's great hymn of thanksgiving for his victory over Saul) survives, it is difficult to pronounce a judgment on the form of the text.[118] It may, however, be noted that our text has in v. 33b the third person, with the Middle Egyptian (M, Gabra) and the Bohairic (B) against the second person of the Sahidic (S, Budge) and the LXX (Rahlfs). Instead of ϲⲁⲩⲧⲉⲛ (v. 34a), the other versions have ϲⲟⲃⲧⲉ/ϲⲟⲃϯ/ϲⲁⲃⲧⲉ. Other examples of this interchange, although not in the same Psalm verse, can be cited.[119] It can be explained by the phonological similarity of ϲⲁⲩⲧⲉⲛ ⲛ- and ϲⲁⲃⲧⲉ ⲛ- and perhaps also by the current use of ϲⲟⲟⲩⲧⲛ/ϲⲁⲩⲧⲉⲛ in semantically related expressions (in particular: "to straighten the way of . . .").

The similarity of the two final verses with Psalms 17:40b-41a (LXX: "You fettered under me all that rose up against me; You gave me the back of my enemies") can only be called approximative, since quite different verbs are used and, instead of the expression "all that rose up against me", a simple "all my enemies" is found. Moreover, the interpretation of the end of the last one (in l. 2 only) is uncertain. If indeed, as is quite plausible, ϩⲁ ⲡⲉϩⲏⲩ should be read (for a more common ϩⲓ ⲡⲁϩⲟⲩ, or the like) one would expect a continuation ⲙⲁⲓ (text: ⲛⲁ-), in conformity with the M version (Gabra) of v. 41a: ϩⲁⲕⲣⲓⲟⲩⲉ ⲛⲛⲁϫⲉϫⲉ ⲛϲⲁ ⲡⲉϩⲟⲩ ⲙⲙⲁⲓ ("You repelled my enemies at the back of me"). Correcting the doubtful word into ϩⲁ ⲡⲉϲⲏⲧ is not my preferred option, but would obtain something quite close to the very differently phrased S version (Budge) of the same verse: ⲁⲕϯ ⲛⲛⲁϫⲁϫⲉ ϩⲁ ⲛⲁⲟⲩⲉⲣⲏⲧⲉ ("You laid down my enemies under my feet").

The language is a variety of the Fayoumi-Sahidic that is commonly found in tenth-eleventh-century documents[120] and inscriptions[121] from the southeastern basin of the Fayoum.[122] As usual, there is no trace of lambdacism, but (F/M) ⲁ occurs normally for (S/B) ⲟ. Also F forms (ⲡⲛⲟⲩⲧⲉⲓ) appear side by side with S ones (ϫⲁϫⲉ). Another characteristic feature is the tendency towards suppression of the initial, pre-consonantal ⲛ-/ⲙ-. Here, ⲙⲁⲥ is found for ⲙⲙⲁⲥ, ⲧⲁⲃ for ⲛⲧⲁⲃ (=ⲛⲧⲁϥ, S ⲛⲧⲟϥ), ⲑⲉ ⲛⲁ- for ⲛⲑⲉ ⲛⲛⲁ-, but note ⲛⲟⲩϭⲁⲙ. (Thanks to the latter feature we can be quite sure that indeed ⲉⲧϲⲁⲩⲧⲉⲛ ⲛⲁ-, not ⲉⲧϲⲁⲩⲧⲉ ⲛⲛⲁ-, is to be read.) Also noteworthy are ⲁⲁ- for possessive ⲧⲁ- (in ⲁⲁϩⲓⲉ)[123] and the three instances of ⲉ for ⲏ (in ϩⲓⲉ, ⲟⲩⲣⲉⲧⲉ, ⲧⲉⲣⲟⲩ).[124] In ⲉϲϲⲟⲩⲁⲃ an intervocalic consonant (ϲ) is doubled, a feature also occurring in the Louvre shawl, discussed below. In spite of some minor slips, the artist has produced a careful and well-readable text in what may tentatively be considered a late-Coptic idiom of the Fayoum. Stylistic similarities with other textiles, produced in Tutun,[125] suggest

JACQUES VAN DER VLIET

that the Antwerp shawl could indeed come from that province, perhaps more particularly from its southeastern part.

2. Paris, Musée du Louvre, inv. E 26792
(old AC 826; Figs. 2.3–2.4)

Two horizontal lines of text, white on a brown background, near the short upper and lower ends of a woolen shawl, present size 102 × 62 cm. The lines almost occupy the full width of the shawl, leaving narrow left- and righthand margins; length of the lines, 60.5 (l. 1) and 59 (l. 2) cm. Both lines are written in the same direction and are not mirrored over the axis of the shawl. Each line is opened and closed by a cross (absent at the end of l. 2). In spite of the moderate overall state of the shawl, the text appears to be reasonably well preserved. There are few brief lacunae only. However, the piece has been crudely sewn upon a piece of modern textile, which gave rise to occasional dislocation of the text, particularly in l. 1. On the photos, the extremities of the lines are still covered by a modern frame that will be removed in the course of restoration.

Script: square and upright uncials, about 8 mm high; carefully and well drawn, although rather stiff; slightly less regular than on the Antwerp piece. Four-stroke square м; asymmetric ү; ϩ with a long and low rightward tail; λ with a dropping head. The strokes above the nomen sacrum π͞ο͞ϭ appear to lack in l. 2.

Both lines contain an essentially identical text. As on the Antwerp shawl (no. 1, above), the language is an informal variety of Fayoumi-Sahidic (Sᶠ) that suggests a late date (roughly tenth-eleventh century). Text and script provide no other clues for dating.

Provenance: from the Fayoum, according to the indications of the former owner, Dr. P.A. Puy-Haubert, an Alexandrian physician; his collection was acquired by the Louvre in 1959;[126] the present piece was accessioned in March 1960.

Bibliography: text unpublished; for the object, see Bourguet 1964: 647, L 2 (as no. AC 826); mentioned in Bénazeth, Durand and Rutschowscaya 1999: 155, n. 60; see furthermore Bénazeth 2006: 122.

(1) ✕ cмoy eπϭ(ωι)c ne̲[ϩϩeм]ϩaλ тнроy eнтн πϭ(ωι)c
(2) ✕ {ɴ}cм̲[oy] eπϭ(ωι)c neϩϩeмϩaλ тнроy eнтн π<ϭ>(ωι)c

(1) neтaϩϩιpaттоy ϩeм пнι eмπϭ(ωι)c ϩeн nay- ✕
(2) <ɴ>eтaϩϩιpaттоy ϩeм пнι eмπϭ(ωι)c ϩeн na<y>λн eмпнι

eπϭ(ωι)c: π͞ο͞ϭ (twice): first π redundant; neϩϩeмϩaλ: l. neϩeмϩaλ; eнтн for more properly eнтe; neтaϩϩιpaттоy: l. neтaϩιpaтоy; eмπнι (l. 2): end of line, but no cross follows.

(Psalms 133:1a) Bless the Lord, all (you) servants of the Lord,
(1b) who stand in the house of the Lord, in the courts of the house (of our God).

48

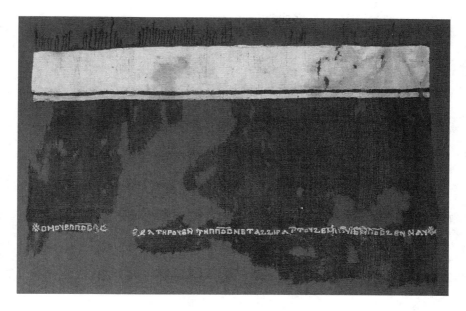

Figure 2.3 Paris, Musée du Louvre, inv. E 26792: upper line of Coptic text
(photo: © Claire Beugnot, 2016)

Figure 2.4 Paris, Musée du Louvre, inv. E 26792: lower line of Coptic text
(photo: © Claire Beugnot, 2016)

The text comprises the major part of Psalms 133:1ab, a verse belonging to the best known of the entire Psalter. In monastic liturgy throughout the Christian world, Psalms 133 has a prominent place as an "invitatorium" in the nocturnal hours.[127] On account of the very outspoken liturgical character of the text, there can be little doubt that this shawl was made for a member of the clergy, most probably a monk (see the discussion above).

The language of the inscription is, again, a variety of S^f, though with slightly different characteristics from that of the Antwerp shawl. No F versions of Psalms 133:1 survive; the S (Budge) version runs thus:

(1a) cмоү епхоеіc нємгаλ тнроү мпхоеіc
(1b) нетагератоү гм пнι мпхоιс гн наүλн мпнι мпенноүте.

This S version does not render the initial ἰδοὺ δὴ ("Ecce nunc") of the LXX (B: гнппе λε; M: геϊпе λн); this element is lacking on the Louvre shawl too. In two other major respects the Louvre text diverges from the S (Budge) version: the B/F compendium п̄о̄с̄ is used[128] and, in v. 1a, the genitival preposition нте (here spelled: ентн) occurs, also a usage favored by B/F (and actually attested in the B version of this verse). Other differences are rather of an orthographical nature and are either well attested in S^f texts or can easily be explained by F influence: schwa is spelled out (гем for гм, etc.); -агерат⸗ has ι for е; нте is spelled ентн (which actually is the pronominal state form). Furthermore, the copyist has a preference for doubling certain (intervocalic) consonants (e.g. нетаггіраттоү for нетагіратоү; cf. ессоүав, for есоүав, in the Antwerp shawl, above). Occasionally, he left out a letter by mistake. On the whole, however, the artist produced a carefully rendered and easily identifiable text, in an idiom that confirms the stated provenance.

Addenda

Cäcilia Fluck kindly drew my attention to two fragmentary silk textiles in the Vatican (Volbach 1942: 20–1, nos. T 19 and T 20). One of these is inscribed with Psalms 109:4 and 5 in Greek, a selection that clearly alludes to priesthood (as Volbach observed), the other one bears Psalms 120:7 in Latin, a quote with a definite apotropaic ring. Note that the provenance and nature of both these fragments is unknown (Volbach hesitatingly attributes them to Rome).

Psalms 44 may be quoted on the rim of the *maphorion* of the Virgin Mary in late Byzantine representations; for discussion, see Babić 1991.

Notes

1   I wish to thank Dominique Bénazeth, Cäcilia Fluck and Antoine De Moor for their generous assistance; Joost Hagen for his biblical scholarship; and Tasha Vorderstrasse and Ewa Zakrzewska for their incisive criticism.

2   Blair 1998, a handsome introduction to Islamic epigraphy, reviews all these categories.

"IN A ROBE OF GOLD"

3   See Fluck 1997: 66–7; cf. Bénazeth, Durand and Rutschowscaya 1999: 148.

4   Gaselee 1924.

5   Pioneer studies are Fluck 1996–1997 and Bénazeth, Durand and Rutschowscaya 1999.

6   See Bénazeth, Durand and Rutschowscaya 1999: 146, who correctly stress their relationship with other kinds of mural decoration.

7   Inv. 9334; cf. Jurriaans-Helle and van Beek 2000: 13 (the reading of the title is uncertain, but it is hardly probable that the person in question was a Roman soldier with the military title of *benificarius*, as the authors suggest).

8   Cited in Elsner 1998: 103.

9   Various collections, see e.g. Forrer 1893: 25, pl. 14, fig. 1; Kendrick 1921: 18, no. 326, pl. 4 (cf. Gaselee 1924: 76–7, no. 6); Shurinova 1967: cat. no. 35, pl. no 33.

10   Kendrick 1921: 19, no. 329, pl. 11.

11   Gaselee 1924: 78–9, no. 8; Kendrick 1921: 19, no. 329 dated this fragment to the fifth or sixth century, but note that if Gaselee's plausible readings are correct, the piece would be non-Christian and its recipient, the pagan god Zeus! [Jacques' copy of *Textile messages* adds: "incorrect!"] For the donation of luxury garments, inscribed with the names of the donors, to churches, see John of Ephesus, *Lives of the Eastern Saints*, ch. 55 ("History of John and Sosiana"; Brooks 1926: 192 [538]–195 [541]; I owe this reference to Tasha Vorderstrasse).

12   I owe this characterization of late antique Egypt to Bagnall 1993: 260, n. 155, who gives further references.

13   See, for the social mechanisms at work in late antiquity, Brown 1992: 35–47 and the chapter on "Art and social life" in Elsner 1998: 91–113.

14   Both pseudo-script (non-letters) and pseudo-text (non-text) are quite common on Egyptian textiles; an example of pseudo-Coptic: Vienna, Papyrussammlung, coll. Tamerit, inv. T 6 (Cat. Vienna 1999: 108–09, no. 101).

15   See the fine study by Beck 2001.

16   A point rightly stressed for medieval inscribed textiles by Bierman 1997: 114–15.

17   Cf. Bénazeth, Durand and Rutschowscaya 1999: 146.

18   Rutschowscaya 1990.

19   See Durand 2006: 83–94.

20   Inv. 6847; see Rutschowscaya 1990: 140 (cf. 137); see also Fluck 2006: 155–56; Mälck 2006, col. fig. 33.

21   As the image has a potency of its own, legends could be dispensed with; see Maguire 1995: 56–7.

22   An *orbiculus* in Washington, Textile Museum, inv. 11.18 (see Rutschowscaya 1990: 143; cf. Maguire 1995: 56); see also Durand 2006: 86 and col. fig. 7.

23   A hanging in Stuttgart, Württembergisches Landesmuseum, inv. 1984–103 (Cat. Hamm 1996: 368–69, no. 421, with further literature).

24   Maguire 1990; see also its sequel: Maguire 1995; for a broader perspective: Dauterman Maguire, Maguire and Duncan-Flowers 1989.

25   Inv. 23.177; see Maguire 1990: 217, fig. 8. An almost identical piece in the collections of the Katoen Natie, Antwerp, inv. 437, is inscribed ⲕⲁⲧⲁⲧⲣⲓⲃⲉ; see De Moor, Verhecken-Lammens and van Strydonck 2006: 223 and fig. 73, and note 73 below.

26   Interesting examples of which are discussed in Van den Hoek, Feissel and Herrmann 1994; see also Dauterman Maguire, Maguire and Duncan-Flowers 1989.

27   *CPG*, no. 2529; Layton 1987: 197–98, no. 163 (4). The Sahidic text is edited, after a late tenth-century manuscript, in Budge 1915: 512–23.

28   Budge 1915: 521 (the italics are my own).

29   Budge 1915: 519–20.

30   See Plumley 1982: 92 (correcting his translation), who already connects the phylactery with the Michael text.

31 Inv. 9993; see now Fluck, Linscheid and Merz 2000: 223–25, no. 155, pl. 15 (with bibliography); Fluck 2006: 164; Mälck 2006, col. fig. 35.

32 For Islamic inscribed textiles and *tirâz* in particular, see the recent introductions in Blair 1997a; Blair 1998: 164–81; Stillman 2000: 120–37; and Stillman and Sanders 2000 (with a rich bibliography).

33 Thus Lombard 1978: 190–99, where a striking description is given.

34 Cf. Bénazeth, Durand and Rutschowscaya 1999: 148, and the discussion below.

35 Also the word *tirâz* found its way into Coptic (ⲁⲧⲧⲓⲣⲁⲥ; see Hasitzka 1998: 34).

36 Now restored to their full glory, see Bolman 2002; Cornu 1997: 56–63, does not include Coptic wall paintings among her Christian iconographical sources for "Islamic" dress.

37 See William Lyster, in Bolman 2002: 109–11 (with 112, pl. 7.14, showing Patriarchs Severus and Dioscorus with *tirâz* bands on their shoulders).

38 See Bolman 2002: 59, pl. 4.25 (cf. 230, Birger A. Pearson); for a thorough discussion of the scene: Van Loon 1999: 103–08, 176–80.

39 See Walters 1989: 194 (with pl. 18), 195 and 206.

40 Even Byzantine dignitaries in eleventh-century Cappadocia can be seen wearing Arabic *tirâz* inscriptions on their sleeves and turban; see Rodley 1985: 199, fig. 38 (f and c). Other examples are mentioned in Cornu 1997: 56–63.

41 Cairo, Museum of Islamic Art, inv. 10846. It is traditionally called the "turban of Samuel ibn-Mûsâ" and dated to June 707 (thus still in Stillman 2000: 125; Stillman and Sanders 2000: 536). However, the date is highly questionable and the father's name definitely reads Murqus (with a *sîn*, which was not uncommon in medieval Egypt), not Mûsâ; see Muḥammad 'Abbâs 1997: 65–6 (but note that his transcription is inaccurate). It should be added that the piece, although made for a Christian, as his name shows, does not bear an overtly Christian character; the text and decoration are religiously neutral.

42 See, preliminarily, Godlewski 2002; the inscriptions are being studied by Gisela Helmecke, Berlin.

43 *Pace* Fluck 1997: 61–2.

44 Thus perhaps in Appendix I, no. 6.

45 Thus in Louvre, inv. E 26798 (Van der Vliet 2000; Cat. Rouen 2002: 129–30, no. 95) and Berlin, Museum für Byzantinische Kunst, inv. 10065 (Fluck 1996: 165–66, pl. 10b, fig. 5; Cat. Hamm 1996: 184–85, no. 323c).

46 For the term "Kufesque", see Blair 1998: 79–80.

47 Apart from crosses and geometrical designs, even figurative motifs of the late antique tradition can still be found, e.g. in Appendix I, nos. 1 and 5. See further Fluck 1997: 59–61. A piece in the Victoria and Albert Museum combines long (pseudo-?) Coptic and Arabic inscriptions with conspicuous traditional imagery, including the apotropaic horseman motif (Kendrick 1922: 11, no. 627, pl. 7; cf. Gaselee 1924: 82, no. 19).

48 See Durand and Rettig 2002; their arguments for dating this group are flimsy, however; see now De Moor, Verhecken-Lammens and van Strydonck 2006: esp. 224–26.

49 The Louvre tunic, inv. E 26798 (quoted above) and our Appendix I, no. 7 (both illustrated in Cat. Rouen 2002: 129–30, no. 95, and 101, fig. 1).

50 Some examples of the latter combination are discussed by Durand and Rettig 2002.

51 See the provisional list of inscribed textiles from the Fayoum or in Fayoumic dialects in Boud'hors and Calament 2004: 468–69, Annexe, nos. 31–40, and our Appendix II.

52 An exception would be the Louvre shawl, inv. E 26793, that appears to bear a ruined Diocletian date; see Bourguet 1964: 647, L 1; Boud'hors and Calament 2004: 468, Annexe, no. 31; see also Bénazeth 2006: 122.

53 Van der Vliet 2000: 242.

54 See Cat. Rouen 2002: 199, no. 165; a similar example is the New York shawl of Pantouleos, Metropolitan Museum of Art, inv. nos. 31.19.13 and 31.19.15. Both are now discussed in Durand and Rettig 2002.

"IN A ROBE OF GOLD"

55 Inv. E.127.1903; Martin 2005: 168–72, no. 115.
56 For Greek and Coptic in medieval Egyptian liturgy, see Quecke 1970: 131–35.
57 Possibly, the Greek selected for this text reflects the high rank of the owner/destinee; see below.
58 See above, note 47.
59 A less known example: Leiden, Rijksmuseum van Oudheden, inv. F 1928/10.2 (Van't Hooft et al. 1994: 183, no. 444, pl. 33). Cf. Fluck 1996: 167–68; Fluck 1997: 65; Fluck 2006: 164, on Berlin, Museum für Byzantinische Kunst, inv. 9304.
60 E.g. Berlin, Museum für Byzantinische Kunst, inv. nos. 9949 and 9992 (Fluck, Linscheid and Merz 2000: 234–35, 232, nos. 167 and 64); for inv. no. 9949 see Fluck 2006: 162 and fig. 33; Louvre, inv. E 26798 (Van der Vliet 2000: 241); cf. Fluck 1996: (170; Bénazeth, Durand and Rutschowscaya 1999: 148.
61 See Goitein 1983: 196. The primary source of inspiration for both Jewish and Christian invocations of this type will rather have been liturgy.
62 Coptic text of the New York shawl of Pantouleos, Metropolitan Museum of Art, inv. nos. 31.19.13 and 31.19.15 (see Durand and Rettig 2002: 168–70; Boud'hors and Calament 2004: 468, Annexe no. 34).
63 Satzinger 1968: 115, no. 387, ll. 1–2; similar prayers occur in Coptic colophons and monumental epigraphy; see Van der Vliet 2002–2003: 143–46 [Study 7]; for parallels in Byzantine epigraphy, see Feissel 1980: 515–16.
64 Thus already Bénazeth, Durand and Rutschowscaya 1999: 148, where further examples from textiles are quoted.
65 The classic discussion of the invocation remains Peterson 1926; among more recent studies, see Diethart 1982 (documents); Van den Hoek, Feissel and Herrmann 1994: 48 and 56–9 (personal adornment).
66 *CANT*, no. 239.
67 Jacques 1969: 202.
68 See Blair 1997b.
69 Inv. 25405 (Cat. Rouen 2002: 199, no. 165); compare the very similar New York shawl of Pantouleos, Metropolitan Museum of Art, inv. nos. 31.19.13 and 31.19.15 (see Durand and Rettig 2002: 168).
70 Only partially read in Fluck 1996: 166; my readings are after the published photo, pl. 11, fig. 6. See also Fluck 2006: 163.
71 A reading "ⲧⲁⲓⲁ: honor" seems less likely.
72 Museum für Byzantinische Kunst, inv. 9948 (Fluck, Linscheid and Merz 2000: 234, no. 166); read: ⲣⲏⳓⲉ, ⲧⲉⲗⲏⲗ, ⲟⲩⲛⲟⲩⲩ (sic).
73 Van Lantschoot 1929: in nos. 2, 15 and 40; similar formulae with ⲥⲉⲓ, "to enjoy", in nos. 53, 68 and 80. The exact Greek equivalent of Coptic ⲉⲓⲧⲉ in this context is the verb κατατρίβω that occurs on a textile fragment at the Katoen Natie in Antwerp, inv. 437 (see De Moor, Verhecken-Lammens and van Strydonck 2006: 223 and fig. 73, and above, note 25); for the use of Greek verbs like φορέω, "to wear", and χράω, "to use", in similar late antique contexts, see Van den Hoek, Feissel and Herrmann 1994: 45–51, with n. 21; note that the wish ⲉⲩⳓⲟⲣⲓ, "flourish!", quoted above after Maguire 1990: 217, is lexically related to φορέω.
74 After Van Lantschoot 1929: 5, no. 2, B, ll. 13–27; the manuscript is now in the Pierpont Morgan Library, New York (Depuydt 1993: 262–63, no. 131).
75 Van Lantschoot 1929, under the nos. quoted, describes the formula as an "acclamation" and is evasive about its interpretation (in his indices, vol. II, p. 148, he appears to derive ⲉⲧ⳿from ⲣⲧⲁⲓ); Depuydt 1993: 263, calls it "possibly a donation" (elsewhere, less correctly, a "memorial").
76 Further possible instances on textile include: Brussels, Koninklijke Musea voor Kunst en Geschiedenis, inv. Tx. 2662, quoted below; Hazorea, Wilfrid Israel Museum for Oriental Art, inv. 1215 (Cat. Jerusalem 1980: no. 272), where . . . ⲅⲉⲙ ⲡ[ⲣ]ⲁⳓⲓ ⲙⲉⲛ

ⲡⲟⲩ[ⲁϥ should be read; Bouvier collection, inv. S 436 (Stauffer 1991: 184, no. 86), also quoted below.

77   See Appendix I, no. 7, that was found *in situ*; for Arabic *tirâz* textiles used as a shroud, see Sokoly 1997.

78   In spite of occasional formulary overlap, I cannot agree with the conclusion of Fluck 1997: 62.

79   Cf. Bénazeth, Durand and Rutschowscaya 1999: 148.

80   As shown by the fragment Berlin, Museum für Byzantinische Kunst, inv. 9948 (Fluck, Linscheid and Merz 2000: 234, no. 166), with a prayer for a Maria, the daughter of Joseph, the son of Samuel; also Brussels, Koninklijke Musea voor Kunst en Geschiedenis, inv. Tx. 2662, quoted below, was made for a female destinee.

81   Berlin, Museum für Byzantinische Kunst, inv. 10065 (Fluck 1996: 165–66, pl. 10, fig. 5; Fluck 2006: 162; Cat. Hamm 1996: 284–85, no. 323c); likewise on a fragment from the Bouvier collection, inv. S 436 (Stauffer 1991: 184, no. 86), with a wish to wear the garment "in gladness: ⲣⲁϣ!".

82   Brussels, Koninklijke Musea voor Kunst en Geschiedenis, inv. Tx. 2662 (Lafontaine-Dosogne and De Jonghe 1988: 14, pl. 40; better: Forrer 1891: 25–6, pl. 8, no. 32; pl. 10, no. 8), probably again with a wish to wear the garment in "gladness" (ⲣⲏϣⲓ).

83   According to an eleventh- or twelfth-century document from the Cairo Geniza, the dowry of a Karaite girl contained a shirt with a *tirâz* embroidered in gold; Olszowy-Schlanger 1998: no. 23, l. 28, cf. p. 229 (I owe this suggestion to Tasha Vorderstrasse).

84   Goitein 1983: 196–97; cf. Stillman 2000: 56.

85   Thus the pieces from Berlin, Museum für Byzantinische Kunst, in Fluck, Linscheid and Merz 2000: 232, 234–35, nos. 164, 166, 167.

86   I fail to understand why some colleagues persist in splitting up this distinct clerical title (Greek παπᾶς) into ⲡ- (det. article-sing.-masc.) + ⲁⲡⲁ; its correct interpretation already in Crum 1939: 13b; cf. Förster 2002: 608, and in particular Derda and Wipszycka 1994: esp. 54–6.

87   Louvre, inv. E 26798 (Van der Vliet 2000; Cat. Rouen 2002: 129–30, no. 95).

88   Biktor: Louvre, inv. E 26793 (Bourguet 1964: 647, L 1; Cat. Lattes 1999: 253–54, no. 77; Boud'hors and Calament 2004: 468, Annexe no. 31); Thoter: Athens, Benaki Museum, inv. 15343 (Cat. Rouen 2002: 199, no. 166; below Appendix I, no. 3).

89   Inv. 143; see Cat. Mariemont 1997: 223, 225, no. 108; "my brother" is a common form of address for a monk.

90   Liturgical in the sense of "réservé à un usage liturgique" (Bénazeth, Durand and Rutschowscaya 1999: 154).

91   Admittedly, the (uninscribed) monastic hood of *papa* Kolthi, Louvre, inv. E 26799 (cf. Cat. Rouen 2002: 129–30, no. 95), appears to be an example of ecclesiastical dress in the technical sense. But I would now hesitate to call the tunic that went with the hood a *liturgical* garment, as I did in Van der Vliet 2000 (on the authority of Cat. Lattes 1999: 265–68, nos. 88–9, where the argument seems to contradict the position taken, in the same volume, by Bénazeth, Durand and Rutschowscaya 1999: 154).

92   See the interesting discussion in Goitein 1983: 196–98.

93   They could thus be counted with the "Occasionalia" ("Gruppe D: Gelegenheitstexte") in the classification of Horn 2000: 104–06.

94   See the detailed discussion of two examples in Appendix II, below.

95   For a traditional but still useful general introduction, see e.g. Oesterley 1937: 185–205; for details, any modern commentary can be consulted.

96   Cf. Horn 2000: 105.

"IN A ROBE OF GOLD"

97 For an amulet with Psalms 109:1 in Fayoumic Coptic, see Stegemann 1934: 25–6 and 62–3, no. 42 [105]; for the magical use of the same Psalm against an enemy, Evelyn White 1932: 390.

98 Widely used, both in Greek and in Coptic, on all kinds of amulets, see in particular Van Haelst 1976: 80–5, no. 183–202; Quecke 1979; Pillinger 2001b.

99 Cf. Quecke 1995: 114.

100 See e.g. Zanetti 1995: 62, 66.

101 See Quecke 1995; the texts in Rahlfs 1907: 238–54, after copies by W. E. Crum.

102 In the West, in the compline; see for the present-day Bohairic hours, Burmester 1967: 104 (compline; cf. 107). For an extensive discussion of the late antique–early medieval practice in the East, see Quecke 1970: 163–70.

103 Again, this does not imply that it is a liturgical garment in the technical sense ("réservé à un usage liturgique", see above). In this particular case, the text is suggestive, although by no means conclusive; whether and how such a shawl could be used during actual worship is a question that can only be decided on the basis of the vestiary habits prevailing in contemporary Egyptian liturgy.

104 Rahlfs 1907: 245.

105 Cf. the introductory remarks, above.

106 Here it may well be noted that the garment bearing Psalms 44:9a showed, according to report, a "gelbfarbigen Seideeinschuss" (Forrer 1891: 26), undoubtedly meant to imitate gold thread; for the royal connotations of golden yellow in the Islamic world, see Lombard 1978: 129–30; cf. on gold threads: ibidem, 190–97, 234–35; Bierman 1997; Stillman 2000: 54.

107 See for a sample of the kind of prayer- and wish-formulae that are employed, Blair 1997b.

108 See Blair 1997a: 98.

109 See, as an introduction, Van Lent 2001, with further bibliography.

110 See e.g. Van Lantschoot 1929: 30, no. 15, ll. 55–7, and *passim*; it should be emphasized that it is formally and functionally a real dating formula, no "acclamation" (*pace* Van Lantschoot).

111 A similar point seems to be made by MacCoull 1989: 203.

112 Judging from the summary description, it belongs to a well-known class of fakes for which ancient materials were used; therefore the Psalm text could be authentic.

113 It was presented by Harald Froschauer during the Berlin conference; cf. Froschauer 2006: 141–2, no. 6 and fig. 28; not enough of the text survives to permit a certain identification.

114 A new publication by Renate Pillinger, Vienna, is in preparation.

115 I was able to study the tunic during the 2003 season of the Polish Mission at Naqlun. Summary information about the burial in question: Godlewski 2001: 160; Godlewski 2002: 103, with a partial photo of the tunic: 101, fig. 1.

116 A set of working photographs were kindly put at my disposal by Cäcilia Fluck.

117 Similar shapes of ϭ occur in the eleventh-century inscriptions of the monastery church at Naqlun.

118 For remains of the Fayoumic Psalter, see Quecke 1979: 332; Boud'hors 1998: 65–75; cf. Depuydt 1993: 523, no. 279; Urbaniak-Walczak 2004: 654–63.

119 See Crum 1939: 371b.

120 See Urbaniak-Walczak 1999: 105–07 (Naqlun); cf. Stern 1885: 26.

121 Cf. Boud'hors and Calament 2004: 447–61 (Tutun); Van der Vliet 2002–2003, 143–46 [Study 7] (Tutun-Bulgusuq-region).

122 For a detailed discussion of the varieties of Fayoum Coptic, see Rodolphe Kasser in Diebner and Kasser 1989: 70–108 ("Diversité dialectale copte en MoyenneÉgypte et au Fayoum").

123 Cf. Kahle 1954: 130–31, par. 111, but also e.g. Boud'hors and Calament 2004: 461: stela New York, Metropolitan Museum of Art, inv. 10.176.38, l. 8 (ⲉⲛⲇⲁⲛ for ⲉⲛⲧⲁⲛ; Tutun, tenth century).
124 Cf. Kahle 1954: 75, par. 34.
125 For which see Durand and Rettig 2002.
126 Cf. Bourguet 1960; Bourguet 1964: 7–8; earlier notices of the collection: Bourguet 1953; Kuentz 1957.
127 See the references given in note 102 above.
128 Cf. Rodolphe Kasser in Diebner and Kasser 1989: 87.

Bibliography

Babić, G. 1991, "Le maphorion de la Vierge et le psaume 44 (45) sur les images du XIVe siècle", in: E. Kypraiou (ed.), *Euphrosunon: Aphierôma ston Manolê Chatzêdakê* I, Athens: Tameio archaiologikon poron kai apallotrioseon, 57–64.

Bagnall, R.S. 1993, *Egypt in late antiquity*, Princeton: Princeton University Press.

Beck, R.M. 2001, *Texte auf Textilien in Ostafrika: Sprichwörtlichkeit als Eigenschaft ambiger Kommunikation*, Cologne: Köppe.

Bénazeth, D. 2006, "Textiles avec inscriptions du premier millenaire, conservés au musée du Louvre (Département des antiquités égyptiennes)", in: C. Fluck and G. Helmecke (eds.), *Textile messages: inscribed fabrics from Roman to Abbasid Egypt*, Leiden/Boston: Brill, 115–29.

Bénazeth, D., M. Durand, and M-H. Rutschowscaya. 1999, "Le rapport de l'objet copte à l'écrit: des textiles aux vêtements liturgiques", in: Cat. Lattes, 145–56.

Bierman, I.A. 1997, "Inscribing the city: Fatimid Cairo", in: H.C. Ackermann (ed.), *Islamische Textilkunst des Mittelalters: Aktuelle Probleme*, Riggisberg: Abegg-Stiftung, 105–14.

Blair, S.S. 1997a, "Inscriptions on medieval Islamic textiles", in: H.C. Ackermann (ed.), *Islamische Textilkunst des Mittelalters: Aktuelle Probleme*, Riggisberg: Abegg-Stiftung, 95–104.

Blair, S.S. 1997b, "A note on the prayers inscribed on several medieval silk textiles in the Abegg Foundation", in: H.C. Ackermann (ed.), *Islamische Textilkunst des Mittelalters: Aktuelle Probleme*, Riggisberg: Abegg-Stiftung, 129–37.

Blair, S.S. 1998, *Islamic inscriptions*, New York: NYU Press.

Bolman, E.S. (ed.) 2002, *Monastic visions: wall paintings in the Monastery of St. Antony at the Red Sea*, Cairo: ARCE/New Haven: Yale University Press.

Boud'hors, A. 1998, *Catalogue des fragments coptes de la Bibliothèque Nationale et Universitaire de Strasbourg* I: *fragments bibliques*, Louvain: Peeters.

Boud'hors, A., and F. Calament. 2004, "Un ensemble de stèles fayoumiques inédites: a propos de la stèle funéraire de Pantoleos de Toutôn", in: M. Immerzeel and J. van der Vliet (eds.), *Coptic studies on the threshold of a new millennium: proceedings of the Seventh International Congress of Coptic Studies, Leiden, 27 August-2 September 2000*, Louvain/Paris/Dudley, MA: Peeters 2004, I, 447–75.

Bourguet, P. du 1953, "Un groupe de tissus coptes d'époque musulmane", *Cahiers de Byrsa* 3: 167–74.

Bourguet, P. du 1960, "Note sur un lot de tissus coptes récemment acquis au Musée du Louvre", *Bulletin de liaison du Centre international d'études des textiles anciens* 12: 7–9.

Bourguet, P. du 1964, *Musée national du Louvre: catalogue des étoffes coptes*, Paris: Editions des Musées Nationaux.

"IN A ROBE OF GOLD"

Brooks, E.W. (ed.) 1926, *John of Ephesus, Lives of the Eastern Saints* III, Paris: Firmin-Didot.

Brown, P. 1992, *Power and persuasion in late antiquity: towards a Christian empire*, Madison: University of Wisconsin Press.

Budge, E.A. Wallis. 1898, *The earliest known Coptic Psalter*, London: K. Paul.

Budge, E.A. Wallis. 1915, *Miscellaneous Coptic texts in the dialect of Upper Egypt*, London: British Museum; reprinted New York: AMS Press 1977.

Burmester, O.H.E. 1967, *The Egyptian or Coptic Church: a detailed description of her liturgical services and the rites and ceremonies observed in the administration of her sacraments*, Cairo: Société d'archéologie copte.

Burmester, O.H.E., and E. Dévaud. 1925, *Psalterii versio memphitica e recognitione Pauli de Lagarde*, Louvain: J.B. Istas.

Cat. Hamm 1996 = *Ägypten. Schätze aus dem Wüstensand: Kunst und Kultur der Christen am Nil*, Wiesbaden: Reichert Verlag 1996 (exhibition catalogue Gustav-Lübcke-Museum, Hamm).

Cat. Jerusalem 1980 = A. Baginski and A. Tidhar 1980, *Textiles from Egypt, 4th–13th centuries C.E.*, Jerusalem: L.A. Mayer Memorial Institute for Islamic Art (exhibition catalogue of the L.A. Mayer Memorial Institute for Islamic Art, Jerusalem).

Cat. Lattes 1999 = N. Bosson and S.H. Aufrère. (eds.) 1999, *Égyptes . . . : L'Égyptien et le copte*, Lattes: Musée archéologique Henri Prades (exhibition catalogue Musée archéologique Henri Prades, Lattes).

Cat. Mariemont 1997 = M-C. Bruwier (ed.) 1997, *Égyptiennes: étoffes coptes du Nil*, Mariemont: Musée royal de Mariemont (exhibition catalogue Musée royal de Mariemont, Mariemont).

Cat. Rouen 2002 = M. Durand and F. Saragoza. (eds.) 2002, *Égypte, la trame de l'histoire: textiles pharaoniques, coptes et islamiques*, Paris: Somogy (exhibition catalogue Musée départemental des Antiquités, Rouen).

Cat. Vienna 1999 = J. Henner, H. Förster, and U. Horak. 1999, *Christliches mit Feder und Faden: Christliches in Texten, Textilien und Alltagsgegenständen aus Ägypten*, Vienna: Österreichische V.-G. (exhibition catalogue Papyrusmuseum der Österreichischen Nationalbibliothek, Vienna).

Cornu, G. 1997, "Sources iconographiques pour l'étude des tissus et costumes islamiques du IXe au XIIIe siècle", in: H.C. Ackermann (ed.), *Islamische Textilkunst des Mittelalters: Aktuelle Probleme*, Riggisberg: Abegg-Stiftung, 53–63.

Crum, W.E. 1939, *A Coptic dictionary*, Oxford: Oxford University Press (and later reprints).

Czaja-Szewzak, B. 2004, "Naqlun 2003: from scraps to tunic", *PAM* 15: 159–64.

Dauterman, M.E., H.P. Maguire, and M.J. Duncan-Flowers. 1989, *Art and holy powers in the early Christian house*, Urbana: Krannert Art Museum/University of Illinois at Urbana-Champaign.

De Moor, A., C. Verhecken-Lammens, and M. van Strydonck. 2006, "Relevance and irrelevance of radiocarbon dating of inscribed textiles", in: C. Fluck and G. Helmecke (eds.), *Textile messages: inscribed fabrics from Roman to Abbasid Egypt*, Leiden/Boston: Brill, 223–31.

Depuydt, L. 1993, *Catalogue of Coptic manuscripts in the Pierpont Morgan Library* I-II, Louvain: Peeters.

Derda, T., and E. Wipszycka. 1994, "L'emploi des titres abba, apa et papas dans l'Egypte byzantine", *JJP* 24: 23–56.

Diebner, B.J., and R. Kasser. 1989, *Hamburger Papyrus bil. 1: Die alttestamentlichen Texte des Papyrus bilinguis 1 der Staats- und Universitätsbibliothek Hamburg*, Geneva: P. Cramer.

Diethart, J.M. 1982, "Κύριε βοήθει in byzantinischen Notarsunterschriften", *ZPE* 49: 79–82.

Durand, M. 2006, "Vers une pseudo-épigraphie textile en langue copte: Diogène, Panopé, Thétis", in: C. Fluck and G. Helmecke (eds.), *Textile messages: inscribed fabrics from Roman to Abbasid Egypt*, Leiden/Boston: Brill, 83–94.

Durand, M., and S. Rettig. 2002, "Un atelier sous contrôle califal identifié dans le Fayoum: le tirâz privé de Tutûn", in: Cat. Rouen 2002, 167–70.

Elsner, J. 1998, *Imperial Rome and Christian triumph: the art of the Roman Empire AD 100–450*, Oxford/New York: Oxford University Press.

Evelyn White, H.G. 1932, *The monasteries of the Wâdi 'n Natrûn II: the history of the monasteries of Nitria and of Scetis*, New York: Metropolitan Museum of Art.

Feissel, D. 1980, "Inscriptions byzantines de Ténos", *Bulletin de Correspondance Hellénique* 104: 477–518.

Fluck, C. 1996, "Koptische Textilien mit Inschriften in Berlin (I)", *BSAC* 35: 161–72.

Fluck, C. 1997, "Koptische Textilien mit Inschriften in Berlin (II)", *BSAC* 36: 59–70.

Fluck, C. 2006, " 'Denkt liebevoll an mich . . .': Textilien mit Inschriften im Museum für Byzantinische Kunst, Berlin", in: C. Fluck and G. Helmecke (eds.), *Textile messages: inscribed fabrics from Roman to Abbasid Egypt*, Leiden/Boston: Brill, 151–71.

Fluck, C., P. Linscheid, and S. Merz. 2000, *Textilien aus Ägypten I: Textilien aus dem Vorbezitz von Theodor Graf, Carl Schmidt und dem Ägyptischen Museum Berlin*, Wiesbaden: Reichert Verlag.

Forrer, R. 1891, *Die Graeber- und Textilfunde von Achmim-Panopolis*, Strasbourg: E. Birkhäuser.

Forrer, R. 1893, *Die frühchristlichen Alterthümer aus dem Gräberfelde von Achmim-Panopolis*, Strasbourg: F.Lohbauer.

Froschauer, H. 2006, "Koptische Textilien mit Inschriften in der Papyrussammlung der Österreichischen Nationalbibliothek", in: C. Fluck and G. Helmecke (eds.), *Textile messages: inscribed fabrics from Roman to Abbasid Egypt*, Leiden/Boston: Brill, 131–50.

Förster, H. 2002, *Wörterbuch der griechischen Wörter in den koptischen dokumentarischen Texten*, Berlin/New York: de Gruyter.

Gabra, G. 1995, *Der Psalter im oxyrhynchitischen (mesokemischen/mittelägyptischen) Dialekt*, Heidelberg: Heidelberger Orientverlag.

Gaselee, S. 1924, "Lettered Egyptian textiles in the Victoria and Albert Museum", *Archaeologia* 23: 73–84.

Godlewski, W. 2001, "Naqlun: excavations, 2000", *PAM* 12: 149–61.

Godlewski, W. 2002, "Les textiles issus des fouilles récentes de Naqlûn", in: Cat. Rouen 2002, 100–04.

Goitein, S.D. 1983, *A Mediterranean society: the Jewish communities of the Arab world as portrayed in the documents of the Cairo Geniza IV: daily life*, Berkeley/Los Angeles/London: Dan Wymann Books.

Hasitzka, M. 1998, "Bekleidung und Textilien auf unedierten koptischen Papyri der Papyrussammlung in Wien: Termini", *GRAFMA Newsletter* 2: 28–34.

Horn, J. 2000, "Die koptische (sahidische) Überlieferung des alttestamentlichen Psalmenbuches: Versuch einer Gruppierung der Textzeugen für die Herstellung des Textes", in: A. Aejmelaeus, and U. Quast (eds.), *Der Septuaginta-Psalter und seine Tochterübersetzungen*, Göttingen: Vandenhoeck & Ruprecht, 97–106.

"IN A ROBE OF GOLD"

Jacques, X. 1969, "Les deux fragments conservés des 'Actes d'André et de Paul'", *Orientalia* 38: 187–213.

Jurriaans-Helle, G., and R. van Beek. 2000, "Van Lineair A tot alfabet", *Mededelingenblad Allard Pierson Museum* 77: 10–14.

Kahle, P.E. 1954, *Bala'izah: Coptic texts from Deir el-Bala'izah in Upper Egypt*, London: Oxford University Press.

Kendrick, A.F. 1921–1922, *Victoria and Albert Museum: catalogue of textiles from buryinggrounds in Egypt*, vols. 2–3, London: His Majesty's stationary office.

Kuentz, C. 1957, "Tissus coptes", in: Z.V. Togan (ed.), *Proceedings of the twenty-second congress of orientalists held in Istanbul, September 15th to 22nd 1951*, II: *communications*, Leiden: Brill, 540–41.

Lafontaine-Dosogne, J., and D. De Jonghe. 1988, *Textiles coptes des Musées royaux d'Art et d'Histoire*, Brussels: Musées royaux d'art et d'histoire.

Layton, B. 1987, *Catalogue of Coptic literary manuscripts in the British Library acquired since the year 1906*, London: British Library.

Lombard, M. 1978, *Études d'économie médiévale* III: *les textiles dans le monde musulman du VIIe au XIIe siècle*, Paris/La Haye/New York: Mouton Éditeur.

MacCoull, L.S.B. 1989, "The Teshlot papyri and the survival of documentary Coptic in the eleventh century", *OCP* 55: 201–06.

Maguire, H. 1990, "Garments pleasing to God: the significance of domestic textile designs in the early Byzantine period", *DOP* 44: 215–24.

Maguire, H. 1995, "Magic and the Christian image", in: H. Maguire (ed.), *Byzantine magic*, Washington, DC: Dumbarton Oaks Research Library, 51–71.

Mälck, K. 2006, "Herstellungstechniken der beschrifteten Textilien des Museums für Byzantinische Kunst, Berlin", in: C. Fluck and G. Helmecke (eds.), *Textile messages: inscribed fabrics from Roman to Abbasid Egypt*, Leiden/Boston: Brill, 239–51.

Martin, G.T. 2005, *Stelae from Egypt and Nubia in the Fitzwilliam Museum, Cambridge, c. 3000 BC-AD 1150*, Cambridge: Cambridge University Press.

Muḥammad, Abbâs Muḥammad Salîm. 1997, "The function of some woven fabrics in Riggisberg", in: H.C. Ackermann (ed.), *Islamische Textilkunst des Mittelalters: Aktuelle Probleme*, Riggisberg: Abegg-Stiftung, 65–9.

Oesterley, W.O.E. 1937, *A fresh approach to the Psalms*, London: Ivor Nicholson and Watson.

Olszowy-Schlanger, J. 1998, *Karaite marriage documents from the Cairo Geniza: legal tradition and community life in mediaeval Egypt and Palestine*, Leiden: Brill.

Peterson, E. 1926, *ΕΙΣ ΘΕΟΣ: Epigraphische, formgeschichtliche und religionsgeschichtliche Untersuchungen*, Göttingen: Vandenhoek & Ruprecht.

Pillinger, R. 2001a, "Drei Amulettarmbänder mit Psalmzitaten", in: U. Horak (ed.), *Realia coptica: Festgabe zum 60. Geburtstag von Hermann Harrauer*, Vienna: Holzhausen, 75–80.

Pillinger, R. 2001b, "Elf Stofffragmente mit Inschriften und christlichen Darstellungen im Museum für angewandte Kunst in Wien", *Mitteilungen zur Christlichen Archäologie* 7: 35–42.

Plumley, J.M. 1982, "Nubian Christian numerical cryptograms: some elucidations", in: P. van Moorsel (ed.), *New discoveries in Nubia: proceedings of the colloquium on Nubian studies, The Hague 1979*, Leiden: NINO, 91–7.

Quecke, H. 1970, *Untersuchungen zum koptischen Stundengebet*, Louvain: Peeters.

Quecke, H. 1979, "Ein Faijumisches Fragment aus Ps 90 (91) (pHeid. Kopt. 184)", in: M. Görg and E. Pusch (eds.), *Festschrift Elmar Edel, 12. März 1979*, Bamberg: Manfred Görg, 332–37.

Quecke, H. 1995, "Psalmverse als 'Hymnen' in der koptischen Liturgie?" in: J-M. Rosenstiehl (ed.), *Christianisme d'Égypte: hommages à René-Georges Coquin*, Louvain: Peeters, 101–14.

Rahlfs, A. 1907, *Septuaginta-Studien* II: *Der Text des Septuaginta-Psalters*, Göttingen: Vandenhoeck & Ruprecht.

Rahlfs, A. 1931, *Psalmi cum Odis*, Göttingen: Vandenhoeck & Ruprecht; reprinted Vandenhoeck & Ruprecht 1967.

Riegl, A. 1889, *Die ägyptischen Textilfunde im K. K. Österreich. Museum: Allgemeine Charakteristik und Katalog*, Vienna: R. v. Waldheim.

Rodley, L. 1985, *Cave monasteries of Byzantine Cappadocia*, Cambridge: Cambridge University Press.

Rutschowscaya, M-H. 1990, *Tissus coptes*, Paris: Adam Biro.

Satzinger, H. 1968, *Koptische Urkunden* III, Berlin: Verlag Bruno Hesling.

Shier, L.A. 1952, "An embroidered hanging with Coptic inscription from Egypt", *American Journal of Archaeology* 56: 176–77.

Shurinova, R. 1967, *Coptic textiles: collection of Coptic textiles, State Pushkin Museum of Fine Arts, Moscow*, Leningrad: Izd. Avrora.

Sokoly, J.A. 1997, "Between life and death: the funerary context of ṭirâz textiles", in: H.C. Ackermann (ed.), *Islamische Textilkunst des Mittelalters: Aktuelle Probleme*, Riggisberg: Abegg-Stiftung, 71–8.

Stauffer, A. 1991, *Textiles d'Egypte de la collection Bouvier/Textilien aus Ägypten aus der Sammlung Bouvier*, Bern: Benteli Verlag.

Stegemann, V. 1934, *Die koptischen Zaubertexte der Sammlung Papyrus Erzherzog Rainer in Wien*, Heidelberg: Carl Winters Universitätsbuchhandlung.

Stern, L. 1885, "Faijumische Papyri im ägyptischen Museum zu Berlin", *ZÄS* 23: 23–44.

Stillman, Y.K. 2000, *Arab dress: a short history, from the dawn of Islam to modern times*, ed. N.A. Stillman, Leiden/Boston/Cologne: Brill.

Stillman, Y.K., and P. Sanders. 2000, "s.v. Ṭirâz", in: *Encyclopaedia of Islam* 10, new edition, Leiden: Brill: 534–8.

Thomson, C. 1954, *The Septuagint Bible: the oldest version of the Old Testament*, revised edition by C.A. Muses, Indian Hills: Falcon's Wing.

Urbaniak-Walczak, K. 1999, "Deir el-Naqlun: Die koptischen Texte aus der Ermitage Nr. 25", *JJP* 29: 93–136.

Urbaniak-Walczak, K. 2004, " 'Hermeneiai'-Fragmente oder den 'Hermeneiai' verwandte Texte aus Deir el-Naqlun (Faijum)", in: M. Immerzeel, and J. van der Vliet (eds.), *Coptic studies on the threshold of a new millennium: proceedings of the Seventh International Congress of Coptic Studies, Leiden, 27 August-2 September 2000*, Louvain/Paris/Dudley, MA: Peeters 2004, I, 647–68.

Van den Hoek, A., D. Feissel, and J.J. Herrmann. 1994, "Lucky wearers: a ring in Boston and a Greek epigraphic tradition of late Roman and Byzantine times", *Journal of the Museum of Fine Arts, Boston* 6: 41–62.

Van der Vliet, J. 2000, "A Naqlûn monk brought home: on the provenance of Louvre inv. E 26798–9", *BSAC* 39: 239–44 [Study 11].

Van der Vliet, J. 2002–2003, "Monumenta fayumica", *Enchoria* 28: 137–46 [Study 7].

Van Haelst, J. 1976, *Catalogue des papyrus littéraires juifs et chrétiens*, Paris: Sorbonne.

Van Lantschoot, A. 1929, *Recueil des colophons des manuscrits chrétiens d'Égypte: les colophons coptes des manuscrits sahidiques* I, Louvain: J.B. Istas; reprinted Milan: Cisalpino- La Goliardica 1973.

"IN A ROBE OF GOLD"

Van Lent, J. 2001, *Koptische apocalypsen uit de tijd na de Arabische verovering van Egypte*, Leiden: Oosters genootschap in Nederland.

Van Loon, G.J.M. 1999, *The gate of heaven: wall paintings with Old Testament scenes in the altar room and the hûrus of Coptic churches*, Leiden: NINO.

Van't Hooft, Ph.P.M., M.J. Raven, E.H.C. van Rooij, and G.M. Vogelsang-Eastwood. 1994, *Pharaonic and early medieval Egyptian textiles*, Leiden: Rijksmuseum van Oudheden.

Volbach, W.F. 1942, *I tessuti del Museo Sacro Vaticano*, Città del Vaticano: Biblioteca Apostolica Vaticana.

Walters, C.C. 1989, "Christian paintings from Tebtynis", *JEA* 75: 191–208.

Zanetti, U. 1995, "Un index liturgique du Monastère Blanc", in: J-M. Rosenstiehl (ed.), *Christianisme d'Égypte: hommages à René-Georges Coquin*, Louvain: Peeters, 55–75.

3

CHRISTUS IMPERAT
An ignored Coptic dating formula

Jacques van der Vliet

Introduction

It is a tenacious myth that "the Copts", understood as the Monophysite majority of Egyptians, welcomed the Arab conquerors as liberators when they invaded Egypt in the autumn of 639. Undeniably, the repressive ecclesiastical politics of Patriarch Cyrus during the latter years of the reign of the emperor Heraclius (d. 641) had provoked vivid opposition among a considerable part of the population. Yet the popular image of "native" Egyptians welcoming the Arab armies as deliverers from "Greek" tyranny is not confirmed by our best sources for this period. The most important of these is the *Chronicle* of John, bishop of Nikiu, a Monophysite who wrote in Greek towards the end of the seventh century, only a few decades after the events he describes. John's *Chronicle*, in spite of its clear anti-Chalcedonian bias, does not betray the slightest sympathy for the Muslim invaders or for the "false Christians" who took their side (chapter CXXI, ed. Zotenberg).[1]

The myth of the Arab liberators is the product of a much later era, when Arab rule had become a permanent reality that needed explanation and justification, and conceptual frames were developed that enabled Egypt's predominantly Christian population to redefine its place under drastically altered circumstances. For the seventh century, however, there is nothing to suggest that John of Nikiu or his contemporaries envisaged a future outside of the familiar framework of the Byzantine Empire, whatever their religious sympathies may have been. In fact, nobody would have been able to realize that the conquerors were there to remain. Yet the impact was dramatic: Egypt was definitively severed from the Hellenistic world to which it had belonged culturally ever since Alexander the Great, and from the Roman empire of which it had been part politically since the reign of Augustus. The Roman empire, moreover, had been a Christian empire ever since Constantine the Great. From an Egyptian point of view, the new rulers belonged to a foreign nation that religiously, culturally and politically fell outside of the framework of the empire. In the course of the following centuries, Egyptians sought to accommodate to the new situation in various ways.

In a very modest way this process of gradual accommodation is reflected in the dating systems and formulae that were used in Greek and Coptic documents from post-conquest Egypt. It is the aim of this homage to Gawdat Gabra, today's foremost Egyptian Copticist, to draw the attention to a medieval Coptic dating formula that has hitherto not been recognized as such.

From empire to caliphate

Ever since Justinian issued his *Novella* 47 in 537, contracts and legal documents had to be dated after the current regnal year of the ruling emperor, the consul of the current year, the indiction year and, finally, the month and day.[2] For Egypt this meant an unintentional return to the practice of dating by regnal years that had been current under the pharaohs. From Mauricius onwards these formulae had to be preceded by an invocation of Christ, "in the name of the Lord and Master Jesus Christ, our God and Savior", which later was expanded into a Trinitarian formula.[3] Reference to the emperor was usually made in the opening colophon of a document or in oath formulae, when such were required. With the surrender of Alexandria in 642, the system of dating after the emperors and consuls of the Byzantine empire became automatically obsolete.

Yet adherence to the empire did not die in 642 or even in 652, when the Arab conquest of Egypt was completed with the pacification of its southern frontier. Whereas in official documents that touched the new regime Hijra dates were introduced very soon,[4] non-Arabic (i.e. Greek and Coptic) documents that did not concern state affairs left more room for scribal conservatism or nostalgia. In a recent paper, Arietta Papaconstantinou was able to quote various instances of the adherence to Byzantine habits precisely in dating formulae.[5] Thus, a Bohairic inscription in one of the monastic settlements of Kellia commemorates the death of a monk on: "the 1st of Thoout, indiction year 9, in the reign of Kostos".[6] The emperor can only be Constans II (641–68) and the date must therefore correspond to A.D. 651 or 666.

Soon, however, even monks had to face the effective end of Byzantine rule in Egypt, and new scribal strategies were adopted that offered acceptable alternatives for Egypt's Christian population. The most durably successful of these was dating according to the Era of Diocletian, which started to occur in normal Greek papyrus documents from A.D. 657/658 onwards and in similar Coptic ones from 720/721 onwards.[7] As an era starting from a fixed year (283/284, the first regnal year of Diocletian) it offered a practical alternative for the Islamic Hijra era, while at the same time retaining the reference to a Roman emperor, albeit one long defunct.[8] Beginning in the eighth century, it was rebaptized as the Era of the Martyrs,[9] and as the properly "Coptic" era it is still used by the Coptic Orthodox Church for official purposes.

Whenever necessary, the actual rulers of the country, the caliphs and their governors, could be mentioned in a more or less oblique way. Some seventh-century inscriptions are dated to the local incipience of Arab rule in a way that still characterizes the Arabs unequivocally as foreign occupants. Thus, an inscription from Kellia dates the death of a monk to "the 19th of Thobi in indiction year 8, the Lami (a local designation of the Arabs) being in Egypt for 24 years and three months".[10] The date situates the text in January 665 and the point of reference of the scribe may have been the so-called Treaty of Babylon of November 641, as has been suggested, or any other event that marked the advent of the Arabs in local memory.[11]

A Sahidic inscription commemorating the restoration of the church of a St. Abraham the Anchorite, from an unknown provenance in Upper Egypt, dates the event:

> in the four hundred and fourteenth year since Diocletian, (i.e.) 414, the most holy Apa Komes being bishop for 18 years, Lord Victor being . . ., the most pious Priest Theon being . . ., N.N. being lector in this holy place for 11 years, the nation (ἔθνος) of the Saracens having power over (ⲁⲙⲁϩⲧⲉ ⲉϫⲛ) the land since 55 years.[12]

This date is furthermore stated to correspond to 13 Phaophi of an eleventh indiction year, that is 10 October 697. The scribe therefore situates the beginning of Arab rule over Egypt in or around the year 642, conceivably at the fall of Alexandria. His terminology is quite explicit in qualifying the Arabs ("Saracens") as a "gentile nation" (ἔθνος) and their rule as a foreign occupation.

A new dating formula

The protocols and oath formulae of the seventh-eighth-century Coptic legal documents from Western Thebes evoke the Muslim rulers preferably as "our lords, the kings that rule over us by virtue of the command of God almighty" (*P.KRU* 1, ll. 41–4, A.D. 750). Such a formula sounds polite enough. Yet, as Arietta Papaconstantinou pointed out, the phrasing may allude to the interpretation of Muslim rule as a divine punishment for the sins of mankind, a view amply developed in paraenetic and apocalyptic literature (see below).[13] Some variants even bring out its temporary character, as if still some change was expected: "our lords, the kings that rule over us *presently* (ⲧⲉⲛⲟⲩ) by virtue of the command of God almighty" (*P.KRU* 4, ll. 25–7, A.D. 749). In a similar context, the Byzantine emperors had been qualified, in a much more positive manner, as "our lords, the Christ-loving kings" (*P.KRU* 105, l. 12, sixth-seventh century).[14]

In the same corpus of Coptic legal documents from Thebes, we find for the first time the dating formula that is the subject of this essay. A sale contract of (probably) A.D. 730 is dated:

ϩⲙ ⲡⲟⲟⲩ ⲛϩⲟⲟⲩ ⲉⲧⲉ ⲥⲟⲩ ϣⲙⲟⲩⲛ ⲙⲡⲉⲃⲁⲧ ⲙⲡⲁⲣⲙϩⲁⲧⲡ ⲛⲧⲉⲣⲟⲙⲡⲉ
[ⲧⲁⲓ ⲛⲧ]ⲣⲓⲥⲕ(ⲁⲓ)<ⲇⲉⲕⲁⲧⲏⲥ> ⲛⲧⲉⲕⲇⲓⲁⲛⲟⲥ [ⲛⲛⲁϩⲣⲛ ⲏⲗⲓ]ⲁⲥ ⲡϣⲏⲣⲉ
[ⲙⲡⲙⲁⲕ(ⲁⲣⲓⲟⲥ) ⲃ]ⲁⲥⲓⲗⲉ ⲙⲛ [ⲡⲉⲧⲣⲟⲥ ⲡϣ]ⲏⲣⲉ ⲙⲡⲙⲁⲕ(ⲁⲣⲓⲟⲥ) ⲕⲱⲙⲟⲥ
ⲉⲩⲁⲣⲭⲉⲓⲥⲑⲁⲓ ⲉⲡⲕⲁⲥⲧⲣⲟⲛ ⲛϫⲏⲙⲉ ⲉⲣⲉ [ⲡⲉⲛ]ϫⲟⲉⲓⲥ ⲓ(ⲏⲥⲟⲩ)ⲥ ⲡⲉⲭ(ⲣⲓⲥⲧⲟ)ⲥ
ⲟ̣ [ⲛⲣⲣⲟ] ⲉϫⲙ ⲡⲕⲟⲥⲙⲟⲥ ⲧⲏⲣϥ

65

on this very day, that is the eighth of the month of Paremhatp, of this very year thirteen (?) of the indiction, during the office of Elias, the son of the blessed Basile, and Petros, the son of the blessed Komos, governing the *kastron* of Jeme, *our Lord Jesus Christ being king over the entire world.*

(*P.KRU* 9, ll. 3–13)

The final phrase clearly echoes others found at the beginning of similar Theban legal documents, such as "(by) the power and the might and the permanence of our lords, the kings, who have authority over the entire country by virtue of the command of God almighty." (*P.KRU* 20, ll. 3–7, A.D. 759). Such phrases refer to the political authority prevailing at the time of writing. Only, instead of referring to worldly authority over the country of Egypt, the formula of *P.KRU* 9 claims Christ's divine authority over the entire *kosmos*.

But is this really a dating formula? A. van Lantschoot, commenting on a later example of the same formula in a book colophon, calls it without further explanation an "acclamation".[15] Whereas a historical link with acclamations can indeed be surmised (see below), there can nevertheless be no doubt that the phrase as it appears in *P.KRU* 9 and various later sources discussed below must be classed as a dating formula according to both formal and functional criteria.

Formally, it follows the pattern of both earlier and later mentions of actually ruling kings or officials when an event or a transaction is situated in their period of office. In such cases Coptic documents and inscriptions frequently use a circumstantial clause with a stative of ⲉⲓⲣⲉ. When a king or emperor is concerned, this would be the stative of the light verb compound ⲣ ⲣⲣⲟ, "to become king, emperor; to reign", equivalent of Greek βασιλεύω.[16] A sixth-century example, in Bohairic, is found in a commemorative *dipinto* at Kellia. It runs:

N.N. ⲁϥⲙⲧⲟⲛ ⲙⲙⲟϥ ⲛⲥⲟⲩ ⲓ︤ⲋ︦ ⲙⲡⲁϣⲟⲛⲥ ϧⲉⲛ ϯⲁⲭⲡ ⲓ︤ⲃ︦ ⲉϥⲟⲓ ⲛⲟⲩⲣⲟ ⲛϫⲉ ⲓⲟⲩⲥⲧⲓⲛⲓⲁⲛⲟⲩ ⲉϥⲟⲓ ⲛⲁⲣⲭⲏⲉⲡⲓⲥⲕⲟⲕⲟⲥ ⲛϫⲉ ⲁⲡⲁ ⲑⲉⲟⲇⲟⲥⲓ ⲉϥⲭⲏ ϧⲉⲛ ϯⲉϫⲱⲣⲓⲥⲧⲓⲁ

N.N. went to rest (i.e. died) on the 16th of Pashons, in the year of the indiction 12, Justinian being king, Apa Theodosi(-os) being archbishop while staying in exile.[17]

The year, a twelfth indiction in the reign of Justinian, must correspond to A.D. 549 or 564.[18]

In Nubia, where Christian kingdoms perpetuated Byzantine patterns of rule till late medieval times, documents and inscriptions continued to be dated after the current regnal year of the ruling king. Circumstantial phrases of the type described would then introduce the various officials that were in office at the time. A Sahidic document recording the sale of a piece of land in the former Dodekaschoinos begins thus:

ϩΡΑΙ ϩΝ ΤΜΝΤΕΡΟ ϩΝ ΤϢΟΡΠ ΝΡΟΜΠΕ ΜΠΕΝΦΙΛΟΧ(ΡΙϹΤΟϹ) ΠΡΡΟ (l. ΝΡΡΟ)
ΜΜΑΙΝΟΥΤΕ ΙϢϩΑΝΝΗϹ ΕΡΕ ΠΑΜΙΑΝΤ(ΟϹ) ΓΕϢΡΓΙΟϹ Ο ΜΠΡΟΤ(Ο)ΜΠΑ`Χ´
(...) (ΜΝ) ΠΡΟΤΟΔΟΜΕϹΤΙΚΟϹ (ΜΝ) ΠΡΟΤΟΜΗΖΙΤΕΡΟϹ

In the reign, in the first year of our Christ-loving (and) pious king John, the spotless George being *proto*-(. . .) [19] and *protodomestikos* and *protomeizoteros*.

(P.Lond.Copt. I 449)

Following the regnal year, a circumstantial clause situates the transaction in the period of office of a certain George, who occupied various high positions in the Makurian state apparatus at some time in the ninth century. Such clauses were part of the official protocol of Nubian documents, but they were similarly used in Egyptian inscriptions and documents as the example of the inscription commemorating the restoration of the church of St. Abraham the Anchorite, quoted above, shows. Apart from serving dating purposes, the protocol features of these texts underscore their official character and confer legitimacy on them even when they are not themselves legal documents, as in the case of commemorative inscriptions.

The formula of the type "Christ being king" serves the same purposes. Obviously, the reference to Christ's kingship cannot be of any immediate practical, chronological use. Yet both *P.KRU* 9 and the various later attestations of the formula unmistakably show that its functional domain is that of the dating clause. In *P.KRU* 9 it is found at the head of the document, immediately preceding the body of the contract, expanding the dating clause that, in addition to the day of the month and the indiction year, states the eponymous administrators of the *kastron* of Jeme in office at the time.

Later examples are found intermittently in inscriptions, book colophons and documents till the beginning of the thirteenth century, that is till the period when the Coptic scribal tradition comes to an end, without being particularly numerous. In all cases, the formula is part of the dating lemma of the text in question, either as an addition to a date according to the Diocletian or Saracene eras or at the head or the end of a series of current office holders.[20] Also in the latter case it is always part of an official protocol that serves to date and legitimize the text. The following brief review of some examples is by no means exhaustive, but is merely meant to give an impression of the genres where the dating formula of the type "Christ being king" occurs and illustrate its wide chronological and geographical range.

Some more examples

Perhaps the southernmost occurrence of the formula is found in the colophon of a Sahidic literary codex written in 1053, in either southern Egypt or northern Nubia, for a Nubian patron. The following date is given: "this very day, 15 of Paone, this very year, 769 of the Era of the Martyrs, 448 of the Saracens, Christ being king over us (ΕΡΕ ΠΕΧ(ΡΙϹΤΟ)Ϲ Ο ΝΕΡΡ`Ο´ ϩΙΧϢΝ). Amen".[21] The earliest piece in van

Lantschoot's collection of Sahidic colophons that is dated with reference to Christ's kingship can be assigned to the end of the ninth century (892/893) and originates from the Fayoum.[22] The latest example in the same collection dates to the early twelfth century,[23] while a further five belong to the ninth and tenth centuries.[24]

A rare occurence in the funerary inscription of a private person, a priest Paese, was found at the monastery of Apa Jeremiah in Saqqara. This would, moreover, be a fairly early epigraphic example, at least if the date, somewhere in the decade 754–63, is read correctly.[25] A more northern origin must be assigned to inscriptions in Bohairic. An inscription of unknown provenance in the Coptic Museum (inv. no. 359) is most likely dedicatory in character and dates to 897/898. The piece has never been properly edited and the following are my readings, after the published photo:

ϧⲉⲛ
ϯⲙⲁϩ ⲭ̅ⲓ̅ⲇ̅ ⲛⲣⲟⲙⲡⲓ ⲛⲧⲉ ⲇⲓⲟⲕⲗ-
ⲏϯⲁⲛⲟⲥ ⲉϥⲟⲓ ⲉⲛⲟⲩⲣⲱ ⲉϩⲣⲏ ⲉ-
ϫⲱⲛ ⲛϫⲉ ⲡⲉⲛⲛⲏⲃ ⲡⲭ(ⲣⲓⲥⲧⲟ)ⲥ

In the 614th year of Diocletian, our lord Christ being king over us (ll. 1–4).[26]

The commemorative stela for St. John Kame and his "spiritual son" Stephanos, from the Wadi al-Natrun, is likewise in Bohairic and from the ninth century (A.D. 858).[27]

A somewhat later example in Sahidic is again from the Fayoum. The great dedicatory inscription in the apse of the church of Naqlun (Dayr al-Malak), executed probably around 1025–1030, dates the refurbishment of the apse to the tenure of a whole series of Church dignitaries who were in office when the work was completed. The formula "our Lord Jesus Christ being king over us" follows a number of identically phrased clauses that mention the reigning archbishop, Zacharias; a local bishop, Severos; the archimandrite Papnoute, the head of the local monastic community; and two more dignitaries attached to the church of Naqlun: an archpriest Kabri (Gabriel) and one other, whose name is lost.[28]

From the same period, the episcopate of archbishop Zacharias (1004–1032), we have the historical note written in a manuscript from the Wadi al-Natrun by a scribe from the Fayoum, the deacon Joseph from Tutun. As he admits, Joseph had fled the Fayoum during the persecution of al-Hakim (996–1021) to find refuge in the Monastery of Saint Macarius. He ends his note stating that it was written: "when Abba Zacharias was archbishop over the entire country of Egypt, in this evil time that we have come upon, Christ being king over us for ever (ⲉⲣⲉ ⲡⲉⲭ(ⲣⲓⲥⲧⲟ)ⲥ ⲱ̄ ⲛⲉⲣⲁ ⲉϫⲱⲛ ϣⲁ ⲉⲓⲛⲏϩ). Amen".[29] The text, written in the Fayoumi brand of Sahidic, primarily records a particularly high Nile and a great earthquake. The latter took place on the twenty-fifth Mesore of the Diocletian year 730, which permits dating the note to A.D. 1014.

The latest example known to me is a marriage contract, probably from Middle Egypt. The body of the text, preceding the signatures of the witnesses, ends thus:

According to the will of God the Father and His only-begotten Son and the Holy Spirit, Amen. On the fourth of Thouth, of the era (year) 925 (?), Christ being king over us, our father the archbishop Abba Joannou (being patriarch) of Rakote, our father Abba Michael (being) patriarch of Antioch, our father and bishop Abba Gabriel being bishop over us in Upper [Hermou-?]polis.

(SB Kopt. I 44. 50–55)

Here, different from the apse inscription at Naqlun, the formula precedes the series of Church dignitaries in office at the time of writing and immediately follows the month and era dates. The latter date corresponds most likely to A.D. 1208.[30]

All occurrences of the dating formula of the type "Christ being king" quoted above are not private texts, but formal utterances, in principle of a public nature. Even the note of Deacon Joseph is not the scribbled expression of an individual's frustration, but a chronicler's historical entry that formally addresses "my holy fathers, the clerics" of all ranks. Colophons of literary codices are more or less public mementoes, proclaiming the munificence of their donors. Both dedicatory and commemorative inscriptions, including the epitaph of the priest Paese from Saqqara and the more ambitious monument erected for St. John Kame, were meant to be seen and recited publicly.[31] Legal contracts have likewise an official and rhetorical function. All examples are in Coptic whereas none in Greek are known to me, but this is perhaps not astonishing, given the late date of most of them. After its first appearance in a Theban document of the early eighth century, the formula is attested till the early thirteenth century. Geographically it is found from the Wadi al-Natrun in the north to Nubia in the south, in both Bohairic and Sahidic (including its Fayoumic variant).

Christ as king under Muslim rule

Given the lack of practical value of the dating formula "our Lord Jesus Christ being king over us", one may wonder what the reason for its use over many centuries may have been. One of the reasons is certainly to be sought in the nature of the colophon type of dating formulae, a way of dating that had its roots in the Byzantine rules regulating the format of a valid legal document, prevalent since the time of Justinian. Even outside of the Byzantine empire this format retained its model function for dating and, more importantly, authorizing and legitimizing written speech acts. As the medieval examples from Nubia show, the protocol specifying the regnal year of the reigning sovereign and the tenure of the highest officials of the kingdom was firmly embedded in the scribal culture of the Christian Nile Valley. As a rhetorical strategy it helped also to shape texts that were not themselves legal documents, but were yet meant to impress their audience and convey official status, such as the two Coptic dedicatory inscriptions commemorating successive restorations of the Church of the Apostles at Faras (I. Khartoum

Copt. 1, from the eighth, and no. 2 from the tenth century). The use of the "Christ being king" formula represents the same rhetorical tradition.

At the same time, the formula adheres to the rhetorics of divine kingship. The concept of Christ's kingship has New Testament roots. Initially situated in an eschatological perspective, it informed later representations of Christ's realized rule in the present. Such representations were part of Byzantine imperial ideology and had spread well beyond the borders of the Empire, for instance to Nubia and Western Europe. An inscription discovered in the 1970s in a tomb in the Alexandrian suburb of Gabbari provides an important link between imperial ideology and the post-conquest dating formula. This late antique *dipinto* shows that a Greek precursor of the famous triple acclamation of the Carolingian *laudes regiae*: *Christus vincit, Christus regnat, Christus imperat*, was already current in fifth- or sixth-century Egypt:

X(ριστ)ὲ ὁ θ(εὸ)ς ἡμῶν, δ<ό>ξα σοι,
X(ριστὸ)ς νικᾷ,
X(ριστὸ)ς βασιλεύει.
Christ, our God, Yours is the glory! Christ prevails! Christ is king![32]

Meant to legitimize the rule of "the Christ-loving kings" in a Byzantine context, the same text acquires a different meaning under Muslim rule. It now denies rather than legitimizes the political present. In its form and rhetorical purpose, the dating formula "Christ being king over us" adheres both to the Byzantine colophon tradition and to the late antique tradition of acclaiming the powers in authority.[33] Only it does so in drastically altered political circumstances.

This brings us to a final reason for the use of the "Christ being king" formula, that is of a political nature. After a few generations, the awareness that Arab rule was not a passing occupation, as the Persian rule had been, began to trigger various reactions. In addition to overtly political reactions, such as protests and revolts, the new situation demanded new ideological orientations. Rethinking the place of a Christian community under non-Christian rule was an important means of coming to grips with the new political reality. In this process, the concept of Christ's kingship proved to be of great help. Conceived in an eschatological perspective, it helped to shape the so-called apocalyptic literature that arose very soon after the Arab conquests, first in Syria and then also in Egypt, where the genre flourished for many centuries.[34] Describing Arab rule as divine punishment for sins past and present, these writings at the same time preached a hopeful message of future political change leading to the final establishment of Christ's rule. The dating formula "Christ being king over us" claims the actual realization of Christ's rule for the present of the community to which the scribe belongs. It does not simply avoid stating the regnal year of some more or less unpopular Muslim ruler, but also positively situates the donation of a manuscript, the restoration of a church, or the execution of a contract and, by the same token, the entire life of the community in the extra-temporal and cosmic reign of Christ and under his ideal suzerainty. As no change of regime could be hoped for any longer, the reflection on Christianity as a polity turned towards biblical and transcendental models rather than political strategies.[35] The Coptic dating formula "Christ being king over us", while drawing upon the forms of Byzantine imperial ideology, affirms Christian citizenship in a kingdom that is "not of this world".

Notes

1 Regrettably, the *Chronicle* of John is transmitted in a defective Ethiopic translation only (Zotenberg 1883); it is best accessible in the English translation of R.H. Charles (1916). For the chapters on the Arab conquest, see Altheim and Stiehl 1971; cf. Sijpesteijn 2007: 439–44.
2 For what follows, see Bagnall and Worp 2004: 43–54; cf. Feissel 1993.
3 Bagnall and Worp 2004: 99–109.
4 Thus, for example, in the well-known bilingual papyrus *SB* VI 9576, dated A.H. 22, or April 643.
5 Papaconstantinou 2009: 454–56.
6 Bridel, Bosson and Sierro 1999: 306, no. 134, ll. 2–3; cf. Luisier 2007: 222.
7 Bagnall and Worp 2004: 64.
8 Papaconstantinou 2009: 454–5.
9 Bagnall and Worp 2004: 67–8.
10 Bridel, Bosson and Sierro 1999: 304, no. 132-bis, ll. 2–5.
11 See Luisier 2007: 222.
12 Alexandria, Greco-Roman Museum, inv. no. A 14529 (not accessible at the time of writing). My readings are based on the mediocre photograph in Brunsch 1994; the text given in *SB Kopt.* III 1584 cannot be used.
13 Papaconstantinou 2009: 456.
14 Note that both Coptic and Greek use one word only for "king" and "emperor" (Sah. ⲣⲣⲟ / Boh. ⲟⲩⲣⲟ / Gr. βασιλεύς); in the translations given here this is consistently rendered as "king", also in cases where "emperor" would be the appropriate English term.
15 Van Lantschoot 1929: 28 *ad* no. XV, F ("une acclamation").
16 Cf. Crum 1939: 299b; this is more precisely an example of noun incorporation as described in Reintges 2004: 232–3.
17 Bridel, Bosson and Sierro 1999: 311, no. 147, ll. 3–8.
18 See Luisier 2007: 221.
19 Reading uncertain. Perhaps "protoeparch"?
20 It is used instead of a reference to the rule of the persecutor Diocletian in the Bohairic Martyrdom of St. Pisoura (Hyvernat 1886: 133, ll. 17–18, after a twelfth/thirteenth-century MS); this example confirms both its character as a dating formula and its political character (for which see below).
21 Van Lantschoot 1929: no. CXXI, r°, ll. 1–4; cf. *P.Lond.Copt.* II 83.
22 Van Lantschoot 1929: no. XV, ll. 54–7.
23 Van Lantschoot 1929: no. LXXX r°, ll. 62–3, with ⲉⲭⲱⲓ for ⲉⲭⲱⲛ.
24 Van Lantschoot 1929: nos. XL, LXXXII, XCI, CI, CIV; similarly: Proverbio 2001: 412 (A.D. 989).
25 Wietheger 1992: 197, no. 191, cf. 198–99.
26 Kamel and Girgis 1987: no. 60, with a photograph and a facsimile that leave much to be desired.
27 Evelyn White 1933: 193–94, pl. LV-B, ll. 22–3; cf. Boud'hors and Delahaye 2008: 110–11.

28 Van der Vliet, in Godlewski forthcoming. Godlewski (2000: 93, fig. 2) shows part of the inscription, which runs over the entire width of the apse, immediately below the feet of the Apostles.

29 Hebbelynck and van Lantschoot 1937: 510–11.

30 According to the *editio princeps*, Möller 1918.

31 For the public function of epitaphs, see Van der Vliet 2011: 177–87.

32 For the text and further discussion, see Heinen 1982 and 2003.

33 For the role of inscribed acclamations in late antiquity, see the seminal essay by Roueché (1984); for acclamations of the "Christus vincit" type, Peterson 1926: 152–63, remains valuable.

34 For the Coptic apocalyptic tradition after the Arab conquest, see Van Lent 2001, with further bibliography.

35 See also Van der Vliet 2006, for a similar phenomenon observed in medieval Coptic *tiraz* inscriptions.

Bibliography

Altheim, F., and R. Stiehl. 1971, "Die arabische Eroberung Ägyptens nach Iohannes von Nikiu", in: F. Altheim and R. Stiehl (eds.), *Christentum am Roten Meer*, Berlin/New York: de Gruyter, vol. I: 356–89.

Bagnall, R.S., and K.A. Worp. 2004, *Chronological systems of Byzantine Egypt: second edition*, Leiden/Boston: Brill.

Boud'hors, A., and G-R. Delahaye. 2008, "Nouvel exemple d'une pierre d'autel remployée: la stèle de Dorotheos", in: A. Boud'hors and C. Louis (eds.), *Études coptes X: Douzième journée d'études (Lyon, 19–21 mai 2005)*, Paris: de Boccard, 103–22.

Bridel, P., N. Boson, and D. Sierro. (eds.) 1999, *Explorations aux Qouçoûr el-Izeila lors des campagnes 1981, 1982, 1984, 1985, 1986, 1989 et 1990*, Louvain: Peeters.

Brunsch, W. 1994, "Koptische und griechische Inschriften aus Alexandrien", *WZKM* 84: 9–33.

Charles, R.H. (transl.) 1916, *The Chronicle of John, bishop of Nikiu*, London/Oxford: Williams & Norgate; reprinted Amsterdam: Philo Press 1981.

Crum, W.E. 1939, *A Coptic dictionary*, Oxford: Oxford University Press.

Evelyn White, H.G. 1933, *The monasteries of the Wâdi 'n Natrûn III: the architecture and archaeology*, New York: Metropolitan Museum of Art.

Feissel, D. 1993, "La réforme chronologique de 537 et son application dans l'épigraphie grecque: années de règnes et dates consulaires de Justinien à Héraclius", *Ktema* 18: 171–88.

Godlewski, W. 2000, "Les peintures de l'église de l'Archange Gabriel à Naqlun", *BSAC* 39: 89–101.

Godlewski, W. forthcoming, *Naqlun: the church of the Archangel Gabriel*, Warsaw.

Hebbelynck, A., and A. van Lantschoot. 1937, *Codices coptici Vaticani, Barberiniani, Borgiani, Rossiani I: codices coptici Vaticani*, Vatican: Biblioteca Apostolica Vaticana.

Heinen, H. 1982, "Eine neue alexandrinische Inschrift und die mittelalterlichen *laudes regiae: Christus vincit, Christus regnat, Christus imperat*", in: G. Wirth (ed.), *Romanitas – Christianitas: Untersuchungen zur Geschichte und Literatur der römischen Kaiserzeit (Festschrift J. Straub)*, Berlin/New York: de Gruyter, 675–701.

Heinen, H. 2003, "Dipinti aus der West-Nekropole Alexandriens. Gabbari", in: J-Y. Empereur and M-D. Nenna (eds.), *Nécropolis 2/2*, Cairo: IFAO, 639–52.

Hyvernat, H. (ed.) 1886, *Les Actes des martyrs de l'Égypte*, Paris: E. Leroux; reprinted Hildesheim/New York: G. Olms 1977.

Kamel, I., and G.D. Girgis. 1987, *Coptic funerary stelae*, Cairo: Organisation égyptienne générale du livre.

Luisier, Ph. 2007, "Les années de l'indiction dans les inscriptions des Kellia", *ZPE* 159: 217–22.

Möller, G. 1918, "Ein koptischer Ehevertrag", *ZÄS* 55: 67–74.

Papaconstantinou, A. 2009, "'What remains behind': Hellenism and Romanitas in Christian Egypt after the Arab conquest", in: H.M. Cotton et al. (eds.), *From Hellenism to Islam: cultural and linguistic change in the Roman Near East*, Cambridge: Cambridge University Press, 447–66.

Peterson, E. 1926, *ΕΙΣ ΘΕΟΣ: Epigraphische, formgeschichtliche und religionsgeschichtliche Untersuchungen*, Göttingen: Vandenhoek & Ruprecht.

Reintges, Chr.H. 2004, *Coptic Egyptian (Sahidic dialect): a learner's grammar*, Cologne: Köppe.

Roueché, C. 1984, "Acclamations in the later Roman Empire: new evidence from Aphrodisias", *JRS* 74: 181–99.

Sijpesteijn, P.M. 2007, "The Arab conquest of Egypt and the beginning of Muslim rule", in: R.S. Bagnall (ed.), *Egypt in the Byzantine world, 300–700*, Cambridge: Cambridge University Press, 437–59.

Van der Vliet, J. 2006, "'In a robe of gold': status, magic and politics on inscribed Christian textiles from Egypt", in: C. Fluck and G. Helmecke (eds.), *Textile messages: inscribed fabrics from Roman to Abbasid Egypt*, Leiden/Boston: Brill, 23–76 [Study 2].

Van der Vliet, J. 2011, "'What is man?' The Nubian tradition of Coptic funerary inscriptions", in: A. Lajtar, and J. van der Vliet (eds.), *Nubian voices: studies in Christian Nubian culture*, Warsaw: University of Warsaw/Raphael Taubenschlag Foundation, 171–224 [Study 31].

Vania Proverbio, D. 2001, "Additamentum Sinuthianum: nuovi frammenti dal Monastero Bianco in un codice copto della Biblioteca Apostolica Vaticana", *Rendiconti Accademia dei Lincei*: 409–17.

Van Lantschoot, A. 1929, *Recueil des colophons des manuscrits chrétiens d'Égypte* I: *les colophons coptes des manuscrits sahidiques*, Louvain: J.B. Istas; reprinted Milan: Cisalpino-La Goliardica 1973.

Van Lent, J. 2001, *Koptische apocalypsen uit de tijd na de Arabische verovering van Egypte*, Leiden: Oosters genootschap in Nederland.

Wietheger, C. 1992, *Das Jeremias-Kloster zu Saqqara unter besonderer Berücksichtigung der Inschriften*, Altenberge: Oros Verlag.

Zotenberg, H. (ed.) 1883, *Chronique de Jean, évêque de Nikiou*, Paris: Imprimerie nationale.

4

PERENNIAL HELLENISM!
László Török and the al-Muʻallaqa lintel
(Coptic Museum inv. no. 753)

Jacques van der Vliet

Introduction

Quite recently, László Török published a sturdy and richly illustrated handbook about the late antique art of Egypt, entitled *Transfigurations of Hellenism* (Török 2005). Although an appraisal of its contribution to the scholarly debate is beyond the scope of the present paper,[1] it must certainly be counted among its merits that it avoids both the traditional misnomer of "Coptic art" for the Christian art of late antique Egypt and the obsolete paradigm that pits the "indigenous" element against the "Greek". Instead, it situates the Egyptian art of this period in a wider Hellenistic and Byzantine context. As an epilogue to the nine chapters of his book, the author offers a discussion of the famous wooden lintel from the Church of the Holy Virgin Mary, surnamed al-Muʻallaqa, in Old Cairo (now in the Coptic Museum under inv. no. 753). This lintel bears a monumental Greek inscription arranged in four lines on top of a sculptured frieze. The inscription, which is fully dated, and the vividly sculptured scenes in the lower register are remarkable for their sophisticated Hellenism, and they have accordingly met with considerable interest from both art historians and epigraphists.[2]

In the epilogue to his book, subtitled "Perennial Hellenism?", Török critically reviews various scholarly opinions about this monument (2005: 351–58). One of his aims in doing so is to bring out the considerable uncertainties that surround the history of Egypt's late antique art. As an example of this, the Cairo lintel could not have been better chosen. Mainly on the basis of style, art historians have dated it very differently, although usually somewhere between the fourth and sixth centuries. Therefore it must have come as a shock for many of them that, in 1986, the much damaged date of the inscription was read with a high degree of likelihood by L.S.B. MacCoull as the Diocletian year 451, which corresponds to A.D. 734/735. Astonishingly, Török, instead of using this information as damning evidence of the fragility of stylistic dating criteria, questions the reading of the date itself. In the final paragraph of his book, he even concludes that all options for dating the lintel seem to be open. Apparently, the readers are free to choose the date that fits them best, whether it be the fifth, the sixth or the eighth century.[3]

75

The disdain for the work of the epigraphist that is apparent here cannot be left unchallenged.

It may be useful to recall the basic facts briefly. After a long liturgically inspired hymn of praise, celebrating Christ as the Godhead incarnate, the inscription concludes (in l. 4) with a brief prayer for the sponsors of the monument: a prelate named Abba Theodore and a deacon and steward called George. This is followed by a dating formula that follows a conventional pattern. In translation, these final phrases read:

> (Lord,) come to the assistance of Abba Theodore, *proedros* (πρόεδρος), and George, deacon and steward (οἰκονόμος)! The 12th of the month Pachon, (year of the) indiction 3, (year since) Diocletian 451 [i.e. May 7, 735 A.D.].

Unfortunately, unlike the indiction year (3),[4] the Diocletian year (451) is hardly legible anymore, and the first editors of the text were even unable to decipher it. Only in 1937 did P. Jouguet succeed in identifying the last two digits as. ṇạ: ?51).[5] In 1986, L.S.B. MacCoull proposed to read the first digit as an *upsilon*; in addition to autoptic study, she was able to use an unpublished early photo of the monument.[6] Her reading ($\overline{\gamma\text{ν}\alpha}$, i.e. 451) was confirmed, again on the basis of autoptic study, by J.-L. Fournet in 1993, who qualified it as "palaeographically the only possible one".[7] This new reading has been accepted by most scholars since, including the latest editor of the inscription, J. Hammerstaedt,[8] who published a thorough discussion of the text, surprisingly ignored by Török. MacCoull,[9] followed by Hammerstaedt,[10] identified the Abba Theodore of the inscription, undoubtedly a bishop, as the homonymous "Monophysite" patriarch of Alexandria. The latter was in office between 731 and 743, which is consistent with the reading of the Diocletian date.

Török cites two arguments which, in his opinion, weaken the eighth-century dating of the monument considerably, although he does not categorically exclude it. First, he denies that the *proedros* Abba Theodore, mentioned in the inscription, can be identified with the like-named patriarch.[11] Secondly, he claims that the damaged Diocletian date might just as well be read as [1]51, i.e. A.D. 434/435. Ignoring Fournet's 1993 article, he considers MacCoull's reading of the date as 451 as uncertain as the alternative reading of 151.[12]

For the identification of Abba Theodore, Török, referring to J.-M. Spieser's study of 1995, asserts that the title *proedros* (πρόεδρος) "may refer to a bishop, but not to a patriarch".[13] This is not only a blatant error in itself, but also an entirely false representation of Spieser's argument. Spieser correctly observed that *proedros*, in addition to its well-known use for patriarchs, may also designate a simple bishop.[14] MacCoull's translation "patriarch"[15] is therefore misleading. In fact, as an ecclesiastical title, *proedros*, literally "president", is a somewhat vague and high-flown epithet that may be used for any church dignitary of episcopal

PERENNIAL HELLENISM!

rank.[16] In the Christian epigraphy of the Nile Valley, the term is quite uncommon.[17] The only other occurrence known to me is found in a near-contemporary (A.D. 707) Coptic foundation inscription from Faras, in Lower Nubia, where it certainly designates a local bishop, Abba Paulos.[18] Since in the Cairo lintel no specific patriarchal titles are given, it would appear that the occurrence of a *proedros* Theodore in the inscription cannot be used as an argument to confirm its date, in the way that it had been done by MacCoull.

Spieser added an interesting observation, however, which had not been made before. A local bishop named Theodore is known from the so-called *History of the Patriarchs of Alexandria* to have been active precisely in the period concerned.[19] This Bishop Theodore of Misr, the new capital of Old Cairo–Fustat that grew up around Babylon of Egypt, participated in the synod that was convened in Misr itself on 28 Mesore of the year 459, i.e. 21 August 743 A.D., in order to elect a new patriarch.[20] He is mentioned several times in the *History of the Patriarchs* and must have been a prominent figure during the episcopate of Patriarch Theodore and his immediate successor, Michael I. He is represented as a close associate of the well-known Bishop Moses of Wasim/Letopolis,[21] and described as "the oldest of the bishops at that time" and "the first of three bishops named Theodore who successively occupied the see of Misr".[22] Some versions of the report about the synod of 743 even call him the metropolitan (*mutranus*) of Misr.[23] In all likelihood, this senior prelate can be identified with the Abba Theodore mentioned eight years earlier on the al-Mu'allaqa lintel. As the reigning bishop of the see, Theodore would naturally be given a place of honor in the inscription, before the deacon and steward George, who may have been the actual donor of the monument.[24] Thus, even if Abba Theodore was not the patriarch, who still resided in Alexandria at the time, but a local bishop, as seems likely, this is not an argument against the eighth-century date of the lintel. On the contrary, as Spieser was the first to point out, the activity of a contemporary Bishop Theodore in Old Cairo–Fustat is confirmed by an independent source, the *History of the Patriarchs*.

In order to question the reading of the Diocletian date as 451, Török[25] refers to an observation by A. Iacobini, who supposes that it could "in theory" also be read as year 151.[26] It is true that the date is seriously damaged. Even from the early photograph published by MacCoull, the first digit cannot be read with ultimate certainty. We have to rely on the acumen of MacCoull and Fournet.[27] Nevertheless, an alternative reading can be ruled out for other reasons as well. When we accept, as Török himself does, the last two digits of the year (51), which were read by Jouguet already in the 1930s, the possibilities for filling in the first digit are extremely limited. Even when the net is cast very wide, there are only three possible Diocletian years +51 that coincide with a third indiction year, as is demanded by the inscription on the lintel, which is unambiguous at this point. These are: 151 (ⲣ̅ⲛ̅ⲁ̅, A.D. 434/435), 451 (ⲩ̅ⲛ̅ⲁ̅, A.D. 734/735) and 751 (ⲯ̅ⲛ̅ⲁ̅, A.D. 1034/1035). Hence, for those who remain suspicious of MacCoull's reading of the first digit, there would be theoretically two, and only two, alternatives: A.D. 435 or 1035.

The second alternative, 1035, seems an unlikely candidate for a variety of reasons, but first of all because the habit of dating by the Diocletian era, as distinct from the era of the Martyrs, began to disappear precisely in the eleventh century. From about the year 1000, the number of attestations declines sharply, and examples come to be virtually confined to Nubian funerary inscriptions and Coptic book colophons. Indiction years went out of use in Christian documents from Egypt more or less simultaneously.[28] For very similar reasons the earlier alternative, A.D. 435, can be excluded.

Although it might seem obvious that the Diocletian era was used by all and sundry ever since the first year of Diocletian, this was certainly not the case. In fact, the originally pagan counting by Diocletian years was adopted by steps and stages, only to become widespread among Egyptian Christians after the Arab conquest of 641. Thus, in ordinary Greek papyrus documents, the Diocletian era is first used only after the middle of the seventh century, possibly in reaction to the introduction of the Hijra or "Saracene" era. In Coptic papyri it appears even somewhat later.[29] The tables in the chronological handbook of Bagnall and Worp graphically show which kind of sources were the first to adopt Diocletian dates.[30] Remarkably enough, the use of the era was initially limited to a few well-defined groups or genres of texts. These were astrological texts and horoscopes; Christian literary texts, mainly patristic and chronographical ones; and pagan inscriptions in Hieratic, Demotic and Greek.[31] This last series ended around the middle of the fifth century, with the petering out of the pagan epigraphical tradition in Philae. But even then the habit was not immediately adopted into Christian epigraphy. Christian inscriptions that are dated after Diocletian began to appear only around the turn of the fifth-sixth centuries, with a series of epitaphs from monasteries in or near Alexandria. The earliest of these epitaphs that is presently known dates from A.D. 491 (*SB* III 6250), but most of them bear sixth-century dates. From about the same time only, double dates as we find in the al-Mu'allaqa lintel, combining Diocletian and indiction years, came into regular use. Therefore, assigning a public Christian inscription, formally dated by a double Diocletian and indiction year, to the first half of the fifth century must be considered an anachronism. Actually, such a date is as unlikely as an eleventh-century date. The evidence of Egyptian epigraphical dating habits practically excludes either of both theoretical alternatives to the date proposed by MacCoull and confirmed by Fournet.[32]

To conclude, Török's doubts about the eighth-century date of the lintel from the al-Mu'allaqa church cannot be substantiated. His plea for an "open" date is not well argued and betrays a remarkable contempt for the epigraphic evidence. This might be considered a negligible error, if it did not formally license the readers of Török's handbook to continue thinking whatever they like about the age of this monument and its historical significance. In fact, the work of Jouguet, MacCoull and Fournet has established the date of the lintel beyond reasonable doubt. It can be safely assigned to the year 735 and to the episcopacy of a well-documented bishop of Old Cairo, Theodore. If this hurts art historians' sensibilities about style, a hazy category at best, so much the better. Hellenism in Egypt did not die in 641.

PERENNIAL HELLENISM!

Also other recent discoveries (for example in the Church of the Virgin in Dayr al-Suryan) show that, in spite of political and denominational frontiers, Christian Egypt remained culturally focused on the Byzantine world for many years and even centuries after the Arab conquest.

Notes

1 See the extensive review by Fluck (2006).
2 For the earlier bibliography see Coquin 1974: 83 (general), and Hammerstaedt 1999: 187 (on the inscription); the most significant art-historical contributions are mentioned by Török (2005: 351–58). Beautiful new photos can be found in Gawdat Gabra and Eaton-Krauss 2007: 206–11, no. 129. As these and other photos show, the inscription with its date cannot have been secondary additions to an existing work of art.
3 Török 2005: 358.
4 The indiction is a fifteen-year cycle used for administrative purposes; see Bagnall and Worp 2004.
5 In Simaika 1937: 27.
6 MacCoull 1986: Tafel XIII.
7 Fournet 1993: 243.
8 Hammerstaedt 1999: 187–99.
9 MacCoull 1986: 233.
10 Hammerstaedt 1999: 197.
11 Török 2005: 354.
12 Török 2005: 354–55.
13 Török 2005: 354.
14 Spieser 1995: 311–12.
15 MacCoull 1986: 232.
16 Spieser refers to Salaville 1930: part. 418–22; patristic references are also cited in Lampe 1961: 1144–45, s.v.; some canonical ones in Jerg 1970: 174, 176, n. 486, 491.
17 In the Greek papyri it appears only as a civil title; in Coptic papyri it seems not to be attested.
18 *I. Khartoum Copt.* 1, l. 9.
19 Spieser 1995: 312, who refers to Fedalto 1988: 614; the same reference in Worp 1994: 297. The ultimate source of all references given by Fedalto is the (Arabic) *History of the Patriarchs*. For the author and background of this part of the *History of the Patriarchs*, see Den Heijer 1989: 8, 118, 145–46.
20 Evetts 1910: 105–06, quoted in Munier 1943: 25 (cf. Seybold 1912: 159). For the rapid development of the new capital in this period, see Raymond 1993: 24–30.
21 See Graf 1944–1953: I, 474–75; Timm 1984–1992: VI, 2988–89.
22 Evetts 1910: 104; cf. Seybold 1912: 158.
23 Evetts 1910: 106; Munier 1943: 25. This may well be an anachronism, though; thus, the title lacks in the early Hamburg manuscript (Seybold 1912: 159).
24 For the relationship between bishops and stewards, see Wipszycka 1972: 135–41; cf. Schmelz 2002: 162–64, who gives further examples of the combination deacon and steward.
25 Török 2005: 254–55.
26 Iacobini 2000: 201–02, n. 61; also Iacobini ignores Fournet's 1993 article.
27 Note, though, that the most recently published photo, in Gawdat Gabra and Eaton-Krauss 2007: 207, shows convincing traces of an *upsilon* as the first digit.
28 See the table in Bagnall and Worp 2004: 80–1; the mostly rather doubtful exceptions to the general patterns discussed here and below have not been taken into consideration.

29 Bagnall and Worp 2004: 64.
30 Bagnall and Worp 2004: 68–81.
31 Bagnall and Worp 2004: 63–4.
32 I deliberately refrained here from using criteria that could be considered subjective (like paleography, or form and contents of the text).

Bibliography

Bagnall, R.S., and K.A. Worp. 2004, *Chronological systems of Byzantine Egypt: second edition*, Leiden/Boston: Brill.

Coquin, C. 1974, *Les édifices chrétiens du Vieux-Caire* I: *bibliographie et topographie historiques*, Cairo: IFAO.

Den Heijer, J. 1989, *Mawhûb ibn Mansûr ibn Mufarriğ et l'historiographie copto-arabe: étude sur la composition de l'Histoire des Patriarches d'Alexandrie*, Louvain: Peeters.

Evetts, B. (ed.) 1910, *History of the Patriarchs of the Coptic Church of Alexandria* III: *Agathon to Michael I (766)*, Paris: Firmin-Didot; reprinted Turnhout: Brepols 1947.

Fedalto, G. 1988, *Hierarchia ecclesiastica orientalis* II: *Patriarchatus Alexandriae, Antiochiae, Hierosolymitanae*, Padova: Messaggero.

Fluck, C. 2006, [Review of Török 2005], *Orientalistische Literaturzeitung* 101: 615–23.

Fournet, J-L. 1993, "L'inscription grecque de l'église al-Mu'allaqa: quelques corrections", *BSAC* 93: 237–44.

Gabra, G., and M. Eaton-Krauss. 2007, *The treasures of Coptic art in the Coptic Museum and Churches of Old Cairo*, Cairo/New York: American University in Cairo Press.

Graf, G. 1944–1953, *Geschichte der christlichen arabischen Literatur*, Vatican City: Bibliotheca Apostolica Vaticana.

Hammerstaedt, J. 1999, *Griechische Anaphorenfragmente aus Ägypten und Nubien*, Opladen: Westdeutscher Verlag.

Iacobini, A. 2000, *Visioni dipinti: immagini della contemplazione negli affreschi di Bâwît*, Rome: Viella.

Jerg, E. 1970, *Vir venerabilis: Untersuchungen zur Titulatur der Bischöfe in den ausserkirchlichen Texten der Spätantike als Beitrag zur Deutung ihrer öffentlichen Stellung*, Vienna: Herder.

Lampe, G.W.H. 1961, *A patristic Greek lexicon*, Oxford: Clarendon Press.

MacCoull, L.S.B. 1986, "Redating the inscription of el-Moallaqa", *ZPE* 64: 230–34.

Munier, H. 1943, *Recueil des listes épiscopales de l'Église copte*, Cairo: Société d'archéologie copte.

Raymond, A. 1993, *Le Caire*, Paris: Fayard.

Salaville, S. 1930, "Le titre ecclésiastique de "proedros" dans les documents byzantins", *Échos d'Orient* 29: 416–36.

Schmelz, G. 2002, *Kirkliche Amtsträger im spätantiken Ägypten nach den Aussagen der griechischen und koptischen Papyri und Ostraka*, München/Leipzig: K.G. Saur.

Seybold, C.F. (ed.) 1912, *Severus ibn al Muqaffa': Alexandrinische Patriarchengeschichte von S. Marcus bis Michael I, 61–767*, Hamburg: L. Gräfe.

Simaika, M.H. 1937, *Guide sommaire du Musée copte et des principales églises du Caire*, Cairo: Imprimerie nationale.

Spieser, J.-M. 1995, "A propos du linteau d'al-Moallaqa", in: *Orbis romanus christianusque ab Diocletiani aetate usque ad Heraclium: travaux sur l'antiquité tardive rassemblés autour des recherches de Noël Duval*, Paris: de Boccard, 311–20; reprinted

PERENNIAL HELLENISM!

in J-M. Spieser. 2001, *Urban and religious spaces in late antiquity and early Byzantium*, Aldershot: Ashgate.

Timm, S. 1984–1992, *Das christlich-koptische Ägypten in arabischer Zeit*, 6 vols., Wiesbaden: Reichert Verlag.

Török, L. 2005, *Transfigurations of Hellenism: aspects of late Antique art in Egypt AD 250–700*, Leiden/Boston: Brill.

Wipszycka, E. 1972, *Les ressources et les activités économiques des églises en Égypte du IVe au VIIIe siècle*, Brussels: Fondation égyptologique Reine Elisabeth.

Worp, K.A. 1994, "A checklist of bishops in Byzantine Egypt (A.D. 325–c. 750)', *ZPE* 100: 283–318.

5

HISTORY THROUGH INSCRIPTIONS
Coptic epigraphy in the Wadi al-Natrun[1]

Jacques van der Vliet

Introduction

The painter Theodore, active in the monastery of Saint Antony near the Red Sea in the thirteenth century, inscribed on one of the walls of the church of this monastery the following proverb: ⲧⲭⲓⲝ ⲛⲁⲧⲁⲕⲟ ⲡⲉⲥⲅⲁ[ⲓ ⲛ]ⲁⲙⲟⲩⲛ ⲉ[ⲃ]ⲟⲗ, that is: "the hand will perish, what is written will remain".[2] It is the almost exact Coptic counterpart of the well-known Latin saying: *vox audita perit, littera scripta manet.* Man writes to survive. The painter Theodore, too, wished to be remembered beyond his earthly existence. The quotation above is from the opening lines of a commemorative inscription which he wrote next to one of his paintings and in which, addressing the readers, he asks for their prayers.

The desire to save from oblivion persons or events was the motivating force behind much of the traditional source material of *epigraphy*: the study and, in particular, the decipherment and interpretation of ancient inscriptions. The scholarly discipline of epigraphy developed within the context of classical studies and was originally concerned primarily with Latin and Greek commemorative and funerary inscriptions. The study, for example, of hieroglyphic Egyptian, Aramaic or Coptic inscriptions is a later development, and even at present these disciplines retain some of the aims and methods of traditional classical epigraphy. The object of Coptic epigraphy in a broad sense can be understood as the surviving inscriptional material from Christian Egypt and Nubia.[3] As such, it comprises much more than just Coptic inscriptions in stone. It is characterized by a wide variety of media (stone, plaster, wood) and purposes (funerary, dedicatory, didactic) as well as by great linguistic diversity (in addition to Greek, Coptic and Arabic, the principal languages of Christian Egypt, Old Nubian, Syriac, Armenian and Ge'ez may also be found). Moreover, much of this Coptic epigraphic material bears a monastic stamp.

Egyptian monasteries, whether still inhabited or long deserted, have produced large ensembles of inscriptions of all kinds. Often, these are an important source for reconstructing the history of a particular establishment. The monastic chronicles of Latin Europe are virtually unknown in Egypt and in quite a number of cases the only written evidence is inscriptions and stray documents. Striking examples of such a predominantly epigraphic documentation are offered by the Monasteries of Jeremiah in Saqqara[4] and Hatre (Hadra) near Aswan.[5] Therefore, the most valuable scholarly contributions in the field of Coptic epigraphy are precisely those publications which do not stop at the more or less satisfactory edition of a handful of inscriptions but try to integrate the epigraphic source material into a broader vision of the general history and culture of a particular site or cluster of sites. Although more recent examples are not lacking, the monumental three-volume set of Hugh G. Evelyn White's *The monasteries of the Wâdi 'n Natrûn* does remain an outstanding example of such an integrative approach.

What, then, would be the role of inscriptions in a future reworking of Evelyn White's panoramatic study? Admittedly, epigraphic material is certainly not the principal source of information for the history of the Wadi al-Natrun. Even if important inscriptions are not lacking, the epigraphic record does not match other sources, either from the Wadi itself or of external origin, for their number and wealth of information. Funerary inscriptions on tombstones, for example, abundant at many other monastic sites in Egypt, are practically absent in the Wadi al-Natrun.[6] This may be connected with local usage: also the site of the Kellia produced only one small funerary stela.[7] It should, moreover, be remembered that excavations are the classical source of inscriptions. The Wadi al-Natrun, although an ancient and a monumental site, is not primarily an archaeological site: it is dominated by four big monasteries which are very much alive and whose communities are much more concerned with the future than with the past. In fact, excavations in the area are generally of a recent date and modest in scale. Also the current Dutch restoration project in Dayr al-Suryan, which greatly increased the number of known mural inscriptions, is modest in scale. Nevertheless, epigraphy does have a lot to offer to the historian of the Wadi al-Natrun and I hope to be able, in the following pages, to give an idea, first, of the variety and richness of the epigraphic sources, and then of their importance.

Categories of inscriptions

As was remarked above, the epigraphic material from Coptic monastic sites is as a rule much more varied than the formal stone inscriptions of classical epigraphy. This variety can be observed in the monasteries of the Wadi al-Natrun too and will be briefly illustrated in the following paragraphs.

Although the area did not produce an extensive corpus of funerary stelae, monumental inscriptions on stone are by no means entirely lacking. It may suffice to mention two examples, both inscribed on marble slabs and both kept at present in Dayr al-Suryan. The first is a Bohairic text which commemorates the demise of Saint John Kame in A.D. 858;[8] the second, a bilingual (Greek–Old Nubian) epitaph of the Nubian King George (Georgi, Georgios) of A.D. 1157.[9] These two monuments provide important historical information, but it is unfortunate that both were detached from their original contexts long ago.[10]

For another group of inscriptions the situation is, in this respect, much more favorable. They would never have survived outside their original contexts, since they are inseparable from the walls and other architectural elements of the monastery buildings themselves. These are the more or less formal mural inscriptions which have been painted upon or scratched into monastery walls from the earliest days of their existence until the present day. They bear a direct physical relationship to the architecture of the holy place to which their fate is linked.

For our purposes, several kinds or genres of mural inscriptions can be distinguished. The most informal and familiar of these is certainly the graffito: a usually short inscription often mentioning merely the name of its author, sometimes accompanied by a prayer.[11] The author may be a visitor, a pilgrim for example, or a local. So-called graffiti may be scratched in (as their name suggests), but they are perhaps even more often painted or, nowadays, written with a ballpoint or felt-tip pen. They can be found everywhere, from pharaonic temples to subway stations, but certain spots are more privileged than others. In the Wadi al-Natrun monasteries, a noteworthy concentration of them can be found, for example, in the chapels on the second floor of the *qasr* [fort] of Dayr Abu Maqar. The clouds of crude graffiti, in Coptic, Syriac, Armenian, Arabic and French, hardly enhance the esthetic effect of the medieval wall paintings in these rooms.[12] Still, they are born not out of mischief or lust for destruction, but out of a desire to remain present, to "survive", within a certain sacred space and thereby to partake of its *baraka* [blessing]. What distinguishes them from other inscriptions is their informal and personal character.

Less informal in character are *dipinti*: full-fledged texts that are painted on the walls with considerable care and usually on a fairly large scale. Most of the recently discovered Syriac inscriptions at Dayr al-Suryan are actually *dipinti*, some of them quite monumental ones.[13] Unlike graffiti, they certainly cannot have been made without the permission of the local authorities. It takes considerable time, skill and technical means to produce them. Their contents, too, are much more varied: they may be commemorative in character, but also, for example, didactic. In spite of being more formal, *dipinti* are nevertheless usually "incidental" in the sense that they do not belong to the planned decoration scheme of a building. In this respect they differ from a third category of painted mural inscriptions: those which intimately belong to the painted decoration of the building, usually a church or chapel.

Ancient and modern wall paintings and even icons may bear a whole series of inscriptions that are functionally related to the picture. In the main, three types can be distinguished: legends, founder's mementos and painter's mementos.[14] The latter two usually contain a prayer, respectively, for the sponsor and the painter of the scene. As regards their form, such prayers often resemble those written, with similar purpose, in the colophons of manuscripts. The legends which accompany so many paintings, although the most familiar as a genre, are actually perhaps the least well understood. They may seem a rather unsophisticated way of identifying, for an ignorant audience, the actors, events or other elements of a painted scene. That is why they are often invaluable for modern scholars when facing damaged, incomplete or indistinct scenes.[15] In fact, the aim of the legends is entirely pedagogical:

while identifying the elements of the picture, they invite its beholder to activate his or her knowledge of the Holy Scriptures, of sacred literature, or even of major theological issues. Thus, the biblical texts on the prophets' scrolls in the Annunciation scene in the Church of the Virgin at Dayr al-Suryan evoke the mystery of the Incarnation as adumbrated in the Old Testament.[16] Rather than attesting to ignorance, legends appeal to erudition. All three elements (legends, a founder's memento and a painter's memento) can be found together in a typical form in a recently discovered but as yet unpublished monument of the Wadi al-Natrun, a painted hermitage in the surroundings of Dayr Abu Maqar, which will be provisionally presented below.

Finally, there are other bearers of inscriptions which can be easily overlooked but which nevertheless have their own importance within monastic epigraphy and particularly in the monasteries of the Wadi al-Natrun. The category in question is that of moveable objects, like furniture and ecclesiastical implements. These too may bear legends and/or commemorative inscriptions. A striking example, again from Dayr al-Suryan, is the bronze flabellum (ceremonial fan) with a Syriac inscription, dated A.D. 1202, now in the Museum of Mariemont in Belgium.[17]

Inscriptions as historical sources

These three categories of inscriptions – the monumental stone inscriptions, those painted upon or scratched into the walls, and finally those adorning moveable objects – constitute the subject of monastic epigraphy in its breadth and variety. As I hope to show, they are a source of information which the historian of the Wadi al-Natrun would be unwise to neglect, even if they do not always write history with a capital letter. In what follows, I will be referring in particular to the Coptic, that is to say Bohairic Coptic, inscriptions that have been discovered since 1990 near Dayr Abu Maqar and in the Church of the Virgin at Dayr al-Suryan,[18] and are still only partly published.[19] These can be approached from various angles, each apt to throw light on a particular aspect of past life in the monasteries concerned. Here I will limit myself to only three aspects of the recent finds, namely their historical, linguistic and cultic implications.

The importance of inscriptions may be purely historical, in the narrow sense of "documenting local history". They produce names and dates for persons or events from the monastery's history. Thus, the conventional prayers accompanying wall paintings may acquaint us with the names of artists and donors. In Dayr al-Suryan, for example, two big ornamental crosses painted on the north wall in the northeastern corner of the nave of the Church of the Virgin (Figure 5.1, no. 2) are the work of an artist Solomon.[20] Two brief and entirely conventional prayers request God's mercy for Solomon, but at present provide no further information beyond his name and his activity as an artist. It is therefore also difficult to say whether the crosses painted by him were made to order, as part of the planned decoration of the Church, or a merely tolerated product of his personal devotion.

The painted panels in a room hewn into the rock to the south of Dayr Abu Maqar and discovered in the autumn of 1990 have certainly been made to order.[21]

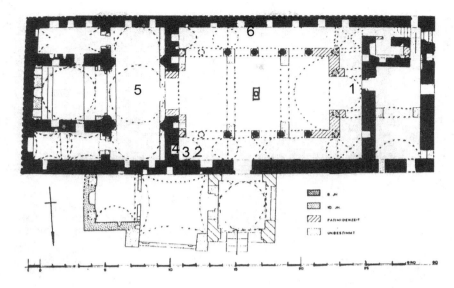

Figure 5.1 The Church of the Virgin, Dayr al-Suryan: plan, after P. Grossmann, showing the location of some of the recently discovered inscriptions

    1   the great Annunciation
    2   Solomon's crosses
    3   graffito of Michael
    4   presumed original position of tenth-century reliquary
    5   commemorative inscription in the dome above the *khurus*
    6   *dipinto* of Jacob

One wall shows Saint Menas with his camels and Christ standing to his right, while another wall shows the Virgin Mary feeding the Logos next to a much damaged representation of Christ Pantokrator upon the Four Living Creatures. The paintings are provided with conventional legends in Greek, whereas colophon-type prayers for the founder and the painter of the ensemble are conceived in Bohairic Coptic. The painter's memento is much damaged and the artist's name lost. Fortunately, enough survives to date his work to the year 943/944 (A.M. 660). According to the other memento, it was an ascetic called Father Mena Panau who ordered and paid for the construction and decoration of the "cave" (it is called thus in the inscription, with the Greek word σπήλαιον) and possibly the adjoining hermitages as well. There is a good chance that the same person is mentioned as the spiritual father of a Deacon Gabriel, who appears as the scribe of a Vatican manuscript of A.D. 978/79.[22] Around the middle of the tenth century, he must have been a person of some importance within the monastic community and we may expect to find more traces of him.

Of an entirely different nature is the great commemorative inscription running beneath the painted decoration adorning the central dome above the *khurus* [room separating the altar room from the nave] of the Church of the Virgin in Dayr al-Suryan (Figure 5.1, no. 5). This text, which has gradually been uncovered since 1999,[23] consists of one line of tall decorative letters and mentions, in addition to a deacon John and two priests, namely *papa* Moses and *papa* Aaron;[24] the former, Moses, is furthermore styled *hegoumenos* [head of a church or monastery] and *oikonomos* [steward]. Judging from similar inscriptions, for example those running underneath the eleventh-century apse composition in the church of the monastery at Naqlun (Dayr al-Malak) in the Fayoum,[25] the persons commemorated here were the monastic authorities in whose period of office and under whose supervision the dome above the *khurus* was refurbished. They may have been the sponsors of the operation as well, but this is by no means necessary and as yet not borne out by the extant remains of the text. We are rather dealing here with the kind of dedicatory inscription which "dates" important construction works according to the ecclesiastical and, sometimes, worldly authorities of the time.[26] Indeed, as my colleague Lucas van Rompay acutely observed, the Moses and Aaron of the inscription have a good chance of being identical to the Moses and Aaron mentioned, as "priests and directors of the monastery of the Syrians", in the undated colophon of a Syriac manuscript from Dayr al-Suryan.[27] Karel Innemée, on the other hand, proposes to identify the Moses of the inscription with the famous Abbot Moses of Nisibis (first half of the tenth century).[28]

Although at the present moment the evidence provided by this inscription does not appear conclusive, it may in the near future enable us to fit the works undertaken in the course of decorating and refurbishing the Church of the Virgin into a broader picture of the life and history of the monastic community some 1000 years ago. Already it raises the question, simply by being conceived in Coptic, of language within the community, a question for which, as becomes more and more evident, no easy and clear-cut answers can be expected.[29] The close proximity of Syriac, Coptic, Greek and Arabic on the walls of the Church of the Virgin poses not so much, to my mind, the question of ethnicity but primarily that of domains or registers and, only secondarily, that of the historical circumstances which may have determined the criteria for code selection and their changes. I prefer not to dwell on this theme but will rather turn to Coptic as it is used in the recently discovered inscriptions of the Church of the Virgin.

Coptic, as is well known, is not a uniform language. Several varieties of Coptic can be recognized, often unhappily styled "dialects", some of which had only a limited, local or regional, currency. Not so Bohairic Coptic, which was used in a wide area which included the Wadi al-Natrun. Bohairic, however, is not a unity either. Again, several varieties of Bohairic can be distinguished. Thus, the variety found in most of the literary manuscripts from Dayr Abu Maqar is traditionally called Nitrian Bohairic. Although well attested thanks to a great number of hagiographical and homiletic texts, it has been little studied.[30] The situation is even worse for non-literary Bohairic,[31] that is Bohairic as it is found, for example, in documents and inscriptions, and which, we may think, must at a certain period have been close to the living language as it was used for everyday purposes in a particular region. In fact, witnesses of non-literary Bohairic are relatively scarce[32] and therefore the recent inscriptional finds in Dayr al-Suryan are, next to the texts discovered by the French and Swiss expeditions on the walls of the Kellia, welcome additions to the repertoire.[33] A few examples may explain the interest of the new Dayr al-Suryan inscriptions.

On the south wall of the nave, next to the later painting of the Three Patriarchs, a *dipinto* of nine lines in Bohairic Coptic has come to light (Figure 5.1, no. 6).[34] The painting partly overlaps the inscription and there is no connection between them. The text is predominantly of a didactic character: epigrammatic exhortations to modesty are followed by a short prayer for the author, a certain Jacob. Similar *dipinti*, which must have been meant to bear out the wit and erudition of their authors, can be found on several monastic sites throughout Egypt.[35] Here we are concerned rather with the state of language reflected in the text. Most remarkably, it shows a clear tendency towards devoicing of the labial /b/ not only in final but in initial position as well. Thus the name of the author, Jacob, is written ιακωπ (l. 7),[36] but also πολ- for βολ- (properly, βελ-) is found (l. 6).[37] The latter phenomenon is quite rare,[38] whereas final devoicing of /b/ occurs occasionally both in mural inscriptions from the Kellia[39] and in Wadi al-Natrun literary manuscripts.[40] Its conspicuous presence in such a short text as the Dayr al-Suryan *dipinto* of Jacob suggests that it may have been a characteristic of non-literary Bohairic of the Western Delta. In the same *dipinto*, it is, in a sense, mirrored by another phenomenon, which is much more common in non-literary Wadi al-Natrun Bohairic, namely the "occlusivefication" of the intervocalic glide /w/ into /b/. Thus we find, in addition to the almost commonplace μεβι for μεγι (l. 9),[41] also εβον for εγον (l. 8).[42] A possibly related phenomenon, that is a doubling of the intervocalic glide /w/, can be observed, be it only once (εγογωνι), in the legends of the great Annunciation in the western semi-dome of the Church of the Virgin in Dayr al-Suryan (Figure 5.1, no. 1).[43] The same double phonological shift that occurs in the *dipinto* of Jacob (/w/ becoming /b/ and /b/ becoming /p/) would appear to be reflected in the spelling βωτεπ for ογωτεβ in the memorial inscription of Saint John Kame, mentioned above (l. 13). This inscription shows, in addition, a spelling ογβωτεπ (l. 4), with a doubling or reinforcement of the labial in word-initial position (single ογ- or β- becoming ογβ-) that occurs more often in texts from the region, as well as in literary "Nitrian" Bohairic.[44] Here a whole pattern of shifts in the notation of labials can be detected which is characteristic of a greater group of texts from this general region and deserves further study.[45]

The texts which, in the Dayr al-Suryan Annunciation, the four Old Testament prophets exhibit on their scrolls offer some additional interest. They correspond neither in their textual nor in their linguistic form to those of the Bohairic Bible. They are nonetheless Bohairic and certainly not "Bashmuric" as has been claimed.[46] However, especially in its so-called "sahidizing" aspect, their language is closer to the local "Nitrian" variety of literary Bohairic than to classical "biblical Bohairic". Thus χι is found for ϭι.[47] This aspect of the texts is perhaps most clearly visible in the treatment of the /h/-sounds, which is remarkable for the total absence of the Bohairic ⲃ. The latter is replaced by either x (once)[48] or ⲉ (twice).[49]

Although the relatively small amount of material makes all theories very fragile, it would appear that the recently discovered inscriptions in the Wadi al-Natrun can afford us a glimpse of a regional variety of "living" Bohairic. In this respect they resemble the colophons and readers' notes found in the Coptic literary manuscripts produced in the Wadi al-Natrun.[50] These "marginal" texts are also by nature and contents close to much of the inscriptional material and deserve to be studied together with it. Both inscriptions and colophons might well appear to represent a similar brand of "subliterary Nitrian" Bohairic, which occasionally also surfaces in the far more regular local form of "literary Nitrian" Bohairic. If this observation would prove to be correct, it would strongly militate against the view that "literary Nitrian" Bohairic is a merely artificial dialect which derives its specific character from heavy contamination with literary Sahidic.[51]

When inscriptions are particularly apt to throw light upon the living local vernacular, it is because they are as a rule firmly tied up with a particular spot or site. They do not travel, as books so easily do. This connection with a certain spot brings us to another important aspect of epigraphic material, to wit its relationship with its surrounding space. The legends of paintings, of course, are – we may say – glued to the picture in which they have their logical place. The position of graffiti and *dipinti* on church walls must correspond to a certain logic as well. Unfortunately, such logic as there once must have been in putting an inscription in a certain place is not always easily reconstructed. To quote just one example, I am unable to suggest why the nine-line *dipinto* of Jacob on the southern wall of the nave in the Church of the Virgin was situated just there and nowhere else. A similar *dipinto* in the Monastery of Saint Phoibammon in Western Thebes was inscribed on a doorpost.[52] What inspiration could have guided the hand of the epigrammist Jacob?

There are, however, also more telling examples. On the eastern part of the northern wall of the nave (near nos. 2–3 in Figure 5.1), a cluster of graffiti and *dipinti* has been discovered, in several layers and in different vernaculars: Syriac, Coptic, Arabic. For this reason, Karel Innemée baptized it "the Palimpsest Wall".[53] Such clustering suggests that this particular spot enjoyed the favor of passers-by and visitors on account of its venerability and *baraka*. One Coptic graffito of an otherwise quite ungratifying appearance permits us to be more precise (Figure 5.1, no. 3).[54] These four short lines written by a certain Michael, who may have been a visitor,[55] comprise a short prayer asking "the God of the Holy James the Persian (ⲡⲁⲅⲓⲟⲥ ⲓⲁ[ⲕⲱⲃⲟⲥ] ⲡⲓⲡⲉⲣⲥⲓⲥ)" to have mercy on the author. In Coptic literature but also in Coptic epigraphy, the invocatory formula "God of Saint so-and-so" is very frequent. One might say that it invokes God while at same time engaging a particular saint as the supplicant's mediator. Therefore the saint in question is never an arbitrary one, but always the one topically most relevant: he or she is at this particular place or in that particular situation the most appropriate mediator.[56]

HISTORY THROUGH INSCRIPTIONS

There can therefore be little doubt that wherever God is invoked as "the God of James the Persian", James the Persian himself must have been near at hand.

In fact, James the Persian, or James the Sawn Asunder, is no foreigner in the monasteries of the Wadi al-Natrun. The only complete Coptic account of his martyrdom and of the translation of his relics is extant in a tenth-century Bohairic manuscript from Dayr Abu Maqar.[57] The same monastery, too, produced a fragment of another copy.[58] That in the late eleventh-century relics of Saint James the Persian were kept in the Wadi al-Natrun is known through Mawhub Ibn Mansur Ibn Mufarrij's notice in the *History of the Patriarchs of Alexandria*.[59] Other sources suggest their presence in Dayr al-Suryan at an even earlier date.

An inscribed piece of church furniture from Dayr al-Suryan, the well-known inlaid reliquary, now in the museum of the monastery, represents among other saints a standing figure in military attire. A legend in Greek calls him "the Holy James: ο ⲁⲅⲓⲟⲥ ⲓⲁⲕⲟⲃⲟⲥ".[60] On account of a far later inventory of the contents of this shrine, which mentions, among other relics, those of Saint James the Persian, already Evelyn White had concluded that the figure called James on the front of the chest represents James the Persian and that from the outset the reliquary was destined to contain the latter's relics.[61] Moreover, the object itself shows a strong family resemblance to the wooden screens which Moses of Nisibis had erected in front of the *haykal* [altar room] and the *khurus* [room separating the altar room and the nave] of the Church of the Virgin in the early years of the tenth century. It can be inferred that the reliquary dates from the same time and had originally been intended for this same church. Evelyn White suggested that its original place might have been in the *khurus*.[62] However, by invoking "the God of James the Persian", the humble Coptic inscription of Michael contradicts the latter point of view. Instead, it confirms what my colleague Karel Innemée had surmised already on the basis of the reliquary's measurements: that it had stood originally in the niche of the short east wall adjacent to the northern wall of the nave (Figure 5.1, no. 4),[63] that is precisely next to the spot where Michael's graffito was uncovered. Here the textual and archaeological evidence combine to pinpoint a local cult of the relics of Saint James the Persian, at least as old as the tenth century, to the northeastern corner of the nave of the Church of the Virgin.

Conclusions

The inscriptions briefly discussed above represent a mere selection from the epigraphic material in Coptic discovered in the Wadi al-Natrun during the last fifteen years. They nevertheless suffice to show two things: First, the Wadi al-Natrun has considerable potential for exciting finds in this field. Secondly, inscriptions, even quite insignificant ones, can be interesting as historical sources on their own account. They are able to shed light on a wide variety of aspects of the life of a monastic community through the ages: on the patronage of art and artists, on local language politics and language variety, on local cults and their "topography". Each of these aspects deserves further study in the years to come. There can be little doubt that in a future rewriting of the history of monasticism in the Wadi al-Natrun, the epigraphic evidence will prove to be as crucial as that of literary, liturgical or archaeological sources.

Notes

1 I wish to thank my colleagues Mat Immerzeel and Brian P. Muhs for their friendly assistance.

2 Painter's memento in the nave; text: R.-G. Coquin and P.-H. Laferrière, "Les inscriptions pariétales de l'ancienne église du monastère de S. Antoine, dans le désert oriental", *BIFAO* 78 (1978), 279–80, no. 14, ll. 3–9; reproduced in P. van Moorsel, *Les peintures du monastère de Saint-Antoine près de la mer Rouge*, Le Caire: IFAO 1995–1997, I, 182, n. 2 (cf. II, pl. 82); the latest edition is by B.A. Pearson, in E.S. Bolman (ed.), *Monastic visions: wall paintings in the Monastery of St. Antony at the Red Sea*, Cairo: ARCE/New Haven: Yale University Press 2002, 229, under N35.3 (cf. 38, with fig. 4.2), who failed to understand both beginning and end of the inscription. For the painter Theodore and his work, see now Bolman's beautiful book. Some further examples of this proverb in Coptic: W.E. Crum, "Inscriptions from Shenoute's monastery", *Journal of Theological Studies* 5 (1904), 555 (no. A 1, ll. 26–7); 561 (no. A 8) with 562, n. 2; 563 (no. A 9); in Arabic: A. Hebbelynck, and A. van Lantschoot, *Codices coptici Vaticani, Barberiniani, Borgiani, Rossiani* I: *Codices coptici Vaticani*, Vatican: Biblioteca Apostolica Vaticana 1937, 427.

3 For a useful overview of Coptic epigraphy, see M. Krause, "Inscriptions", in: *CoptEnc* 4, 1290–99.

4 See C. Wietheger, *Das Jeremias-Kloster zu Saqqara unter besonderer Berücksichtigung der Inschriften*, Altenberge: Oros 1992.

5 Cf. S. Timm, *Das christlich-koptische Ägypten in arabischer Zeit*, Wiesbaden: Reichert Verlag 1984–1992, II, 664–67.

6 Cf. H.G. Evelyn White, *The monasteries of the Wâdi 'n Natrûn*, New York: Metropolitan Museum of Art 1926–1933, III, 53.

7 R. Kasser et al., *Kellia: topographie*, Genève: Georg 1972, 82, no. 58, 5/20, fig. 33.

8 Evelyn White, *Monasteries* III, 193–94 (with an incorrect date A.D.), pl. LV, B; presently mounted in a case in the *khurus* of the Church of the Virgin Mary.

9 Published in F.L. Griffith, "Christian documents from Nubia", *PBA* 14 (1928), 118–28, pl. I; cf. White, *Monasteries* III, 215–17, pl. LXXVI; D.A. Welsby, *The Medieval kingdoms of Nubia: Pagans, Christians and Muslims along the Middle Nile*, London: British Museum Press 2002, 89–90, 260 (all give an incorrect date A.D.); presently exhibited in the museum of the monastery.

10 Presumably both were brought in from neighboring monastic sites; see White, *Monasteries* III, 193 and 217.

11 On ancient Christian graffiti, see now W. Eck, "Graffiti an Pilgerorten im spätrömischen Reich", in: *Akten des XII. internationalen Kongresses für christliche Archäologie*, Münster: Aschendorffsche Verlagsbuchhandlung 1995; for a representative ensemble from a Coptic monastery, S.H. Griffith's chapter in Bolman, *Monastic visions*, 185–93.

12 For these graffiti, see White, *Monasteries* III, 75–6, 79–80, pl. XVI; J. Leroy, *Les peintures des couvents du Ouadi Natroun*, Le Caire: IFAO, 45–9, pl. 91–103.

13 See, for example, K.D. Jenner and L. van Rompay, in K. Innemée et al., "New discoveries in the al-'Adrâ' Church of Dayr as-Suryân in the Wâdî al-Naṭrûn", *Mitteilungen zur christlichen Archäologie* 4 (1998), 96–103; K. Innemée and L. van Rompay, "La présence des Syriens dans le Wadi al-Natrun (Égypte): à propos des découvertes

HISTORY THROUGH INSCRIPTIONS

récentes de peintures et de textes muraux dans l'Église de la Vierge du Couvent des Syriens", *Parole de l'Orient* 23 (1998), 174–80.

14 See the slightly different classification by Pearson, in Bolman, *Monastic Visions*, 217–19.

15 See, for example, K. Innemée and L. Van Rompay, "Deir al-Surian (Egypt): new discoveries of 2001–2002", *Hugoye: Journal of Syriac Studies* 5/2 (2002), 13–14 and 28–40 [www.bethmardutho.org/index.php/hugoye/volume-index.html] on the scenes with Constantine and Abgar in the *khurus* of the Church of the Virgin Dayr al-Suryan and their Syriac legends.

16 Cf. P.P.V. van Moorsel, "La grande annonciation de Deir es-Sourian", in: *Called to Egypt: collected studies on painting in Christian Egypt*, Leiden: NINO 2000, 211.

17 Published by J. Leroy, "Un flabellum syriaque daté du Deir Souriani (Egypte)", *Les Cahiers de Mariemont* 5–6 (1974–75), 31–9.

18 For the latter, see the accompanying plan.

19 The new inscriptions from the Church of the Virgin will be published in full in the final publication of the Leiden Dayr al-Suryan project; for preliminary publications and discussions, see the footnotes and the bibliography in the Appendix.

20 Both crosses are visible on the photo in L. van Rompay, and A.B. Schmidt, "Takritans in the Egyptian desert: the Monastery of the Syrians in the ninth century", *Journal of the Canadian Society for Syriac Studies* 1 (2001), 60, to the right. More about this wall below.

21 The ensemble was recorded immediately after discovery by the late Professor Paul van Moorsel and his team; this documentation is kept in Leiden and its publication is forthcoming.

22 Scribal colophon edited by Hebbelynck and van Lantschoot, *Codices coptici*, 472–73; reproduced in H. Hyvernat, *Album de paléographie copte*, Paris: E. Leroux/Rome: Spithoever 1888, reprinted Osnabrück: Zeller Verlag 1972, pl. XL, 2. Note, however, that there is a two-letter lacuna in the name (ⲙⲏⲛⲁ ⲡ[ⲁⲛ]ⲁⲩ is my own reconstruction) and that the manuscript, although acquired at Dayr Abu Maqar, was originally written for a church outside the Wadi al-Natrun.

23 See K. Innemée, L. van Rompay, and E. Sobczynski, "Deir al-Surian (Egypt): its wall-paintings, wall-texts, and manuscripts", *Hugoye: Journal of Syriac Studies* 2/2 (1999) (www.bethmardutho.org/index.php/hugoye/volume-index.html), 8 (with ill. 3 and 4; cf. 37.e and 46); L. van Rompay, "Deir al-Surian: miscellaneous reflections", *Essays on Christian art and culture in the Middle East* 3 (2000), 87, pl. 3; it is not yet fully uncovered; for the most recent discoveries in the *khurus*, see now Innemée in Innemée and van Rompay, "Deir al-Surian (Egypt): new discoveries of 2001–2002".

24 For the title *papa*, see T. Derda and E. Wipszycka, "L'emploi des titres *abba*, *apa* et *papas* dans l'Egypte byzantine", *JJP* 24 (1994), 54–6.

25 Partly visible in W. Godlewski, "Les peintures de l'église de l'Archange Gabriel à Naqlun", *BSAC* 39 (2000), 93 and fig. 2.

26 Compare the Naqlun inscription just mentioned, which dates the refurbishment of the apse to the periods of office of Archbishop Zacharias of Alexandria, of Bishop Severos of al-Fayoum, and of a local abbot (archimandrite) Papnoute, a local archpriest Kabri, and two more local dignitaries whose names and functions are lost (upper line of text; edition by the present author forthcoming in W. Godlewski's publication of the church of the Naqlun monastery).

27 See Van Rompay, "Deir al-Surian", 84.

28 K. Innemée, "Deir al-Surian (Egypt): conservation work of Autumn 2000", *Hugoye: Journal of Syriac Studies* 4/2 (2001), 10, www.bethmardutho.org/index.php/hugoye/volume-index.html; Innemée and van Rompay, "Deir al-Surian (Egypt): new discoveries of 2001–2002", 12.

29 See the cautious remarks by Van Rompay, in Innemée, Van Rompay and Sobczynski, "Deir al-Surian (Egypt)", 44–50, and Van Rompay, "Deir al-Surian".

30 See, however, the provisional remarks by L.Th. Lefort, "Littérature bohairique", *Le Muséon* 44 (1931), 115–35, and A. Shisha-Halevy, "Bohairic", in: *CoptEnc* 8, 58.

31 Cf. Shisha-Halevy, "Bohairic", 58: "nonliterary Bohairic is still a complete mystery".

32 Although the list of Bohairic inscriptions given by Roquet a quarter of a century ago (G. Roquet, "Inscriptions bohaïriques de Dayr Abû Maqâr", *BIFAO* 77 (1977), 163–64) can be considerably expanded now, the corpus is still not impressive.

33 For the Bohairic of the Kellia inscriptions, see now R. Kasser, "L'épigraphie copte aux Kellia et l'information qu'elle donne sur l'importance de la langue bohairique *B5*", *BSAC* 37 (1998), 15–48, and idem, "Langue copte bohairique: son attestation par les inscriptions des Kellia et leur évaluation linguistique", in: S. Emmel et al. (eds.), *Ägypten und Nubien in spätantiker und christlicher Zeit: Akten des 6. Internationalen Koptologenkongresses, Münster, 20.-26. Juli 1996*, Wiesbaden: Reichert Verlag 1999, II, 335–46.

34 For the inscriptions on this wall, see Van Rompay, in K. Innemée and L. van Rompay, "Deir al-Surian (Egypt): new discoveries of January 2000", *Hugoye: Journal of Syriac Studies* 3/2 (2000), 17–32, www.bethmardutho.org/index.php/hugoye/volume-index.html. The present text is no. 1 in fig. 1; cf. furthermore Innemée and van Rompay, "La présence des Syriens"; 171; Innemée, Van Rompay and Sobczynski, "Deir al-Surian (Egypt)", 40.f.

35 For a very similar *dipinto* in the Monastery of Saint Phoibammon in Western Thebes, see H.E. Winlock and W.E. Crum, *The monastery of Epiphanius at Thebes*, New York: Metropolitan Museum of Art 1926, I, 12–13; W. Godlewski, *Le monastère de St Phoibammon*, Warsaw: PWN–Éditions scientifiques de Pologne 1986, 150–51, no. 27 (who dates it to the seventh–eighth century). The typical exhortation "may a wise man (σοφός) solve this saying" (here l. 6) recurs in the *dipinti* which a certain Ananias left on the west wall of the church of the Red Monastery, Sohag (undated and unpublished).

36 Note also ⲕⲱⲡ for ⲕⲱⲃ (l. 5); ϫⲱⲡ for ϫⲱⲃ (l. 7).

37 The text offers no real examples of initial ⲃ-. For the vocalization of the nominal state form, see Shisha-Halevy, "Bohairic", 58, who refers to H.J. Polotsky, *Collected papers*, Jerusalem: Magnes Press 1971, 344.

38 Some scarce Sahidic examples in W.E. Crum, *A Coptic dictionary*, Oxford: Oxford University Press 1939, 27.a, and P.E. Kahle, *Bala'izah: Coptic texts from Deir el-Bala'izah in Upper Egypt*, London: Oxford University Press 1954, I, 93, under 65A.

39 Thus ϫⲱⲡ/ϫⲟⲡ for ϫⲱⲃ, see Kasser, "L'épigraphie copte", 41 s.v.

40 Several examples in Crum, *Coptic dictionary*, 27.a. Especially ⲕⲱⲡ for ⲕⲱⲃ is not rare, see ibidem, 98–9, also e.g. a memento of A.D. 1025 in codex Vat. copt. 584, f. 35v (Hebbelynck and van Lantschoot, *codices coptici*, 390); it may have influenced the spelling ⲓⲁⲕⲱⲡ here.

41 ⲙⲉⲃⲓ also occurs in the Kellia inscriptions (Kasser, "L'épigraphie copte", 36 and 30); other examples in the Church of the Virgin at Dayr al-Suryan are a graffito below the Galaktotrophousa in the *khurus* (unpublished) and a graffito on the capital of a column in the southern nave (see Innemée and van Rompay, "Deir al-Surian (Egypt): New discoveries of January 2000", 33–5; Van Rompay, "Deir al-Surian", 87, pl. 4–5; cf. 84).

42 ⲃⲟⲛ for ⲟⲩⲟⲛ also once in the Kellia (Kasser, "L'épigraphie copte", 39 and 30). Both ⲙⲉⲃⲓ and ⲃⲟⲛ occur passim in the colophons and readers' notes of the Vatican Bohairic codices from Dayr Abu Maqar.

43 Van Moorsel, "La grande annonciation", 214, l. 10; 218, fig. 6.

44 Thus in the Kellia: ⲟⲩⲃⲟⲛ for ⲟⲩⲟⲛ (cf. Kasser, "L'épigraphie copte", 39 and 30: "cumul"). More frequently in "Nitrian" literary texts: Crum, *Coptic dictionary*, 467.a

HISTORY THROUGH INSCRIPTIONS

(oy "supplemented" by в and conversely); Lefort, "Littérature bohairique", 123 (under I.c), and e.g. oyвєрⲱ)єноⲩϥ1 (spelling of the proper name Barsanuphius; cf. Van der Vliet, "Un évêché fantôme: ⲛ1оⲩвєрⲱ)єноⲩϥ1", *GM* 120 (1991), 109–11; codex Vat. copt. 6212, Hebbelynck and van Lantschoot, *Codices coptici*, 445–46).

45 See, provisionally, Kasser's remarks ("L'épigraphie copte", 30 and 31–2) on part of the Kellia material, although I would prefer to avoid the misnomer "Bashmuric".

46 So-called "Bashmuric" is supposed to be characterized first of all by the absence of specifically Coptic (Demotic) graphemes (cf. R. Kasser, "Bashmuric", in: *CoptEnc* 8, 47–8; R. Kasser and A. Shisha-Halevy, "Dialect G (or Bashmuric or Mansuric)", in: *CoptEnc* 8, 74–6); this is not the case here, as it is in occasional reader's notes in Wadi al-Natrun literary manuscripts (e.g. codex Vat. copt. 591, f. 29v; Hebbelynck and van Lantschoot, *Codices coptici*, 401).

47 Van Moorsel, "La grande annonciation", 214, l. 10; 218, fig. 6 (ⲋ only occurs in the compendium ⲡⲟⲥ). Cf. Shisha-Halevy, "Bohairic", 58.

48 Van Moorsel, "La grande annonciation", 214, l. 8; 218, fig. 5: ⲛхнтϥ; contrast next example.

49 Ibidem, 213, l. 6; 217, fig. 4: єⲅоⲩⲛ ⲛⲅнтϥ. Cf. Lefort, "Littérature bohairique", 124 (under II); Shisha-Halevy, "Bohairic", 58.

50 Although it is to be regretted that the Bohairic sequel to A. van Lantschoot's collection of Sahidic colophons never appeared, rich material can be gleaned from Hebbelynck and van Lantschoot, *Codices coptici*.

51 The classic statement of this view is Lefort, "Littérature bohairique".

52 See above, note 35.

53 K. Innemée, "Recent discoveries of wall-paintings in Deir Al-Surian", *Hugoye: Journal of Syriac Studies* 1/2 (1998), 21, www.bethmardutho.org/index.php/hugoye/volume-index.html. See furthermore idem, "The iconographical program of paintings in the church of al-'Adra in Deir al-Sourian: some preliminary observations", in: M. Krause and S. Schaten (eds.), *ΘΕΜΕΛΙΑ: spätantike und koptologische Studien Peter Grossmann zum 65. Geburtstag*, Wiesbaden: Reichert Verlag 1998, 148; Innemée and van Rompay, "La présence des Syriens", 177–78; Innemée, Van Rompay and Sobczynski, "Deir al-Surian (Egypt)", 40.c-d.

54 Not on fig. 2 in Innemée and van Rompay, "La présence des Syriens", 178, but partly and vaguely visible on fig. 5 in Van Rompay and Schmidt, "Takritans", 60 (extreme lower righthand corner).

55 The inscription mentions his place of origin (end of l. 3, now largely lost), which, however, does not necessarily imply that he did not live in the Wadi al-Natrun.

56 Cf. J. van der Vliet, "The Church of the Twelve Apostles: the earliest cathedral of Faras?" *Orientalia* 68 (1999), 93 [Study 26].

57 Codex Vat. copt. 591, ed. I. Balestri and H. Hyvernat, *Acta Martyrum* II, Paris: Typographeo Reipublicae1924; reprinted Louvain: L. Durbecq 1953, 24–61; an annotation would seem to date the manuscript to A.D. 883/84, but see Hebbelynck and van Lantschoot, *Codices coptici*, 402. The interesting translation story (Balestri and Hyvernat, *Acta*, 50–60) was studied by O. von Lemm, *Iberica*, St. Petersburg: Academy of Sciences 1906, 2–19; it links the Saint with the Oxyrhynchite nome, not (yet?) with the Wadi al-Natrun.

58 See White, *Monasteries* I, 75, no. XIV.

59 Discussed by White, *Monasteries* II, 363–64.

60 The reliquary is decribed and discussed in White, *Monasteries* III, 194–96, with pl. LXIII. On the figure of Saint James ("attired in a short, skirted tunic"), ibidem, 195. Cf. Innemée, Van Rompay and Sobczynski, "Deir al-Surian (Egypt)", 33.

61 White, *Monasteries* III, 196, where also the document in question ("a paper formerly affixed to the wall of the church") is translated.

62 See ibidem, 194, 196, and caption to pl. LXIII ("from the choir (?)").
63 Cf. Innemée, "The iconographical program", 148; idem, "Recent discoveries of wall-paintings", 21.

Appendix

Bibliography of publications relating to Wadi al-Natrun epigraphy that have appeared after 1924 (date of demise of H.G. Evelyn White): Although fairly complete as far as major publications are concerned, the list does not pretend to be exhaustive. Brief annotations are meant to be indicative of the epigraphic relevance of each title.

[N.B. All articles published in the electronic journal *Hugoye: Journal of Syriac Studies*, originally available at http://syrcom.cua.edu/hugoye, are now accessible at www.bethmardutho.org/index.php/hugoye/volume-index.html.]

Evelyn White, H.G. 1926–1933, *The monasteries of the Wâdi 'n Natrûn* I-III, New York: Metropolitan Museum of Art: vol. III quotes and discusses the epigraphic sources available to the author.

Griffith, F.L. 1928, "Christian documents from Nubia", *PBA* 14: 117–46: edition of the epitaph of King George (Dayr al-Suryan: 118–28).

Hunt, L.-A. 1993, "Eternal light and life: a thirteenth-century icon from the Monastery of the Syrians, Egypt, and the Jerusalem paschal liturgy", *Jahrbuch der österreichischen Byzantinistik* 43: 349–74; reprinted in: *Byzantium, Eastern Christendom and Islam* II, London: The Pindar Press 2000, 127–52: brief Greek legends (reprint, 128).

Hunt, L.-A., "The fine incense of virginity: a late twelfth century wallpainting of the Annunciation at the Monastery of the Syrians, Egypt", *Byzantine and Modern Greek Studies* 19 (1995): 182–232; reprinted in: *Byzantium, Eastern Christendom and Islam* I, London: The Pindar Press 1998, 158–204: due attention is paid to the legends (but see now Van Moorsel, "La grande annonciation", below).

Immerzeel, M., "Discovery of wall-paintings in Deir anba Bishoi (Wadi'n Natrun)", *Newsletter/Bulletin d'information International Association for Coptic Studies* 30 (1992): 8–11: mentions (unpublished) Coptic inscriptions ("Chapel of Benjamin").

[Innemée, K.C., "Dayr al-Suryan: new discoveries", *Claremont Coptic encyclopedia* (2016) (http://ccdl.libraries.claremont.edu/cdm/singleitem/collection/cce/id/2137/rec/1)].

Innemée, K.C., "Deir al-Surian (Egypt): conservation work of Autumn 2000", *Hugoye: Journal of Syriac Studies* 4/2 (2001): discusses i.a. the Coptic inscription in the dome above the *khurus*, Church of the Virgin (10).

Innemée, K.C., "The iconographical program of paintings in the church of al-'Adra in Deir al-Sourian: some preliminary observations", in: M. Krause and S. Schaten (eds.), *ΘΕΜΕΛΙΑ: Spätantike und koptologische Studien Peter Grossmann zum 65. Geburtstag*, Wiesbaden: Reichert Verlag 1998, 143–53: refers *passim* to legends and other inscriptions.

Innemée, K.C., "New discoveries at Deir al-Sourian, Wadi al-Natrun", in: S. Emmel et al. (eds.), *Ägypten und Nubien in spätantiker und christlicher Zeit: Akten des 6. internationalen Koptologenkongresses, Münster, 20.-26. Juli 1996*, Wiesbaden: Reichert Verlag

1999, I: 213–22: refers to legends and other inscriptions (Coptic, Greek, Syriac) in the Church of the Virgin.

Innemée, K.C., "Recent discoveries of wall-paintings in Deir Al-Surian", *Hugoye: Journal of Syriac Studies* 1/2 (1998): refers to legends and other inscriptions (Coptic, Greek, Syriac) in the Church of the Virgin.

Innemée, K.C., P. Grossmann, K.D. Jenner, and L. van Rompay, "New discoveries in the al-'Aḏrâ' Church of Dayr as-Suryân in the Wâdî al-Naṭrûn", *Mitteilungen zur christlichen Archäologie* 4 (1998): 79–103: occasional discussion of Greek legends (Innemée, 79–90); four Syriac *dipinti* (Jenner and van Rompay, 96–103).

Innemée, K.C., and L. van Rompay, "Deir al-Surian (Egypt): new discoveries of January 2000", *Hugoye: Journal of Syriac Studies* 3/2 (2000): mural inscriptions in Coptic and Syriac, Church of the Virgin.

Innemée, K.C., and L. van Rompay, "Deir al-Surian (Egypt): new discoveries of 2001–2002", *Hugoye: Journal of Syriac Studies* 5/2 (2002): Coptic and Syriac legends; Syriac graffiti, Church of the Virgin.

Innemée, K.C., and L. van Rompay, "La présence des Syriens dans le Wadi al-Natrun (Égypte): à propos des découvertes récentes de peintures et de textes muraux dans l'Église de la Vierge du Couvent des Syriens", *Parole de l'Orient* 23 (1998): 167–202: chiefly on the Syriac texts (174–80).

Innemée, K.C., L. Van Rompay, and E. Sobczynski, "Deir al-Surian (Egypt): its wall-paintings, wall-texts, and manuscripts", *Hugoye: Journal of Syriac Studies* 2/2 (1999): references to legends and other inscriptions (section I, Innemée); discussion of Syriac and Coptic inscriptions in the Church of the Virgin (section II, Van Rompay).

Leroy, J., "Le décor de l'église du couvent des Syriens au Ouady Natroun (Égypte)", *CArch* 23 (1974): 151–67: quotes legends (Syriac, Greek, Coptic); facsimile copies of Syriac dedicatory texts (154).

Leroy, J., *Les peintures des couvents du Ouadi Natroun*, Le Caire: IFAO 1982: quotes passim legends and other inscriptions.

Leroy, J., "Un flabellum syriaque daté du Deir Souriani (Egypte)", *Les Cahiers de Mariemont* 5–6 (1974–75): 31–9: Syriac inscription on a flabellum (A.D. 1202).

Martin, M.J., "A Syriac inscription from Deir al-Surian", *Hugoye: Journal of Syriac Studies* 5/2 (2002): dedicatory text on a marble column.

Martyros El-Souriany, "The youngest layer of plaster in the Church of the Holy Virgin Mary in El-Sourian Monastery", in: *Seventh International Congress of Coptic Studies: abstracts of papers*, Leiden: Faculty of Arts/Leiden University 2000, 73: the full paper, which is partly based upon Arabic graffiti, [did not appear in M. Immerzeel and J. van der Vliet (eds.), *Coptic studies on the threshold of a new millennium: proceedings of the Seventh International Congress of Coptic Studies, Leiden, 27 August-2 September 2000*, Louvain/Paris/Dudley, MA: Peeters 2004].

Meinardus, O.F.A., "The collection of coptica in the Monastery of St. Macarius", *BSAC* 19 (1967–1968): 235–48: several inscribed objects; some texts are given.

Meinardus, O.F.A., "The museum of Dair as-Surîân also known as the Monastery of the Holy Virgin and St. John Kame", *BSAC* 17 (1963–1964): 225–34: several inscribed objects, most of them previously known.

Monneret de Villard, U., *Les églises du monastère des Syriens au Wâdî en-Naṭrûn*, Milan: privately printed 1928: quotes legends from the screens of Moses of Nisibis (30–1; cf. Evelyn White, *Monasteries* III, 180 ff.; Leroy, "Le décor de l'église").

Roquet, G., "Inscriptions bohaïriques de Dayr Abû Maqâr", *BIFAO* 77 (1977): 163–79: dedicatory inscriptions in the "Haykal of Benjamin"; cf. Leroy, *Les peintures des couvents*, 124–25, pl. III-IV, and *SB Kopt.* I 495–97.

Van Loon, G.J.M., *The gate of heaven: wall paintings with Old Testament scenes in the altar room and the ḫûrus of Coptic churches*, Leiden: NINO 1999: quotes passim the legends of the paintings discussed, many of them from the Wadi al-Natrun monasteries.

Van Moorsel, P.P.V. "La grande annonciation de Deir es-Sourian", *BIFAO* 95 (1995), 517–37: replaced by citation below.

Van Moorsel, P.P.V., "La grande annonciation de Deir es-Sourian", in: *Called to Egypt: collected studies on painting in Christian Egypt*, Leiden: NINO 2000 203–24: text of the legends (Greek and Bohairic; 213–14); replaces citation above.

Van Rompay, L., "Art and material culture of the Christian Syriac tradition: some current projects", *Hugoye: Journal of Syriac Studies* 1/1 (1998): preliminary remarks on Syriac text finds in Dayr al-Suryan, Church of the Virgin (7–8).

Van Rompay, L., "Deir al-Surian: miscellaneous reflections", *Essays on Christian art and culture in the Middle East* 3 (2000), 80–7: includes discussion of mural inscriptions in Coptic and Syriac, Church of the Virgin.

Van Rompay, L., and A.B. Schmidt, "A new Syriac inscription in Deir al-Surian (Egypt)", *Hugoye: Journal of Syriac Studies* 4/1 (2001): wooden beam, A.D. 1285/86.

Van Rompay, L., and A.B. Schmidt, "Takritans in the Egyptian desert: the Monastery of the Syrians in the ninth century", *Journal of the Canadian Society for Syriac Studies* 1 (2001): 41–60: includes discussion of the new mural inscriptions in Syriac, Church of the Virgin (49–51).

6

RECONSTRUCTING THE LANDSCAPE
Epigraphic sources for the Christian Fayoum

Jacques van der Vliet

Introduction

Inscriptions are a rich and potentially important source for the history of Christianity in the Fayoum. A wide variety of inscriptions survives – from different places and periods and in different genres and languages – in Greek, in Coptic and to a minor degree in Arabic. Writing a history of the Fayoum on the basis of these inscriptions alone, however, would hardly be possible. The epigraphic record is incomplete and full of lacunae, and scattered over time and space. The inscriptions from the Fayoum may be seen as small pieces of a great jigsaw puzzle that have to be combined with other pieces to yield any sense at all. In order to allow inscriptions to take their proper place within the picture, it is necessary not only that they are available, in the form of reliable editions, but also that they can be restored to their original context. Isolated artifacts are rarely helpful for any purpose.

The present paper will be devoted to the elementary task of mapping the epigraphic evidence. This involves, first, simple stocktaking, collecting the artifacts, and, secondly, an attempt to situate them in space and time. Of course, this task can here only be undertaken in a provisional and cursory way, but it is a necessary first step before inscriptions can be used as historical sources in anything other than a haphazard manner. Such a survey may bring out both the limits and the potential interest of epigraphic sources.[1]

One of the most striking characteristics of the epigraphic evidence from the Fayoum is its diversity. For the present purpose, I distinguish four large groups of artifacts, according to a classification that, like every classification, is arbitrary and allows some overlap between the various categories.[2] The first group of artifacts consists of funerary inscriptions; the second of monumental dedicatory inscriptions; the third of ensembles of mural inscriptions; and, finally, inscriptions on various *objets d'art*.

Funerary inscriptions

Funerary stelae are the most widely known category of inscriptions from the Fayoum. Sofia Schaten [(2005)] discusses them in detail and here a few general remarks may suffice. Those most characteristic of the Fayoum are the elaborate limestone stelae of a "Hellenistic" type, with crosses and praying figures in a rich architectural decor, that can be found in almost every collection of late antique Egyptian art.[3] They are inscribed in Greek or, less frequently, Fayoumic Coptic. Familiar as they seem, very little is actually known about them. Their provenance, for instance, can be ascertained only in rare cases. Most of them may come from anywhere in the Fayoum or even from elsewhere. What is worse, to the best of my knowledge, not one item has ever been recorded *in situ*. It can therefore only be inferred how these stelae were conceived functionally and how they related architectonically to the tomb or the cemetery.[4]

Situating these monuments in time is hardly easier than situating them on the map. Absolute dates, i.e. Diocletian years, are rare and generally do not occur in Fayoum sources before the middle of the seventh century.[5] The very few dated stelae known show that the traditional sculptured types were current until well into the eighth century, but also suggest that with time a development in the textual formulae had taken place. Whereas the earlier texts are very brief and simple, longer formulae with clear liturgical echoes had made their appearance. The most common of them is the well-known prayer for being received in the bosom of Abraham, Isaac and Jacob.[6] Stylistic criteria also suggest a development over a longer period of time that may have comprised the sixth to eighth centuries.[7]

Therefore, although these sculptured stone stelae, decorated in a style considered typical of the Fayoum, make up a sizeable corpus, their usefulness for anything other than prosopographical listing or iconographical comparison is limited. Their chronology can be established only approximately, their provenance is but rarely known and, from an archaeological point of view, their role and function as funerary monuments are undocumented. There is a total loss of context that hampers any effort at interpretation.

Apart from this familiar class of stelae, other types of funerary inscriptions exist, some of them also discussed by Schaten [(2005)]. Those presenting pre-Christian textual or formal features may predate the heyday of the richly sculptured stelae. They can be stone slabs with minimal decoration[8] or wooden tablets continuing the style of the earlier "mummy-labels",[9] and are always inscribed in Greek. Simple stone slabs remained in use long afterward. A relatively late example, from the Louvre, is inscribed in Coptic and bears a date corresponding to A.D. 913.[10] Here, several differences with the older classes of stelae can be observed. The language is no longer Greek or Fayoumic, but a form of Fayoumi-Sahidic (Sahidic strongly tainted by Fayoumic in the vowel system, but usually without lambdacism)[11] that is typical of tenth-eleventh-century sources from the region, and the text exhibits the more developed, liturgically inspired formulae that enjoyed popularity from the seventh century onward.

Liturgical inspiration is also characteristic of another group of late funerary inscriptions, recently identified by Anne Boud'hors and Florence Calament (2004). Three closely related wooden stelae, inscribed again in Fayoumi-Sahidic, all open with the formula "God of the spirits and Lord of all flesh." They bear dates in the second quarter of the tenth century and can be assigned to the neighborhood of Tutun, the successor settlement to Tebtynis. Formally they exhibit a modified cross-form. This group can be compared to other wooden funerary crosses from the Fayoum, which are also inscribed with liturgically inspired formulae.[12]

It may provisionally be concluded that the domain of Fayoum funerary inscriptions saw several developments through time. After ill-defined beginnings in Greek, the heyday of the elaborately sculptured stone stelae, inscribed in Greek and Fayoumic Coptic, coincides most probably with the sixth to eighth centuries. From the seventh century, the inscriptions become more elaborate: they witness the influence of liturgical formulae and occasionally provide absolute dates. Sometime between the eighth and the tenth centuries the sculptured architectonical stelae appear to have gone out of fashion. Instead, a certain preference for a freestanding cross-form and for wood as a material can be observed. Linguistically, Coptic became more widespread than Greek and, ousting Fayoumic proper, a local brand of Sahidic, called here Fayoumi-Sahidic, rose to prominence.

Monumental dedicatory inscriptions

Few monumental inscriptions from Christian times survive, except those that are part of larger ensembles of painted church decoration, which will be discussed later. The Greek limestone inscription found in the neighborhood of al-Nazlah, which commemorates the revetment of a church of St. Menas, is perhaps the best known of them. It has been plausibly dated to the second half of the seventh century (perhaps A.D. 672).[13] This inscription is at the same time the only evidence for Christianity in al-Nazlah in pre-modern times and the only witness to the existence of a church of Saint Menas in or near that town.

A later group of monumental lintels, inscribed in Coptic, originates probably from a more southeastern part of the Fayoum.[14] The five pieces that are now assigned to this group share a common formulary – a short prayer asking for God's protection – a common idiom (usually Fayoumi-Sahidic), and a common layout, the text being arranged symmetrically around a central cross. They record the efforts, generally not merely of an individual, but of an entire family (man, wife, children), in the building of a "house"[15] or "living quarter."[16] Judging from the formulae used, these must have been pious foundations. Most of the lintels bear tenth-century dates and are made of wood. Boud'hors and Calament rightly noted the relationship of these lintels with the wooden stelae from the Tutun region published by them (2004: 454–55). Their observation is confirmed by the recorded provenance of one of these lintels, the only one in stone, which has been found in Abu Hamid,[17] one of the hamlets that make up the village now known as Qasr al-Basil, somewhat to the east of Tebtynis, on the Bahr al-Gharaq, in the same southeastern corner of the Fayoum as Tutun. S. Björnesjö has proposed identifying the site of the present-day Qasr al-Basil with the medieval village of Peljisok/Bulgusuq (Björnesjö 1993: 239–40). The latter is known from quite a number of Coptic and Arabic documentary texts from about the tenth to twelfth centuries, but has never been located with certainty.[18] Various written sources, among which is an inscribed silver chalice in the Louvre, show that it had a church dedicated to the Virgin Mary.[19] One of the wooden lintels belonging to the same tenth-century group as the stone lintel from Abu Hamid/Qasr al-Basil also mentions a church of the Virgin.[20] Although churches of the Virgin must have been common,[21] this lends substance to the idea that this entire group of tenth-century lintels originate from the basin of the Bahr al-Gharak.[22] Both documentary texts and epigraphic data suggest that, around the year 1000, this region was an important Christian center.

Mural inscriptions

Perhaps more than any other kind, mural inscriptions are likely to provide information of substance and density. Not only because they often make up ensembles, but also because they become less easily detached from their original context than moveable objects; with some luck they stick to the wall.

In addition to the graffiti at the isolated site of the so-called Dayr Abu Lifa (Timm 1984–1992: II, 588–89), two important ensembles of inscribed mural paintings are known: one from the now-lost church at Tebtynis, edited in 1989 by C.C. Walters after the documentation recorded by the Grenfell and Hunt expedition over a century ago, and the other in the church of the Monastery of the Archangel Gabriel at Naqlun, in recent years discovered and restored by the mission of the Polish Centre of Mediterranean Archaeology of the University of Warsaw.[23] Within such a decorated church, various kinds of inscriptions can be expected: legends accompanying the wall paintings; dedicatory inscriptions commemorating patrons, donors and artists; and, finally, inscriptions and graffiti commemorating visitors.

Both churches were apparently decorated according to a kind of program: the Tebtynis church in the tenth century, the one in Naqlun in the eleventh century. Donors – well-to-do persons or groups of persons – provided the means for executing the individual parts of the scheme. Prayers for their well-being were inscribed next to the paintings they donated. Regrettably, the church in Tebtynis was only hastily recorded and it must be feared that much valuable information has been lost forever. Out of several prayers for donors, only one has come down to us in a complete form. It asks for Christ's blessing for "our brother Papas, son of Mercury," who in A.D. 952 or 953 paid for a painting of an unidentified archangel, perhaps St. Michael, that apparently did not itself survive.[24] To judge from the title of this "brother" Papas, he was a monk and this lends support to the idea that the church in question was a monastery church,[25] like the one at Naqlun.

The inscriptions from the Naqlun church are more numerous and much richer in information. They combine with the archaeological, literary and documentary material from the same site to create a picture of the monastic community as it flourished in the eleventh century. At that time, the apse was refurbished and the entire church redecorated. These works of renovation naturally focused on the liturgical center of the church, i.e. on the altar area in the central apse. The inscriptions in this part of the church, as far as they survive, are the decorative expression of an entire hierarchy of roles and functions. In this respect, the Naqlun ensemble is probably unique in Egypt.

The *auctor intellectualis* of the refurbishment and redecoration of the church was undoubtedly the archimandrite Papnoute, the head of the monastic community. He is commemorated in a window grill that occupies an elevated and central position in the apse, but in fact is placed so high that it cannot be read well from floor level (Godlewski 2000: 92). Still central, but in a lower position and in relatively small script, is a long but badly damaged inscription with a prayer for a group of donors. The most conspicuous inscriptions, however, are those that run below the Apostles and that are written in a big decorative script that is readable even from a distance (Godlewski 2000: 93, fig. 2). The uppermost of them is a long foundation inscription that situates the building's operation in terms of time and jurisdiction in the period of tenure of a whole series of officials: first the reigning archbishop Zacharias (died 1032), then the local bishop, a certain Severos,[26] the archimandrite Papnoute, and finally three more dignitaries who are said to be connected to "this church": an archpriest Kabri and two more whose names are lost. Below this line there is a second line in even bigger script containing a prayer exclusively for the local bishop, stating *inter alia* that "[it is he] who provides the churches within his see with everything they (text: you) need".[27]

The layout of this ensemble of inscriptions reflects clear hierarchical principles of organization. The place of honor reverts to the local bishop whose care for the churches in his diocese is emphatically praised. An inscription on a somewhat smaller scale but in a more elevated position situates the building operations within a larger hierarchical frame that includes the heads of the monastery, among them the archimandrite Papnoute. As the probable patron of the whole operation, his name returns in a privileged position, in a window grill, where it is central although almost unreadable. The role of the donors is not less central, but they have to be satisfied with the usual prayer inscribed in a panel below the window.

Such inscribed panels, commemorating the donors and the painters of individual paintings, are found not only near the altar, but also throughout the church.[28] The impression conveyed by these prayers is that the donors were mostly well-to-do monks, priests and deacons from the "mountain of Naqlun", where they had their dwellings. Apparently they belonged to the wealthy elite of the monastery, who did not live in the cenobium, but rather in the hermitages behind it.[29] The surviving mementos for painters are generally much damaged. In the two cases where names survive, they appear to be deacons; one of them, a certain John, is said to be from Aqfahs, in the Nile Valley, which suggests that here, as elsewhere, the artists were not members of the community, but recruited from outside.[30]

A different category of inscriptions is that of the graffiti and *dipinti* commemorating visitors. A large number of them can be found on the north wall, in Arabic and Coptic, most of them now nearly unreadable. Some of these bear witness to the connections of the monastery with the outside world, in particular with the neighboring districts of the Nile Valley.[31] The most interesting of these is no doubt an inscription near the altar that records the visit of a Bishop James of Atfih in the year 1033, not long after the redecoration of the apse (Godlewski 2000: 94–5). According to the long text he left behind, he was accompanied by a cortège of three priests who were *hegoumenoi* and superiors of various monasteries and churches. On what occasion this important delegation from Atfih visited the monastery must remain a matter of speculation.

As may be seen from this quick review, the inscriptions in the church of Naqlun do not merely provide dates and legends for the wall paintings, but actually situate these paintings within the social and ecclesiastical life of the Fayoum in the eleventh century.

Inscriptions on *objets d'art*

The final category of inscriptions to be considered are those on various *objets d'art*. The nature of these may vary widely, from pottery to icons.[32] Indeed, as mentioned above, an inscribed silver chalice in the Louvre is one of the rare surviving sources for the existence of a church of the Virgin in Peljisok/Bulgusuq.[33] Here, the focus will be on Christian inscribed textiles, for which the Fayoum appears to have been an important center.[34]

In the centuries following the Arab conquest in A.D. 642, Egyptian Christians did not only adopt the originally Islamic, so-called *tiraz* style of dress, characterized by extensive inscriptions in Arabic, but they also developed what may be called a Christian *tiraz* style. This is characterized by inscriptions with an overtly Christian character, often enhanced with crosses and Christian monograms. The language used may be Greek or Coptic in addition to Arabic. As for the contents, they usually contain short personalized prayers giving the name and sometimes the filiation of the owner or beneficiary. In addition, quotations from the Greek or Coptic Psalter may be found. For textiles exhibiting this Christian *tiraz* style, the Fayoum must have been an important center. Remarkably, the evidence again appears to point, geographically, toward the southern part of the Fayoum (more precisely the basin of the Bahr al-Gharak), and, chronologically, toward a comparatively late period, around the year 1000. Recent research by French colleagues has shown that a characteristic group of these Christian inscribed textiles can be traced to a private *tiraz* workshop at Tutun.[35] This group is characterized by light inscriptions on a dark background and a combination of Kufic or Kufesque lines of Arabic text with lines of Coptic, usually the typical Fayoumi-Sahidic of the period.

In addition to this Tutun group, it is again the Monastery of the Archangel Gabriel that provides us with different examples of Christian *tiraz* textiles. During the recent campaigns of the Polish mission, a cemetery was uncovered that produced examples of inscribed textiles both in the Arabo-Islamic and in the typically Christian *tiraz* styles (Godlewski 2002). Among the examples of the latter style is a beautiful tunic inscribed with Psalms 46:2–3 in Sahidic Coptic.[36] Another tunic, closely resembling it in general style and outlook, but without a Psalm quote, is kept in the Louvre. Thanks to the inscriptions on its sleeves, we know that its owner, a priest Kolthi, was an inmate of this monastery (Van der Vliet 2000 [Study 11]). These inscribed clothes are difficult to date, but they must belong to a period around about the eleventh century. Whether or not they were mainly or even exclusively worn by members of the clergy is a matter of further research. In any case, the examples show that around the year 1000 the Fayoum still supported a Christian population that was sufficiently wealthy, literate and self-conscious to order and wear these typically Christian pieces of luxury dress.

Conclusion

Within the literate society of the late antique and early medieval Fayoum, inscriptions remained an important means of communication, even if they seem less numerous and prominent than in earlier periods. In spite of their fragmentary and often isolated nature, the epigraphic sources briefly presented here permit some general conclusions about trends and developments. Thus, in the period between the eighth and the tenth centuries, the sculptured stone monuments that continued late antique traditions gave way to a much wider range of inscribed texts involving a far greater variety of materials, including wood, plastered church walls, textiles and *objets d'art*. Linguistic diversity increased at the same time as material diversity: whereas earlier documents were inscribed in Greek and occasionally in Fayoumic Coptic, later ones show a preference for Coptic (in addition to Greek and Arabic), but now the Coptic is usually some local brand of Sahidic. Geographically, the apparently growing importance of the basin of the Bahr al-Gharak is quite remarkable. Around the year 1000, Tutun, Naqlun and Bulgusuq appear as major sources of inscribed material. Is this a meaningful shift or merely an accidental one, representing the chances of preservation and discovery?

Before an attempt can be made to analyze these developments in terms of the social and cultural history of the Fayoum, an important preparatory task awaits: that of trying to reintegrate the scattered and isolated pieces of evidence into their original context. In my opinion, this should take place along two lines: first, a reintegration in terms of space and time by combining field archaeology with museum archaeology, tracing artifacts scattered in museums back to their provenance after which, through comparison, more may be said about their chronology. Secondly, the epigraphic evidence from a particular place or region should be studied together and in combination with the archaeological and literary sources available for that particular place or region. This is a process of reintegration in terms of the source material. In classical scholarship, epigraphy counts as a *Hilfswissenschaft* [auxiliary science]. Only the double process of reintegration advocated here can make the study of inscriptions a really helpful tool for writing the history of Christianity in the Fayoum.

Notes

1 The three volumes of the *I. Fayoum*, edited by É. Bernand between 1975 and 1981, can serve as a model even if they cover almost exclusively pre-Christian inscriptions in Greek. Boud'hors and Calament include a valuable "Annexe" listing Christian inscribed material (2004: 461–69), but it is limited linguistically to Coptic. The Greek Christian inscriptions from the Fayoum known at the time were assembled in Lefebvre 1907.

2 The classification in Boud'hors and Calament 2004 is more restrictive (stelae, lintels, textiles).

3 Particularly impressive ensembles are in the Coptic Museum, Cairo and the Staatliche Museen, Berlin, for which see respectively Crum 1902 (reprinted 1975) and Wulff 1909.

4 Cf. Thomas (2000: 10–11 and 14–15) on the lack of information on provenance and original emplacement of the stelae.

5 See the tables in Bagnall and Worp 2004: 68 ff.; note that most of the Arsinoite documents mentioned are papyri and that the presumed Diocletian date 306 (A.D. 590) on a (hidden) stela (at p. 72; ed. Boyaval 1966: 369–70), although accepted as *SB* X 10705 and *SEG* XXIV 1204, is spurious.

6 For which see, in general, Staerk 1950; cf. Lefebvre 1907: xxx. For a Fayoum stela dated A.D. 703 of the orans type bearing this prayer, see Von Falck 1996: no. 67 (= Wulff 1909: no. 77; Lefebvre 1907: no. 790).

7 For an attempt to classify these stelae stylistically see Zuntz 1932: 27–33; see also Schaten 2005.

8 Thus perhaps a small stela, dated by its editor to the fourth to fifth centuries; see Nachtergael 1994: 143–46; (cf. *SEG* XLIV 1447B); its form is reminiscent of the triangular stelae of Akhmim (Panopolis) and it bears a simple ankh-cross surrounded by the characters "alpha" and "omega" at the top. Note, however, that neither date nor provenance of this piece is fully certain.

9 One such wooden tablet, of a physician named Phoibammon, perhaps from Naqlun, is preserved at Dayr al-'Azab; see the publication by Derda and van der Vliet 2006.

10 Louvre E 27220; see Coquin 1983: 103–05; Coquin and Rutschowscaya 1994: 107–08. The "Bishop Victor" mentioned in the text (line 12) can be plausibly identified (*pace* Coquin and Rutschowscaya 1994: 108, n. 6) with a Victor, Bishop of the Fayoum, known from several others sources written around the year 900; see Depuydt 1993, vol. 1: 244–45 (with further references) and 621.

11 Crum's *Sf*; I avoid using more specific "dialectological" sigla. That the language of the stela is a kind of Fayoumic rather than Bohairic (thus Coquin) was already suggested by Boud'hors and Calament (2004: 467, sub b).

12 One from Berlin, again in Fayoumi-Sahidic, is dated to A.D. 934, another, fragmentary one, from the Polish excavations in Naqlun is inscribed in Greek. Here, no date survives, but its editor, A. Łajtar (Łajtar 1994: 265–69), was certainly correct in dating this piece to about the ninth or tenth century. See further Boud'hors and Calament 2004: 453.

13 Bernand 1975–1981, no. 131; see also Timm 1984–1992: IV, 1758–59; Schaten 1999: 307–08 (cf. *SEG* XLIX 2205; Bingen 2000: 576, no. 709); Papaconstantinou 2001: 147.

14 First identified by Roquet (1978). See also Schaten 1998, who adds another one (at 306, n. 7); Boud'hors and Calament 2004: 467–68; Van der Vliet 2002–2003: 143–46.

15 "Houses" (plur.) in lintel no. 1 published by Roquet 1978; singular in his no. 2.

16 Thus in the Berlin lintel, for which see Van der Vliet 2002–2003: 143–46; the other two lintels, now in the British Museum and Cairo (?), suffered major damage at this point.

RECONSTRUCTING THE LANDSCAPE

17　The lintel in Cairo (?), see Lefebvre 1910: 59.

18　Cf. Abbott 1937: 16, n. 39; Timm 1984–1992: I, 430–31.

19　See Boud'hors 1999 (silver chalice); Depuydt 1993, I: 367.68, no. 182; II: pl. 404 (ninth-to-tenth-century colophon); 'Abd al-Masih 1941 (letter, eleventh-twelfth century).

20　Viz. the Berlin lintel, see Van der Vliet 2002–2003: 143–46.

21　Papaconstantinou (2000: 83) mentions only one for the Fayoum, but she is concerned with a slightly earlier period.

22　Cf. Schaten 1998: 306: "Anzunehmen ist . . . daß die Herkunft dieser Türstürze auf eine begrenzte Region, vielleicht sogar auf einen Ort zurückzuführen ist".

23　The inscriptions will be edited in full by the present author in a volume devoted to the church, to appear under the editorship of Prof. W. Godlewski. See provisionally Urbaniak-Walczak 1993 (*editio princeps* of three inscriptions on the west wall); Godlewski 2000 (comments *passim* on the inscriptions).

24　Note that Walters (1989: 205) incorrectly interpreted this inscription; it does not support his idea that the monastery bore the name of St. Michael.

25　Cf. Walters 1989: 207.

26　The name of a Bishop Severos of the Fayoum, perhaps the same person, is also found inscribed on the fragment of an amphora discovered in Naqlun during the 2001 season of the Polish excavations.

27　There is something of a riddle connected with this bishop, since his name is given as John. Therefore he cannot have been the same person as the Bishop Severos mentioned in the upper line of the text. A change of tenure may have occurred during the decoration works on the apse.

28　For a representative example of a donor inscription, see Urbaniak-Walczak 1993: 163–66 (no. II). Similar memorials to donors and painters can be observed in other medieval monastic churches; see, e.g., Bolman 2002: 38 and Pearson 2002: 218–19.

29　Although the terminology may be deceptive, some differentiation seems to be intended by "the (holy) mountain of Neklone [(Naqlun)]", on the one hand, and "this church", on the other.

30　See Urbaniak-Walczak 1993: 161–63 (no. I); cf. Godlewski 2000: 96. The same painter was active elsewhere on the west wall. On the north wall a painter Theodoulos left his name.

31　E.g., the Coptic graffito of a Deacon Markos, son of Paulos of Aqfahs (north wall).

32　For an amphora inscribed with the name of a bishop of the Fayoum, see note 26 above. An icon with a Greek-Coptic inscription in the Benaki Museum, Athens (inv. no. 8953), can on account of the dialect be ascribed to the Fayoum (see Chatzidakis 1967: fig. 19, 204).

33　See note 19 above.

34　For what follows I refer in particular to Bénazeth, Durand and Rutschowscaya 1999; various contributions in Durand and Saragoza 2002; and my own article, Van der Vliet 2006.

35　See in particular Durand and Rettig 2002.

36　See Godlewski 2002: 103, with a partial photo of the tunic at 101, fig. 1; cf. Van der Vliet, "In a robe of gold", Appendix I, no. 7 [Study 2].

Bibliography

Abbott, N. 1937, *The monasteries of the Fayyûm*, Chicago: University of Chicago Press.

Abd al-Masîh, Y. 1941, "Letter from a Bishop of al-Fayyûm", *BSAC* 7: 15–18.

Bagnall, R.S., and K.A. Worp. 2004, *Chronological systems of Byzantine Egypt: second edition*, Leiden/Boston: Brill.

Bénazeth, D., M. Durand, and M-H. Rutschowscaya. 1999. "Le rapport de l'objet copte à l'écrit: des textiles aux vêtements liturgiques", in: N. Bosson, and S.H. Aufrère (eds.), *Égyptes . . . l'égyptien et le copte: catalogue de l'exposition*, Lattes: Musée archéologique Henri Prades, 145–56.

Bernand, É. 1975–1981, *Recueil des inscriptions grecques du Fayoum* I, Leiden; vols. 2 and 3, Cairo: IFAO.

Bingen, J. 2000, "Bulletin épigraphique: Égypte et Nubie", *Revue d'études grecques* 113: 572–80.

Björnesjö, S. 1993, "Toponymie de Tebtynis à l'époque islamique", *Annales islamologiques* 27: 233–43.

Bolman, E.S. (ed.) 2002, *Monastic visions: wall paintings in the Monastery of St. Antony at the Red Sea*, Cairo: ARCE/New Haven: Yale University Press.

Boud'hors, A. 1999, "127. Calice", in: N. Bosson and S.H. Aufrère (eds.), *Égyptes . . . l'égyptien et le copte: catalogue de l'exposition*, Lattes: Musée archéologique Henri Prades, 297–98.

Boud'hors, A., and F. Calament. 2004, "Un ensemble de stèles fayoumiques inédites: à propos de la stèle funéraire de Pantoleos de Toutôn", in: M. Immerzeel and J. van der Vliet (eds.), *Coptic studies on the threshold of a new millennium*, Leuven: Peeters, I, 447–75.

Boyaval, B. 1966, "Huit stèles inédites d'Égypte", *Chronique d'Égypte* 41: 361–70.

Chatzidakis, M. 1967, "An encaustic icon of Christ at Sinai", *ArtB* 49: 197–208.

Coquin, R-G. 1983, "Deux stèles funéraires coptes", *BIFAO* 83: 101–05.

Coquin, R-G., and M-H. Rutschowscaya. 1994, "Les stèles coptes du Département des antiquités égyptiennes du Louvre", *BIFAO* 94: 107–31.

Crum, W.E. 1902, *Coptic monuments*, Cairo: IFAO; reprinted Osnabrück: Otto Zeller Verlag 1975.

Depuydt, L. 1993, *Catalogue of Coptic manuscripts in the Pierpont Morgan Library* I-II, Louvain: Peeters.

Derda, T., and J. van der Vliet. 2006, "Four Christian funerary inscriptions from the Fayum (I. Deir el-'Azab 1–4)", *JJP* 36: 21–33 [Study 9].

Durand, M., and S. Rettig. 2002, "Un atelier sous contrôle califal identifié dans le Fayoum: le tirâz privé de Tutûn", in: M. Durand, and F. Saragoza (eds.), *Égypte, la trame de l'histoire: textiles pharaoniques, coptes et islamiques*, Paris: Somogy, 167–70.

Durand, M., and F. Saragoza. (eds.) 2002, *Égypte, la trame de l'histoire: textiles pharaoniques, coptes et islamiques*, Paris: Somogy.

Godlewski, W. 2000, "Les peintures de l'église de l'Archange Gabriel à Naqlun", *BSAC* 39: 89–101.

Godlewski, W. 2002, "Les textiles issus des fouilles récentes de Naqlûn", in: M. Durand and F. Saragoza (eds.), *Égypte, la trame de l'histoire: textiles pharaoniques, coptes et islamiques*, Paris: Somogy, 100–04.

Łajtar, A. 1994, "Two Greek inscriptions from Deir el-Naqlun", *Nubica* 3/1: 265–274.

Lefebvre, G. 1907, *Recueil des inscriptions grecques-chrétiennes d'Égypte*, Cairo: IFAO; reprinted Chicago: Ares 1978.

Lefebvre, G. 1910, "Égypte chrétienne II", *ASAE* 10: 50–65.

Nachtergael, G. 1994, "Deux inscriptions grecques d'Égypte (ancienne Collection Eid)", *Chronique d'Égypte* 69: 140–46.

Papaconstantinou, A. 2000, "Les sanctuaires de la Vierge dans l'Égypte byzantine et omeyyade", *JJP* 30: 81–94.

Papaconstantinou, A. 2001, *Le culte des saints en Égypte des Byzantins aux Abbasides: l'apport des inscriptions et des papyrus grecs et coptes*, Paris: Éditions du CNRS.

Pearson, B.A. 2002, "The Coptic inscriptions in the Church of St. Antony", in: E.S. Bolman (ed.), *Monastic visions: wall paintings in the Monastery of St. Antony at the Red Sea*, Cairo: ARCE /New Haven: Yale University Press, 217–39.

Roquet, G. 1978, "Linteaux commémoratifs en dialecte fayoumique", *BIFAO* 78: 339–45.

Schaten, S. 1998, "Inschriften auf Türstürzen", in: M. Krause, and S. Schaten (eds.), *ΘΕΜΕΛΙΑ: Spätantike und koptologische Studien Peter Grossmann zum 65. Geburtstag*, Wiesbaden: Reichert Verlag, 305–15.

Schaten, S. 1999, "Griechische und koptische Bauinschriften", in: S. Emmel et al. (eds.), *Ägypten und Nubien in spätantiker und christlicher Zeit: Akten des 6. internationalen Koptologenkongresses, Münster, 20–26. Juli 1996*, Wiesbaden: Reichert Verlag 1999, II: 305–14.

Schaten, S. 2005. "Christian funerary stelae from the Fayoum", in: G. Gabra (ed.), *Christianity and monasticism in the Fayoum Oasis*, Cairo/New York: American University in Cairo Press, 257–63.

Staerk, W. 1950, "Abrahams Schoss", *RAC* 1: 27–28.

Thomas, T.K. 2000, *Late antique Egyptian funerary sculpture: images for this world and the next*, Princeton: Princeton University Press.

Timm, S. 1984–1992, *Das christlich-koptische Ägypten in arabischer Zeit*, 6 vols., Wiesbaden: Reichert Verlag.

Urbaniak-Walczak, K. 1993, "Drei Inschriften aus der Kirche des Erzengels Gabriel in Deir an-Naqlun im Faijum", *BSAC* 32: 161–69.

Van der Vliet, J. 2000, "A Naqlûn monk brought home: on the provenance of Louvre inv. E 26798–9", *BSAC* 39: 239–44 [Study 11].

Van der Vliet, J. 2002–2003, "Monumenta fayumica", *Enchoria* 28: 137–46 [Study 7].

Van der Vliet, J. 2006, " 'In a robe of gold': status, magic and politics on inscribed Christian textiles from Egypt", in: C. Fluck and G. Helmecke (eds.), *Textile messages: inscribed fabrics from Roman to Abbasid Egypt*, Leiden/Boston: Brill, 23–67 [Study 2].

Von Falck, M. (ed.) 1996, *Ägypten: Schätze aus dem Wüstensand: Kunst und Kultur der Christen am Nil*, Wiesbaden: Reichert Verlag.

Walters, C.C. 1989, "Christian paintings from Tebtunis", *JEA* 75: 191–208.

Wulff, O. 1909, *Altchristliche und mittelalterliche byzantinische und italienische Bildwerke*, Berlin: Reimer.

Zuntz, D. 1932, "Koptische Grabstelen: Ihre zeitliche und örtliche Einordnung", *MDAIK* 2: 22–38.

7

MONUMENTA FAYUMICA[1]

Jacques van der Vliet

I. Epitaph of Anni (Amsterdam, Allard Pierson Museum, inv. no. 7804)

Among the gems of the Coptic collection of the Allard Pierson Museum, the archaeological museum of the University of Amsterdam, counts a small funerary stela of bright beige limestone, measuring 52 × 21.5 × 5.5 cm.[2] Its elegant sculpted decoration is dominated by the architectural motif of an *aedicula*, a niche or small chapel, with two columns surmounted by capitals that support a conch surrounded by a vine and slightly damaged *acroteria* [the pointed ornaments on both sides of the conch]. The *aedicula* encloses a Maltese cross on a pedestal. This pedestal extends the high plinth of the stela, the upper part of which, just below the *aedicula*, bears an epitaph inscribed across the entire width of the stone.

Although the stela has been known for a long time and has often been reproduced and exhibited, the inscription was never published.[3] It comprises four lines of Coptic text, which are separated from the lower part of the undecorated plinth by a simple horizontal line underneath. A Greek cross marks the beginning of the text. The writing is not very elegant, without being clumsy. The letters, shallowly carved uncials, are slightly inclined to the right and about 1 cm high. The asymmetrical ⲁ is drawn with straight lines, so that it resembles a ⲗ (otherwise absent from the text); the ⲙ is quite broad and has straight, short legs. There is a ligature at the end of l. 3. The text is slightly worn and difficult to read in some places.

> + ⲡⲛⲟⲩⲧⲓ ⲙⲙⲉⲉⲓ ⲁⲗⲓ ⲟⲩⲛⲉⲓ
> ⲙⲛ ⲧⲉⲯⲩⲭⲏ ⲛⲁⲛⲛⲓ
> ⲁⲥⲕⲉ ⲥⲱⲙⲁ ⲉⲣⲁⲏⲓ ⲛⲥⲟⲩ
> 4. ⲓ̅ⲁ̅ ⲙⲡⲁⲣⲉⲙⲣⲁⲧⲡⲉ ⲓⲛ(ⲇⲓⲕⲧⲓⲱⲛⲟⲥ) ⲍ̅

3: ⲛⲥⲟⲩ: -ⲟⲩ ligatured and with a final loop ‖ 4: ⲓ̅ⲁ̅: possibly ⲓ̅ⲝ̅ | ⲙⲡⲁⲣⲉⲙⲣⲁⲧⲡⲉ: partly effaced.

+ True God, have mercy upon the soul of Anni.
She laid down the body on the 11th of Paremhatpe (Phamenoth) of the seventh indiction (year).

111

Figure 7.1 Stela Allard Pierson Museum inv. no. 7804
(photo: © Allard Pierson Museum)

The language of the text is a regular Fayoumic Coptic. If the readings given above are correct, only the spelling Paremhatpe (with a final -ⲉ) would be unusual. The formulary of the epitaph, characterized by the prayer ⲁⲗⲓ ⲟⲩⲛⲉⲓ ⲙⲛ ("have mercy upon") and the death formula ⲕⲉ ⲥⲱⲙⲁ ⲉⲣⲗⲏⲓ ("to lay down the body"), is widespread in Coptic funerary inscriptions, in particular from the Fayoum and Middle Egypt.[4] The opening acclamation ("True God"), undoubtedly an echo of the Nicene Creed,[5] is less common. It is found, however, with some variation (including the use of the Greek loanword ἀληθινός instead of ⲙⲙⲉⲉⲓ) in several later Coptic inscriptions from the Fayoum and elsewhere.[6] The name of the deceased, Anni, is a diminutive of the biblical proper name Anna (Hannah).[7] The text tells us nothing about this woman or girl except the date of her demise, 7 March.

Since a more elaborate discussion of the stela in Amsterdam is due to appear elsewhere,[8] a few words about its dating and provenance will suffice. Judging from its iconography, which features a cross inside a niche or chapel in a classical style,[9] this small monument belongs to a significant group of Byzantine stelae that are believed to come from the Fayoum. The Fayoumic dialect of the present epitaph confirms this general provenance. A more accurate localization is difficult, however. Almost without exception, the stelae of this type, which are found in all the main museums of Egypt and Europe, were acquired in commerce at the end of the nineteenth or the beginning of the twentieth centuries. The same holds for the Amsterdam stela, which was part of the private collection of the German Egyptologist Friedrich Wilhelm, Freiherr von Bissing (1873–1956), until 1921.[10] We know nothing about its provenance or original context, although it is likely that von Bissing bought it from a dealer in Cairo or Giza in the early twentieth century.[11] Only for very few Fayoumic stelae of the *aedicula*-cross type do we have more precise information. A small group of similar monuments was discovered in Naqlun, at the site of the Monastery of the Archangel Gabriel,[12] whereas an isolated stela was kept in the church of Sinnuris (published as Figure 7.2 below).[13] Regrettably, in its style and general appearance, the stela in Amsterdam is neither very close to the Naqlun stelae, which constitute a very homogeneous group, nor to the single item from Sinnuris. Therefore, it seems impossible at present to link the monument to a specific place or region.

The question of its date is hardly easier to solve. The great majority of Greek and Coptic epitaphs from the Fayoum are dated, in addition to the day and month of death, by indiction years only, as is the case here. As far as can be judged, the few exceptions that contain an absolute date are never very close to the general type represented by the stela in Amsterdam.[14] Nevertheless, its iconography, which will not be studied here, suggests a date later than the last quarter of the sixth century.[15]

II. Epitaph of Damian, a pastry baker (formerly Sinnuris, church of Saint Michael)

In a monograph devoted to the monasteries and churches of the Fayoum, published in 1998, Fathy Khurshid published a limestone funerary stela, which he had discovered in the *haykal* [altar room] of the ancient church of Sinnuris, a town north of Fayoum city, ancient Psineuris.[16] His book provides good photos, a facsimile and a brief description of the monument, which is exceptional in that its provenance is certain. This documentation is priceless since the ancient church of Sinnuris, dedicated to the Archangel Michael, has been entirely demolished and replaced by a new one in the 1990s and the present location of the stela is unknown.[17] Since Fathy Khurshid, an archaeologist by training, did not decipher the inscription, an epitaph in Greek, and his book knew a limited distribution, even in Egypt, the text is edited here on the basis of the published documentation.

The richly sculpted limestone stela measures $57 \times 24 \times 10$ cm. The monument is virtually complete, except for some surface damage that affects part of the text as well as the upper and lower ends of the stone. The Greek text consists of eight lines of coarse uncials, engraved above and around a small conch in a half-round sunken field that is surrounded by a decorative frame. This field surmounts a niche filled with an elaborate cross, flanked by two columns. The script is clear but not very regular. The shape of the σ is normal in l. 4, whereas it resembles a ſ in ll. 2 and 5. The scribe wrote an A with a broken bar, except at l. 4, where it takes the form ⲁ. The μ at the end of l. 4 looks more like a Coptic ⳙ and the υ has the shape of a V throughout. The ψ (l. 3) is written as a ϯ, a form that is common in Greek stelae from the Fayoum.[18] A Greek cross marks the beginning of the text; at the end of the text, below l. 8, two more are carved on either side of the small conch.

```
   + Κ(ύρι)ε, ἀνά-
     παυσον τὴ-
     ν ψυχὴν τοῦ τ-
4. ούλου σου Ταμ-
     ιανο[ῦ] σιλικνιαρίο-
     υ. Ἐκυμήθη Ἐπίπι
     η΄ ἀρχῇ ἐννάτ-
8. ης [ἰ]ν(δικτίωνος)
```

3-4. δούλου || 4-5. Δαμιανοῦ || 5-6. σιλιγνιαρίου || 6. ἐκοιμήθη || 8. [ἰ]νs stone (interrupted by the small conch).

+ Lord, grant rest to the soul of your servant Tamanios (Damian), the pastry baker. He fell asleep (on) Epipi (Epeiph) 8, at the beginning of the ninth indiction (year).

Figure 7.2 Stela from Sinnuris (after the original publication)

The common formulary is characteristic of the Greek-Christian epigraphy of the Fayoum.[19] The name of the pastry baker who is commemorated by the stela from Sinnuris, Damian, is one of the most common in the Christian Arsinoite, whereas his profession is well documented in the papyri from the Fayoum.[20] However, as far as I know, it is the first appearance of a σιλιγνιάριος in an inscription from Egypt. The death formula (ll. 6–8) states that the date of Epeiph 8 (July 2) is at the beginning of the indiction year, which is a chronological peculiarity of the Fayoum region.[21] The orthographic variants reported in the apparatus, τ for initial δ (l. 3; again in l. 4, in the name of the deceased), κ for γ (l. 5) and υ for οι (l. 6), are unremarkable.[22]

The stela from Sinnuris is quite similar to the one in Amsterdam that was studied above (no. I). In both cases it is a carved stela, the main decoration of which consists of a cross inside an *aedicula*. There are, however, differences in the style and the arrangement of constituent elements. On the Amsterdam monument, the epitaph is inscribed below the *aedicula*; here, it is at the top of the stela: the two columns of the *aedicula* support a curved space that contains the inscription, which is arranged above and around a conch of modest dimensions. The vine and *acroteria* are missing here, as is, by the way, the high pedestal, which lends the Amsterdam monument a certain elegance. By contrast, the cross and the surrounding space are abundantly decorated. Although these two stelae represent the same general type, which is characteristic of the Fayoum, there is no doubt that they belong to two distinct local traditions. Indeed, the particular decoration and style of the stela from Dayr Sinnuris enable us to compare it with a number of very similar stelae that belong to the same *aedicula*-cross type and originate from the Fayoum as well.[23] It remains to be seen whether these should also be assigned to the Sinnuris region.

As for the dating of the stela, which does not have an absolute date, it can be observed that, from a stylistic point of view, it is very close to the monument of the *meizoteros* Apa Ôl, from Damanhur/Hermopolis Parva.[24] The latter monument is dated to 409 of the Diocletian area (A.D. 693), which would place the stela of the pastry baker Damian with all likelihood in the seventh-eighth centuries.

III. Commemorative lintel of a Deacon Severus (Berlin, Skulpturensammlung und Museum für Byzantinische Kunst, inv. no. 9898)

Among the epigraphic monuments of Christian antiquity, lintels occupy a privileged place.[25] Usually surmounting the monumental entrance to a building, their role is far from being exclusively decorative. Their inscriptions often commemorate the founders or even the occupants of the building, while the nature of their decoration and the textual formulae used reveal their apotropaic function. In an article published in 1978, G. Roquet drew the attention to an important set of lintels inscribed in Fayoumic or Fayoumi-Sahidic Coptic.[26] These five lintels, all of which are dated to the tenth century, bear a very close formal resemblance

JACQUES VAN DER VLIET

and must originate from the Fayoum or neighboring regions, judging from their language. One of them, a beautiful specimen of sycamore wood, is now kept in the Skulpturensammlung und Museum für Byzantinische Kunst in Berlin, the former Kaiser-Friedrich-Museum. It was published in 1949 by Maria Cramer.[27] Regrettably, the editor failed to solve all problems posed by the text in a satisfactory manner. Although Roquet has proposed a new translation that corrects the worst mistakes of the first one,[28] the false readings of the *editio princeps* and the importance of the monument justify, in my opinion, the following re-edition of the text.

The lintel measures 14 × 133 cm (according to the *editio princeps* that records no thickness). The text counts eight lines, grouped on either side of a sculptured cross inscribed in a circular frame (ll. 1–4 to the left, ll. 5–8 to the right of the cross). The writing is quite careful, even if sporadically a bit stiff. The scribe uses quite a number of abbreviations and ligatures as well as a monogram (in l. 4). The following transcription is based on the photograph of the *editio princeps*, which is perfectly legible apart from one or two details.

> + ΠϬ(ⲱI)ⲥ Ⲓ(ⲎⲤⲞⲨ)ⲥ ⲠⲉⲬⲢ(ⲒⲤⲦⲞ)ⲥ ⲠⲉⲚⲀⲖⲓⲐⲓⲚⲱ(ⲥ)
> ⲚⲞⲨϮ ⲤⲘⲞⲨ ⲀⲨⲱ ϨⲀⲢⲉϨ Π-
> 3. ⲱⲚⲀϨ ⲠⲉⲔϨⲉⲘϨⲀⲖ ⲠⲀⲓ(ⲀⲔⲱⲚ)
> ⲤⲉⲨⲎⲢⲱⲤ ΠϢⲎⲢⲒ ⲠⲀⲓ(ⲀⲔⲱⲚ) ⲦΠⲀⲢ-
> (ⲐⲉⲚⲞⲤ) ⲬⲀⲎⲖ ΠⲉϤⲒⲱⲦ ⲘⲉⲚ ⲦⲀⲔⲀ-
> 6. ⲐⲱⲚ ⲦⲉϤⲤⲒⲘⲒ Ⲙⲉ ⲚⲉⲂϢⲎⲢⲒ
> Ⲭⲉ ⲚⲦⲀϤ ⲀϤⲔⲀⲦ ΠⲒⲘⲀⲚϢⲱ(Πⲉ)
> Ϩⲉ ⲚⲉϤϨⲒⲤⲒ ⲘⲒⲚⲉ ⲘⲀϤ Ⲁ(ⲓⲞ)Ⲕ(ⲖⲎⲦⲒⲀⲚⲞⲨ) ⲬⲚⲐ

1–2. ⲠⲉⲚⲀⲖⲓⲐⲓ ⲚⲞ͞Ⲛ͞Ⲭ Cramer ‖ 2. ⲞⲨ twice as a ligature resembling Ⲭ ‖ Ϯ ⲤⲘⲬ Cramer ‖ 4–5. ⲦΠⲀⲢϩ: ⲦΠⲀⲢ written as a monogram, ⲦⲀⲢ Cramer, ⲦⲀⲎϩ for ⲦⲀⲎⲨ or ⲦⲀⲒⲎⲨ? Roquet ‖ 5. ΠⲉϤⲒⲱⲦ: Πⲉ ligatured, Π resembling Ⲅ ‖ 6. Ⲙⲉ ligatured | ⲚⲉⲂϢⲎⲢⲒ: Ⲛⲉ ligatured, Ⲛ resembling Ⲙ, ⲚⲉϤϢⲎⲢⲒ Cramer ‖ 8. ⲚⲉϤϨⲒⲤⲒ ⲘⲒⲚⲉ ⲘⲀϤ: Ⲛⲉ twice ligatured, ⲚⲉϤ ϨⲒⲤⲒⲘⲒ ⲚⲉⲘⲀϤ Cramer.

O Lord Jesus Christ, our true God, bless and protect the life of your servant, the deacon Severus, the son of the deacon of (the church of) the Virgin, Chael, his father, as well as Takathon, his wife, and his children, for he built this dwelling by his own efforts, (in the year) of Diocletian 659 (= A.D. 942/943).

The language of the text is a Sahidic that is strongly influenced by Fayoumic with regard to the quality of the vowels. This is, in fact, the common idiom of inscriptions and documents from the Fayoum in the tenth and eleventh centuries. Note in particular /a/ for /o/ in closed syllables (ⲚⲦⲀϤ, ⲔⲀⲦ, ⲘⲀϤ) and /i/ for final /e/ (ⲚⲞⲨϮ, ϢⲎⲢⲒ, ⲤⲒⲘⲒ, ϨⲒⲤⲒ). Other characteristic spellings include ⲱⲚⲀϨ (Sah. ⲱⲚϨ), ⲤⲒⲘⲒ (Sah. ⲤϨⲒⲘⲉ),[29] as well as the compendium Π͞Ⲟ͞Ⲥ (in l. 1). On the other hand, there is no question of lambdacism as in Fayoumic proper. As is quite normal in this late idiom, the phonetic value of the prepositions is strongly reduced in the present text, to such an extent that ⲉ- and Ⲛ-/Ⲙ- are written nowhere and that ⲘⲉⲚ

116

(ⲙⲛ) and ⲅⲉⲛ (ⲉⲛ) become ⲙⲉ and ⲅⲉ before ⲛ in *anlaut* (but note in l. 5: ⲙⲉⲛ before ⲧ in *anlaut*).

As for the structure and wording of the text, the Berlin lintel conforms to the four other monuments belonging to the same group and, beyond these, to a broader category of dedicatory prayers, including among others prayers in colophons, dedications of paintings, etc.[30] It is, therefore, a prayer. For the form of the acclamation at the beginning, with its echo of the Nicene Creed, it can be compared to the Amsterdam stela edited above. The text invokes the protection of God for a deacon Severus and his family,[31] who erected a building called "this dwelling" (ⲡⲓⲙⲁⲛϣⲱ(ⲡⲉ)), "by his own efforts", that is "by his own means". It is difficult to specify the nature of this foundation. In any case, as the use of the word ⲙⲁⲛϣⲱⲡⲉ clearly shows, it was not a tomb.[32] Did the Berlin lintel mark the entrance of a private house inhabited by Severus and his family or, for instance, that of a pious foundation, of which Severus was the benefactor? The characteristic form of the text that connects it with the prayers in colophons and other dedicatory texts favors the second hypothesis.

Like the other lintels from the group, the specimen in Berlin provides the filiation of the founder. Severus' father, Chael, was a deacon like him, but his title is specified by an element, the meaning of which escaped the previous editors. This monogram at the end of l. 4 should be read neither as ⲧⲁⲣ (Cramer) nor as ⲧⲁⲏ (Roquet), but as ⲧⲡⲁⲣ; it is, moreover, an abbreviation, as is shown by the abbreviation sign at the beginning of l. 5. It should be resolved as ⲧⲡⲁⲣ(ⲑⲉⲛⲟⲥ), which is clearly the name of the church where Chael served as a deacon.[33] Unfortunately, this church of the Virgin cannot be located with certainty. The explicit nomenclature suggests a locality where there were several churches that needed to be distinguished by name and, therefore, perhaps a town or a sizeable village. On the other hand, the origin of the Berlin lintel is unknown[34] and there were many churches dedicated to the Virgin in the towns and villages of the Fayoum.[35]

As regards the origin of the four other lintels that are part of the same group, the data are hardly more abundant.[36] There is only the lintel in Cairo, which is known to have been found, according to its editor, at Abou Hamed (Fayoum).[37] This village is situated on the Bahr al-Gharak, a few kilometers east of Tebtynis-Tutun, in other words, in the southern part of the Fayoum, the former *Polemônos meris*.[38] Given the homogeneity of the whole, one would be tempted to assume that the five lintels, including the one in Berlin, all originate from this same region.[39] An additional clue is provided by a small series of four wooden funerary stelae, dated to the tenth century, which are very similar to the lintels in several respects and must originate from the same area of Tebtynis-Tutun.[40] We will recall that this southern part of the Fayoum, which includes the basin bypassed by Bahr al-Gharak, was much more important in the tenth century than it is today, and that it had considerable Christian centers, like the Monastery of Naqlun, Peljisok-Bulgusuq[41] and, of course, Tebtynis-Tutun. For the Christian history of the region at the time we have a rich documentation, in Coptic and Arabic. Unless new data demonstrate the contrary, there is every reason to add both the Berlin lintel and the epigraphic sources that are connected with it.

Notes

1 I warmly thank Sofia Schaten, Harry Blonk and Klaas A. Worp as well as the staff of the Allard Pierson Museum for their invaluable help.
2 Provenance: collection of Friedrich Wilhelm, Freiherr von Bissing (see below); from 1921 onward, Museum Scheurleer (a private museum, founded by C.W. Lunsingh Scheurleer, at Carnegielaan 12, The Hague; cf. W.R. Dawson et al., *Who was who in Egyptology*, London: Egypt Exploration Society 1995, 377), inv. no. S. 553; the Allard Pierson Museum acquired the stela by purchase in 1934.
3 The stela has a rich bibliography: *Beknopte gids van de Egyptische en Grieksche verzamelingen, Museum Carnegielaan 12*, The Hague: Museum Carnegielaan 12 1924, 25, no. 36; *Algemeene gids Allard Pierson Museum*, Amsterdam: Universiteit van Amsterdam 1937 (reprinted Universiteit van Amsterdam 1956), 112, no. 948; exhibition catalogue *Uit de schatkamers der oudheid*, Amsterdam: Stedelijk Museum Amsterdam 1938, no. 296; R.A. Lunsingh Scheurleer, *Egypte, geschenk van de Nijl*, Steenwijk: Concept en Design B.V. 1992, 192, ill. 164, cf. pp. 194 and 210; exhibition catalogue *Kopten: christelijke cultuur in Egypte = Mededelingenblad Vereniging van Vrienden Allard Pierson Museum* 72–3 (November 1998), no. 78, 11, ill. 22, cf. p. 12 (R.A. Lunsingh Scheurleer); A. Boud'hors and F. Calament, "Un ensemble de stèles fayoumiques inédites: à propos de la stèle funéraire de Pantoleos de Toutôn", in: M. Immerzeel and J. van der Vliet (eds.), *Coptic studies on the threshold of a new millennium: proceedings of the Seventh International Congress of Coptic Studies, Leiden, 27 August-2 September 2000*, Louvain/Paris/Dudley, MA: Peeters 2004, I, 447–75, mentioned in an appendix "Épigraphie copte fayoumique", under no. 20.
4 See C. Wietheger, *Das Jeremias-Kloster zu Saqqara unter besonderer Berücksichtigung der Inschriften*, Altenberge: Oros 1992, 146–48; for the Coptic epigraphy of the Fayoum in general, see Boud'hors and Calament, "Ensemble de stèles fayoumiques", in particular their appendix.
5 Cf. C.M. Kaufmann, *Handbuch der altchristlichen Epigraphik*, Freiburg im Breisgau: Herder 1917, 144.
6 E.g. in the inscription re-edited below, under no. III, ll. 1–2 (tenth century). Also from the Fayoum, on a textile in New York, ninth-tenth century (see Boud'hors and Calament, "Ensemble de stèles fayoumiques", no. 34 of their appendix).
7 See F. Preisigke, *Namenbuch*, Heidelberg: privately printed 1922, col. 32; for an epigraphic example from Saqqara, Wietheger, *Das Jeremias-Kloster*, no. 32 (= *SB Kopt.* I 694).
8 A.A. Gzula and J. van der Vliet, "Een Koptische grafsteen uit de Fayoem", in: *Mededelingenblad Vereniging van Vrienden Allard Pierson Museum* (forthcoming). The stela will be included in the general catalogue of the Coptic collections of the museum.
9 For an essay on the iconographic typology of the stelae from the Fayoum, see D. Zuntz, "Koptische Grabstelen: Ihre zeitliche und örtliche Einordnung", *MDAIK* 2 (1932), 27–33.
10 For von Bissing, see the notice in Dawson, *Who was who in Egyptology*, 46–7.
11 Cf. perhaps F.W. von Bissing, "Funde in Ägypten", *JDAI* 16 (1901), *Archäologischer Anzeiger*, 59, where the author regrets, in an account of the new Greco-Roman

MONUMENTA FAYUMICA

discoveries, that the merchants merely offered "einigen Grabstelen, deren Erwerbungen mich gelang (some funerary stelae that I was able to acquire)".

12 For an example known since 1930, but lost since, see the edition by A. Łajtar, "Two Greek inscriptions from Deir el-Naqlun", *Nubica* 3/1 (1994), 269–74; similar ones were found during recent excavations on site.

13 For other pieces, which might well belong to the same class of monuments, insufficient documentation is available to judge their style and iconography; e.g. G. Lefebvre, *Recueil des inscriptions grecques-chrétiennes d'Égypte*, Cairo: IFAO 1907, no. 75 (a stela acquired at Tell el-Gurob); id., "Égypte chrétienne II", *ASAE* 10 (1910), 276–77 (two stelae from "El Gebeli, ten minutes from Kimân Farès", included in *I. Fayoum* I, as nos. 30–1).

14 E.g. Lefebvre, *Recueil*, no. 790 (a stela with an orans figure, A.D. 703); *SEG* 24, no. 1204 (A.D. 590; without decoration?). The monuments studied by Boud'hors and Calament, "Ensemble de stèles fayoumiques", are certainly much later (tenth century).

15 See the study announced above (in n. 8); cf. the stela Berlin no. 8827, which is fairly close to the Amsterdam stela in its style and which Zuntz, "Koptische Grabstelen", 32, pl. IV.d, would date to the beginning of the seventh century.

16 *The churches and monasteries of the Fayoum province, from the spread of Christianity until the end of the Ottoman period* (in Arabic), Cairo: Supreme Council of Antiquities 1998, pl. 54–8, cf. pl. 50 and p. 225, n. 560 (on the stela); pp. 135–38 (on the history and the architecture of the church). I owe the photograph reproduced here to the generosity of Sofia Schaten, who obtained it from the late Anba Samuel, Bishop of Shibin al-Qanatir.

17 For all information on Christian Sinnuris and its church, see now Study 8.

18 See e.g. W.E. Crum, *Coptic monuments*, Cairo: IFAO 1902; reprinted Osnabrück: Otto Zeller Verlag 1975, no. 8584, pl. XXXII, above (= Lefebvre, *Recueil*, no. 80); no. 8604, pl. XXXVI, l. 2 (= *ibid.*, no. 81); Lefebvre, "Égypte chrétienne II", 276–77, no. 819, l. 2; exhibition catalogue *Ägypten: Schätze aus dem Wüstensand*, Wiesbaden: Reichert Verlag 1996, no. 67, l. 3 (= Lefebvre, *Recueil*, no. 790).

19 Cf. Lefebvre, *Recueil*, XXVII; Zuntz, "Koptische Grabstelen", 23–4.

20 J.M. Diethart, *Prosopographia Arsinoitica* I, Vienna: Hollinek 1980, lists for the Fayoum about fifty Damians (nos. 1571–21; all from the sixth-eighth centuries); among them, there are two bakers (ἀρτοκόπος: nos. 1589 and 1615), but no σιλιγνιάριος; for the σιλιγνιάριος (F. Preisigke, *Wörterbuch* II, col. 459: "Feinbäcker"), there are thirteen attestations (*Prosopographia Arsinoitica* I, 386, s.v.); for the present spelling (with κ), cf. *ibid.*, no. 4228. On this profession in the papyri: E. Battaglia, *'Artos': Il lessico della panificazione nei papiri greci*, Milan: Vita e Pensiero 1989, 198–99.

21 For the formula ἀρχῇ ἰνδικτίωνος, see the discussion by R.S. Bagnall and K.A. Worp, *The chronological systems of Byzantine Egypt*, Zutphen: Terra Publishing Co. 1978, 17–29, cf. 55–60 [idem, *Chronological systems of Byzantine Egypt: second edition*, Leiden: Brill 2004, 22–35, 110–15]; and the same authors, "Chronological notes on Byzantine documents, III", *BASP* 16 (1979), 239–43.

22 Cf. F.T. Gignac, *A grammar of the Greek papyri of the Roman and Byzantine periods* I: *phonology*, Milan: Cisalpino-La Goliardica 1976, 50, 76–80 and 197–98.

23 Cf. Crum, *Coptic monuments*, no. 8589, pl. XXXIII (= Lefebvre, *Recueil*, no. 94); no. 8590, id. (= *ibid.*, no. 95); no. 8604, pl. XXXVI (= *ibid.*, no. 81) and especially no. 8598, pl. XXXIV (= *ibid.*, no. 107); W. Brunsch, "Koptische und griechische Inschriften aus Alexandria", *WZKM* 84 (1994), no. A 11954 (cf. A. Łajtar, "Die griechischen und koptischen Inschriften im Griechisch-römischen Museum in Alexandria: Hinweis auf eine neue Veröffentlichung", *JJP* 26 (1996), 62–3); exhibition catalogue *Ägypten: Schätze aus dem Wüstensand*, no. 81 (= Lefebvre, *Recueil*, no. 793). The group is discussed by Zuntz, "Koptische Grabstelen", 31–2.

24 Crum, *Coptic monuments*, no. 8599, pl. XXXV (= Lefebvre, *Recueil*, no. 62); cf. Zuntz, "Koptische Grabstelen", 31–2.

25 See Kaufmann, *Handbuch*, especially 410–17.

26 "Linteaux commémoratifs en dialecte fayoumique", *BIFAO* 78 (1978), 339–45; cf. Boud'hors and Calament, "Ensemble de stèles fayoumiques", nos. 27–30 in their appendix. To the group of four pieces known to Roquet, S. Schaten, "Inschriften auf Türstürzen", in: *ΘΕΜΕΛΙΑ: Spätantike und koptologische Studien Peter Grossmann zum 65. Geburtstag*, Wiesbaden: Reichert Verlag 1998, 306, n. 7, was able to add a fifth one, British Museum no. 54040 (unedited; photograph in W.V. Davies, *Egyptian hieroglyphs*, London: British Museum Press 1987, 27, ill. 33) [cf. Study 10].

27 *Koptische Inschriften im Kaiser-Friedrich-Museum zu Berlin*, Cairo: Société d'archéologie copte 1949, 41–2, no. 9898 (20 791), pl. VII.

28 "Linteaux", 342, no. 4, with important corrections in the notes to his translation.

29 Cf. Roquet, "Linteaux", 340, n. e.

30 For the formula on the Berlin lintel Roquet ("Linteaux", 342, no. 4, n. b) already refers to the prayers in the colophons collected by A. van Lantschoot, *Recueil des colophons des manuscrits chrétiens d'Égypte* I: *les colophons coptes des manuscrits sahidiques*, Louvain: J.B. Istas 1929.

31 The name of his wife, Tagathon, is unusual. Perhaps, this is the (masculine) name Ἀγάθων (Preisigke, *Namenbuch*, col. 5), marked as feminine by the Coptic morpheme ⲧ-. The other names are common.

32 As the first editor already assumed; cf. Schaten, "Inschriften auf Türstürzen", 306–07. The two lintels published by Roquet ("Linteaux", 339 and 341) identify the building in question as "these houses" (no. 1, l. 5: ⲛⲓⲏⲓ) or "this house" (no. 2, l. 3: ⲡⲏ<ⲓ>). For the interpretation of ⲙⲁⲛϣⲱⲡⲉ on the lintels from Bawit, see M. Krause, "Die Inschriften auf den Türsturzbalken des Apa-Apollon-Klosters von Bawit", in: *Mélanges Antoine Guillaumont: contributions à l'étude des christianismes orientaux*, Genève 1988, 111–20, in particular 113.

33 Apart from being written as a monogram, the spelling is almost identical to the habitual abbreviation of ⲧⲡⲁⲣⲑⲉⲛⲟⲥ, viz. ⲧⲡⲁⲣ with a raised ⲑ, as found for instance in L. Depuydt, *Catalogue of Coptic manuscripts in the Pierpont Morgan Library*, Louvain: Peeters 1993, II, pl. 10.

34 Cf. Cramer, *Koptische Inschriften*, 41: "Von Prof. Wreszinski (W. Wreszinski, 1880–1935) und Dr. Pieper (M. Pieper, 1882–1941) in Kairo gekauft".

35 A. Papaconstantinou, "Les sanctuaires de la Vierge dans l'Égypte byzantine et omeyyade", *JJP* 30 (2000), 81–94, lists just one church in the city of Arsinoe (p. 83, nos. 1 and 1bis), but the data for later periods are much richer; on the evolution of the nomenclature, see her discussion on pp. 92–3.

36 It is true that Roquet recognized a toponym "Petêb: ⲡⲉⲧⲏⲃ" on one of the lintels published by him (no. 1, l. 5). However, the interpretation and identification that he proposes seem hardly certain to me ("Linteaux", 340, n. f; cf. S. Timm, *Das christlich-koptische Ägypten in arabischer Zeit*, Wiesbaden: Reichert Verlag 1984–1992, IV, 1907). [For the solution, see Study 10: J. van der Vliet, and A. Jeudy, "A lintel from the Fayoum in the British Museum", n. 12.]

37 G. Lefebvre, "Égypte chrétienne III", *ASAE* 10 (1910), 59–60.

38 For the medieval topography of the region, see S. Björnesjö, "Toponymie de Tebtynis à l'époque islamique", *Annales Islamologiques* 27 (1993), 233–43; add: Timm, *Ägypten* VI, 2887–92; Depuydt, *Catalogue* I, CIII–CXVI.

39 Cf. Schaten, "Inschriften auf Türstürzen", 306: "Anzunehmen ist . . . daß die Herkunft dieser Türstürze auf eine begrenzte Region, vielleicht sogar auf einen Ort zurückzuführen ist".

MONUMENTA FAYUMICA

40 Collected and edited by Boud'hors and Calament, "Ensemble de stèles fayoumiques", who emphasize the relationship between the two groups of monuments.

41 It must be observed that some authors searched for this hamlet, the location of which is uncertain, exactly to the east of Tebtynis (see Björnesjö, "Toponymie", 239–40); at the time, there was a church of the Virgin, see Timm, *Ägypten* I, 430–31; Depuydt, *Catalogue* I, 367–68, no. 182; II, pl. 404.

8

MONUMENTS OF CHRISTIAN SINNURIS (FAYOUM, EGYPT)

Peter Grossmann, Tomasz Derda and Jacques van der Vliet

Introduction

Following an earthquake in the early 1990s, the ancient sanctuary of the Archangel Michael (Kanisat al-Malak, Dayr al-Malak Mikhail) in Sinnuris (locally pronounced Sinauris), a district town about twelve kilometers north of Fayoum city, was demolished and replaced by a glittering new church. Today, hardly any remains from Sinnuris' past survive *in situ*, and very little is known about the Christian history of the town, which boasts a flourishing Coptic Orthodox parish as well as a protestant church.[1]

For late antique and early medieval Sinnuris (Ψενῦρις, Ψινεῦρις), a certain amount of papyrological documentation exists in Greek and Coptic, which still awaits systematic study.[2] These sources suggest that the town had its share in the considerable economic activity that characterized the Fayoum in the Byzantine and Early Islamic periods. It was, in fact, sufficiently well known in the fifth and sixth centuries to be mentioned as an Αἰγυπτία κώμη in the *Ethnica* of Stephen of Byzantium (s.v. Ψενῦρις). This impression of prosperity is confirmed by the beautiful stone monuments from this period that were preserved until recently in the Church of St. Michael: the Greek funerary stela of the baker (*siligniarios*) Damian, published earlier by one of the present authors,[3] and the votive relief of Phib and Phoibammon, republished below.

For a somewhat later period, we have the description of the Fayoum written around the year 1245 by a Muslim author known as Uthman al-Nabulusi. He reports that the quite considerable town of Sinnuris numbered two churches, one of which was functioning, whereas the other had fallen into disuse, in addition to a monastery to the west of the town, which he designates as Dayr Sinnuris.[4] Fathy Khurshid, writing in the 1980s, applies the same name to the modern church complex. This complex, however, is situated in the northeastern part of the town center and is otherwise always referred to by the name of its patron saint. Already in the seventeenth century, the early orientalist Vansleb (J.M. Wansleben) described it as a church dedicated to the Archangel Michael.[5] It cannot, therefore, be identified out of hand with al-Nabulusi's western monastery.[6] It seems more likely that it was a successor to one of the two town churches mentioned by al-Nabulusi,

123

but even this cannot be verified for lack of further information. In any case, in September 2003 nothing remained of the earlier churches of Sinnuris other than a ruinous altar screen with some Coptic and Arabic inscriptions, dumped in a narrow corridor to the east of the new church. In the literature only scarce documentation is found about the church's predecessors (see below).

The present article may be described as a form of salvage archaeology on paper. The first part consists of Peter Grossmann's notes on the medieval church of Sinnuris, elements of which were still visible in 1977, although much had been changed and built over. In the second part, Tomasz Derda and Jacques van der Vliet republish the late antique votive relief of Phib and Phoibammon, which they recorded in September 2003, adding an iconographical commentary by van der Vliet, who also acted as the general editor of the article.[7] In an appendix, van der Vliet briefly presents the inscriptions of the altar screen mentioned above.

I. The ancient Church of the Archangel Michael in Sinnuris

The Church of St. Michael in Sinnuris, as it was visible in the 1970s (Figure 8.1), preserved only meager vestiges of its original architecture.[8] Merely the apse and some pillars in the nave appeared to represent an earlier building phase (Figure 8.2).[9] It is unknown when the church to which these older elements belonged was erected. Our discussion below suggests that it may have been of medieval date, yet it may

Figure 8.1 Interior of the Church of St. Michael in Sinnuris in 1977
(photo: Peter Grossmann)

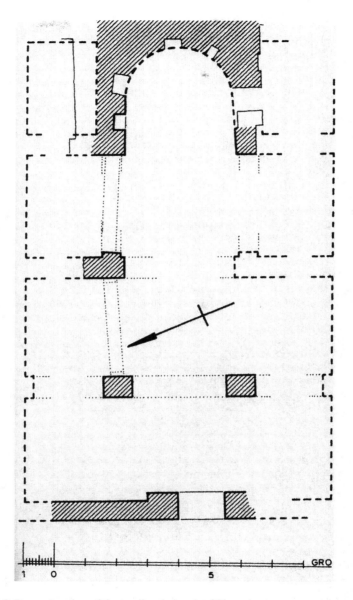

Figure 8.2 Reconstruction of the medieval church of Sinnuris
(drawing: Peter Grossmann)

have replaced a late antique predecessor in this location.[10] The lateral outer walls of this earlier, possibly medieval church were demolished already in the nineteenth century, during reconstruction and enlargement works. The structure was expanded on its south side with a three-aisled basilical church which had three altar rooms (*haykals*). A panel in this South Church, inscribed in Arabic, recorded that it was built in 1890.[11] The one-aisled northern annex, with only one altar room, was not dated but may have been somewhat older than the South Church (Figure 8.4). Nothing of all these constructions remains today.

Fortunately, in 1977, before the church was demolished, those elements of its architecture that appeared to belong to its earlier, possibly medieval building phase could be recorded. These seem to point to a basilical structure with rectangular pillars that are placed in a strangely transversal alignment. A western pair of such pillars, integrated into later constructions, survived *in situ*.[12] In the light of the general proportions of the church, it seems unlikely that there were originally more of these pillars to the west. The strong west wall that was incorporated into the more recent northern annex may have preserved parts of the original western outer wall. Only the situation of the entrance, placed out of the axis of the building, is a curious feature. The phenomenon is not without parallels elsewhere: in the Church of the Mother of God in the Monastery of St. Catherine in the Sinai, for instance, dating from the reign of Justinian, where the displacement of the main entrance is probably due to irregularities in the soil.[13] In the Sinnuris church, however, this entrance does not seem to have been part of the original plan.

To the east of this pair of transversal pillars, and situated at about the same distance as to the western entrance wall, one finds another pillar belonging to this same building phase. This latter pillar shows a more or less westward oriented T-shaped profile, but with legs of unequal length on all sides, as can often be found in pillars of transeptal churches.[14] In the present case, however, the pillar is most probably the remainder of an originally continuous transverse wall that separated the former *khurus* from the nave.[15] Its longest leg, which points towards the north, has an irregular northern end, clearly indicating that the masonry was broken through and, therefore, must once have continued northwards. The very short legs projecting south and east, on the other hand, may be original. They served as support for arches stretching in a southern and eastern direction. At the time of our visit, these arches were no longer present, but the broken surface of the masonry on the south face still clearly shows at which height the southern arch sprang from the pillar (Figure 8.1). On the east side, the springing point of the original broad arch could still be recognized, but the latter was replaced by a more modern narrow arch that, moreover, did not run exactly parallel to the axis of the church (also visible on Figure 8.1).

Another arch that connected this remnant of the partition wall of the *khurus* with the next pillar to the west was likewise not original. It sprang from an oddly narrow and shapeless console that had been inserted secondarily. It is doubtful whether there was originally a similar arch at this point. The southern arch that sprang from the T-shaped pillar closed the great middle gate of the *khurus* and must therefore have been the first or primary triumphal arch. The eastern arch, springing from the same pillar, must have served to articulate the structure of the vaults above the *khurus*.

Besides these three pillars and the remains of the west wall, only parts of the apse of the earlier building phase survived. The apse had a stilted ground plan and was provided with several wall niches. Yet only the first niche on the left can be considered original. It had smooth lateral jambs and was the only one to be suitably positioned in relation to the ground plan of the apse. All other niches were inserted at a later stage, as is shown by their haphazard placement and mediocre workmanship. The vault of the apse also no longer reflects the original situation. The frontal arch of the apse, which could be called the second triumphal arch, shows a pointed shape, which has no tradition in medieval Christian architecture in Egypt. The few examples that are extant elsewhere have a clearly more truncated profile.[16]

In the south wall of the apse, furthermore, a wide door was inserted that connects the apse with the northern *haykal* of the newly built South Church of 1890. It is doubtful whether there was always a door at this place. The inner door opening, as visible thirty years ago, nowhere showed an area that could be considered original. The original southern *pastophorion* [apse side room] must have been accessible only through the *khurus*. The same was true for the northern *pastophorion*, which had no extant opening to the apse.

Any further comments on the architecture of the earlier church must be considered speculative. As no remains of lateral outer walls survive, it is impossible to say how wide the aisles and the church as a whole may have been. The conjectural dotted lines added in our Figure 8.2 are purely hypothetical and based upon other similarly proportioned medieval churches in Egypt.[17] Nevertheless, it can be assumed that the pair of transversally oriented rectangular pillars were positioned more or less in the middle between the partition wall of the *khurus* and the west wall, and that the two bays of the nave were of equal length. As was already pointed out above, the west wall may have preserved parts of the original outer wall, even though the entrance that was inserted in this wall, out of the axis of the building, does not seem to belong to the original plan.

Typologically the older building phase of the Sinnuris church resembles to some extent the medieval reconstruction phase of the church of Dayr Abu Hennis, south of Antinoopolis.[18] There, the middle of the building was marked by a series of domes, which may also have been the case in Sinnuris (Figure 8.3). The Sinnuris church would then number among the classical examples of the Upper-Egyptian elongated domed churches ("Langhauskuppelkirchen") that were developed in Egypt and are practically only found there.[19] This type of church apparently arose when, after the Fatimids, the size of newly built churches diminished, a phenomenon that can also be observed in the Byzantine Empire in about the same period.[20] As a result of this development, the habit grew up of covering churches with massive vaulted constructions instead of saddle roofs built of timber beams. Difficulties in procuring the necessary wood may have played a minor role in this decision, though, in particular for the poorer communities, such difficulties should not be underestimated.[21] Rather, as contemporary authors already pointed out, the massive vaults were more resistent against fire and vermin than wooden roof

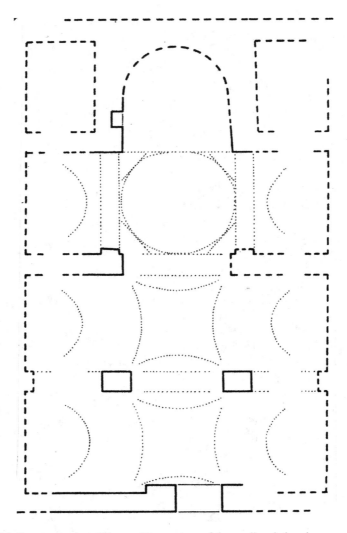

Figure 8.3 Reconstruction of the vaulting system of the medieval church
(drawing: Peter Grossmann)

constructions.[22] At the time, building with vaults was considered a more modern way of engineering. In the same period and for the same reasons, roofing churches with massive barrel vaults had become common in Lower Egypt and in particular in the monasteries of the Wadi al-Natrun, the ancient Scetis. Unlike the church of Dayr Abu Hennis, which was only secondarily transformed into an elongated domed church, the earlier church in Sinnuris may have been conceived in this way from the beginning. The lateral compartments of the bays may have been roofed

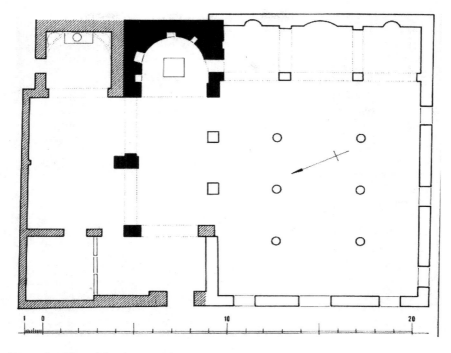

Figure 8.4 Plan of the church with its extensions in 1977
(drawing: Jacek Kościuk)

with transversal barrel vaults,[23] which did not rule out introducing a hanging dome in the middle. Egyptian church architecture also knows spatial forms that can be qualified as single axis buildings with a central plan, covered with domed vaults.[24]

Regrettably nothing is known about the date of this earlier building phase of the church of Sinnuris. No datable sculptural or tectonic elements have survived – if they ever existed – and an attempt at dating the church can only be based on considerations of a general nature. An important criterion is the evident presence of a *khurus* situated between the nave and the actual sanctuary. Although the introduction of the *khurus* may date back as far as the seventh century, probably following a decision of Patriarch Benjamin I (627–665), it was initially apparently limited to monastic churches.[25] Its adoption in the architecture of secular churches can hardly predate the ninth century,[26] and even then the church at Sinnuris would probably not have been among the first churches in which this innovation was introduced. The spatial proportions of the *khurus* in Sinnuris also argue against an early date, as its depth corresponds practically to that of the individual bays of the nave, an arrangement that can only be attributed to a later development.

On the other hand, this earlier church of Sinnuris very probably had no lateral altar rooms. The spatial proportions show that the rooms next to the apse were almost certainly the normal *pastophoria* demanded by the liturgy, and not secondary altar rooms. The latter only appeared in the Mamluk period, when, under the influence of a change in religious politics, the previously prevailing tolerance towards the Christian population was abandoned.[27] The first Mamluk rulers were al-Malik az-Zahir Baybars (1260–77) and Qalawun (1279–90), but the changing religious climate was perhaps not felt immediately. In the *History of the Patriarchs*, the change becomes apparent for the first time in the Life of Patriarch John VII (1262–93), in an incident that took place already in 1264/1265 but could still be settled without bloodshed.[28] On the basis of typological criteria, therefore, we would propose a date for the construction of the earlier church of Sinnuris somewhere before the thirteenth century.

II. Mary and the angels: an inscribed stone relief from Sinnuris

The stone monument studied here has actually been known since the end of the seventeenth century. It was first described by Vansleb, who had discovered it in the church of Sinnuris in July 1672. According to his account, the clergy of the church had moved the relief some time before his visit from the *khurus* ("le Choeur") to a more remote part of the church ("dans un coin de l'Eglise"), in order to prevent the faithful from venerating the sculptured figures.[29] This may explain why, until quite recently, the relief was walled into the northern wall of the *haykal* of the northern extension, a part of the church that was not publicly accessible. There it was found in February 1966 by Father G. Giamberardini,[30] who published the monument in a rather summary way in his monograph on the cult of the Virgin Mary in Egypt. Its original location is likely to remain unknown. Since it is definitely not a funerary monument (see below), it can hardly have come, as Giamberardini was told, from "un vecchio cimitero".[31] In 1977, it was spotted and photographed by Grossmann. In September 2003, Derda and van der Vliet rediscovered the monument in one of the back rooms of the new church, where Father Cherubim, the parish priest, kindly allowed them to record it. Regrettably, by their second visit, in early February 2004, the stone had vanished.

The inscription

The monument consists of a rectangular slab of fine-quality limestone[32] or possibly marble (Figures 8.5–8.6).[33] The dimensions of the slab are 65.5 × 57 × 5–6.5 cm. Its front side shows, in shallow relief, a standing figure of the Virgin Mary with the infant Jesus on her arm, flanked by the Archangels Michael (on her right; left for the viewer) and Gabriel (on her left; right for the viewer). The pictorial field is surrounded on all sides by a raised frame. The piece is complete, although its surface is rather worn and in places disfigured by chips and scratches (some very recent). The outer rim bears traces of whitewash and mortar that betray the stone's

Figure 8.5 The stone relief of Phib and Phoibammon in 1977
(photo: Peter Grossmann)

former position, when it was built into a wall. Whatever polychromy it may have borne originally is lost now.

A dedicatory text consisting of four short lines of Greek is incised between the feet of Michael and the Virgin (ll. 1–3) and below Gabriel (ll. 3–4); l. 3 is interrupted by the feet of the Virgin and continues on l. 4 for lack of space. A fifth line

Figure 8.6 The stone relief of Phib and Phoibammon in 2003 (photo: Jacques van der Vliet)

of text, containing the legends, runs over the lower frame of the stone. The letters are uncials of a late antique type, competently incised by a single hand. A is symmetrical, with a low-dropping broken bar; M, three-stroke with a low saddle. The height of the letters is approximately 2 cm. Paleography suggests a date between the sixth and the eighth centuries.

Bibliography: Giamberardini 1974–1978: I, 197–98, no. 13, fig. 21 (photo, brief description, partial reading of the text after the stone); briefly described or mentioned: Vansleb 1677: 265–66; Timm 1984–1992: V, 2355–56 (after Vansleb); Samuel al Syriani and Badii Habib 1990: 147; Fathy Khurshid 1998: 135–36.

MONUMENTS OF CHRISTIAN SINNURIS

Sinnuris (Fayoum), *ca.* sixth-eighth century

[pictorial field]
    Εὐχά-
    ριον
    Φῖβ ‖ (καὶ) Φοιβάμμων-
4. ος
[lower frame]
    Μιχαήλ, ἡ ἁιγία Μαρίας, Γαβριήλ.

3. (καὶ): ϛ, not read by Giamberardini ‖ 4. Giamberardini ⲇ̣ⲟ ‖ 5. Giamberardini ⲙⲁⲣⲓⲁ; l. ἡ ἁγία Μαρία.

Votive offering of Phib and Phoibammon.
Michael, the Holy Mary, Gabriel.

The stone's indubitable εὐχάριον (ll. 1–2) is absent from the dictionaries.[34] At first sight, the format of the text could suggest an abbreviation of εὐχαριστήριον, "thank-offering". As in the present case, Christian votive inscriptions with εὐχαριστήριον typically consist of proper names immediately following the opening word εὐχαριστήριον.[35] Such inscriptions occur throughout Egypt and Nubia, even if they are not very common.[36] Yet the word εὐχάριον on the Sinnuris relief is not marked as an abbreviation, whereas in a Nubian inscription that does abbreviate εὐχαριστήριον the stone reads ⲉⲩⲭⲁⲣϛ, with an abbreviation mark.[37] The fairly skillful workmanship of the stone from Sinnuris argues against a spelling error (by haplography). Moreover, since εὐχάριον occurs elsewhere, we prefer to leave the word as it stands.

In addition to the present inscription, the word εὐχάριον appears in a similar context in two pre-Christian inscriptions from present-day Bulgaria.[38] Both are dedications of marble sculptures of traditional gods: the Thracian horseman, and Zeus and Hera, respectively. A further occurrence in a Christian stela from Pontus is more doubtful.[39] In each of these cases, εὐχάριον can be interpreted as a diminutive derived from εὐχή, analogous to ψυχάριον from ψυχή, for instance. The word εὐχή itself is frequently found, in similar contexts, as a word for "votive offering", on a par with προσφορά or εὐχαριστήριον;[40] examples from Egypt are not very common, however.[41]

The continuation of the inscription within the pictorial field identifies the two dedicators of the piece. Their names, Phib and Phoibammon, are typically Egyptian. Phib (Φῖβις, "the Ibis")[42] has in the past been considered a form of Phoibammon, but these are clearly two distinct names.[43] Both were extremely common in the Fayoum in Byzantine and early Islamic times.[44] The genitival case ending of the second name, Phoibammon, in l. 4, was read by Giamberardini erroneously as ⲇ̣ⲟ, which he interpreted as an abbreviation of ⲇⲓⲟⲕⲗⲏⲧⲓⲁⲛⲟⲥ. He accordingly proposed to read the group ϥⲓⲃ in l. 3 as a Diocletian year 512, corresponding to A.D. 796.[45] The stone does not bear a date, however.

133

The line of text on the frame below identifies the saints represented. This manner of exhibiting the legends is strongly reminiscent of the well-known, possibly sixth-century tapestry from Egypt, now in the Cleveland Museum of Art, which represents Mary enthroned with the infant Jesus between Michael and Gabriel. There, an architrave above the figures bears an almost identical inscription.[46] In the name of the Virgin, the spelling ἁιγία for ἁγία is an example of a rarely attested interchange of α and αι.[47] In the present word it also occurs once in a Coptic environment: πϩⲁⲓⲅⲓⲟⲥ ïⲱϩⲁⲛⲛⲏⲥ, "the Holy John".[48] It is uncertain whether these spellings could in some way be influenced by the shift of the following γ to a fricative /j/ that had already occurred in antiquity (compare the modern Greek pronunciation: *aya*).[49] The genitival form Μαρίας must be a scribal slip, perhaps influenced by inscriptions where a covert noun such as "picture, image" or "tomb-stone" demands a following genitive.[50] It may finally be noted that the angel on the left of the Virgin is not Raphael, as was stated by Vansleb and some later authors following him, but Gabriel.[51]

The image

Although strictly speaking four persons are depicted, the composition follows an essentially triadic pattern. The central and most prominent element is the standing Virgin Mary, who holds the infant Jesus in an attitude corresponding to the type commonly known as the Virgin Hodegetria.[52] Mary carries Jesus on her left arm, holding him with her other arm. The child stretches out one hand in a gesture of blessing, with two fingers raised; in his other hand, he probably holds the traditional scroll, although this can no longer be established with certainty. The asymmetry of the Hodegetria type is underlined by the position of the Virgin's feet, which peep out from under her robe. In spite of the flatness of the relief, the angels on either side of Mary are both depicted as slightly turned towards the central group of the Virgin and the child. Asymmetrically, both angels carry a scepter, crowned by a knob inscribed with a cross, in their right hand, and an orb with a cross in their left hand. They are winged and wear a tunic and a long cloak (*himation*), flowing from their right shoulder; their tunics are tucked up so as to end shortly below the knee and are decorated with *clavi* and roundels. Their attire closely resembles that of the two angels flanking the Virgin with child in one of the better-known apse paintings from Bawit.[53] All four figures bear a nimbus, and in the case of Jesus, it is inscribed with a cross.

It is well known that the Egyptian tradition shows a marked preference for representing the Virgin Mary with the child enthroned.[54] The Sinnuris relief, as well as examples from the Wadi al-Natrun[55] and Dayr al-Shaykha (al-Qubbaniya),[56] all of which show the standing Hodegetria type, prove that this preference is by no means an absolute rule. Yet it is striking that the closest parallel for the general composition of the relief is the apse mosaic in the church of the Panagia Angeloktistos in Kiti on Cyprus.[57] The mosaic, which is usually dated to the sixth or seventh century, is definitely more dynamic: the two archangels are depicted in a symmetrical movement, offering their orbs to the Virgin and the child. The overall composition is nearly identical, however, including identical legends above the figures, more or less as in the Cleveland tapestry mentioned above. The only differences are that in Kiti the Virgin is standing on a low dais and the angels are dressed differently. Although contacts between Egypt and Cyprus were quite lively in the seventh century,[58] it seems far-fetched to postulate influence either way, as we are simply dealing with an iconographic type that had rapidly become popular in the entire Mediterranean world. In any case, the Sinnuris relief is undoubtedly a product of local workmanship, as the native Egyptian names of the donors demonstrate.

Compositions showing the Virgin with the child – most often enthroned – between two angels have a long history in Christian Egypt, where they appear in various contexts and media.[59] A first group occurs in liturgical settings. The best known among these representations are those that are part of a painted apse-composition.[60] This tradition can be traced back to the late antique oratories of Bawit[61] and Saqqara,[62] and is found as well in the great "programs" of the medieval monastic churches,[63] such as the old church of the Monastery of St. Antony on the Red Sea.[64] Not far from Egypt, on Cyprus, the mosaic at Kiti also adorns the apse of a church. The possibly sixth-century Cleveland tapestry has been compared to contemporary apse compositions with regard to its iconographical program.[65] A medieval icon beam from Old Cairo, which shows the motif on its central panel, likewise belongs in a liturgical context.[66]

A second group occurs in what may be called a liminal context, marking a boundary between sacred and non-sacred space.[67] Thus a monumental version of the motif, which can be dated with considerable certainty to the early sixth century, was found painted in the courtyard of an Alexandrian house.[68] A far later wall painting from the Fayoum itself, in the monastery church of Naqlun (Dayr al-Malak, eleventh century), can be usefully compared to the Alexandrian painting because of its liminal setting. Although situated within the church, it is not part of an apse composition, but one of a series of large-scale paintings on the west wall (i.e. the entrance wall) of the Church of St. Gabriel. The scene occupies a large panel in the southern part of the upper register, next to a centrally located symmetrical pair of equestrian saints.[69] Even two late ninth-century frontispieces in Coptic manuscripts from the scriptorium of Touton in the Fayoum may be called liminal as they are situated at the beginning of a book.[70]

In addition to these predominantly painted representations, a small series of stone reliefs from Egypt show the same motif on various scales and with a number of variations.[71] They usually lack a known provenance and survive isolated from their original context. Although the loss of context seriously hampers the interpretation of these reliefs, they are certainly non-funerary in character and appear to belong to the greater class of Byzantine sculptured icons.[72] The Sinnuris stone must be numbered among this latter group of rather neglected monuments.

The original architectural setting of the Sinnuris relief is unknown. As Grossmann's analysis shows, the *khurus* of the old church, where the stone was exhibited prior to Vansleb's visit, was of much later date than the stone itself. It can merely be assumed that the stone once adorned part of an ancient predecessor of the medieval church. This negative observation leaves only two elements to provide a framework for the interpretation of the relief, the text and the iconography of the scene.

Whereas most other Egyptian stone icons that depict the Virgin between angels are anepigraphic, the inscription of the Sinnuris relief characterizes it as an *ex-voto*, offered in fulfillment of a vow, which provides an immediate clue to its interpretation. Texts such as the present one testify to a reciprocal relationship between donors and saints.[73] The monument is an expression of gratitude felt by two explicitly named individuals, Phib and Phoibammon, towards the saints depicted in the relief and identified in the inscription below, for favors prayed for and received. The text, furthermore, formally excludes a liturgical or funerary interpretation. Given its votive nature, the monument must originally have occupied a position where it could be seen and – presumably – touched by everybody. Its worn and glossy surface may bear witness to its veneration as an icon through the centuries, as indeed Vansleb's report confirms.

The presence of text links the Sinnuris relief to another sculptured icon from the Fayoum that likewise shows a triple composition with the Virgin Mary as its central character. The Greek inscription of this more or less contemporary stone relief, sculptured in a far more provincial style, was recently republished by the late Georges Nachtergael.[74] The stone was found in al-Lahun, ancient Ptolemais Hormou, in 1961 and depicts the Virgin standing, without the child, in a praying pose (the Virgin *deomene, orans*) between two anonymous winged angels, equally depicted in praying position.[75] The resulting triadic image is a powerful icon of intercessory prayer, the message of which is reinforced by the written prayer that surrounds the Virgin: "Lord, help (βοήθησον) Elias, administrator". The dedicatory formula below the scene identifies the Virgin as "Our Lady, the Mother of God". To the honor of the Virgin, the *oikonomos* Elias dedicated either the relief itself or – as seems more likely – its lost architectural setting: a church or chapel (or part of it).[76] The element of prayer connects both icons, but the Sinnuris relief, besides being more sophisticated in style, shows a different iconography that conveys an altogether different range of associations.

In interpreting the rather diverse compositions that depict the Virgin between two angels, scholars usually focus on the central figure of the Mother of God, on the various ways in which she and the infant Jesus are represented, or on her place within a larger iconographical program. It is obvious, however, that the triadic structure of the composition, far from being a casual arrangement, represents an established iconographical motif in its own right, with its own significance. Underlining the triadic structure of the scene, the inscription of the Sinnuris relief identifies only three saints: Michael, Mary and Gabriel. Angelic beings, in particular Michael and Gabriel, situated to the right and left of a central character, occur in very different late antique contexts. The two main sources cited for this configuration are usually Judeo-Christian angelology, on the one hand,[77] and the Byzantine court, on the other.[78] These two options are not mutually exclusive, however; rather they illustrate the interchangeability of terrestrial and celestial visualizations of power.[79] The idea of a heavenly court shaped after the imperial court on earth – or *vice versa* – is most clearly reflected in the visual arts, where both Christ and the Mother of God, usually carrying the infant Jesus, can be surrounded by angels in courtly dress.[80] This is true for the entire Mediterranean world, including Egypt. As recent studies rightly emphasize, several of the aforementioned Egyptian compositions depicting the Virgin between two angels display a markedly hieratic and courtly character.[81] What in its core may seem an idyllic scene, a mother holding or even feeding her child, is in reality an icon of supernatural power drawing upon images of the highest authority on earth.

Documentary evidence is the best source from which to gauge the general impact that the visual motif of – specifically – the Virgin flanked by the Archangels Michael and Gabriel may have had on its late antique or early medieval Egyptian beholder. Two significant examples may suffice here. Mainly in the monastic centers of Middle Egypt, a triad that links the Virgin with the archangels is a frequent and prominent ingredient of the common inscriptions of the so-called "litany type".[82] Such inscriptions open with the invocation of a series of saints, in formulas of varying length and composition. Often, the texts first invoke the Holy Trinity, then Michael, Gabriel and Mary, usually in that order, only rarely with Mary preceding the angels.[83] That this configuration is actually meant to reflect a spatially conceived triadic arrangement is neatly demonstrated by the relatively frequent cases where the "logical" order, which pairs the archangels, is replaced by the "illogical" (marked) order Michael-Mary-Gabriel.[84] Here, the order of the names is undoubtedly "iconic" and appears to be directly informed by the visual experience of a triad consisting of the Virgin flanked by the archangels. In all such inscriptions, the saints invoked are expected to intercede either for the dead or for the living, and there can be no doubt that the role of the "Marian triad" in these litanies is also primarily intercessory.

The second example is taken from a "magical" ritual that, in its explicit verbal reproduction of the image, spells out its potency even more clearly. The text belongs to the type known as the "Prayer of the Virgin *ad Bartos*", attested in Coptic, Ethiopic and Arabic, but quoted here in its Sahidic Coptic form.[85] Within a ritual setting, it combines a set of prayers addressing Jesus and the Archangels Michael and Gabriel, put into the mouth of the Virgin, with a narrative framework that serves to lend authority to the prayers. First, the Virgin Mary, licensed by her son, addresses Jesus in a prayer that He himself is alleged to have communicated to her in order that "you (scil. Mary) heal the sick through it and the deranged and those who are suffering in the prisons and everyone who is tormented by the impure spirits".[86] The invocations of this prayer emphatically bring out the unity of Christ under his double aspect, as God and Son of God and as man, son of Mary:

> Hail, firstborn of his father and firstborn of my womb! . . .
> Hail, mouth that drew milk from my virgin breasts!
> Hail, hand that created our father Adam! . . .
> Then, as the Virgin said these things, she looked to her right and, seeing Michael, and Gabriel at her left, she was alarmed.
> Immediately Gabriel said to her:
> Do not be afraid, Mary.
> I am Gabriel, who brought <you> the happy tidings of your prime (?).
>
> <I> have come to you to fulfil your request and whatever you are asking for!
> And Mary said:
> Who is he with the golden scepter in his hand?
> He said to her:
> That is Michael, the greatest among the entire host of the angels.
> And she adopted a sweet voice and said . . .

What follows is, first, a prayer to Michael, which briefly evokes the story of his elevation to the rank of commander-in-chief of the heavenly hosts, in lieu of Mastema (Satan), and then a prayer to Gabriel. In both largely analogous prayers, the angels are asked to empower water and oil so that they may be used to heal the sick and the suffering, and ward off the attacks of sorcerers and demons. Not only the actors in this ritual text – Mary, Jesus and the two archangels – but also their spatial disposition, with Michael and his scepter to the right, and Gabriel to the left of the Virgin, correspond to those found in the pictorial tradition.

The "verbal icon" of the Virgin Mary flanked by Michael and Gabriel, as found in a Coptic popular ritual for healing and protection, conveys a fairly precise idea of how this conventional triadic composition was perceived by its contemporary beholder. Not primarily as a theological statement, although this element is emphatically present even in the magical ritual, but as a potent image of intercessory power that evokes the ability of Mary, the Mother of God, and the two leaders of the heavenly hosts to intervene on behalf of the believers who appeal to them, and to heal and rescue them.[87] Its inherent visual power explains the positioning of the motif of the Virgin and child enthroned between angels in liminal contexts, such as the west wall of the church in Naqlun, close to the undoubtedly apotropaic double image of the equestrian saints above the entrance.[88] An overt example of the prophylactic use of the same motif is offered by a gold medallion from Cyprus, dated to the seventh century, which bears the prayer "Christ our God, help (βοήθισον) us", inscribed around the triadic group of the Virgin with the Child and a pair of angels.[89] The image itself mediates divine assistance and protection. In the case of the Sinnuris relief, the monument is explicitly referred to as a "votive offering", a token of gratitude for answered prayer. There can be no doubt that Phib and Phoibammon, two well-to-do citizens of Byzantine or early Islamic Sinnuris, dedicated this monument to celebrate some profound experience of healing or deliverance.

Appendix

The old altar screen

During their first visit to Sinnuris, in September 2003, Derda and van der Vliet came across a discarded altar screen with inlaid Coptic and Arabic inscriptions, lying exposed to the elements in a narrow corridor at the east side of the present church. Due to the awkward position in which the screen was found, it could not be properly recorded. It was only possible to take some rather unsatisfactory photographs of the inscriptions; these are published here together with the text of the inscriptions themselves. In spite of the deficient state of our documentation, the publication of this brief note seemed justified in view of the recent interest in similar medieval and post-medieval altar screens and their inscriptions.[90]

The Sinnuris altar screen belongs to a widespread class of objects made of woodwork inlaid with crosses, decorative elements and inscriptions in bone or ivory. Such screens separate the altar room (*haykal*) or altar rooms from the nave (or from the *khurus*, if there is one), while at the same time providing acces to the sanctuary through the doorway at the center of the screen.[91] They are still today produced in more or less the same style and, though they may be crowned by epistyles carrying series of icons, such screens are not themselves an iconostasis in the modern Greek sense.

The text consists of two lines, inlaid in bone or ivory and symetrically arranged above the central doorway (Figures 8.7–8.9). The upper line, in Arabic, is divided over three panels; the lower line gives the same text twice within one panel, once in Arabic (right), and once in Bohairic Coptic (left).

[upper line]

برسم الملاك ميخائيل بسنورس (middle)
Intended for (the church of) the Angel Michael in Sinnuris.

من له تعب(left) (right) عوض يا رب
Remunerate, o Lord, ‖ him who toiled.

[lower line]

Arabic
السلام لهيكل الله

Bohairic Coptic
ϫⲉⲣⲉ ⲡⲓⲉⲣⲫⲉⲓ ⲛ̄ⲧⲉ ⲫ(ⲛⲟⲩ)ϯ
Hail, temple of God!

Figure 8.7 The inscriptions on the altar screen: middle
(photo: J. van der Vliet)

Figure 8.8 The inscriptions on the altar screen: right
(photo: J. van der Vliet)

Figure 8.9 The inscriptions on the altar screen: left
(photo: J. van der Vliet)

The upper line of text is a common prayer for the anonymous donor or donors of the screen, mentioning the name of the church for which it was destined.[92] The lower line of text, immediately above the doorway, consists of an acclamation that brings out the character of the altar room as sacred space, God's dwelling on earth. Other texts in a similar position are frequently quotations from the Psalms that refer to the screen's function as a gate. The present text is found more often on Coptic altar screens, sometimes in the slightly expanded form "Hail, temple of God the Father!"[93] The decoration of the screen, and the phrasing and setting of the inscriptions, are reminiscent of eighteenth-century examples from the Monastery of St. Anthony near the Red Sea.[94] The Sinnuris screen bears no date but may likewise date to the eighteenth century, when many Coptic churches were refurbished.[95] It is not visible on the photograph of Figure 8.1, and must therefore have adorned the South Church, until this was demolished in the 1990s.

Notes

1. The Orthodox church is no. 19 in the list of churches of the (then) diocese of Fayoum and Giza published in Somers Clarke 1912: 205; it is erroneously mentioned under the lemma "Sînarû" in Timm 1984–1992: V, 2354–55. For the protestant church, see Meinardus 1977: 579 (under Baptist Evangelical Church); Timm 1979: 135.
2. See Wessely 1904: 163–65, 167–68; Timm 1984–1992: IV, 2034–35 (s.v. Psenyris); V: 2355–56 (s.v. Sinauris); Derda 2008: 143, *ad* no. 27, l. 7, with full references that

are not repeated here. Contrary to Wessely, we consider both forms of the toponym to refer to the same place.

3 Van der Vliet 2002–2003: 140–42, Tafel 22, right [Study 7]. In the 1980s, this stela could still be seen in the old church, see Fathy Khurshid 1998: pl. 50; it has disappeared since then.

4 Ed. Moritz 1899: 107, ll. 15–16; Salmon 1901: 50–1; cf. Keenan 2005.

5 Vansleb 1677: 265; we thank Gertrud van Loon for kindly providing copies of the corresponding pages from Vansleb's Italian diary of 1676 (Bibliothèque nationale de France, ms. italien 435, 51–2); for Vansleb, see now Hamilton 2006: 142–51.

6 *Pace* Van der Vliet 2002–2003: 140.

7 The authors gratefully acknowledge the assistance of Clara ten Hacken, Leiden, for the Arabic texts. The article profited from critical remarks by Gertrud van Loon, Zuzana Skalova, and Ewa D. Zakrzewska.

8 Vansleb, who visited Sinnuris on July 31, 1672, must have seen more, but he was far from impressed. He characterized the church as "fort pauvre", without giving further details (Vansleb 1677: 265); he showed more interest in the stone relief, described below.

9 Samuel al Syriani and Badii Habib 1990: 146–47, fig. 190 (ground plan), pl. 188; cf. Fathy Khurshid 1998: 135–38, pl. 8 (ground plan) and 46–58; a brief notice with a photo of the interior also in Leclant 1979: 364, pl. XIII, fig. 18 (here reproduced again as Figure 8.1).

10 For the late antique and early medieval sources on Sinnuris, see the references given in the introduction.

11 The panel is reproduced in Fathy Khurshid 1998: pl. 46; cf. Samuel al Syriani and Badii Habib 1990: 147. It situates the renovation of the church under the administration of a *mu'allim* Shihat Abd al-Sayyid and in the episcopacy of the famous Bishop Abraham, bishop of the Fayoum between 1881 and 1914, but does not provide further details. For a plan of this South Church, see Samuel al Syriani and Badii Habib 1990: 147, and Fathy Khurshid 1998: pl. 8.

12 Unwarrantedly considered modern by Samuel al Syriani and Badii Habib 1990: 147; that they had been integrated into the design of the modern South Church does not contradict their original character. The next two pillars to the east are not original, however; their position corresponds exactly to that of the inner piers of the South Church and they must have been inserted together with these.

13 See Grossmann 1990: 34, fig. 3; also in the city church of the Holy Anargyroi at Pharan (Sinai) the outer entrance to the narthex as well as the entrance to the nave are clearly placed out of the axe of the church, see Grossmann 1998: 66–74, with fig. 5.

14 For transeptal churches in the Mediterranean, see Lemerle 1953, in particular 687–94, and recently Grossmann 2008: esp. 99–108, figs. 1–4.

15 On the significance of the *khurus* in Egyptian church architecture, see Grossmann 2002: 71–6.

16 A characteristic example is the primary triumphal arch of the so-called Upper Church of Dayr Abu Fana, cf. Buschhausen 1988: esp. 354–56; see also the large triumphal arch in the partition wall of the *khurus* of the Shenoute monastery near Sohag, cf. Grossmann 2002: 528–36, pl. Xb.

17 Examples include the church of Dayr al-Maymun, south of Atfih, ancient Aphroditopolis (see Grossmann 1982: 178–80, fig. 58; Samuel al Syriani and Badii Habib 1990: 150–51, fig. 198), and that of Dayr al-Hammam, Fayoum (see Samuel al Syriani and Badii Habib 1990: 142–43, fig. 183).

18 See Grossmann 2002: 522–23, fig. 141.

19 Grossmann 1982: 165–69.

20 Deichmann 1937: 49–53.

MONUMENTS OF CHRISTIAN SINNURIS

21 Grossmann 1982: 158–61; on the value of wood for building, in particular 159, n. 688.
22 Abu 'l-Makarim, ed. Evetts 1895: 95–6 (fol. 27b), 106 (fol. 31a), 120–22 (fol. 37a-b), 127 (fol. 39a), 186 (fol. 63b).
23 In Dayr Abu Hennis, the lateral compartments bore transversal conchate vaults; see Grossmann 1982: 121, fig. 50.
24 Grossmann 1982: 86–7, with fig. 31, with reference to Chapel 25 in al-Bagawat.
25 For the development of the *khurus*, see Grossmann 2002: 72–6.
26 Grossmann 2002: 78–9.
27 Grossmann 1982: 229–30, 234–35.
28 *History of the Patriarchs* III: 3, ed. Khater and Burmester 1970: 229–30 (fol. 237r).
29 Vansleb 1677: 265; his Italian diary (p. 52; see n. 5, above) gives a more roundabout version of the story than the printed French text, though without adding further details of interest.
30 Giamberardini 1974–1978, I, 197 ("infissa nella parete nord-ovest"); see also Samuel al Syriani and Badii Habib 1990: 147; Fathy Khurshid 1998: 136.
31 Giamberardini 1974–1978: I, 197, who consequently mistook it for a tombstone.
32 Thus also Giamberardini 1974–1978: I, 197.
33 According to Samuel al Syriani and Badii Habib 1990: 147. Our second (fruitless) visit to Sinnuris, in February 2004, was made to verify the type of stone used.
34 Giamberardini 1974–1978: I, 198, who believed that the inscription was funerary, took it for a proper name ("Eucarione").
35 Łajtar 2004: 92; cf. Vikan 1995: 570–72. For εὐχαριστήριον in liturgical texts and diptychs, see Engberding 1967.
36 Though less rare than Łajtar (2004: 92) suggests.
37 Łajtar 2004: 89–94 (referring to the photo published by C.M. Firth in 1927; cf. *SEG* LIV 1773).
38 *IGBulg* I2, 277; *IGBulg* III/2 1817.
39 *Studia Pontica* III, 54. The editor, H. Grégoire, reads the word as a feminine proper name, formally neuter and incorrectly in the nominative instead of the genitive. His interpretation is questionable, however.
40 See H.G. Liddell, R. Scott and H.S. Jones, *A Greek-English lexicon*, Oxford: Clarendon Press 1968 (and later impressions), Suppl., 66, s.v. εὐχή; Łajtar 2004: 92.
41 A Jewish example, probably from Alexandria: Horbury and Noy 1992: no. 19; ὑπέρ εὐχῆς in a Christian inscription from Panopolis: van der Vliet 2008: 153–55 [Study 16].
42 See Vergote 1954: 11, no. 44.
43 For Phoibammon, see Crum, in Winlock and Crum 1926: 110; Vycichl 1983: 10 and 245.
44 See Diethart 1980: nos 5262–5326 (Phib); 5481–5679 (Phoibammon). Given the commonness of the names and the absence of a link with Sinnuris, a Phib, son of Phoibammon, from a place named Mouei (Diethart 1980: no. 5309, seventh cent.; for Mouei, see Timm 1984–1992: IV, 1688), cannot be identified out of hand with the present dedicators.
45 Giamberardini 1974–1978: II, 198.
46 See Rutschowscaya 1990: 134–37. For the persistance of the designation of the Virgin as "Holy Mary" in Egypt, see Papaconstantinou 2000: 92–3.
47 See Gignac 1976: 194–95.
48 Crum 1902, no. 8328, a Sahidic funerary inscription, quoted in Förster 2002: 10.
49 For this shift, see Gignac 1976: 71–5. It is clearly apparent on the Cleveland tapestry mentioned above, which spells: ἡ ἁήα (for ἁγία) Μαρία.
50 For an example from the Fayoum, see Derda and van der Vliet 2006: 27–8 ("[monument] of Papas" [Study 9]; cf. *SEG* LVI 1977).
51 Correct in Giamberardini 1974–1978: I, 197, and Samuel al Syriani and Badii Habib 1990: 147.

52 The literature on the Virgin Hodegetria is immense, see e.g. Babić 1994; Bacci 1996; Angelidi and Papamastorakis 2000, mainly focused on the Constantinopolitan icon; cf. Snelders and Immerzeel 2004: 119–22; for the early history of the type, also Wellen 1961: 176–78 (cf. 208–09). Recently, Bissera Pentcheva (2006: 110–17) argued for a distinction between pre-iconoclastic representations of the Hodegetria type, which stress the intimacy between mother and child, and later images of the Hodegetria proper, which focus on the intercessory prayer of the Virgin. The present representation would belong to the former group.

53 Chapel no. XXVIII: Clédat 1904: pls 96 and 98, cf. p. 154; the angels are identified in Greek as "angel of God" (left) and "angel of the Lord" (right). Compare the angels in similar scenes at Saqqara (see below note 62) and the relief Coptic Museum 7814 (Pahor Labib 1948–49: pl. I; Beckwith 1963: fig. 111; D. Bénazeth in Catalogue Paris 2000: 187, no. 202). For the iconography of angels in general, see Leroy 1974: 205–07 (Egypt); Maguire 1995: esp. 65–6 (Byzantium).

54 See e.g. Lazarev 1995 (originally 1938): 198–99 (on the Galaktotrophousa), 229–30 (on the Hodegetria); Snelders and Immerzeel 2004: 122–23 (Hodegetria); various examples are quoted below.

55 A book illumination, a secondary addition to a ninth-tenth century Bohairic manuscript; see Leroy 1974: 108–10, pl. 98, 1.

56 A much deteriorated wall painting, difficult to date; see Junker 1922: 49–50, pl. 6, fig. 10. We owe this reference to Renate Dekker, Leiden.

57 See e.g. Volbach and Lafontaine-Dosogne 1963: pl. 9.

58 Cf. Gascou 2008: 71–2.

59 In the following, we do not intend to study the iconography of the Virgin in late antique and medieval Egypt in any detail; for further examples and references, see Giamberardini 1974–1978; Leroy 1974: 202–05 and idem 1975: 52–4; Rutschowscaya 2000; Snelders and Immerzeel 2004: 122–23; Bolman 2004 and 2005.

60 For the place of the Virgin in such compositions, see Van Moorsel 1989: 127–31; reprinted 2000: 188–91).

61 Most notably Chapel no. XXVIII, already quoted above n. 53.

62 See Van Moorsel and Huijbers 1981: 129–31, 150–64; cf. Wietheger 1992: 50–4.

63 See Van Loon 1999: 196–201.

64 Van Moorsel 1995–1997: I, 45–8; Bolman 2002: 64–5, cf. 70–1. See also Esna (Dayr al-Shuhada): Leroy 1975: 6–10, cf. 53–4, and Coquin 1975: 245 (C).

65 Rutschowscaya 1990: 134–37; Rutschowscaya 2000: 222–25.

66 Skalova and Gabra 2003: 180–83.

67 Cf. Krueger 2011.

68 Rodziewicz 1984: 194–206; a reconstruction of this very fragmentary painting: fig. 236.

69 See Godlewski 1993: 190–93; idem 1997: 130, fig. 9–10. The subject is identified (in Coptic) as ⲧⲉⲓ[ⲉ̅]ⲁⲅⲓⲁ [ⲙⲁⲣⲓⲁ ⲙⲏ] ⲛⲓⲁⲣⲭⲁⲛⲅⲉⲗⲟⲥ ⲉⲧⲟ[ⲩⲁ]ⲁⲃ: this (image of the) Holy Mary (ἁγία Μαρία) and these Holy Archangels (van der Vliet's reconstruction of the dedicatory prayer provisionally published in Urbaniak-Walczak 1993: 166).

70 Depuydt 1993: II, pl. 11 (M574, a liturgical collection) and 12 (M612, Ps.-John Chrysostom, *Encomium on the Four Bodiless Creatures*); cf. Leroy 1974: 94–7, pls 31 and 34.

71 See Pahor Labib 1948–1949; Giamberardini 1974–1978: I, 155–56, 196–97, figs. 3, 19–20; Beckwith 1963: 25–6; D. Bénazeth in Catalogue Paris 2000: 186–87, nos 201–02; cf. Loverdou-Tsigarida 2000: 238–39. An additional example is discussed below.

72 See, in general, Lange 1964; for the Virgin in relief icons, Loverdou-Tsigarida 2000; Paribeni 2008.

73 Compare Vikan 1995.

74 Nachtergael 2006 (cf. *SEG* LVI 1985); the earlier publication is Giamberardini 1963: 66–7, no. 15, pl. VIII. Nachtergael dates the relief to the sixth-seventh centuries on paleographical grounds.

75 Probably under the influence of Giamberardini's erroneous interpretation of the stone as a tombstone, Nachtergael (2006) identifies the central character of the scene as the (male) donor of the relief; it is clearly a woman, however. The Virgin is also the central character in the inscription below the scene; the angels remain unnamed.

76 See the commentary by Nachtergael 2006: 12–13.

77 See e.g. Testa 1983, in particular 286–89 and 290–91. For a triad with a plausible background in traditional Egyptian religion, see van der Vliet 2000.

78 Cf. Maltese 1990: in particular 116–21.

79 See Maguire 1997.

80 See Stuhlfauth 1897: 207–33 ("Die Engel als repräsentative Umgebung Christi"); 203–07 ("Die Engel als repräsentative Umgebung der Maria"); cf. Klauser 1962: 301–04; Maguire 1995: 65–6; Jolivet-Lévy 1998.

81 Thus Godlewski 1993: 190 (on the Naqlun painting); D. Bénazeth in Catalogue Paris 2000: 186–87 (on the limestone reliefs, nos 201–02); Rutschowscaya 2000: 225 (on the Cleveland hanging); cf. Leroy 1975: 53.

82 See Wietheger 1992: 210–42; Papaconstantinou 2001: 22, 387–402.

83 The latter (hierarchically correct) order can also be found in the properly liturgical litanies that must have served as a model; see, for an example from the present-day Bohairic Psalmodia, ⲡⲭⲱⲙ ⲛ̅ⲧⲉ ϯⲯⲁⲗⲙⲟⲇⲓⲁ ⲉⲑⲟⲩⲁⲃ ⲛ̅ⲧⲉⲙⲣⲟⲙⲡⲓ, ed. Cairo 1984: 65 (Bohairic-Arabic), and Brogi 1962: 20 (Italian translation).

84 Wietheger 1992: 210–42; Lefebvre 1907: no. 228 (Ashmunayn).

85 After the edition in Crum 1897: 210–18, from a manuscript to be dated "not before the 11th century" (210); cf. Crum 1905: no. 368; Kropp 1930–1931: II, 127–35, no. XXXIX. For the genre of "prayers of Mary" in Coptic magic, and a review of the most important parallels, see Meyer 1996: 57–9; cf. Bolman 2005; 16–17.

86 This and the following quotation follow the text in Crum 1897: 213, omitting some phrases for the sake of brevity; the translations are by van der Vliet.

87 A similar conclusion in Bolman 2005: 16–17.

88 For the apotropaic role of equestrian saints, see Snelders and Jeudy 2006.

89 See Maguire 2005: 184 and 191 (Fig. 16.1).

90 See in particular Hunt 1987 (with MacCoull 1989); Rutschowscaya and Boutros 2004; Skalova 2004: 1529–38; Bolman 2006: esp. 93–103; Jeudy 2009: 53–61.

91 For general descriptions that are not repeated here, see Butler 1884: I, 28–32; Burmester 1967: 19; Rutschowscaya and Boutros 2004: 1502–03.

92 For similar texts, see Rutschowscaya and Boutros 2004: 1504–05.

93 See Rutschowscaya and Boutros 2004: 1503–04.

94 Illustrated in Rutschowscaya and Boutros 2004: figs. 15–21.

95 See Guirguis 2008: 39–44.

Bibliography

Angelidi, Chr., and T. Papamastorakis. 2000, "The veneration of the Virgin Hodegetria and the Hodegon monastery", in: Catalogue Athens 2000, 373–87.

Babić, G. 1994, "Les images byzantines et leurs degrés de signification: l'exemple de l'Hodigitria", in: A. Guillou and J. Durand (eds.), *Byzance et les images*, Paris: La Documentation française, 189–222.

Bacci, M. 1996, "La Panayia Hodighitria e la Madonna de Constantinopoli", *Arte cristiana* 84: 3–12.

PETER GROSSMANN ET AL.

Beckwith, J. 1963, *Coptic sculpture, 300–1300*, London: A. Tiranti.

Bolman, E.S. 2002, *Monastic visions: wall paintings in the Monastery of St. Antony at the red sea*, Cairo: ARCE/New Haven: Yale University Press.

Bolman, E.S. 2004, "The Coptic Galaktotrophousa revisited", in: M. Immerzeel and J. van der Vliet (eds.), *Coptic studies on the threshold of a new millennium: proceedings of the Seventh International Congress of Coptic Studies, Leiden, 27 August-2 September 2000*, Louvain/Paris/Dudley, MA: Peeters 2004, II, 1173–84.

Bolman, E.S. 2005, "The enigmatic Coptic Galaktotrophousa and the cult of the Virgin Mary in Egypt", in: M. Vassilaki (ed.), *Images of the Mother of God: perceptions of the Theotokos in Byzantium*, Aldershot/Burlington: Routledge 2005, 13–22.

Bolman, E.S. 2006, "Veiling sanctity in Christian Egypt: visual and spatial solutions", in: S.E.J. Gerstel (ed.), *Thresholds of the sacred: architectural, art historical, liturgical, and theological perspectives on religious screens, East and West*, Washington, DC: Dumbarton Oaks, 72–104.

Brogi, M. 1962, *La santa Salmodia annuale della Chiesa copta*, Cairo: Edizioni del Centro francescano di studi orientali cristiani.

Burmester, O.H.E. 1967, *The Egyptian or Coptic Church: a detailed description of her liturgical services and the rites and ceremonies observed in the administration of her sacraments*, Cairo: Société d´archéologie copte.

Buschhausen, H. 1988, "Die Ausgrabungen von Dayr Abû Fâna in Mittelägypten im Jahre 1987", *Jahrbuch der österreichischen Byzantinistik* 38: 353–62.

Butler, A.J. 1884, *The ancient Coptic churches of Egypt*, Oxford: Clarendon Press; reprinted Oxford: Oxford University Press 1970.

Catalogue Athens. 2000, (M. Vassilaki, ed.), *Mother of God: representations of the Virgin in Byzantine art*, Milan: Skira Editore.

Catalogue Paris. 2000, *L'art copte en Égypte: 2000 ans de christianisme*, Paris: Gallimard.

Clarke, S. 1912, *Christian antiquities in the Nile Valley: a contribution towards the study of the ancient churches*, Oxford: Clarendon Press.

Clédat, J. 1904. *Le monastère et la nécropole de Baouît*, Cairo: IFAO.

Coquin, R-G. 1975, "Les inscriptions pariétales des monastères d'Esna: Dayr al-Šuhadâ' – Dayr al-Faḫûrî", *BIFAO* 75: 241–84.

Crum, W.E. 1897, "A Coptic palimpsest: I. Prayer of the Virgin in 'Bartos'. II. Fragment of a patriarchal history", *Proceedings of the Society of Biblical Archaeology* 19: 210–22.

Crum, W.E. 1902, *Coptic monuments*, Cairo: IFAO; reprinted Osnabrück: Otto Zeller Verlag 1975.

Crum, W.E. 1905, *Catalogue of the Coptic manuscripts in the British Museum*, London: British Museum.

Deichmann, F.W. 1937, *Versuch einer Darstellung der Grundrisstypen des Kirchenbaues in frühchristlicher und byzantinischer Zeit im Morgenlande auf kunstgeographischer Grundlage*, Ph.D dissertation, Halle; published in Würzburg: K. Triltsch 1937.

Depuydt, L. 1993, *Catalogue of Coptic manuscripts in the Pierpont Morgan Library* I-II, Louvain: Peeters.

Derda, T. 2008, *Deir el-Naqlun: the Greek papyri* II, Warsaw: University of Warsaw/ Raphael Taubenschlag Foundation.

Derda, T., and J. van der Vliet. 2006, "Four Christian funerary inscriptions from the Fayum (I. Deir al-'Azab 1–4)", *JJP* 36: 21–33 [Study 9].

Diethart, J.M. 1980, *Prosopographia Arsinoitica* I, Vienna: Hollinek.

Engberding, H. 1967, "Eucharisterion in ägyptischen liturgischen Texten", in: P. Wirth (ed.), *Polychordia: Festschrift Franz Dölger,* II, Amsterdam: A.M. Hakkert, 148–61.

Evetts, B.T.A. (ed.) 1895, *The churches and monasteries of Egypt and some neighbouring countries attributed to Abû Sâlih, the Armenian,* Oxford: Clarendon Press; reprinted Oxford: Butler & Tanner 1969; Piscataway: Gorgias Press 2001.

Förster, H. 2002, *Wörterbuch der griechischen Wörter in den koptischen dokumentarischen Texten,* Berlin/New York: de Gruyter.

Gascou, J. 2008, "Religion et identité communautaire à Alexandrie à la fin de l'époque byzantine d'après les *Miracles des saints Cyr et Jean*", in: J-Y. Empereur and Chr. Décobert (eds.), *Alexandrie médiévale* 3, Cairo: IFAO, 69–88.

Giamberardini, G. 1963, "La preghiera nella Chiesa copta", *SOC, Collectanea* 8: 3–77.

Giamberardini, G. 1974–1978, *Il culto mariano in Egitto* I–III, Jerusalem: Franciscan Printing Press.

Gignac, F. T. 1976, *A grammar of the Greek papyri of the Roman and Byzantine periods* I: *Phonology,* Milan: Cisalpino- La Goliardica.

Godlewski, W. 1993, "Deir el Naqlun 1989–1992", in: D.W. Johnson and T. Orlandi (eds.), *Acts of the Fifth International Congress of Coptic Studies: Washington, 12–15 August 1992,* Rome: Centro Italiano Microfiches, II/1, 183–95.

Godlewski, W. 1997, "Deir el Naqlun: topography and tentative history", in: *Archeologia e papiri nel Fayyum: storia della ricerca, problemi e prospettive,* Siracusa: Istituto internazionale del papiro, 123–45.

Grossmann, P. 1982, *Mittelalterliche Langhauskuppelkirchen und verwandte Typen in Oberägypten,* Glückstadt: J.J. Augustin.

Grossmann, P. 1990, "Architecture", in: K.A. Manafis (ed.), *Sinai: treasures of the Monastery of Saint Catherine,* Athens: Ekdotike Athenon, 29–39.

Grossmann, P. 1998, *Die antike Stadt Pharan: Ein archäologischer Führer,* Cairo: Archbishopric of Sinai and the Womens Convent of Firan.

Grossmann, P. 2002, *Christliche Architektur in Ägypten,* Leiden: Brill.

Grossmann, P. 2008, "Zur Typologie des Transepts im frühchristlichen Kirchenbau", *JbAC* 51: 97–136.

Guirguis, Magdi. 2008, *An Armenian artist in Ottoman Egypt: Yuhanna al-Armani and his Coptic icons,* Cairo/New York: American University in Cairo Press.

Hamilton, A. 2006, *The Copts and the West, 1439–1822: the European discovery of the Egyptian Church,* Oxford: Oxford University Press.

Horbury, W., and D. Noy. 1992, *Jewish inscriptions of Graeco-Roman Egypt,* Cambridge: Cambridge University Press.

Hunt, L-A. 1987, "Iconic and aniconic: unknown thirteenth and fourteenth century Byzantine icons from Cairo in their woodwork settings", in: A. Berger et al. (eds.), *Varia* II, Bonn: Habelt, 31–48.

Jeudy, A. 2009, "Elite civil et 'mécénat': le rôle du commanditaire dans le développement des arts et des lettres en Egypte chez les coptes du xe au xive siècle", *ECA* 6: 51–65.

Jolivet-Lévy, C. 1998, "Note sur la représentation des archanges en costume impérial dans l'iconographie byzantine", *CArch* 199: 121–28.

Junker, H. 1922, *Das Kloster am Isisberg: Bericht über die Grabungen der Akademie der Wissenschaften in Wien bei El-Kubanieh, Winter 1910–1911,* III, Vienna/Leipzig: Hölder.

Keenan, J.G. 2005, "Landscape and memory: al-Nabulsi's *Ta'rikh al-Fayyum*", *BASP* 42: 203–12.

Khater, A., and O.H.E. Burmester. (eds.) 1970, *The history of the Patriarchs of the Egyptian Church* III/3, Cairo: Société d'archéologie copte.

Khurshid, F. 1998, *The churches and monasteries of the Fayyum province, from the spread of Christianity until the end of the Ottoman period* [in Arabic], Cairo: Supreme Council of Antiquities.

Klauser, T. 1962, "Engel X (in der Kunst)", *RAC* 5: 258–322.

Kropp, A.M. 1930–31, *Ausgewählte koptische Zaubertexte*, Brussels: Édition de la Fondation égyptologique Reine Elisabeth.

Krueger, D. 2011, "Mary at the threshold: the Mother of God as guardian in seventh-century Palestinian miracle accounts", in: L. Brubaker and M.B. Cunningham (eds.), *The cult of the Mother of God in Byzantium: texts and images*, Farnham/Burlington, VT: Ashgate, 31–8.

Labib, Pahor. 1948–1949, "The virgin and child Jesus: reliefs in the Coptic Museum", *BSAC* 13: 191–92.

Łajtar, A. 2004, "Varia Nubica VIII-IX", *JJP* 34: 87–94.

Lange, R. 1964, *Die byzantinische Reliefikone*, Recklinghausen: Verlag Aurel Bongers.

Lazarev, V. 1995, "Studies in the iconography of the Virgin", in: *Studies in Byzantine painting*, London: Pindar Press, 196–248; published originally, with author's name V. Lasareff, in *ArtB* 20 (1938): 26–65.

Leclant, J. 1979, "Fouilles et travaux en Égypte et au Soudan, 1977–1978", *Orientalia* 48: 340–412.

Lefebvre, G. 1907, *Recueil des inscriptions grecques-chrétiennes d'Égypte*, Cairo: IFAO; reprinted Chicago: Ares 1978.

Lemerle, P. 1953, "Saint Démétrius de Thessalonique et les problèmes du martyrion et du transept", *BCH* 77: 660–94.

Leroy, J. 1974, *Les manuscrits coptes et coptes-arabes illustrés*, Paris: IFAO.

Leroy J. 1975, *Les peintures des couvents du désert d'Esna*, Cairo: IFAO.

Loon, G.J.M. van. 1999, *The gate of heaven: wall paintings with Old Testament scenes in the altar room and the ḫûrus of Coptic Churches*, Leiden: NINO.

Loverdou-Tsigarida, K. 2000, "The Mother of God in sculpture", in: Vassilaki 2005, 237–49.

MacCoull, L.S.B. 1989, "An inscription in Haret Zuwaila", *ZPE* 79: 270.

Maguire, H. 1995, "A murderer among the angels: the frontispiece miniatures of Paris Gr. 510 and the iconography of the archangels in Byzantine art", in: R. Ousterhout and L. Brubaker (eds.), *The sacred image: East and West*, Urbana/Chicago: University of Illinois Press, 63–71.

Maguire, H. 1997, "The heavenly court", in: H. Maguire (ed.), *Byzantine court culture from 829 to 1204*, Washington, DC: Dumbarton Oaks, 247–58.

Maguire, H. 2005, "Byzantine domestic art as evidence for the early cult of the Virgin", in: Vassilaki 2005, 183–93.

Maltese, E.V. 1990, "Gli angeli in terra: sull'immaginario dell'angelo bizantino", *Materiali e discussioni per l'analisi dei testi classici* 24: 111–32; reprinted in: Maltese, E.V. 1995, *Dimensioni bizantine: donne, angeli e demoni nel medievo greco*, Turin: Paravia, 69–92.

Meinardus, O.F.A. 1977, *Christian Egypt, ancient and modern*, second revised edition, Cairo: American University in Cairo Press.

Meyer, M. 1996, *The magical book of Mary and the Angels (P. Heid. Inv. Kopt. 685)*, Heidelberg: Universitatsverlag C. Winter.

MONUMENTS OF CHRISTIAN SINNURIS

Moorsel, P.P.V. van 1989, "Forerunners of the Lord: saints of the Old Testament in medieval Coptic church decoration", *CArch* 37: 119–33; reprinted in P.P.V. van Moorsel 2000, *Called to Egypt: collected studies on painting in Christian Egypt*, Leiden: NINO, 179–202.

Moorsel, P.P.V. van 1995–1997, *Les peintures du monastère de Saint-Antoine près de la mer Rouge*, Cairo: IFAO.

Moorsel, P. van, and M. Huijbers. 1981, "Repertory of the preserved wallpaintings from the Monastery of Apa Jeremiah at Saqqara", *Acta ad archaeologiam et artium historiam pertinentia* 9: 125–86.

Moritz, B. (ed.) 1899, *Description du Faiyoum au viime siècle de l'hégire par Abou 'Osmân il Naboulsi il Safadi*, Cairo: Imprimerie nationale; reprinted in Sezgin 1992.

Nachtergael, G. 2006, "Dédicace d'un monument à la Vierge Marie", *Ricerche di egittologia e di antichità copte* 8: 11–14.

Papaconstantinou, A. 2000, "Les sanctuaires de la Vierge dans l'Égypte byzantine et omeyyade: l'apport des textes documentaires", *JJP* 30: 81–94.

Papaconstantinou, A. 2001, *Le culte des saints en Égypte des Byzantins aux Abbasides: l'apport des inscriptions et des papyrus grecs et coptes*, Paris: Éditions du CNRS.

Paribeni, A. 2008, "I rilievi in marmo rappresentanti la Vergine e altri personaggi religiosi: considerazione sulla cronologia e sul loro ruolo nella liturgia", in: C. Pennas, and C. Vanderheyde (eds.), *La sculpture byzantine, viie–xiie siècles: actes du colloque international, 6–8 septembre 2000*, Athens: École Française d'Athènes, 561–75.

Pentcheva, B.V. 2006, *Icons and power: the Mother of God in Byzantium*, University Park, PA: Pennsylvania State University Press.

Rodziewicz, M. 1984, *Les habitations romaines tardives d'Alexandrie à la lumière des fouilles polonaises à Kôm el-Dikka*, Warsaw: Editions scientifiques de Pologne.

Rutschowscaya, M-H. 1990, *Tissus coptes*, Paris: Adam Biro.

Rutschowscaya, M-H. 2000, "The Mother of God in Coptic textiles", in: Catalogue Athens 2000, 219–25.

Rutschowscaya, M-H., and R.W. Boutros. 2004, "Deux arcatures en bois de l'époque ottomane au musée du Louvre", in: M. Immerzeel and J. van der Vliet (eds.), *Coptic studies on the threshold of a new millennium: proceedings of the Seventh International Congress of Coptic Studies, Leiden, 27 August-2 September 2000*, Louvain/Paris/Dudley, MA: Peeters 2004, II, 1499–1524.

Salmon, G. 1901, "Répertoire géographique de la province du Fayyoûm d'après le Kitâb târîkh al-Fayyoûm d'an-Nâboulsî", *BIFAO* 1: 29–77; reprinted in Sezgin 1992.

Samuel al Syriani, and Badii Habib. 1990, *Guide to ancient Coptic churches & monasteries in Upper Egypt*, Cairo: Institute of Coptic Studies, Department of Coptic Architecture.

Sezgin, F. (ed.) 1992, *Studies on the Faiyûm together with Târîh al-Faiyûm wa-bilâdihî by Abû 'Utmân an-Nâbulusî (d. 1261)*, Frankfurt am Main: Institute for the History of Arabic-Islamic Science.

Skalova, Z. 2004, "Five 13th-century great Deesis portraits in the Wadi Natrun: their origin", in: M. Immerzeel and J. van der Vliet (eds.), *Coptic studies on the threshold of a new millennium: proceedings of the Seventh International Congress of Coptic Studies, Leiden, 27 August-2 September 2000*, Louvain/Paris/Dudley, MA: Peeters 2004, II, 1525–50.

Skalova, Z., and Gawdat Gabra. 2003, *Icons of the Nile Valley*, Cairo: Egyptian International Publishing Company Longman.

149

Snelders, B., and M. Immerzeel. 2004, "The thirteenth – century flabellum from Deir al-Surian in the Musée royal de Mariemont (Morlanwelz, Belgium)", *ECA* 1: 113–39.

Snelders, B., and A. Jeudy 2006, "Guarding the entrances: equestrian saints in Egypt and North Mesopotamia", *ECA* 3: 105–42.

Stuhlfauth, G. 1897, *Die Engel in der altchristlichen Kunst*, Freiburg im Breisgau: J.C.B. Mohr.

Testa, E. 1983, "L'angelologia dei giudeo-cristiani", *Liber Annuus* 33: 273–302.

Thomas, T.K. 2000, *Late antique Egyptian funerary sculpture: images for this world and the next*, Princeton: Princeton University Press.

Timm, S. 1979, *Christliche Stätten in Ägypten*, Wiesbaden: Reichert Verlag.

Timm, S. 1984–1992, *Das christlich-koptische Ägypten in arabischer Zeit*, 6 vols., Wiesbaden: Reichert Verlag.

Urbaniak-Walczak, K. 1993, "Drei Inschriften aus der Kirche des Erzengels Gabriel in Deir an-Naqlun im Faijum", *BSAC* 32: 161–69.

Vansleb, [J.M.] 1677, *Nouvelle relation en forme de journal d'un voyage fait en Egypte . . . en 1672 & 1673*, Paris: Estienne Michallet.

Vassilaki, M. (ed.) 2005, *Images of the Mother of God: perceptions of the Theotokos in Byzantium*, Aldershot/Burlington: Routledge.

Vergote, J. 1954, *Les noms propres du P. Bruxelles inv. E. 7616: essai d'interprétation*, Leiden: Brill.

Vikan, G. 1995, "Icons and icon piety in early Byzantium", in: Chr. Moss and K. Kieffer (eds.), *Byzantine East, Latin West: art-historical studies in honor of Kurt Weitzmann*, Princeton, NJ: Dept. of Art and Archaeology, Princeton University, 569–78.

Vliet, J. van der 2000, "Les anges du Soleil: à propos d'un texte magique copte récemment découvert à Deir en-Naqloun (N. 45/95)", in: N. Bosson (ed.), *Études coptes VII: Neuvième journée d'études, Montpellier 3–4 juin 1999*, Louvain/Paris: Peeters, 319–38.

Vliet, J. van der 2002–2003, "Monumenta fayumica", *Enchoria* 28: 137–46 [Study 7].

Vliet, J. van der 2008, "*Parerga*: notes on Christian inscriptions from Egypt and Nubia", *ZPE* 164: 153–58 [Study 16].

Volbach, W.F., and J. Lafontaine-Dosogne. (eds.) 1963, *Byzanz und der christliche Osten*, Berlin: Propylaen.

Vycichl, W. 1983, *Dictionnaire étymologique de la langue copte*, Leuven: Peeters.

Wellen, G.A. 1961, *Theotokos: eine ikonographische Abhandlung über das Gottesmutterbild in frühchristlicher Zeit*, Utrecht/Antwerpen: Het Spectrum.

Wessely, C. 1904, *Topographie des Faijûm (Arsinoites Nomus) in griechischer Zeit*, Wien: C. Gerold's Sohn; reprinted Milan: Cisalpino-Goliardica 1975.

Wietheger, C. 1992, *Das Jeremias-Kloster zu Saqqara unter besonderer Berücksichtigung der Inschriften*, Altenberge: Oros Verlag.

Winlock, H.E., and W.E. Crum. 1926, *The monastery of Epiphanius at Thebes*, New York: Metropolitan Museum of Art.

9

FOUR CHRISTIAN FUNERARY INSCRIPTIONS FROM THE FAYOUM (I. DAYR AL-'AZAB 1–4)

Tomasz Derda and Jacques van der Vliet

Introduction

The four funerary monuments which are published here for the first time are part of the collections of Dayr al-'Azab, now better known as the Monastery of Anba Abraam, a large ecclesiastical center situated a few kilometers to the southeast of Medinet al-Fayoum, on the main road to the Nile Valley.[1] We are grateful to His Grace Anba Abraam, Bishop of the Fayoum, for his permission to study and publish them.[2]

There is no documentation available to show the exact provenance of these four objects. Nevertheless, they can be assumed to come from the Fayoum, which is generally confirmed by their stylistic and textual characteristics. Further, internal criteria allow stela no. 1 to be assigned to one of the Christian cemeteries of the town of Arsinoe. For wooden tablet no. 4, Dayr al-Malak (Naqlun), a few kilometers to the southeast of Dayr al-'Azab, has been cited as a plausible provenance, but this remains uncertain. All four monuments are in Greek and can be dated only very imprecisely to the Byzantine–early Islamic period.[3] In addition to their prosopographical and art-historical interest, they illustrate the prosperity and the high cultural level of the Fayoum in this period.

1. Tombstone of Menas and Gerontios

Upper righthand corner of a tombstone; limestone, 23 × 35 × 9 cm. The lower half of the stone (which must have contained a sculptured motif, either a cross or a praying figure) is lacking. The surviving upper part is irregularly broken at the lefthand side, and at the lower right margin another piece of the stone is broken away. Surface damage occurs at several places. The original width of the stela may have been about 50–51 cm, and while its original height cannot be reconstructed exactly, it may have been over 75 cm.

Remains of thirteen lines of Greek text are preserved, though the beginnings of each line (about one-third of the original length) are missing. Lines 10–13 are interrupted in the middle by a sculptured conch.

The text is written in crudely drawn and rather irregular incised uncials, about 2–2.5 cm high, badly ruled. Characteristic letter forms include the A with a broken bar and a long upward stroke towards the left at the top (also notable in Λ), while the letter B is plump with a narrow base (l. 6).

Figure 9.1 Dayr al-'Azab 1
(photo: Jacques van der Vliet)

FOUR INSCRIPTIONS FROM THE FAYOUM

Arsinoe (Fayoum town),   seventh-eighth century

[+ κ(ύρι)ε Ἰ(ησοῦ)ς ὁ Χ(ριστὸ)ς] ὁ θ(εὸ)ς τῶν πα[τέρων ἡμ-]
ῶν Ἀβρὰμ (καὶ) Ἰ]σὰ[κ] (καὶ) Ἰακ[ὼβ ἀνάπ]αυσ-
ον τὴν ψυχὴ]ν τοῦ δούλου σου Μηνᾶ
4.   [- - -]χς, υἱοῦ Τιμοθέου
     [- - -]τς. ἐκοιμήθη ἥν κ(ύρι)ῳ
     [μ(ηνὶ) .... ἀ]ρχ(ῇ) β ἰν(δικτίωνος). ἀμήν. +
     [ἀνάπαυσον τὴν] ψυχὴν τοῦ δούλου
8.   [σου Γερ]οντίου, υἱοῦ Μαρνι-
     [τᾶ πραιπο]σίτου, [π]ρ(εσβυτέρου) μεγάλ(ης)
     [ἐκκλ(ησίας). ἐκοιμ]ή‖θη ἐν εἰρή-
     [νη, μ(ηνὶ) Πα]‖ῦνι λ,
12. [τέλει] ‖ πρώτη[ς]
     [ἰν(δικτίωνος).] ‖ ἀμήν. [+]

1. filling in of first lacuna e.g.; θ(εο)ς: stone ΘϹ ‖ 2. (καὶ): stone ς (twice) ‖ 4-beginning and 5-beginning: end of abbreviated profession name or title ‖ 5. κ(ύρι)ῳ: stone ΚΩ ‖ 6. ἀ]ρχ(ῇ): stone] ΡΧς ‖ 7. δούλου: final -ου ligatured ‖ 8. Γερ]οντίου: low o after lacuna suggests preceding ρ ‖ 9. [π]ρ(εσβυτέρου) μεγάλ(ης): stone [Π]Ρς ΜΕΓΑΛ/; for the small π in the lacuna, cf. l. 12 ‖ 12. lacuna: τέλει, rather than ἀρχῇ (see below).

+ O Lord Jesus-Christ, God of our fathers Abraham and Isaac and Jacob, rest the soul of your servant Menas, [. . .], son of Timotheos, [. . .]. He fell asleep in the Lord in the month [. . .], (day) [. . .], in the beginning of the second (year of the) indiction. Amen. + Rest the soul of your servant Gerontios, son of Marnitas, praepositus, priest of the Great Church. He fell asleep in peace, in the month Pauni, (day) 30, in the end of the first (year of the) indiction. Amen. +

The text is an epitaph for two men, Menas and Gerontios,[4] whose relationship to each other is not clear from the text as preserved. Both are given titles and a filiation. Gerontios was a priest of the "Great Church" (ll. 9–10), and his father, Marnitas, was most probably a *praepositus*, "prefect" (ll. 8–9).[5] Since the stone originates from the Fayoum, the "Great Church", where Gerontios served as a priest, can be identified as the cathedral church of Arsinoe, well known through a variety of documents, mainly from the early Islamic period.[6] Among the many persons named Gerontios that are listed in the *Prosopographia Arsinoitica* there are in fact one or two priests (both from the sixth-seventh century).[7] It is impossible, however, to link any of these men to the one commemorated here. The name of his father, Marnitas, is extremely rare in Egypt and the two examples mentioned in the repertories are both from the Fayoum.[8] Again, however, it is not possible to identify the father of Gerontios with either of these. The titles or professions of Menas and his father, Timotheos, mentioned in ll. 4 and 5, are now lost but for their final letters. Presumably, both were also men of some status.

   Several other funerary inscriptions commemorating multiple people are known from the Fayoum.[9] The present one is textually unusual in that, after a shared

153

opening invocation (ll. 1–2), it presents a complete epitaph for each of both deceased. Both formulae follow a similar pattern that is common in the Fayoum (κύριε ἀνάπαυσον τὴν ψυχὴν + name + ἐκοιμήθη + date).[10] Some variation occurs in the dating lemmata: whereas Menas died "in the Lord" (l. 5), Gerontios died "in peace" (ll. 10–11). The long opening invocation of ll. 1–2, with its reference to the three patriarchs, is another remarkable feature.[11] It well illustrates the transition from very brief, summary epitaphs (such as no. 2 below) to more developed texts that show a strong liturgical inspiration.[12]

The method of dating observed in the present stela, by "beginnings" and "ends" of indiction years, is again particularly common in documents and inscriptions from the Arsinoite.[13] The reconstruction of τέλει in l. 12 is uncertain, of course, but the association of the end of the indiction with the month of Pauni is in accordance with the predominant statistical evidence for the region.[14] It would indeed be logical if both men had died shortly after each other, Gerontios at the end of a first indiction year (ll. 11–13), Menas in the beginning of the following, second year (l. 6). No absolute dates are given, but the present epitaph can be profitably compared to the stela of Pousei and Kosmas, now in Berlin, but also originally from the Fayoum.[15] This is also a double epitaph, with an extensive text surmounting the pictorial field (lost in the present piece). It likewise opens with the invocation of the "God of our fathers" and, in referring to the patriarchs (ll. 5–9), shows a similar liturgical inspiration.[16] The Berlin stela is dated to Diocletian year 419 (A.D. 703), and the present stela must belong to approximately the same period.

2. Tombstone of Papas

A complete and well-preserved funerary monument, carved from one piece of limestone, measuring 18.5 × 17 × 17 cm. It consists of an otherwise undecorated square block surmounted by a sculptured Maltese cross within a laurel wreath; the space between the arms of the cross is filled in with leaf-like decoration.

The front of the square basis bears an epitaph of four lines in Greek, which leaves the lower third part of the surface empty. The text is written in square incised uncials, about 1.8 cm. high. Characteristic letter forms include a broken-bar symmetrical A.

Fayoum,   *ca.* sixth-eighth century?

    + ἐν ἡρήνη.
    τ(οῦ) ἀναπαυ-
    σαμέν(ου) ἐν
4.   κ(υρί)ῳ Παπᾶ.

1. ἡρήνη, l. εἰρήνη; -ην- ligatured ‖ 2 and 3. abbreviations by a slash ‖ 4. κ(υρί)ῳ: ΚΩ.

+ In peace. (Monument) of Papas, who went to rest in the Lord.

Figure 9.2 Dayr al-'Azab 2
(photo: Jacques van der Vliet)

The peculiar shape of this monument is as yet unparalleled, although crosses within laurel wreaths are common enough in Christian sculpture, funerary and non-funerary, from all over Egypt. Another funerary monument from the Fayoum, now in Cairo, is particularly close in style and appearance to the present piece.[17] It bears an identical textual formula (partly lost), inscribed in a *tabula ansata* [a tablet with handles] below the laurel wreath, and the similarity between both monuments suggests that they might share a common, though unknown, provenance.

Beyond the name of the deceased, Papas,[18] no information can be gained from the epitaph – it even lacks the usual date of demise. This very concise text, characterized by ἐν εἰρήνῃ, is typical of an entire class of Fayoum monuments.[19] Its brevity and the laconic form of the monument could be indicative of a (typologically) early date.

3. Tombstone of N.N.

The upper part of a limestone funerary stela, presently measuring 32 × 23 × 6–8 cm. The lower part, with a sculptured decoration of a cross within a niche,

Figure 9.3 Dayr al-'Azab 3
(photo: Jacques van der Vliet)

has for the greater part broken away; the surviving upper part, including the epigraphic field, is severely damaged.

The remains of five lines of Greek text are still visible, but are so much obliterated as to be almost entirely illegible. Incised uncials, about 1.5 cm high.

Fayoum, *ca.* seventh-eighth century?

+ κ(ύρι)ε, ἀγάπ[αυσον τὴν
ψυχ[ὴν - - -]
ΥΕ. [- - -]
4. ΜΕÇΩΡ[- - -]
μ(ηνὶ) Ἐπείφ δ. . α' ἰν(δικτίωνος) +

1. κ(ύρι)ε: ΚΕ ‖ 3. mere traces ‖ 4. or ΜΕCΧΡ [? ‖ 5. μ(ηνὶ): Μ*s*; α': slightly raised, uncertain whether something precedes; ἰν(δικτίωνος): ΙΝ*s*.

FOUR INSCRIPTIONS FROM THE FAYOUM

+ Lord, grant rest to the soul [of your servant N.N. He/she fell asleep?] in the month Epeiph, (day) 4 (?), of the first (or eleventh) year of the indiction. +

This monument belongs to the group of conventional Fayoum stelae characterized by a big sculptured cross within an *aedicula*.[20] Here, the epigraphic field is squeezed in between the niche containing the cross and the tympanum *plus* conch that would logically crown the niche.[21] Apart from this detail, the style and iconography as well as the very common textual formula, virtually identical to that of the double epitaph no. 1, above, are close to those of the stelae found at Naqlun. There, however, the text is arranged within the niche, around the cross.[22]

4. Wooden funerary tablet of Phoibammon

A *tabula ansata* inscribed with three lines of Greek text, 7 × 23 × 3 cm; dark brown wood of low density and mediocre quality, undoubtedly of local provenance. Near the ends, two holes have been drilled for attaching the *tabula* to a mummy or, perhaps, a coffin.

The text consists of three lines of slightly sprawling incised uncials, varying in height from about 3 cm (in l. 1) to 1.5 cm (in ll. 2–3). Letter forms include a broken-bar symmetric A and a three-stroke M with a low saddle.

Naqlun (?), Fayoum,   *ca.* sixth-eighth century?

    + Φοιβά-
2. μμων, ἰα-
    τρ(ός).

2–3. ἰατρ(ός): IATPς.

+ Phoibammon, physician.

Figure 9.4 Dayr al-'Azab 4
(photo: Jacques van der Vliet)

Wooden tablets, often in the shape of a *tabula ansata*, as in the present case, are a traditional element of funerary equipment in Greco-Roman Egypt, meant to be attached to the mummy of the deceased for the purpose of identification.[23] Conventionally known as mummy labels, they remained in use until well into Christian times.[24] Since they are inscribed with the name of the deceased, usually accompanied by a filiation or a title, they may also serve as a memorial, analogous to grave stelae, for which they offer a cheap alternative.[25] In late antiquity, stone funerary monuments often also incorporate *tabulae ansatae*, typically as a feature exhibiting (part of) the epitaph.[26] In one of these, a Fayoum monument for a children's tomb, the stone *tabula* shows two (fake) holes for attachment, which indicates that it was meant to imitate a wooden tablet like the present one.[27] Transposed into a more costly material and fitted into the symbolic framework of a high-status object, the simple *tabula* had apparently retained its privileged role as a text bearer in a funerary context. By contrast, undoubtedly Christian examples in wood, like the one published here, are quite scarce.[28]

In sources from Christian Egypt, the name Phoibammon is as ubiquitous as the profession of doctor. Tombstones of physicians abound,[29] and from one of the monasteries of Western Thebes another Phoibammon is known who was a physician as well as a deacon.[30] As for the Fayoum, no doctors with this name can be identified with certainty.[31] A possibly fifth-century namesake, who seems to have been a kind of general practitioner, may have inhabited hermitage no. 44, behind the monastery at Naqlun.[32] Since the provenance suggested for the Dayr al-'Azab tablet is no more than conjectural, it cannot be ascertained whether its owner, in addition to being a doctor, was a monk as well.

Notes

1   See Sameh Adli, "Several churches in Upper Egypt", *MDAIK* 36 (1980), 1–14 at 4; S. Timm, *Das christlich-koptische Ägypten in arabischer Zeit*, Wiesbaden: Reichert Verlag 1984–1992, II, 681–82; Samuel al Syriani and Badii Habib, *Guide to ancient Coptic churches & monasteries in Upper Egypt*, Cairo: Institute of Coptic Studies, Department of Coptic Architecture 1990, 146, no. 189; Fathy Khurshid, *The churches and monasteries of the Fayum province: from the spread of Christianity until the end of the Ottoman period* (in Arabic), Cairo: Supreme Council of Antiquities 1998, 127–31; R.N. Hewison, *The Fayoum: history and guide*, Cairo/New York: American University in Cairo Press 2013, 66–7. Locally, a series of illustrated booklets on the monastery are sold, produced by the Coptic Orthodox archdiocese of the Fayoum; one of these, no. 3 in the series, is devoted to its collections.

2   We furthermore thank Sister Irene and Mr. Ezzat Salib for their assistance in Dayr al-'Azab, and Dr. Brian Muhs of the Leiden Papyrological Institute for correcting our English.

3   For recent reviews of the Christian epigraphic record of the Fayoum, see A. Boud'hors, and F. Calament, "Un ensemble de stèles fayoumiques inédites: à propos de la stèle funéraire de Pantoleos de Toutôn", in: M. Immerzeel and J. van der Vliet (eds.), *Coptic studies on the threshold of a new millennium: proceedings of the Seventh International Congress of Coptic Studies, Leiden, 27 August-2 September 2000*, Louvain/Paris/Dudley, MA: Peeters 2004, I, 447–75 (Coptic only; with an appendix:

FOUR INSCRIPTIONS FROM THE FAYOUM

"Epigraphie fayoumique: addenda et corrigenda", *JCoptS* 7 (2005), 131–35); S. Schaten, "Christian funerary stelae from the Fayoum", in: G. Gabra (ed.), *Christianity and monasticism in the Fayoum Oasis*, Cairo/New York: American University in Cairo Press 2005, 257–63; J. van der Vliet, "Reconstructing the landscape: epigraphic sources for the Christian Fayoum", ibidem, 79–89 [Study 6].

4 The latter name is damaged, but from the traces its reading highly plausible. Gerontios was a very common name in the Christian Fayoum, see J.M. Diethart, *Prosopographia Arsinoitica* I, Vienna: Hollinek 1980, nos. 1261–1319; for its wider distribution, see W.J. Cherf, "What's in a name? the Gerontii of the later Roman empire", *ZPE* 100 (1994), 145–74.

5 See F. Preisigke, *Wörterbuch der griechischen Papyrusurkunden* III, Berlin: privately published 1931, 143 (civil officials) and 217 (military); S. Daris, *Il lessico latino nel greco d'Egitto*, Barcelona: Papyrologica Castroctaviana 1912, s.v. *The Prosopographia Arsinoitica* mentions no *praepositi*.

6 See E. Wipszycka, "Καθολική et les autres épithètes qualifiant le nom ἐκκλησία: contribution à l'étude de l'ordre hiérarchique des églises dans l'Égypte byzantine", *JJP* 24 (1994), 191–212 at 200–02 (= eadem, *Études sur le christianisme dans l'Égypte de l'antiquité tardive*, Rome: Institutum Patristicum Augustinianum 1996, 157–75 at 165–66); Timm, *Ägypten* IV, 1511.

7 Diethart, *Prosopographia Arsinoitica*, nos. 1289 (sixth-seventh century) and 1305 (seventh century).

8 Diethart, *Prosopographia Arsinoitica*, nos. 3429 (sixth century) and 3430 (A.D. 663). See also *BGU* II 675 = *SPP* VIII 1139 (with *BL* VI, 12): δι' ἐμοῦ Μαρνιτου ἀντ(ι) γ(εού)χ(ου). The document can be paleographically dated to the seventh century, (see the photograph in A. Bataille, *Les papyrus*, Paris 1955, pl. XIII). The form Μαρνιτου would imply the nominative Μαρνίτης, which is by no way surprising since the names ending in -ᾶς (gen. -ᾶ) may be interchanged with those ending in -ης (gen. -ου).

9 See W. Godlewski, and A. Łajtar, "Grave stelae from Deir el-Naqlun", *JJP* 36 (2006), 43–62, no. 1.

10 Type no. 1 in D. Zuntz, "Koptische Grabstelen: Ihre zeitliche und örtliche Einordnung", *MDAIK* 2 (1932), 22–38 at 23. The same formulary is found in no. 3 below.

11 For its background, see M. Rist, "The God of Abraham, Isaac, and Jacob: a liturgical and magical formula", *Journal of Biblical Literature* 57 (1938), 289–303.

12 For this development in Fayoum epitaphs, see Van der Vliet, "Reconstructing the landscape", 80–1.

13 See now the fully documented discussion in R.S. Bagnall and K.A. Worp, *Chronological systems of Byzantine Egypt: second edition*, Leiden: Brill 2004, 22–35.

14 See Bagnall and Worp, *Chronological systems*, especially 25.

15 Skulpturensammlung und Museum für Byzantinische Kunst, inv. no. 4477: Lefebvre, *Recueil*, no. 790; cat. exhib. *Ägypten: Schätze aus dem Wüstensand*, Wiesbaden: Reichert Verlag 1996, no. 67.

16 It seems as if, in funerary epigraphy, the invocation "God of Abraham (etc.)" (here ll. 1–2) is interchangeable with the well-known prayer for rest "in the bossom of Abraham (etc.)" (found in the Berlin stela); cf. *I. Khartoum Copt.* 46, with commentary.

17 W.E. Crum, *Coptic monuments*, Cairo: IFAO 1902, no. 8543; Lefebvre, *Recueil*, no. 79. Crum connects this piece with his no. 8710 (p. 145, n. 3).

18 Diethart, *Prosopographia Arsinoitica*, mentions only two persons of this name: nos. 4145 (A.D. 531) and 4146 (sixth-seventh century).

19 See Zuntz, "Koptische Grabstelen", 23, types no. 2 and 3; A. Łajtar, "Bemerkungen zu griechischen christlichen Inschriften aus dem Koptischen Museum in Kairo", *ZPE* 97 (1993), 227–35 at 230–31 (with further references).

20 See, most recently, S. Schaten, "Christian funerary stelae", 258–61.

21 Several other examples of this arrangement are known from the Fayoum; e.g. cat. exhib. *Ägypten: Schätze aus dem Wüstensand*, no. 81.

22 For the Naqlun stelae, see A. Łajtar, "Two Greek inscriptions from Deir el-Naqlun", *Nubica* 3/1 (1989–1993), 265–74 at 269–74; in recent years, some fragments of others have been found, for which see now W. Godlewski and A. Łajtar, ["Grave Stelae from Deir el-Naqlun"].

23 For an introduction, see J. Quaegebeur, "Mummy labels: an orientation", in: E. Boswinkel and P.W. Pestman (eds.), *Textes grecs, démotiques et bilingues*, Leiden: Brill 1978, 232–59; on the *tabula ansata*-form: 235–36; Fayoum provenances: 244.

24 For Christian mummy labels, see now H. Froschauer, "Tradition im koptischen Bestattungswesen: Ein christliches Mumientäfelchen aus den Beständen Tamerit in der Papyrussammlung der Österreichischen Nationalbibliothek", *Eirene* 40 (2004), 91–100, and A. Delattre, "Une étiquette de momie copte de l'ancienne collection G.A. Michaelidès", *CdÉ* 80 (2005), 373–74.

25 Quaegebeur, "Mummy labels", 237.

26 For some typical examples from the Fayoum, see Crum, *Coptic monuments*, nos. 8543, 8590; cat. exhib. *Ägypten: Schätze aus dem Wüstensand*, nos. 64, 65, 84, but the phenomenon was common all over Egypt.

27 Berlin, Skulpturensammlung und Museum für Byzantinische Kunst, inv. no. 4709: Lefebvre, *Recueil*, no. 794; cat. exhib. *Ägypten: Schätze aus dem Wüstensand*, no. 84.

28 See Delattre, "Une étiquette de momie copte", who provides a full bibliography.

29 For some examples, see Lefebvre, *Recueil*, nos. 135, 190, 496, 799; Kamal Sabri Kolta, "Namen christlicher Ärzte der koptischen Zeit in Ägypten", *Die Welt des Orients* 14 (1983), 189–95; G. Nachtergael, "Lecture de quelques noms de métiers dans des inscriptions grecques d'Égypte", *CdÉ* 74 (1999), 148–55 at 152–53, no. 3; for the Coptic stela of a female doctor, *I. Lyon Copt.* 1 (*Bulletin des musées et monuments lyonnais* 2001, no. 1, 74–5).

30 R. Rémondon, Yassâ 'Abd al-Masîh, W.C. Till, and O.H.E. KHS-Burmester, *Le Monastère de Phoebammon dans la Thébaïde* II, Cairo: Société d'archéologie copte 1965, 71, no. 116b.

31 Diethart, *Prosopographia Arsinoitica*, lists four physicians and two "archphysicians", none of them called Phoibammon, though.

32 See W. Godlewski, "Excavating the ancient monastery at Naqlun", in: Gabra (ed.), *Christianity and monasticism*, 155–71 at 157–58.

10

A LINTEL FROM THE FAYOUM IN THE BRITISH MUSEUM

Jacques van der Vliet and Adeline Jeudy

Introduction

The lintel published here is a witness of the Christian culture of the Fayoum around the year 1000, when it knew a last revival before the decline of the Middle Ages set in. As Sofia Schaten was the first to observe, this piece can be joined to a small series of similar monuments, previously identified by Gérard Roquet.[1] These are all wooden lintels, now five in number, that contain a Coptic text inscribed on either side of the central motif of a sculpted cross. The inscriptions display the same formulary and include dates that situate them in the tenth century. In the present study, the *editio princeps* of the text by Jacques van der Vliet is followed by a discussion of the figurative decoration by Adeline Jeudy.[2]

The text

The lintel consists of a wooden beam, 16.5 cm high, 192.5 cm wide, and 12 cm thick. The back partly preserves the contours of the trunk of the tree that was used for producing the lintel. At both ends, the beam was crafted so as to be suspended in a support. The front is adorned with three panels with figurative decoration, separated by two panels inscribed with text. Traces of polychrome paint are preserved in several places. The piece is complete and in a fairly good condition; only in the right half, the inscription and the decoration underwent some deterioration, particularly at the expense of ll. 4–5 of the text.

The inscription contains seven lines of Coptic text, three to the left and four to the right of the central panel. On the lefthand panel, the text is incised; on the right panel, it is sculpted in raised relief. The script is very carefully executed and consists of beautiful uncials, tall and elegant on the left, where the letters are provided with apices, somewhat more thick and roundish in the text on the right. The height of the letters is 2.5–3 cm (left panel) and 2.5 cm (right panel). Two large stylized characters on the central panel, on either side of the cross, are part of the symbolism of the sculpted decoration. The language is a "Fayoumi-Sahidic", described below.

The piece was purchased from the Cairene antiquities dealer Panayotis Kyticas in 1915 (accession number 1915/4–10/75; on Kyticas, see Dawson and Uphill 1995: 233).

Unpublished. Reproduced: Davies 1987: 27, pl. 33; Enß 2005: pl. 84, fig. 437c (detail); cf. Schaten 1998: 306, n. 7.

British Museum, inv. no. EA 54040　　　　　　　　　Fayoum, A.D. 920/921

[central panel]
　　　　A　　ⲱ
[left side]
  + ⲡϭ(ⲱⲓ)ⲥ ⲓ(ⲏⲥⲟⲩ)ⲥ ⲡⲉⲭ(ⲣⲓⲥⲧⲟ)ⲥ ϭⲟⲓⲑ(ⲓ) ⲉⲕⲓ-
  ⲣⲓ ⲭⲁⲏⲗ ⲩⲓ(ⲟⲥ) ⲡⲙⲁⲕⲁⲣⲓ-
  ⲟⲥ ⲓⲱⲕⲓⲙ ⲙⲏⲛ ⲧⲉϥⲥⲓⲙⲓ
[right side]
4.　ⲙⲏ[ⲛ ⲛⲉϥϣⲏⲣ]ⲓ ϫⲉ ⲛⲧⲁϥ
　　ⲁϥϣⲁⲡ ⲡⲓⲏⲓ ⲁϥⲕⲁⲧϥ ϩⲉⲛ
　　ⲛⲏϥϩⲓⲥⲓ ⲙⲙⲓⲛ ⲙⲙⲁϥ ϩⲁⲙⲏ-
　　ⲛ ⲉⲥⲉϣⲟⲡⲓ ⲇⲓⲟⲕ(ⲗⲏⲧⲓⲁⲛⲟⲩ) ⲭⲗⲍ

1. ⲡϭ(ⲱⲓ)ⲥ: ⲡⲟ̄ⲥ̄ doubled by an abbreviation stroke | ⲓ̄ⲥ̄ ⲡⲉⲭ̄ⲥ̄ | ϭⲟⲓ˙θ˙: βοηθέω ‖ 1–2. κύριος ‖ 2. ⲩⲓ(ⲟⲥ): ϥ ̗, ligature of a cursive type, υἱός ‖ 2–3. μακάριος ‖ 3. ⲙⲏⲛ: ligatured ‖ 6. ⲙⲙⲁϥ: ⲙⲙ- ligatured | ϩⲁⲙⲏ-: -ⲙⲏ- ligatured ‖ 7. ⲇⲓ̄ⲟ̄ⲕ̄.

Alpha-Omega.
+ O Lord Jesus Christ, help Lord Chael, son of the blessed Iokim, and his wife and his children, for he bought this house and built it by his own efforts. Amen, so be it.
(The year) of Diocletian 637 (= A.D. 920/921).

As with the other lintels from the same series, the inscription is a dedicatory text.³ It takes the form of a prayer that invokes divine protection for a certain Chael (a shortened form of Michael) and his family, on the occasion of the foundation of a house. The characteristic format of the text links it to a common genre of prayers for the sake of benefactors of monasteries and churches. Very similar prayers can be found in the colophons that commemorate the donation of a manuscript⁴ or the dedicatory inscriptions that honor the donors of a wall painting.⁵

Figure 10.1 Lintel British Museum EA 54040: overview
　　　　　(photo: © Courtesy of the Trustees of the British Museum)

In the opening acclamation, the use of the verb βοηθέω, "to help" (l. 1), which is almost standard in such prayers, brings out the apotropaic character of the text, in agreement with the general nature of the lintel's decoration.[6]

The foundation itself is described in l. 5. Despite the damage that affects the inscription in this line, the readings are hardly doubtful. In the analogous text on the lintels Roquet nos. 1 and 2, the foundation in question is identified as "this house" (ⲡⲏⲓ̄<ⲓ>, in no. 2) and "these houses" (ⲛⲓⲏⲓ, in no. 1). Only in the inscription of Roquet no. 4, it is designated as "this dwelling", ⲡⲓⲙⲁⲛϣⲱ(ⲡⲉ).[7] Here, the space in l. 5 and the remaining traces require the reading ⲡⲓⲏⲓ, "this house". As the deictic article ⲡⲓ- shows, the lintel must have been part of the house, exhibited in a visible place, most likely above its principal entrance.

The foundation act is evoked by two brief parallel phrases containing forms of the verbs ϣⲱⲡ, "to buy", and ⲕⲱⲧ, "to build, rebuild". The first verb (ⲁϥϣⲁⲡ, "he bought") is damaged, but the adopted reading is indubitable, even though nowhere else in the corpus of known Coptic monumental inscriptions, the purchase of a building prior to its construction (or reconstruction) appears to be mentioned. Likewise, the lintel Roquet no. 3, published earlier by G. Lefebvre, bears a double formula with in the second position ⲁϥⲕⲁⲧϥ, "(and) he built it", but the corresponding first phrase falls in a lacuna.[8] Book colophons, however, often mention the purchase of the manuscript previous to its donation, using the same verb ϣⲱⲡ, "to buy".[9] Here again, the text of the lintel resembles prayers in colophons, an analogy that suggests that the house in question was a pious foundation rather than a private home.[10] It is perhaps possible to determine its nature more precisely. The word "dwelling", ⲙⲁⲛϣⲱⲡⲉ, which occurs on the lintel Roquet no. 4, is frequently used in monastic contexts, to designate a "monk's cell" or "group of cells", for instance on lintels from the monastery of Bawit.[11] Like most inscribed lintels from Christian Egypt, the one in London in all likelihood surmounted the entrance to a "house" that belonged to a monastic complex.

Chael bought this house and built it "by his own efforts" (ll. 5–6), that is "by his proper means", as the corresponding phrase ⲅⲉ ⲡⲉⲧⲏⲃ, literally "from his property", on the lintel Roquet no. 1 brings out more explicitly.[12] Chael's title, κύριος, "Lord, Sir" (ll. 1–2), and the rich decoration of the monument suggest that he belonged to the social elite. The donor of the lintel Roquet no. 3 bears the same title,[13] whereas the one of no. 4 is a deacon, just like his father, another Chael.[14] The father of our Chael, Iokim, bears no title, apart from the epithet "blessed", which shows that he is deceased (ll. 2–3). His unusual name represents a variant of Joachim. As in the other lintels from the group, the prayer of the text does not just concern the person of the donor; the filiation and the formal inclusion of a wife and children (here ll. 3–4) extend it to his kin.[15] Obviously, the wealthy families of the tenth-century Fayoum were proud to show their munificence and perpetuate their memory in this way.

The coherence of this group of lintels and their origin from the same region of Egypt, the Fayoum, are not only guaranteed by their quasi-identical formulary, but also by their language. It represents what one might call "Fayoumi-Sahidic", a form of late Sahidic that betrays the Fayoumic milieu in a variable degree, particularly in its vocalism.[16] It is the idiom commonly used in non-literary Coptic texts from the Fayoum in the tenth-twelfth centuries, shortly before the final Arabization of the region in the course of the eleventh-twelfth centuries.[17] It is worth mentioning that this Fayoumi-Sahidic was used around A.D. 1000 to produce monumental inscriptions not just on lintels, but in a variety of contexts: in the decoration of churches (in Naqlun, for example), on funerary stelae and elements of formal dress, not to mention the numerous administrative documents.[18] Despite its fluctuating orthography and its almost complete absence from the strictly literary domain, it seems to have acquired an almost official status as the language used among the Christians of the Fayoum in the public domain.

The text on the London lintel is a new witness to this idiom. Its principal particularities, familiar from the entire group of Fayoum lintels, are: /a/ for /o/ in closed syllables (for example in ⲛⲧⲁϥ, l. 4, and -ⲕⲁⲧϥ, l. 5), /i/ for final /e/ (for example ⲉⲓⲥⲓ, l. 6), the shift between ⲃ and ϥ (ϥⲟⲓⲟⲓ, l. 1) and between ⲟ and ⲱ (-ⲟⲡⲓ, l. 7), as well as spellings like ⲙⲏⲛ (l. 3, for ⲙⲛ or ⲙⲉⲛ), ⲛⲏϥ- (l. 6, for ⲛⲉϥ-) and ⲥⲓⲙⲓ (for Sahidic ⲥϩⲓⲙⲉ, in l. 3).[19] Another characteristic of the idiom, the extreme instability of the spelling, even within a single, rather brief text, is illustrated here by the variants ⲧⲉϥ- (l. 3) and ⲛⲏϥ- (l. 6).[20] On the other hand, the lambdacism that characterizes Fayoumic proper is absent; within this small corpus, it is only attested in the lintel Roquet no. 1, which writes ϣⲏⲗⲓ, for Sahidic ϣⲏⲣⲉ.

If its Fayoum provenance seems certain in a general sense, it may be that the origin of the London lintel can be determined more accurately. Among the five lintels that constitute the corpus, one has a provenance that is both precise and certain.[21] At the beginning of the twentieth century, the lintel Roquet no. 3 was discovered in Abu Hamid.[22] This name, which is little used nowadays, refers to one of several hamlets that make up the modern village of Qasr al-Basil, located not far east of Tebtynis, on the Bahr al-Gharaq in the southeastern part of the Fayoum.[23] Another indication is provided by a small series of wooden funerary stelae from the tenth century, which are similar to the group of lintels in several respects, and must originate from the area of Tebtynis-Tutun, in the same region.[24] These indications, however meager they may seem, argue in favor of a provenance in the basin of the Bahr al-Gharaq, viz. the southern province of the Fayoum, which counted several prosperous Christian centers at the time, including the town of Tebtynis-Tutun and the monastery of Naqlun.

The decoration

Our lintel is composed of five horizontal panels that alternate figuration and engraved or carved inscriptions. The central panel is framed by two panels that bear the inscriptions, followed by two panels with figurative motifs. Each element is placed within a medallion. A floral background composed of rosettes and small leaves is visible continuously on either side of the lintel. In the central panel, a Greek cross is flanked by the Alpha and the Omega. The two panels at both ends of the lintel each have two medallions containing griffins, a fish head and a rosette within a star. The two griffins are symmetrically opposed at the extremities of the lintel. The ornamental background in the medallions containing the Alpha and the Omega consists of bunches of grapes, trilobed leaves and small rosettes. A palmette decorates the Omega; it is integrated into the composition, so that the center of the Omega forms the stem of the palmette.

As was emphasized in the study of the text, this lintel is part of a larger group of commemorative lintels from the Fayoum, dated to the tenth century and initially published by Gérard Roquet.[25] In a general way, the decoration of the lintels has an apotropaic quality, given the fact that they are placed at the entrance of a building (church or house).[26] This characteristic of the decoration is strikingly complementary to the text inscribed on our lintel, a prayer invoking protection for a donor and his family. When the cross, as a sign of victory, is placed in a medallion, on a lintel, a tympanum or a tombstone, it is given the role of a barrier against the demonic.[27] This very role was emphasized in a study on the lintels from Syria that decorate the entrance of houses.[28] The Alpha and the Omega complement the cross as protective motifs.[29] In the same category of motifs we find equestrian saints, which are not represented on our lintel, but depicted on other friezes from the sixth century onwards. For example, the lintel British Museum EA 1276, which originates from Middle Egypt and was republished by Sofia Schaten, presents two equestrian saints facing each other.[30]

The other motifs on our lintel are two griffins, a fish and a rosette. Each of them is placed in a medallion. The association of a griffin with a fish is rare. However, many (complete) fish are represented on friezes with a Nilotic theme, the obvious Christian symbolism of which, mixed with Egyptian tradition, is well known. On the lintel in the British Museum, the scales and the fin of the fish are meticulously carved. It is placed in an eight-pointed star formed by two interlocking squares, and only its head has been depicted, facing left towards the griffin. The direction in which the fish is turned, and the fact that only its head is represented, are extraordinary details. However, a lack of parallels does not allow us at present to interpret this way of depicting the fish. The motif that symmetrically opposes it, a rosette, rather belongs to a purely decorative repertoire, but the eight-pointed star that encloses it seems to be important insofar as it is symmetrical to the star in which the fish is placed. This type of star is a very common motif on Islamic friezes, such as the wooden friezes from the Western Palace in Cairo or fragments from Fustat, kept in the Islamic Museum.[31] One fragment presents precisely a griffin in an eight-pointed star.[32] It seems that the star was similarly used for its apotropaic quality, both in Coptic and Islamic art.[33] A Coptic amulet from the Fayoum indeed comprises "on the left side, three pentagrams; on the right side, two eight-pointed stars (or wheels) ending in circles", a description that recalls the motif depicted on our lintel.[34]

By contrast, the griffin is rather uncommon in Coptic iconography. It is found on two other friezes of unknown origin in New York, which seem to be contemporary to our lintel.[35] In addition, two griffins are depicted among the Islamic-inspired decoration on the sanctuary screen of the Church of Sitt Barbara (Old Cairo), dated to the eleventh century.[36] Like the sphinx or the harpy, the griffin belongs to the same tradition of mythical creatures with a protective character or an astrological significance that have become part of the bestiary of Islamic art.[37] The known prototypes of these motifs are to be found in pre-Islamic Iranian art, which inspired the Muslims, who presumably took over their significance as well.[38] Within this range of meanings, the griffin – like the sphinx – can be interpreted as the guardian of the sanctuary.[39] It should also be remembered that the protective griffin is already known in Greco-Roman antiquity. As mythological creatures dedicated to Apollo, griffins protected the treasure of the god in the lands of the Hyperboreans against the Arimaspi. This protective motif is adopted in funerary art, where the griffin appears as the guardian of the tomb.[40] Thus, on a child's sarcophagus at the Lateran Museum, two griffins are depicted on the sides of the coffin.[41] Be that as it may, Coptic funerary stelae from late antiquity do not adopt this motif; they rather display eagles, peacocks or lions.[42]

The griffin on our lintel is characterized by its pointed, erect wing and by a head with dressed ears. The griffin in Islamic art is not necessarily composed of the body of a lion and the head of an eagle, but shows several variations, of which the griffin on our lintel presents an example, for its lion-like body is combined here with an animal head, thus resembling a winged quadruped.[43] It recalls the griffins that decorate the façade of the Mschatta palace in Jordan (ca. 743), which drink in opposing pairs from vases from which a vine grows. Two other friezes of Coptic origin, mentioned in the preceding paragraph, are decorated with griffins that are similar to those on our lintel: the frieze 28.12, kept in the Metropolitan Museum of New York, and the frieze 86.80 in the Newark Museum.[44] As for these two friezes, which were published by Elisabeth Enß, the first presents three medallions containing a central cross flanked by two griffins, and the second three medallions containing a griffin and two quadrupeds without wings (probably a fourth medallion is missing). All of these motifs are embellished with ornamental elements, including palm leaves, but no inscription seems to have been carved ever. The provenance of these two friezes is unknown, but Enß observes that the similarity of their style indicates that they must have come from the same workshop. The author compares them stylistically to our lintel at the British Museum, but she does not explain the significance of the motif of the griffin (which she calls a "winged quadruped"), or the function of the frieze, which she considers to be architectural. The hypothesis that the friezes in New York and Newark originate from the Fayoum seems plausible as well.

In addition, griffins are depicted on a small number of textiles kept in the Louvre, which are contemporary to our lintel and originate from the Fayoum, Antinoe,

Bawit, al-Bersha and Fustat.[45] On these textile fragments, the griffin, compared by Pierre du Bourguet to a "winged horse" from the Sassanid tradition, is placed in a medallion, sometimes surrounded by putti and, in one case, by dancers. The images of griffins are more or less stylized, but the model of the griffin that is so popular in Islamic art, on ceramics, textiles and wooden friezes, is well recognizable on these textiles. Another piece of fabric, discovered at the Monastery of the Archangel Gabriel (Naqlun) in the Fayoum during the excavation of Cemetery A in 2002, features stylized griffins with slender, curved wings in medallions that alternate birds and griffins.[46] Most of the preserved textile fragments belonged to funerary tunics or shrouds, the decoration of which was often enhanced by inscriptions in Arabic.[47] In the Fayoum, a region known for its weaving workshops, one workshop in particular was subject to caliphal control: the private *tirâz* located at Tutun (or Touton, ancient Tebtynis), which is mentioned in inscriptions that enrich the decoration of Fayoum shawls from the ninth-tenth centuries.[48] The presence of Islamic models is thus well documented in the oasis, from the first two centuries after the conquest onwards. It is therefore possible that the griffin depicted on our lintel from the Fayoum was inspired by the decoration of textiles produced in local workshops, textiles with which the sculptor and the commissioner of the lintel must have been familiar. And as was mentioned earlier in the study of the text, our lintel could well originate from a village in the area of Tebtynis. The presence of the griffin on several lintels that probably originate from the Fayoum therefore bears witness to the wide circulation of iconographic models, which were applied to different kinds of support, both wood and textile. As was previously observed, the griffin could acquire a protective character. For this reason, choosing a griffin to enrich the iconography of a lintel is not insignificant. It highlights the protective quality of the motif, thus continuing an ancient tradition.

Finally, the Omega on our lintel is adorned with a palmette, the style of which recalls the stuccos of Samarra and, geographically closer, those of the Mosque of Ibn Tulun in Cairo. The palmette on the lintel calls to mind a similar pattern depicted on the upper panels of the sanctuary doors of the Church of al-Adra in Dayr al-Suryan (Wadi al-Natrun).[49] Each of these two panels contains a saintly figure framed by two palmettes. The style of the palmettes on the doors in Dayr al-Suryan is also reminiscent of the stuccos of Samarra and Ibn Tulun, and the interior of the sanctuary of the church is precisely decorated with similar stuccos.[50] In the case of the doors of Dayr al-Suryan, M. Frazer interprets this motif as the stylized version of a tree that recalls Paradise.[51] With due caution, we could perhaps suggest a similar interpretation for the palmette depicted on the lintel in the British Museum.

In conclusion, whereas the decoration and the text of our lintel contain all the characteristic elements that give it an apotropaic quality, its decoration equally bears witness to the gradual transformation of the Coptic iconographic repertoire into a Copto-Arabic repertoire, mainly by the introduction of Islamic-style griffins. The protective quality of the griffin itself can be situated in a long tradition dating back to antiquity.

Notes

1 See Roquet 1978; Schaten 1998: 306, n. 7.

2 The authors would like to thank the curators of the Department of Egyptian Antiquities of the British Museum, in particular M. Marée and J. Taylor, for their invaluable help. Renate Dekker kindly translated our French text.

3 These other Fayoum lintels will be referred to here by the numbers of Roquet 1978: 339–42; the text of Roquet nos. 1–2 is reprinted in *SB Kopt.* I 400–01; no. 4 is reedited in Van der Vliet 2002–2003: 143–46. In addition to the titles by Roquet and Schaten, quoted previously, the following publications on inscribed lintels from Christian Egypt should be mentioned: Hammerstaedt 1999: 187–99; Krause 1988: 111–20; Rutschowscaya 1986: 150–53; Wietheger 1992: 100–02.

4 Numerous examples in the corpus of van Lantschoot 1929.

5 In the Fayoum, for instance, in a church at Tebtynis (tenth century; now destroyed, see Walters 1989: 205) or in the church of the monastery of Naqlun (eleventh century; see J. van der Vliet, "The inscriptions", in Godlewski forthcoming).

6 Three other lintels in the group feature an acclamation with βοηθέω as well (Roquet nos. 1–3). On the history of this acclamation, see the classical study by Peterson 1926; for its use in medieval Coptic epigraphy, Van der Vliet 2006: 36–7.

7 Van der Vliet 2002–2003: 145.

8 Lefebvre 1910: 60, ll. 4–6. The publication, which provides neither a photograph nor a facsimile, forbids a reconstruction of the text. In any case, the lost phrase was much longer than the corresponding group (ⲁϥϣⲁⲡ) on the London lintel.

9 Examples include van Lantschoot 1929: no. LXXX, vo, ll. 72–6; no. CII, vo, ll. 3–6 and ll. 16–18.

10 See Krause 1991: 1292; it was not a tomb either, see Schaten 1998: 306–07.

11 See Krause 1988: 113; cf. Crum 1939: 580a-b.

12 This is what the phrase ⲅⲉ ⲡⲉⲧⲏⲃ (for more correctly ⲅⲙ ⲡⲉⲧⲉⲛⲧⲏϥ, in proper Sahidic ⲅⲙ ⲡⲉⲧⲛⲧⲁϥ, as in van Lantschoot 1929: no. CII, vo, ll. 16–18) says; there is no question of a toponym ⲡⲉⲧⲏⲃ (*contra* Roquet 1978: 340, who was followed by Timm 1984–1992: IV, 1907, and *SB Kopt.* I 400).

13 Lefebvre 1910: 60, l. 1: ⲕⲩⲣⲱ; for variations in the spelling, see Förster 2002: 453–54.

14 Cf. Van der Vliet 2002–2003: 143–46.

15 Only in the lintel Roquet no. 2 no family is mentioned. Similar mentions of kin occur in other dedicatory inscriptions, for instance in *SB Kopt.* I 302 (commemorating the foundation of a workshop belonging to the Church of the Holy Virgin at Philae; eighth century).

16 It corresponds most often to the Sf of Crum 1939. For a discussion of the Coptic idioms of the Fayoum, see R. Kasser in Diebner and Kasser 1989: 70–108.

17 Note that for literary texts (biblical, patristic and hagiographic), transmitted in codices, proper Fayoumic was replaced by Sahidic; see Boud'hors 2005: 22–4. For the stages of Arabization of the Fayoum, see Gaubert and Mouton 2004: 505–17; Van der Vliet 2005b: 191–94.

18 See Van der Vliet 2005a: 77–89, for a discussion of these different kinds of monuments.

19 The lintels Roquet nos. 1 and 4 likewise read ⲥⲓⲙⲓ, whereas only the lintel Roquet no. 3 has the Sahidic orthography ⲥⲅⲓⲙⲉ. Note as well the use of the compendium ⲡⲟⲥ̅ (l. 1), see Roquet 1978: 341.

20 The same uncertainty in the inscriptions of the Church of Naqlun (eleventh century), see Van der Vliet, "The inscriptions", in Godlewski, *Naqlun* (forthcoming), no. W.4.

21 As was observed in n. 12, the putative toponym ⲡⲉⲧⲏⲃ, read by Roquet (1978, 30) in his lintel no. 1, is a ghost.

22 Lefebvre 1910: 59, adds that it was transported to the Egyptian Museum in Cairo in December 1908. According to Roquet, this lintel would be of stone, but Lefebvre's edition does not state the material.

A LINTEL FROM THE FAYOUM IN THE BRITISH MUSEUM

23 See the discussion in Van der Vliet 2002–2003: 146; 2005a: 82–3.
24 Collected by Boud'hors and Calament 2004 (see their appendix, 466–68), who emphasize the affinity between the two groups of monuments.
25 Roquet 1978.
26 Roquet 1978: 344; Van der Vliet 2002–2003: 143.
27 Thierry 1999: 235; Grabar 1970: 25.
28 Engemann 1975.
29 Horsley 1986: 35; Wietheger 1992: 201–02.
30 Schaten 1998: 307–10; cf. Hall 1905: 132, pl. 91. They are identified as Saints Pakene and Victor by inscriptions.
31 Pauty 1931: pl. XXXIV, no. 4790 and pl. XXXVI, no. 6342. The Western Palace was built at the end of the tenth century by the caliph Al-'Aziz for his daughter Sitt al-Mulk. Al-Mustansir completed it in 1058. Afterwards, the Mamluk sultan Qalawum bought it in 1283 and built his hospital on top of the ruined palace. See Fu'âd Sayyid 1998: 300.
32 *Trésors fatimides*: 91, no. 7.
33 Winkler 1930: 119–27.
34 Crum 1922: 543, cited by Winkler 1930: 125, n. 5.
35 Enß 2005: pl. 84.
36 Pauty 1930: pl. IX.
37 Baer 1965: 54.
38 Baer 1965: 82.
39 They can also be considered as the guardians of the "tree of life". Baer 1965: 56–7; Khazai 1978: 1–34.
40 Cumont 1942: 170.
41 Benndorf and Schöne 1867: no. 438.
42 Crum 1902: pls. XLIV-XLI and XLVII-XLVIII.
43 Gelfer-Jørgensen 1986: 122–23 and pl. 37b.
44 Enß 2005: 66.
45 Du Bourguet 1964: nos. H16, H19, H20, H24, H25, H45, H46, H47, H65. The author dates all these tissues to the tenth century.
46 Godlewski 2003: 167, fig. 3a. The alternation between griffins and other motifs is a common pattern: many Islamic ceramics and fabrics display sphinxes chased by griffins, and vice versa; see Baer 1965: 25.
47 Czaja-Szewczak 2003: 184, fig. 9.
48 M. Durand, in Bosson and Aufrère 1999: 259–61; Durand and Rettig 2002: 167–70.
49 Evelyn White 1933: pl. LXIV.
50 Immerzeel 2004a; Immerzeel 2004b.
51 Frazer 1973: 151.

Bibliography

Baer, E. 1965, *Sphinxes and harpies in medieval Islamic art*, Jerusalem: Israel Oriental Society.

Benndorf, O., and R. Schöne. 1867, *Die antiken Bildwerke des Lateranensischen Museums*, Leipzig: Breitkopf & Härtel.

Bosson, N., and S.H. Aufrère. (eds.) 1999, *Egyptes . . . L'égyptien et le copte: catalogue de l'exposition au musée archéologique Henri Prades à Lattes*, Lattes: Musée archéologique Henri Prades.

Boud'hors, A. 2005, "Manuscripts and literature in Fayoumic Coptic", in: Gawdat Gabra (ed.), *Christianity and monasticism in the Fayoum Oasis*, Cairo: American University in Cairo Press, 21–31.

JACQUES VAN DER VLIET AND ADELINE JEUDY

Boud'hors, A., and F. Calament. 2004, "Un ensemble de stèles fayoumiques inédites: à propos de la stèle funéraire de Pantoleos de Toutôn", in: M. Immerzeel and J. van der Vliet (eds.), *Coptic studies on the threshold of a new millennium: proceedings of the Seventh International Congress of Coptic Studies, Leiden, 27 August-2 September 2000*, Louvain/Paris/Dudley, MA: Peeters 2004, I, 447–75.

Crum, W.E. 1902, *Coptic monuments*, Cairo: IFAO; reprint Osnabrück: Otto Zeller Verlag 1975.

Crum, W.E. 1922, "La magie copte: nouveaux textes", in: *Recueil d'études égyptologiques dédiées à la mémoire de Jean-François Champollion*, Paris: E. Champion, 537–44.

Crum, W.E. 1939, *A Coptic dictionary*, Oxford: Oxford University Press (several reprints).

Cumont, F. 1942, *Recherches sur le symbolisme funéraire des Romains*, Paris: P. Geuthner; reprinted Paris: P. Geuthner 1966.

Czaja-Szewczak, B. 2003, "Burial tunics from Naqlun", *PAM* 14: 177–84.

Davies, W.V. 1987, *Egyptian hieroglyphs*, London: British Museum Press (several reprints).

Dawson, W.R., and E.P. Uphill. 1995, *Who was who in Egyptology*, third edition revised by M. Bierbrier, London: Egypt Exploration Society.

Diebner, B.J., and R. Kasser. 1989, *Hamburger Papyrus bil. 1: Die alttestamentlichen Texte des Papyrus bilinguis 1 der Staats- und Universitätsbibliothek Hamburg*, Geneva: P. Cramer.

Du Bourguet, P. 1964, *Etoffes coptes du Louvre*, Paris: Éditions des Musées nationaux.

Durand, M., and S. Rettig. 2002, "Un atelier sous contrôle califal identifié dans le Fayyoum: le tirâz privé de Tutûn", in: M. Durand and F. Saragoza (eds.), *Egypte, la trame de l'histoire. Textiles pharaoniques, coptes et islamiques*, Paris: Somogy, 167–70.

Engemann, J. 1975, "Zur Verbreitung magischer Übelabwehr in der nichtchristlichen und christlichen Spätantike", *JbAC* 18: 22–48.

Enß, E. 2005, *Holzschnitzereien der spätantiken bis frühislamischen Zeit aus Ägypten: Funktion und Dekor*, Wiesbaden: Reichert Verlag.

Evelyn White, H.G. 1933, *The monasteries of the Wâdi 'n Natrûn* III: *the architecture and archaeology*, New York: Metropolitan Museum of Art.

Förster, H. 2002, *Wörterbuch der griechischen Wörter in den koptischen dokumentarischen Texten*, Berlin/New York: de Gruyter.

Frazer, M. 1973, "Church doors and the gates of paradise: Byzantine bronze doors in Italy", *DOP* 27: 145–62.

Fu'âd Sayyid, A. 1998, *La capitale de l'Egypte jusqu'à l'époque fatimide: al-Qâhira et al-Fustât; essai de reconstitution topographique*, Beirut/Stuttgart: Franz Steiner Verlag.

Gaubert, C., and J-M. Mouton. 2004, "Présentation des archives d'une famille copte du Fayoum à l'époque fatimide", in: M. Immerzeel and J. van der Vliet (eds.), *Coptic studies on the threshold of a new millennium: proceedings of the Seventh International Congress of Coptic Studies, Leiden, 27 August-2 September 2000*, Louvain/Paris/Dudley, MA: Peeters 2004, I, 505–17.

Gelfer-Jørgensen, M. 1986, *Medieval Islamic symbolism and the paintings in the Cefalù Cathedral*, Leiden: Brill.

Godlewski, W. 2003, "Naqlun excavations, 2002", *PAM* 14: 163–71.

Godlewski, W. forthcoming, *Naqlun: the Church of Saint Gabriel*, Warsaw.

Grabar, A. 1970, "Deux portails sculptés paléochrétiens d'Egypte et d'Asie Mineure", *CArch* 20: 15–28.

Hall, H.R. 1905, *Coptic and Greek texts of the Christian period from ostraka, stelae, etc. in the British Museum*, London: British Museum.

A LINTEL FROM THE FAYOUM IN THE BRITISH MUSEUM

Hammerstaedt, J. 1999, *Griechische Anaphorenfragmente aus Ägypten und Nubien*, Opladen: Westdeutscher Verlag.

Horsley, G.H.R. 1981, *New documents illustrating early Christianity* I, North Ride, New South Wales: Macquarie University, 67.

Immerzeel, M. 2004a, "The Stuccoes of Dayr al-Surian: a *waqf* of the Taqritans in Fustat?" in: M. Immerzeel and J. van der Vliet (eds.), *Coptic studies on the threshold of a new millennium: proceedings of the Seventh International Congress of Coptic Studies, Leiden, 27 August-2 September 2000*, Louvain/Paris/Dudley, MA: Peeters 2004, II, 1303–20.

Immerzeel, M. 2004b, "A play of light and shadow: the stuccoes of Dayr al-Suryan and their historical context", *Coptica* 3: 104–29.

Khazai, K. 1978, "L'évolution et la signification du griffon dans l'iconographie iranienne", *Iranica Antiqua* 13: 1–34.

Krause, M. 1988, "Die Inschriften auf den Türsturzbalken des Apa-Apollon-Klosters von Bawit", in: *Mélanges Antoine Guillaumont: contributions à l'étude des christianismes orientaux*, Genève: P. Cramer, 111–20.

Krause, M. 1991, art. "Inscriptions", in: *CoptEnc* 4, 1290–99.

Lefebvre, G. 1910, "Égypte chrétienne II", *ASAE* 10: 50–65.

Pauty, E. 1930, *Bois sculptés d'églises coptes (époque fatimide)*, Le Caire: IFAO.

Pauty, E. 1931, *Catalogue général des bois du Musée Arabe du Caire: Les bois sculptés jusqu'à l'époque ayyoubide*, Le Caire: IFAO.

Peterson, E. 1926, *ΕΙΣ ΘΕΟΣ: Epigraphische, formgeschichtliche und religionsgeschichtliche Untersuchungen*, Göttingen: Vandenhoek & Ruprecht.

Roquet, G. 1978, "Linteaux commémoratifs en dialecte fayoumique", *BIFAO* 78: 339–45.

Rutschowscaya, M-H. 1986, *Catalogue des bois de l'Égypte copte*, Paris: Musée du Louvre.

Schaten, S. 1998, "Inschriften auf Türstürzen", in: M. Krause and S. Schaten (eds.), *ΘΕΜΕΛΙΑ: Spätantike und koptologische Studien Peter Grossmann zum 65. Geburtstag*, Wiesbaden: Reichert Verlag, 305–15.

Schiemenz, G.P. 1986, "Kreuz, Orans und heiliger Reiter im Kampf gegen das Böse", in: G. Koch (ed.), *Studien zur frühchristlichen Kunst* II, Wiesbaden: Harrassowitz, 305–15.

Thierry, N. 1999, "Aux limites du sacré et du magique: un programme d'entrée d'une église en Cappadoce", *Res Orientales* 12: 233–47.

Timm, S. 1984–1992, *Das christlich-koptische Ägypten in arabischer Zeit*, 6 vols., Wiesbaden: Reichert Verlag.

Trésors fatimides du Caire: catalogue de l'exposition présentée à l'Institut du Monde Arabe du 28 avril au 30 août 1998, Paris: Institut du monde arabe.

Van der Vliet, J. 2002–2003, "Monumenta fayumica", *Enchoria* 28: 137–46 [Study 7].

Van der Vliet, J. 2005a, "Reconstructing the landscape: epigraphic sources for the Christian Fayoum", in: Gawdat Gabra (éd.), *Christianity and monasticism in the Fayoum Oasis*, Cairo: American University in Cairo Press, 77–89.

Van der Vliet, J. 2005b, "Naqlun: preliminary observations on the Coptic texts found during seasons 2003 and 2004", *PAM* 16: 191–94.

Van der Vliet, J. 2006, " 'In a robe of gold': status, magic and politics on inscribed Christian textiles from Egypt", in: C. Fluck and G. Helmeke (eds.), *Textile messages: inscribed fabrics from Roman to Abbasid Egypt*, Leiden/Boston: Brill, 23–67 [Study 2].

Van Lantschoot, A. 1929, *Recueil des colophons des manuscrits chrétiens d'Égypte* I: *les colophons coptes des manuscrits sahidiques*, Louvain: J.B. Istas; reprinted Milan: Cisalpino-La Goliardica 1973.

171

Walters, C.C. 1989, "Christian paintings from Tebtunis", *JEA* 75: 191–208.

Wietheger, C. 1992, *Das Jeremias-Kloster zu Saqqara unter besonderer Berücksichtigung der Inschriften*, Altenberge: Oros Verlag.

Winkler, H.A. 1930, *Siegel und Charaktere in der muhammedanischen Zauberei*, Berlin/Leipzig: de Gruyter.

11

A NAQLUN MONK BROUGHT HOME
On the provenance of Louvre inv. E 26798–9

Jacques van der Vliet

Introduction

Before the start, in 1986, of the Polish excavations in Dayr al-Malak, the Monastery of the Archangel Gabriel at Jabal an-Naqlun (southeastern Fayoum), very few ancient documents were known which could be said with certainty to originate from this important site.[1] One exception is the funerary stela of a Christodoros, which has an inscription in Greek and can be dated tentatively to the seventh century. It was discovered on the spot by Johann Georg, Duke of Saxony, in the twenties and recently re-edited by A. Łajtar.[2] This sculptured monument, which bears the representation of a cross within an *aedicula*, belongs to a characteristic group of late antique stelae commonly ascribed to the Fayoum, but only rarely traceable – as here – to a precise place of origin. Interestingly, a much damaged fragment of a similar stela was found during the 1995 excavation campaign in Naqlun.[3]

Slightly more doubtful is the case of a Bohairic letter on paper, presently kept in the British Library under number Or 4720 (31).[4] It addresses a deacon Makari (Macarius), apparently living in Naqlun (ⲛⲉⲕⲗⲱⲛⲓ; ll. 4–5). Plausibly, but by no means necessarily, the letter was delivered to its addressee in Naqlun and, subsequently, remained there until its discovery in recent times. If indeed this reconstruction of events is correct, it would open an interesting perspective. From the seventies of the nineteenth century onwards, several large collections of Greek, Coptic and Arabic manuscripts, purportedly from the Fayoum, appeared on the market to find their way into various European museums.[5] The Bohairic letter in the British Library belongs to one such collection, acquired from the famous Austrian dealer Th. Graf.[6] The number Or 4720, under which it is catalogued, covers a group of Coptic letters, classified by W.E. Crum as "Middle Egyptian" in the widest possible acceptance, all originating from Graf. The contents of most of them testify to a monastic background, whereas some show points of resemblance with documents from Dayr al-Hammam, not far north of Naqlun.[7] Although it should be emphasized that the Graf collection in the British Library is a mixed lot,[8] one feels tempted to suppose that it holds more than one single manuscript from Naqlun.

To this meager file of "early" documents from Naqlun a quite exceptional item can now be added. In 1960, the Musée du Louvre acquired a complete liturgical tunic, a so-called *sticharion*, and a hood belonging to it, which bear inventory numbers E 26798 and 26799.[9] Both had been part of the collection Puy-Haubert. In a recent and very thorough analysis, M. Durand dates these pieces to the tenth or eleventh century. On stylistic grounds, he correctly assumes them to originate from the Fayoum province. However, both the name of their owner and their precise provenience went unrecognized to date.

The Louvre tunic from Naqlun

The ornamentation of the Louvre tunic is quite unlike that of the well-known "Coptic" garments of late antiquity, but rather follows Muslim fashions.[10] This is most clearly apparent in the lines of text, of the so-called *tirâz* type, which decorate the ends of the sleeves.[11] Each sleeve carries one line of Coptic, which was woven into the fabric, so not embroidered. Like their Muslim models, these Christian *tirâz* inscriptions, next to enhancing the status of their bearers, had an apotropaic and amuletic function, warding off evil and bringing good luck. In this respect they do continue an antique tradition.[12] The same apotropaic bias is conspicuously present in the sleeve texts of the Louvre tunic too, both of which will be briefly reconsidered here.

The text on the lefthand sleeve has been transcribed by its editors as follows: ⲡⲣⲁⲛ ⲉⲡⲛⲟⲩⲧⲉ ⲛϣⲁⲣⲡ ⲛϩⲱⲃ ⲛⲓⲙ ⲓⲥⲃⲩ.[13] This they translate as: "The name of God is first in everything . . .", which Durand takes to be "une formule de profession de foi".[14] However, comparison with similar formulae, both on textiles[15] and at the head of Coptic letters,[16] would suggest that it must be reconstructed as: <ϩⲙ> ⲡⲣⲁⲛ ⲉⲡⲛⲟⲩⲧⲉ ⲛϣⲁⲣⲡ ⲛϩⲱⲃ ⲛⲓⲙ. . . . That is: "In the name of God, first of all . . .", presumably followed by some kind of supplication, the beginning of which may be hidden in the obscure group ⲓⲥⲃⲩ (if thus read correctly, perhaps an abbreviation of ⲓ̅ⲥ̅ ⲃⲱⲏⲑⲉⲓ, "Jesus, help . . ."?). The apotropaic value of such an invocation of the name of God is evident beyond doubt.[17] As a "dialectical" feature, ϣⲁⲣⲡ (standard Sahidic: ϣⲟⲣⲡ) may be noted. The orthography is suggestive of the particular brand of Sahidic with strong Fayoumic coloring, which was predominant in the Naqlun region around the turn of the millennium.[18]

The inscription on the righthand sleeve proves to be much more informative. Since the readings of the earlier editors can be partly verified and corrected after the published photograph,[19] the text can be established with considerable certainty as follows: ⲡⲛⲟⲩⲧⲉ ⲃⲱⲓⲉⲓ ⲉⲡ<ⲁ>ⲡⲁ ⲕⲱⲗⲑⲓ ⲡⲁ ⲧⲙⲟⲩⲛⲏ ⲛⲉⲕⲗⲱⲛⲓ ϩ[ⲁⲙⲏⲛ], "God, help (βοηθέω) Father (παπᾶς) Kolthi of the Monastery (μονή) Nekloni (Naqlun). Amen".[20] The age-old βοήθει-acclamation is a familiar feature of apotropaic or protective formulae in all kinds of contexts, including textile decoration.[21]

The righthand sleeve text of the Louvre tunic, therefore, identifies its owner as a priest (ⲡⲁⲡⲁ, παπᾶς),[22] named Kolthi, a hypocoristic form of Kollouthos.[23] Moreover, the latter is explicitly designated as an inmate of the Monastery of Naqlun, which is called here for the first time, as far as I am aware, a μονή, literally "abode", instead of the habitual ⲧⲟⲟⲩ, literally "mountain" (Greek: ὄρος).[24] No conclusions can be drawn from this onomastic variant, since both terms are common and essentially synonymous designations of a monastery.[25] The Coptic spelling ⲙⲟⲩⲛⲏ is well attested, e.g. at nearby Hamuli.[26]

Although the tunic's earlier acquisition history is unknown to the present author, we can safely assume that it was discovered during illegal diggings near Dayr al-Malak. One particular circumstance permits us to be slightly more precise. There can be no doubt that this originally liturgical garment ended up as the owner's shroud.[27] From this we can surmise that it originated from the ancient cemetery of the Naqlun community, which appears to be situated to the west of the present-day monastery. Already the *Legend of Aur*, which pre-dates the thirteenth century, locates the tomb of the monastery's founder "dans le sol à l'ouest de l'église".[28] Even today, the sloping area to the west of the modern monastic buildings reveals a great number of shallow pits indicating the presence of tombs. Over 180 have been identified, but to date only one of these has been excavated by the Polish expedition.[29] Predictably, it proved to be plundered. Nevertheless, it did produce an interesting fragment of decorated textile,[30] thus confirming the cemetery's potential for similar discoveries.

Finally, the Louvre tunic from Naqlun is not an isolated piece. Stylistically it belongs, as Durand's recent analysis makes clear, to a far larger family of textiles with *tirâz*-type inscriptions from the Fayoum. Some of these, characterized by a dark blue dye, can now be traced to a workshop in Tutun, a few kilometers from the Naqlun monastery.[31] Kolthi's tunic, from undyed material, does not really fit in this group, though it resembles several other Fayoum garments made for priests (ⲡⲁⲡⲁ's).[32] Further technical, stylistic and epigraphical research may be able to demonstrate, for some pieces from this latter group, a common origin in some other nearby atelier. Whether any of these, too, were discovered in the cemetery of Naqlun must remain an open question.

Conclusion

The importance of the inscriptions on the Louvre tunic-*cum*-hood, inv. nos. E 26798–9, appears to be twofold. First of all, they identify its owner, a priest Kolthi, as an inmate of the Monastery of Naqlun. Thereby, this unique ecclesiastical vestment becomes one of the relatively rare pieces of Coptic textile, the provenience of which can be established with certainty and precision. Secondly, they add one more to the small number of objects from Naqlun which came to light before independent of modern scholarly excavations. For the heuristics of Christian sources from the Fayoum, this is a capital observation. It now cannot be doubted anymore that the site of Naqlun was among the tributaries of the stream of Coptic antiquities from the Fayoum which reached Europe in the late nineteenth and early twentieth centuries. For the twenty-first century, which the present issue of this admirable *Bulletin* (i.e. *BSAC*) inaugurates, we may expect more discoveries to throw light upon the world of *papa* Kolthi, not only in the field but in museum storerooms as well.

Notes

1 For reviews of previously known sources, see S. Timm, *Das christlich-koptische Ägypten in arabischer Zeit*, Wiesbaden: Reichert Verlag 1984, II, 762–67; T. Derda, *Deir el-Naqlun: the Greek papyri* (*P.Naqlun* I), Warsaw: University of Warsaw/Raphael Taubenschlag Foundation 1995, 20–3. Additionally, Naqlun occurs in an eleventh-century Fayyumic account book (British Library, formerly coll. G. Michaelides), cf. S. Clackson, "The Michaelides Coptic manuscript collection in the Cambridge University Library and British Library", in: D.W. Johnson and T. Orlandi (eds.), *Acts of the Fifth International Congress of Coptic Studies: Washington, 12–15 August 1992*, Rome: Centro Italiano Microfiches 1993, II/1, 123–38, there 130–31; the text is to be edited by Dr. Clackson.

2 "Two Greek inscriptions from Deir el-Naqlun", *Nubica* 3/1 (1989–1993), 265–74 at 269–74, with full bibliography. Its present whereabouts are unknown.

3 Inv. no. N. 24/95; unpublished.

4 Partly edited by W.E. Crum, *Catalogue of the Coptic manuscripts in the British Museum*, London: British Museum 1905, no. 590, 281–82; discussed i.a. by N. Abbott, *The monasteries of the Fayyûm*, Chicago: University of Chicago Press 1937, 47; Derda, *P.Naqlun* I, 22, Crum's summary description suggests that it may date from about the ninth or tenth century.

5 For an early report, see J. Karabacek, *Der Papyrusfund von el-Faijûm*, Vienna: C. Gerold's sohn 1882 (equals *Denkschriften kaiserl. Akademie der Wissenschaften Wien*, phil.-hist. Klasse, Wien: Rohrer, 33, 207–42).

6 For Theodor, Ritter von Graf, 1840–1903, and his activities, succinctly: W.R. Dawson et al., *Who was who in Egyptology*, London: Egypt Exploration Society 1995, 173; see furthermore: H. Hunger, *Aus der Vorgeschichte der Papyrussammlung der Österreichischen Nationalbibliothek*, Vienna: Hollinek 1962; H. Loebenstein, "Vom 'Papyrus Erzherzog Rainer' zur Papyrussammlung der Österreichischen Nationalbibliothek: 100 Jahre Sammeln, Bewahren, Edieren", in: *Festschrift Papyrus Erzherzog Rainer* (*P.RainerCent.*), Vienna: Hollinek 1983, 1, 3–39, there esp. 3–7 and 18–19.

7 See Crum, *Catalogue British Museum*, nos. 529–37, 540–41, 544–50, 578–655, 669; 261–303; for the Dayr al-Hammam manuscripts, see W.E. Crum, *Coptic manuscripts brought from the Fayyum*, London: David Nutt 1893.

8 Cf. Crum's cautious remark, *Catalogue British Museum*, XXIII, n. 1 ("declared by the native sellers to come from the Fayyûm and Aḥmîm; but . . . bought in Cairo").

9 First published by P. du Bourguet, *Musée national du Louvre: catalogue des étoffes coptes* I, Paris: Éditions des Musées nationaux 1964, nos. L 8–9 (AC 832–33), 652–53 (cf. 36); most recently by M. Durand, "Tunique liturgique ornée, et son capuchon", in: N. Bosson and S.H. Aufrère (eds.), catalogue exhibition *Égyptes. . . : L'Égyptien et le copte*, Lattes: Musée archéologique Henri Prades 1999, nos. 88–9, 265–68.

10 On Coptic weavers under Muslim rule: G. Cornu, "Les tisserands coptes après la conquête islamique", in: M. Martiniani-Reber and C. Ritschard (eds.), *Tissus coptes*, Genève: Réunion des Musées nationaux 1991, I, 28–31. [See now Study 2].

A NAQLUN MONK BROUGHT HOME

11 For the term *tirâz*, see Y.K. Stillmann, and P. Sanders "Ṭirâz", *Encyclopaedia of Islam* 10, Leiden: Brill 1999, 534–38; for *tirâz*-type Coptic inscriptions, see C. Fluck, "Koptische Textilien mit Inschriften in Berlin (I)", *BSAC* 35 (1996), 161–72; "Koptische Textilien mit Inschriften in Berlin (II)", *BSAC* 36 (1997), 59–70; cf. D. Bénazeth et al., "Le rapport de l'objet copte à l'écrit: des textiles aux vêtements liturgiques", cat. *Égyptes*, 148.

12 For which, see H. Maguire, "Garments pleasing to God: the significance of domestic textile designs in the early Byzantine period", *DOP* 44 (1990), 215–24, a study of fundamental importance for the interpretation of "Coptic" textile designs.

13 Thus Du Bourguet, *Catalogue*, 652; Durand, cat. *Égyptes*, 266, omits ⲛⲓⲙ, apparently by mistake. The published photographs do not permit verification of their readings.

14 Cat. *Égyptes*, 266.

15 See particularly Fluck, *BSAC* 35, 164–65, nos. 9949 and 9992.

16 Common in several variants, see e.g. W.E. Crum, *A Coptic dictionary*, Oxford: Oxford University Press 1939, 587b; P.E. Kahle, *Bala'izah: Coptic Texts from Deir el-Bala'izah in Upper Egypt*, London: Oxford University Press 1954, II, 639; A. Biedenkopf-Ziehner, *Untersuchungen zum koptischen Briefformular unter Berücksichtigung ägyptischer und griechischer Parallelen*, Würzburg: G. Zauzich 1983, 42 (type i).

17 Compare e.g. a recently edited Coptic ritual for protection: "in order that your (*scil.* God's) name may become my helper (βοηθός) and life, during the whole of the day as well as during the whole of the night", M. Meyer, *The magical book of Mary and the angels* (*P. Heid. Inv. Kopt. 685*), Heidelberg: Universitatsverlag C. Winter 1996, 12, ll. 8–11 (and *passim*); cf. biblical passages like Ps 78: 9, 123:8 (LXX).

18 Of course, ⲉ- for ⲛ-/ⲙ- is ubiquitous, cf. Kahle, *Bala'izah* I, 114, *sub* d.

19 Du Bourguet, *Catalogue*, 652, upper right.

20 Some notes: ⲡ<ⲁ>ⲡⲁ: written in full in similar inscriptions (cf. M. Durand, "Fragment de toile de lin barrée de bandes décoratives", in: cat. *Égyptes*, no. 77, 253–54); ⲕⲱⲗⲉ: eventually, ⲕⲱⲗⲉⲏ might be read (with ⲏ and following ⲡ ligatured); ⲡⲁ ⲧⲙⲟⲩⲛⲏ: ⲁ (thus with Durand) rather resembles ⲁ in ⲡ<ⲁ>ⲡⲁ than ⲉ (Du Bourguet). Durand reads the part of the text between ⲃⲱⲓⲉⲉⲓ and ⲧⲙⲟⲩⲛⲏ as: ⲉⲡⲡⲁⲕⲙⲁⲉⲏⲧⲁ, which he interprets as "ton disciple"; what follows, he finds "incompréhensible". Final "Amen" uncertain: Du Bourguet reads two unclear signs, Durand only one ϩ (adding: "le rabat de la manche mord sur la suite du texte"); on the photo, nothing beyond the first, ϩ-like sign is visible.

21 See E. Peterson, *ΕΙΣ ΘΕΟΣ: Epigraphische, formgeschichtliche und religionsgeschichtliche Untersuchungen*, Göttingen: Vandenhoek & Ruprecht 1926, 2–4 and *passim*; Th. Klauser, art. "Akklamation", *RAC* 1 (1950), col. 216–33, there 228, no. 8; Maguire, *DOP* 44, 220–21; Bénazeth et al., cat. *Égyptes*, 148 (examples from Egyptian textiles).

22 On the titel ⲡⲁⲡⲁ, see T. Derda and E. Wipszycka, "L'emploi des titres *abba*, *apa* et *papas* dans l'Egypte byzantine", *JJP* 24 (1994), 23–56, there 54–6. The title is frequent at Naqlun, both in documents and in inscriptions.

23 Cf. W.E. Crum, "Colluthus: the martyr and his name", *Byzantinische Zeitschrift* 30 (1929–1930), 323–27.

24 For Coptic designations of Naqlun, see Derda, *P.Naqlun* I, 24, 26 (with references).

25 For μονή, see e.g. F. Preisigke, *Wörterbuch der griechischen Papyrusurkunden* III, Berlin: privately published 1931, 262 s.v.; G.W.H. Lampe, *A patristic Greek lexicon*, Oxford: Clarendon Press 1961, 880 s.v. (*sub* 5).

26 Cf. A. van Lantschoot, *Recueil des colophons des manuscrits chrétiens d'Égypte* I: *les colophons coptes des manuscrits sahidiques*, Louvain: J.B. Istas 1929, fasc. 2, 7–8 (note *ad* no. I, l. 8); L. Depuydt, *Catalogue of Coptic manuscripts in the Pierpont Morgan Library* I, Louvain: Peeters 1993, CIV–CVII.

27 Cf. Durand, cat. *Égyptes*, 266.

28 E. Amélineau, *Contes et romans de l'Égypte chrétienne* I, Paris: E. Leroux 1888, 139–40; his translation of the Arabic text, a critical edition of which is now being prepared by Mrs. Clara ten Hacken [*The legend of Saint Aūr and the Monastery of Naqlūn: the Copto-Arabic texts*, Ph.D dissertation, Leiden University 2015; to be published by Peeters Publishers].

29 Cf. W. Godlewski et al., "Deir el Naqlun (Nekloni), 1988–1989: second preliminary report", *Nubica* 3/1 (1989–1993), 201–63, there 216; J. Leclant, G. Clerc, "Fouilles et travaux en Égypte et au Soudan, 1989–1990", *Orientalia* 60 (1991), 196–97; J. Dobrowolski, "The monastic complex of Naqlun: topography of the site", in: *Orbis Aethiopicus* II, Albstadt: Karl Schuler 1992, 309–25, there 318; Derda, *P.Naqlun* I, 39.

30 Illustrated: Godlewski, *Nubica* 3/1, 216, fig. 12; Leclant and Clerc, *Orientalia* 60, fig. 40.

31 Thanks to S. Rettig's improved reading of the Arabic text of Louvre inv. E. 25405, *apud* M. Durand, "Châle en toile de laine et bandes de tapisserie à inscriptions de tirâz bilingues, coptes et arabes", in: cat. *Égyptes*, no. 82, 259; cf. Durand's commentary, ibidem, 259–61.

32 Cf. Durand, ibidem, 254 (on no. 77) and 268.

12

I. VARSOVIE
Graeco-Coptica

Jacques van der Vliet

Introduction

Except in its early days, the epigraphy on stone of Christian Egypt was profoundly marked by the coexistence of Coptic and Greek until about the eleventh century, when the traditions inherited from late antiquity faded out.[1] Considering this bilingualism, which is apparent even in very short inscriptions, it is hardly surprising that modern epigraphists hesitate from time to time about the language of a text, especially when facing a damaged or obscure text.[2] Therefore, the authors of the beautiful *Catalogue des inscriptions grecques du Musée National de Varsovie*, Adam Łajtar and Alfred Twardecki,[3] cannot be blamed for including in their collection two pieces that are, in my opinion, at least partly written in Coptic. If I take the trouble to restore these two pieces, which may seem insignificant, to the Coptic domain, it is not to find fault with the editors, but to underline the necessity of treating the Christian epigraphy of the Nile Valley in a way that does justice to its distinctive peculiarities, one of which is precisely its bilingualism.[4]

I. Varsovie 118

This fragment of a funerary stela made of white limestone, of which Alfred Twardecki provides the *editio princeps* in the above-mentioned catalogue, belongs to the so-called litany type. In the following, I present a hypothetical reconstruction of the epitaph, in order to illustrate the nature of the text. In the lacunae in lines 1–2, common formulae have been supplemented by way of example; this reconstruction presupposes that we are dealing in l. 1 with the beginning of the text, which is very uncertain, however.

Middle Egypt (Saqqara?),   ca. sixth-eighth centuries?

```
   [+ пιω]τ [пϣнρε пεпн(εγм)а ετογа-]
   [ав пε]νιω[τ адам τενмаау z-]
   [ωн] νενιοτε [ετογααв апа]
4. [вικ]τωρ апа [ . . . . . . . апа]
   [хω]ϣρε [ - - - ]
   [ - - - ]
```

179

JACQUES VAN DER VLIET

3.]. viotε [ed. pr. || 5.]. ϙε [ed. pr.

+ The Father, the Son, the Holy Spirit. Our father Adam, our mother Zoe, our holy fathers, Apa Victor, Apa [N.N.], Apa Joore [- - -].

1. The stone shows the lower part of a vertical sign, unidentified in the *ed. pr.*; a т is suggested by the decorative dots at both sides of the shaft (cf. l. 3).

Similar "litanies", which first invoke the Holy Trinity, then a series of sometimes obscure saints, before mentioning the deceased (singular or plural), appear throughout Middle and Upper Egypt. Funerary stelae of the type represented by the stone from Warsaw are usually associated with the large monastic centers of Middle Egypt.[5] Since form and length of the litanies show much variation, even at a single site, the present example is too fragmentary to establish the composition and contents with more certainty. Whereas the name ϫⲱⲱⲣⲉ (l. 5) seems rather hypothetical, that of a saint called Victor (l. 4) is almost certain. In fact, saints named Victor are quite often attested in epitaphs of this type, from Saqqara to Esna. It is impossible to establish whether the name refers to the known martyr or a local saint.[6]

Uncertain as it may be, the name ϫⲱⲱⲣⲉ, "Strong One" (l. 5), brings to mind another martyr, much venerated in Upper Egypt (*BHO* 326–27).[7] This martyr saint, however, does not appear in the known litanies.[8] By contrast, a certain Abraham the Strong (ⲁⲃⲣⲁⲅⲁⲙ ⲡϫⲱⲱⲣⲉ), most probably a local saint, is known from two litanies from Saqqara.[9] If this name is to be supplied in the lacuna of ll. 4–5, the fragment in Warsaw could originate from the Monastery of Apa Jeremiah in Saqqara.

I. Varsovie, appendix I, A8

The discussion of a second Coptic (or rather, Greek-Coptic) epitaph is more delicate. It is one of the inscriptions from the collection of the former Lyceum Hosianum in Braunsberg (former East Prussia) that did not end up in the National Museum of Warsaw, as most of the collection did, and must be considered lost now. We have no more than a sketchy facsimile published in 1913 by its first editor, Seymour de Ricci. Yet this enables us to establish the nature of the text and the monument, the upper part of a funerary stela made of sandstone. In fact, the monument is incomplete, as de Ricci already supposed. The lower part, i.e. about two-thirds of the original height, is lost (compare the similar stelae in W.E. Crum, *Coptic Monuments*, Cairo: IFAO 1902, pl. X *et passim*). The catalogue of the *I. Varsovie*, while adhering to the new interpretation of the text given by Adam Łajtar in 1993, did not re-edit the text. I reproduce it here after de Ricci's facsimile (Figure 12.1).

Bibliography: S. de Ricci, "Inscriptions grecques d'Égypte à Braunsberg et à Saint-Pétersbourg", *Revue épigraphique* 1 (1913), 141–64 at 151–52, no. 19

I. VARSOVIE

Figure 12.1 I. Varsovie, Appendix I, A8
(drawing: de Ricci; adapted by Renate Dekker)

(facsimile, edition, brief commentary); A. Łajtar, "Zwei griechische christliche Inschriften", *ZPE* 95 (1993), 246–48 at 247–48 (revised reading); cf. *SEG* XLIII 1114.

Hermonthis, ca. sixth-eighth centuries?

    Εἷς θεός ὠ β-
    οηθῶν. Ἀμην.
    ϲαρρα ϫη ν-
4. ϥϲιϫ(ωροϲ).

1–2. ὠ βοηθῶν: ὁ βοηθῶν || 3–4. Σαρραχηνυς ιδ´: de Ricci, ed.; Σάρρα Δηνυσία (= Διονυσία) Łajtar *post corr.* || 4. ϥϲιϫ(ωροϲ): ϥϲιϫ de Ricci, facs., ϥϲιϫ Łajtar

One God who helps. Amen.
Sarra (Sarah), the daughter of Ysidoros (?).

As Łajtar already argued in his 1993 article, the formal and textual characteristics of the Braunsberg monument assign it unambiguously to the region of Hermonthis.[10] As for the inscription, although the interpretation of ll. 3–4 proposed by de Ricci (a name Sarradjenus = Saracenus, followed by the number 14) is to be rejected,[11] it is beyond doubt that he read the Coptic ϫ in l. 3 correctly. ϫH N-/ϫe N- is a perfectly common formula for feminine filiation (an orthographic variant of TϣH N-/Tϣe N-, "daughter of").[12] In Coptic funerary epigraphy it appears particularly often in the epitaphs of women from Hermonthis, both for indicating real patronyms and as part of proper names.[13] A much similar filiation is found for instance on another stela from Hermonthis, presently in Cairo, viz. that of capa ϫe mπanicke, "Sarah, the daughter of Paniskos".[14]

In the Braunsberg fragment, only the name of the father poses a problem. Supposing that the text copied by de Ricci is correct and complete as it is, the father's name reads Ysid, which can be plausibly interpreted as an abbreviation of the popular name Isidoros, a solution adopted here. An abbreviation Ἰσίδ(ωρος) is actually attested in the papyri.[15] Should one prefer to read, with Łajtar, a final -ⲁ instead of the -ⲇ, two letters that are often hardly distinguishable, one might assume ycia to be a deformation of the name Isaiah (Hcaïac).[16] The latter name is little attested in Coptic Egypt, however.

Would this lost stela have been the only Coptic monument in the former Braunsberg collection? For the answer, we have to await the publication of the *I. Varsovie coptes*.

Notes

1 For the fairly late appearance of Coptic in this domain, see J. Bingen, "L'épigraphie grecque de l'Égypte post-constantinienne", in: *XI Congresso internazionale di epigrafia greca e latina: Atti* II, Rome: Edizioni Quasar 1999, 613–24, especially 613–14.

2 A. Łajtar, "Notationes legentis", *JJP* 33 (2003), 184–85, lists another example, *I. Lyon* 41 (= Lefebvre, *Recueil*, no. 807), which is Coptic, not Greek.

3 [*I. Varsovie*].

4 For some methodological reflections, see my paper "Épigraphie chrétienne d'Égypte et de Nubie: bilan et perspectives", in: A. Boud'hors and D. Vaillancourt (eds.), *Huitième congrès international d'études coptes (Paris 2004)* I: *bilans et perspectives 2000–2004*, Paris: de Boccard 2006, 303–20 [Study 1].

5 See C. Wietheger, *Das Jeremias-Kloster zu Saqqara unter besonderer Berücksichtigung der Inschriften*, Altenberge: Oros 1992, 210–19; A. Papaconstantinou, *Le culte des saints en Égypte des Byzantins aux Abbasides: l'apport des inscriptions et des papyrus grecs et coptes*, Paris: Éditions du CNRS 2001, 22 and 387–402.

6 Cf. Wietheger, *Jeremias-Kloster*, 223; Papaconstantinou, *Le culte des saints*, 62–8 et 389.

7 See De Lacy O'Leary, *The saints of Egypt*, London: Society for Promoting Christian Knowledge 1937, 125; R-G. Coquin, "Le catalogue de la bibliothèque du couvent de saint Élie 'du rocher'", *BIFAO* 75 (1975), 207–39 at 230–31; T. Orlandi, art. "Joore", in: *CoptEnc* 5, 1370.

8 Analyzed by Papaconstantinou, *Le culte des saints*, 387–402.

9 See Wietheger, *Jeremias-Kloster*, 219.

10 Łajtar, "Zwei griechische christliche Inschriften", 248; cf. de Ricci: "haute Thébaïde".

11 See Łajtar, "Zwei griechische christliche Inschriften", 247.

12 See W.E. Crum, *Coptic dictionary*, Oxford: Oxford University Press 1939, s.v. ϣHpe; S. Sauneron and R-G. Coquin, "Catalogue provisoire des stèles funéraires coptes

I. VARSOVIE

d'Esna", in: J. Vercoutter (ed.), *Livre du centenaire de l'IFAO*, Cairo: IFAO 1980, 239–77 at 240.

13 E.g. W.E. Crum, *Coptic monuments*, Cairo: IFAO 1902, nos. 8445, 8461, 8484, 8500.

14 Crum, *Coptic monuments*, no. 8608.

15 P.E. Kahle, *Bala'izah*, London: Oxford University Press 1954, II, no. 299, l. 4; H. Satzinger, *Koptische Urkunden* III, Berlin: Verlag Bruno Hesling 1968, no. 401, vo, l. 14, twice in Greek text.

16 Cf. D. Foraboschi, *Onomasticon alterum*, Milano: Istituto Editoriale Cisalpino 1967, 334: (Φλαούιος) Ἐισιὰς?

13

A COPTIC FUNERARY STELA IN THE MONTREAL MUSEUM OF FINE ARTS

Jitse Dijkstra and Jacques van der Vliet

Introduction

It is not generally known that the Montreal Museum of Fine Arts houses a collection of more than one hundred Coptic textiles. In order to create a museographic context for these pieces, the Museum acquired three late Egyptian objects, donated by Jacques and Lise Simard, in 2008: a funerary stela from the Roman period, representing a woman reclining on a *kline* (inv. 2008.147), an architectural fragment decorated with a head surrounded by a nimbus, dating to late antiquity (inv. 2008.145), and the Coptic funerary stela published here (inv. 2008.146). The precise provenance of these pieces is unknown, as is the history of their acquisition. Through this first publication of one of them we hope to draw attention to the Coptic antiquities in the Museum's collections.[1]

Description

The small monument consists of a trapezoidal limestone slab of about 21–24 cm wide, which narrows slightly towards the bottom; its height is 36 cm and its thickness 6.5 cm. There are two inscriptions, engraved by the same hand and concerning the same person. On the front side, a Coptic text is laid out around a cross with forked endings on a pedestal, carved in raised relief.[2] The top of the stela bears a brief text that repeats the name of the deceased and part of the date of death; the back side is blank. The stela is complete, except for the lower corners, but has suffered damage, which makes the text at times difficult to read, particularly in lines 5–6 on the front side.

The upper side contains two lines in "Copto-Greek", the front side fourteen lines in Sahidic Coptic (lines 1–6 are engraved above, lines 7–14 below the horizontal bar of the cross). The text is crudely carved and there are several spelling errors. The stonecutter was clearly rather poorly skilled and sometimes misunderstood his more cursive model (see in particular the characteristic reading errors in lines 11 and 12). With a few exceptions, for instance in line 4, the ⲁ is written as an A with an almost triangular broken crossbar. The ⲡ most often assumes a cursive form (Λ);[3] the c in line 13 and the one in line 2 on the upper side are extended towards the right.

Inv. 2008.146 (Figs. 13.1 and 13.2) Middle Egypt
Ca. seventh century? (region of Antinoopolis?)

Figure 13.1 Coptic stela, Montreal Museum of Fine Arts, front side (photo: J.H.F. Dijkstra)

Figure 13.2 Coptic stela, Montreal Museum of Fine Arts, upper side (photo: J.H.F. Dijkstra)

[Front side]

    + ϩⲛ ⲡⲣ‖ⲁⲛ ⲛ-
    ⲛⲉⲓⲱⲧ ‖ ⲙⲛ ⲡⲱ-
    ⲏⲣⲉ ⲙⲛ ‖ ⲡⲉⲡⲛⲁ-
4.   ⲩⲙⲁ ⲉ‖ⲧⲟⲩⲁⲁ-
    ⲃ ⲉⲕⲛⲁ‖ⲣ ⲟⲩⲛⲁ
    ⲙⲛ [ⲡ]‖ⲙⲁⲕⲁ(ⲣⲓⲟⲥ)
    ==========
    ⲓⲱⲥⲏⲫ ‖ ⲛ̄ⲧⲁϥⲙ-
8.   ⲧⲟⲛ ⲛ̄‖ⲙⲟϥ ϩ-
    ⲛ ⲥⲟⲩ ‖ ⲙⲉⲩⲧ
    ⲛⲡⲉ‖ⲃⲟⲧ
    ⲁⲑⲟⲣ ‖ ϩⲙⲉⲛ-
12. ⲣⲟⲙⲡⲉ ‖ ⲧⲛ ϩⲉ-
    ⲃⲧⲟⲙ‖ⲏⲥ {ⲓⲛ}
    ⲓⲛⲁ(ⲓⲕⲧⲓⲱⲛⲟⲥ) ‖

2. ⲛⲉⲓⲱⲧ: read ⲡⲉⲓⲱⲧ ‖ 2–3. ⲡⲱⲏⲣⲉ: read ⲡϣⲏⲣⲉ ‖ 3–4. ⲡⲉⲡⲛⲁⲩⲙⲁ: read ⲡⲉⲡⲛⲉⲩⲙⲁ ‖ 6. [ⲡ]ⲙⲁⲕⲁ(ⲣⲓⲟⲥ): read ⲡⲙⲁⲕⲁ(ⲣⲓⲟⲥ) ‖ 9. ⲙⲉⲩⲧ: read ⲙⲏⲧ ‖ 11. ϩⲙⲉⲛ-: read ϩⲛ ⲧⲉⲓ- ‖ 12. ⲧⲛ: read ⲧⲁⲓ

[Upper side]

    + ⲟ ⲙⲁⲕⲁ(ⲣⲓⲟⲥ) ⲓⲱⲥⲏⲫ ϩⲉⲃ-
    ⲇⲟⲙⲏⲥ

[Front side]

+ In the name of the Father, the Son and the Holy Spirit. May you have mercy on the blessed Joseph. He went to rest on the tenth day of the month Hathor of this very year, the seventh indiction.

[Upper side]

+ The blessed Joseph, seventh (indiction).

JITSE DIJKSTRA AND JACQUES VAN DER VLIET

[Front side]

1–5. An invocation of the Holy Trinity that is often found (with variations) at the beginning of Christian epitaphs from Egypt and Nubia, cf. C. Wietheger, *Das Jeremias-Kloster zu Saqqara unter besonderer Berücksichtigung der Inschriften*, Altenberge: Oros Verlag 1992, 132, where, however, the author's remark on the geographical distribution of the formula needs to be modified: for Aswan, see e.g. the epitaph of Abba Pousi, bishop of Philae (*SB Kopt*. I 789; for an improved text, see S.G. Richter, *Studien zur Christianisierung Nubiens*, Wiesbaden: Reichert Verlag 2002, 119–21; eighth century?);[4] for Nubia, see *I. Khartoum Copt*., p. 26. It appears in the heading of official documents from the reign of the Emperor Phocas (A.D. 602–10) onwards, see R.S. Bagnall and K.A. Worp, *Chronological systems of Byzantine Egypt: second edition*, Leiden: Brill 2004, 99–109 and 290–99, where the present formula bears the siglum 2J.

1–2. The spelling ⲛⲛⲉⲓⲱⲧ is difficult to explain. Perhaps we are dealing here with a simple copying error (the second ⲛ for a cursive ⲡ)?

2–3. ⲡϣⲏⲣⲉ: the stonecutter forgot the descending stroke of the ϣ, which is a rather common error, cf. e.g. the epitaphs *SB Kopt*. I 425.1, II 1142.1 and 1154.1, all in the same formula.

3–4. The spelling ⲡⲉⲡⲛⲁⲩⲙⲁ instead of ⲡⲉⲡⲛⲉⲩⲙⲁ is due to forward vowel assimilation, see F.T. Gignac, *A grammar of the Greek papyri of the Roman and Byzantine periods* I, Milan: Cisalpino-La Goliardica 1976, 233–34; H. Förster, *Wörterbuch der griechischen Wörter in den koptischen dokumentarischen Texten*, Berlin: de Gruyter 2002, s.v. It is found e.g. on the funerary stela *SB Kopt*. I 723.2 (Manqabad, north of Asyut), in the same formula.

5–6. ⲉⲕⲛⲁⲣ ⲟⲩⲛⲁ ⲙⲛ "may you have mercy on": this prayer, addressing God, introduces the name of the deceased. The formula, characterized by the use of the composite verb ⲣ ⲟⲩⲛⲁ ⲙⲛ "to have mercy on" is quite common (as far south as Nubia), but the attestations are most numerous in Middle Egypt, see Wietheger, *Jeremias-Kloster*, 147; cf. *SB Kopt*. I-III indices, s.v. ⲛⲁ. Usually, the name of the deceased is preceded by ⲧⲉⲯⲩⲭⲏ ⲛ-/ⲙ- "the soul of", but the shorter formula (without ⲧⲉⲯⲩⲭⲏ) is well attested in other epitaphs, e.g. *SB Kopt*. I 686.3–5 (Antinoopolis): ⲉⲣ ⲟⲩⲛⲁ ⲙⲛ ⲡⲙⲁⲕⲁⲣⲓⲟⲥ ⲥⲟⲩⲏⲣⲟⲥ "have mercy on the blessed Severos".

5. The scribe uses the Second Future instead of the more common Third Future, as e.g. in G. Lefebvre, "Égypte chrétienne IV", *ASAE* 11 (1911), 238–50 at 243–45 (nos. 4.7 and 5 vº 3–4), two stelae that probably originate from Akhmim. For the "directive" use of the Second Future, see G. Steindorff, *Lehrbuch der koptischen Grammatik*, Chicago: University of Chicago Press 1951, 150–51; B. Layton, *A Coptic grammar*, Wiesbaden: Harrassowitz 2000, 267 (§ 339); C.H. Reintges, *Coptic Egyptian (Sahidic dialect)*, Cologne: Köppe 2004, 267.

6. [ⲡ]ⲙⲁⲕⲁ(ⲣⲓⲟⲥ): the abbreviation mark after the ⲕ shows that the stonecutter first wanted to write the usual abbreviation for ⲙⲁⲕ(ⲁⲣⲓⲟⲥ) "blessed". Then, in order to fill the empty space at the end of the line, he added the fourth letter of ⲙⲁⲕⲁⲣⲓⲟⲥ, the ⲁ, which can be read without difficulty.

7. ⲓ̈ⲱⲥⲏϥ: a biblical name, common in Christian Egypt; cf. *SB Kopt*. I-III indices and *NB Kopt*. s.v.

7–8. ⲛⲧⲁϥⲙⲧⲟⲛ ⲙⲙⲟϥ "he went to rest": a standard formula introducing the date of death. The verb denoting death (ⲙⲧⲟⲛ ⲙⲙⲟ⸗, Greek ἀναπαύομαι) is not a

188

euphemism, but refers to a central notion of Christian eschatology, see J. Helderman, *Die Anapausis im Evangelium Veritatis*, Leiden: Brill 1984, 47–70. The Second Perfect highlights the date of death, for the importance of which, see J. van der Vliet, "'What is man?' The Nubian tradition of Coptic funerary inscriptions", in: A. Łajtar and J. van der Vliet (eds.), *Nubian voices: studies in Christian Nubian culture*, Warsaw: University of Warsaw/Raphael Taubenschlag Foundation 2011, 171–224 at 195–97 [Study 32].

8. ⲛⲙⲟϥ: as elsewhere in the text, the assimilation ⲛ > ⲙ is not realized; for the spelling ⲛⲙⲟϥ, cf. e.g. the stelae from Antinoopolis *SB Kopt.* I 761.4, 766.6 and 773.7 (ⲛⲙⲟⲟⲩ).

8–14. The dating formula concludes the text. It consists of a tripartite structure listing first the day (ⲍⲛ ⲥⲟⲩ ⲙⲉⲩⲧ, i.e. ⲍⲛ ⲥⲟⲩ ⲙⲏⲧ "on the tenth day"), then the month (ⲛⲡⲉⲃⲟⲧ ⲁⲑⲟⲣ, i.e. ⲙⲡⲉⲃⲟⲧ ⲍⲁⲑⲱⲣ "of the month Hathor") and finally the year (ⲍⲙⲉⲛⲣⲟⲙⲡⲉ ⲧⲛ ⲍⲉⲃⲧⲟⲙⲏⲥ {ⲓⲛ} ⲓⲛⲇ(ⲓⲕⲧⲓⲱⲛⲟⲥ), i.e. ⲍⲛ ⲧⲉⲓⲣⲟⲙⲡⲉ ⲧⲁⲓ ⲍⲉⲃⲇⲟⲙⲏⲥ ⲓⲛⲇⲓⲕⲧⲓⲱⲛⲟⲥ "of this very year, the seventh indiction"). It is common in Middle Egypt to record the indiction year in Greek (ἑβδόμης ἰνδικτίωνος).[5]

9. ⲙⲉⲩⲧ: this variant of ⲙⲏⲧ would be an unusual spelling, but the reading is far from certain.

11. ⲍⲙⲉⲛⲣⲟⲙⲡⲉ: probably a reading error for ⲍⲛ ⲧⲉⲓⲣⲟⲙⲡⲉ; the model of the stonecutter must have included some sort of ligature for the last four letters of the sequence ⲍⲛ ⲧⲉⲓ-.

12. ⲧⲛ: another reading error (for ⲧⲁⲓ), resulting from the visual resemblance of ⲛ and ⲁⲓ; see e.g. *P.Naqlun* II 20 rº 8 (ⲛⲛ for ⲛⲁⲓ).

12–13. ⲍⲉⲃⲧⲟⲙⲏⲥ: the confusion between ⲧ and ⲇ is a normal phenomenon in this period, particularly in the Greek vocabulary of Coptic texts, see Gignac, *Grammar*, I, 63; P.E. Kahle, *Bala'izah: Coptic texts from Deir el-Bala'iza in Upper Egypt* I, London: Oxford University Press 1954, 95, no. 68; note that the correct spelling, ⲍⲉⲃⲇⲟⲙⲏⲥ, is read on the upper side, lines 1–2.

13–14. {ⲓⲛ} ⲓⲛⲇ(ⲓⲕⲧⲓⲱⲛⲟⲥ): the stonecutter wrote ⲓⲛ at the end of line 13, but restarted the word at the beginning of line 14. One would expect the abbreviation ⲓⲛⲇ(ⲓⲕⲧⲓⲱⲛⲟⲥ), with an abbreviation mark after the ⲇ, but this reading is not without difficulty: the ⲇ is quite rounded, whereas the upper side displays a perfectly triangular ⲇ.

[Upper side]

1–2. The use of the article ὁ shows that the scribe intended to write this part of the inscription in Greek. Nevertheless, on account of the Coptic ⲍ at the beginning of the word ἑβδόμης and the rudimentary character of the text, we transcribed it here in Coptic uncials. Otherwise, the meaning of this addition escapes us. Is this a "tag" used for identifying the owner during the production process of the stela?[6]

Conclusion: provenance and date

There can be little doubt about the provenance of this epitaph. The overall form of the text and the individual formulae that constitute its elements, most notably the prayer with ⲣ ⲟⲩⲛⲁ ⲙⲛ (lines 5–6) and the way in which the indiction year is noted (lines 11–14), all point to Middle Egypt and recall in particular the numerous Coptic stelae discovered in the vicinity of Antinoopolis by Albert Gayet and Count Jean de Beaucorps, and published by Seymour de Ricci and Gustave Lefebvre at the beginning of the twentieth century.[7]

The language of the main text is Sahidic Coptic, but it must be observed that the scribe never realizes the assimilation ⲛ > ⲙ, not even in ⲛ̄ⲙⲟⳓ in line 8. This is a fairly widespread dialectal feature, but could be another indication of a Middle Egyptian provenance.[8]

The date of the stela is more difficult to establish. One of the Coptic stelae found by de Beaucorps at Antinoopolis presents a formulaic structure that is similar to the text of the stela from Montreal (invocation of the Holy Trinity, prayer with ⲣ ⲟⲩⲛⲁ ⲙⲛ, dating formula with a day of the month in Coptic and an indiction year in Greek). It is, in addition, dated to year 336 of Diocletian, which corresponds to year 619/620 of the Dionysian era.[9] With due caution, we therefore propose to date the stela from Montreal, which is of much inferior craftsmanship, to the seventh century as well. The use of Coptic does not allow for an earlier date; a later date cannot be entirely excluded, but the paleography definitely places the stela before the tenth century.

Notes

1 We sincerely thank Dr. John M. Fossey, curator emeritus of Mediterranean archaeology at the Museum, for encouraging us to publish this stela, for providing us with valuable information, and for his assistance in the entire publication process, in particular during Jitse Dijkstra's visit to the Museum on 15 June 2011, when the stela was photographed and studied.

2 A cross "fourchée" according to the terminology of J.H.F. Dijkstra, *Syene I: The figural and textual graffiti from the temple of Isis at Aswan*, Darmstadt/Mainz: Philipp von Zabern 2012, 81.

3 For such a "cursive" ⲡ in an inscription, see J.H.F. Dijkstra and J. van der Vliet, " 'In year one of King Zachari': evidence of a new Nubian King from the Monastery of St. Simeon at Aswan", *BSF* 8 (2003), 31–9 at 33 [Study 19].

4 On this text, see most recently J.H.F. Dijkstra, *Philae and the end of ancient Egyptian religion: a regional study of religious transformation (298–642 CE)*, Louvain/Paris/ Dudley, MA: Peeters 2008, 325–26, who refers to *SB Kopt.* I 564 (A.D. 733) for the date.

5 Among the funerary stelae of which the provenance is known and which are included in *SB Kopt.* I-III, the following ones have the same dating formula: *SB Kopt.* I 760.6–9; *SB Kopt.* II 1070.6–9, 1071.9–12, 1072.8–11, 1073.5–8 (Antinoopolis); *SB Kopt.* III 1588.10–12, 1589.6–8 (Dayr Abu Hennis). We can add the stelae *SB Kopt.* I 685.10–13 and 687.13–16, which *SB Kopt.* attributes to "Upper Egypt", following the vague provenance given in the *editio princeps* merely on the basis of the fact that they are written in Sahidic (G. Biondi, "Inscriptions coptes", *ASAE* 8 (1907), 77–96, 161–83 at 77). In fact, these two stelae come from Antinoopolis (see below, n. 7). In addition, *SB Kopt.* I 686.6–9 can be mentioned, which has a similar dating formula (only lacking the element ⲙⲡⲉⲃⲟⲧ) and is from Antinoopolis as well. Cf. J. van der Vliet, "*Parerga*: notes on Christian inscriptions from Egypt and Nubia", *ZPE* 164 (2008), 153–58 (on *SB Kopt.* III 1596.14–17, which does not originate from Abu Mina but Middle Egypt) [Study 16].

A COPTIC FUNERARY STELA IN MONTREAL

6   It is extremely rare to find inscriptions on the sides of funerary stelae, which are usually hidden from view. Three stelae from Akhmim published by Lefebvre, "Égypte chrétienne IV", 241–45 (nos. 2–3, 5) are likewise inscribed on the upper side (no. 3 also including the name of the deceased), but these objects were intended to be visible from both sides, which is apparently not the case with the stela from Montreal.

7   On these excavations, see F. Calament, *La révélation d'Antinoé par Albert Gayet: histoire, archéologie, muséographie* (Bibliothèque d'études coptes 18), Cairo: IFAO 2005. The main editions of Coptic stelae are: G. Lefebvre, "Inscriptions chrétiennes du Musée du Caire", *BIFAO* 3 (1903), 69–95 at 87–90 (nos. 35–40) = *SB Kopt.* I 439–43, 686; S. de Ricci, "Inscriptions grecques et coptes", *Annales du Musée Guimet* 30 (1903), 141–43 at 142–43 (nos. 12–28; nos. 15, 24, 26 are nos. 35, 39, 36 in Lefebvre, "Inscriptions chrétiennes"; therefore, his no. 35 comes from Antinoopolis); Biondi, "Inscriptions coptes", 83 (no. 6), 84–7 (nos. 9–13), 88–9 (nos. 15–16, 18), 161 (no. 32), 162–63 (nos. 34, 38) = *SB Kopt.* I 681, 684–88, 690–91, 693, 699, 441, 439 (nos. 11, 34, 38 are nos. 37–8, 35 in Lefebvre, "Inscriptions chrétiennes"; nos. 6, 9–10, 12–13, 15–16, 18, 32 are nos. 22, 14, 12, 16, 21, 19, 25, 13, 17 in de Ricci, "Inscriptions grecques et coptes"; all these stelae therefore originate from Antinoopolis); G. Lefebvre, "Égypte chrétienne III", *ASAE* 10 (1910), 260–84 at 273–75 (nos. 4–7) = *SB Kopt.* I 760–63, and "Égypte chrétienne V", *ASAE* 15 (1915), 113–39 at 114–25 (nos. 1–15) = *SB Kopt.* I 764–78. It is a pity that *SB Kopt.* did not attempt to bring some order in this bibliographical jumble. For the most recent publication of a stela from Gayet's excavations in Antinoopolis, see R.-G. Coquin and M.-H. Rutschowscaya, "Les stèles coptes du Département des antiquités égyptiennes du Louvre", *BIFAO* 94 (1994), 107–31 at 115 (no. 8), fig. 4 = *SB Kopt.* II 1098. Since it is impossible to present all the material, it suffices to mention here the Coptic stelae that were discovered during the later excavations by E. Breccia and S. Donadoni at Antinoopolis in 1936, most of which were edited by H. Munier, "Stèles chrétiennes d'Antinoé", *Aegyptus* 29 (1949), 126–36 at 129–35 (nos. 1–10) = *SB Kopt.* II 1069–78.

8   Cf. Kahle, *Bala'izah*, 99–100.

9   Lefebvre, "Égypte chrétienne V", 113 and 117–19 (no. 5), pl. I = *SB Kopt.* I 768, cf. Bagnall and Worp, *Chronological systems of Byzantine Egypt: Second edition*, Leiden: Brill 2004, 72, n. 37. See also Coquin and Rutschowscaya, "Stèles coptes", 115 (no. 8), fig. 4 = *SB Kopt.* II 1098, a stela from Antinoopolis that includes the same elements (except for the invocation of the Holy Trinity) and is dated to A.D. 610.

14

SNIPPETS FROM THE PAST
Two ancient sites in the Asyut region: Dayr al-Gabrawi and Dayr al-'Izam

Jacques van der Vliet

Introduction: Asyut/Lykopolis in late antiquity

There can be little doubt about the importance of the city of Asyut in the Christian history of Egypt. Even if it is unlikely that Alexander of Lykopolis, the fourth-century author of a witty treatise against the Manichaeans, was really a bishop of that city – as is still occasionally maintained – Melitius, the originator of the Melitian schism, also in the fourth century, certainly was. In the early seventh century, Constantine of Asyut, a contemporary of Patriarch Damian, was one of the prominent bishops of the nascent Severan ("miaphysite") Church, the precursor of the present-day Coptic Orthodox Church. In the period of the persecutions, it was the scene of many martyrdoms and in the fourth-fifth centuries a renowned center of monasticism with which the names of Paul of Tamma and John of Lykopolis remain associated. Still in the early seventh century, John Moschos' *Spiritual Meadow* bears witness to the reputation of its monasteries.

Asyut was a major Christian center, as it still is today, and in late antique and early medieval times its hinterland was covered with monastic settlements and sanctuaries devoted to the cult of the martyrs, many of them centers of pilgrimage. Indeed, historical maps of the region show a remarkable density and concentration of Christian sites, stretching from the region of Abnub on the east bank, via the ancient necropoleis of Asyut itself, mainly on the west bank, to the region of Shotb, south of the city (see, for instance, Wipszycka 2009: 142). Yet all these ancient sites left surprisingly few traces in the modern scholarly record. Thus, when one consults the Brussels database of Coptic documentary papyri, the turnout for Asyut is almost nil (see Delattre 2012, under "provenance: Assiout"). And even though several of the ancient monasteries around Asyut are honored by an article in the *Coptic Encyclopedia*, they appear to have little to contribute to our knowledge of Coptic Egypt. Hence, while Melitius, Constantine and John of Lykopolis continue to attract scholarly attention, our picture of the broader Christian landscape of Asyut in pre-modern times remains hazy and patchy.

The background of this somewhat sad state of affairs is to be sought in the ruthless exploitation of the region by the would-be archaeologists who, in the nineteenth and twentieth centuries, plundered the sites around Asyut, in particular the rich necropoleis of the Old and Middle Kingdoms. These dilettanti and treasure hunters were exclusively interested in pharaonic antiquities and hardly ever in more recent material. For the Christian antiquities of the region the loss was disastrous. Architectural remains were destroyed and poorly recorded, if at all. Other evidence, apart from some inscriptions, was sold on the antiquities market to join the mass of objects and documents for which no certain provenance is known and which – as a result – became virtually worthless for historical research.

In the following pages I intend to illustrate these points by briefly evoking the example of two sites that in recent decades have received a certain amount of scholarly interest: Dayr al-Gabrawi (north of Asyut, near Abnub) and Dayr al-'Izam (west of Asyut).

Dayr al-Gabrawi: the Martyrium of St. Victor the General?

Dayr al-Gabrawi is famous mostly for its Old Kingdom rock-cut tombs dating from the Sixth Dynasty.[1] These were published in 1902 by Norman de Garis Davies, whose team also copied some of the Coptic inscriptions that are witness to the monastic re-use of the tombs (summarily edited by W.E. Crum in Davies 1902: II, 45–6). The tombs were republished in more recent years (Kanawati 2005–13), but this new publication is exclusively focused on the Old Kingdom reliefs. In addition to the tombs, there is a Christian village, sometimes called Dayr Abu Buqtur al-Gabrawi, after the patron of its church, St. Victor, a martyr from the era of Diocletian.[2] The present village church appears to be a fairly recent building, but in the immediate neighborhood various remains of presumably late antique date have been identified. Among these are a structure described as a Roman basilica and sites locally known as the "chapel of Sitt Barbara" and the "monastery of Sitt Barbara", designations that appear to have very little historical foundation.

The various ancient and medieval remains of Dayr al-Gabrawi were the object of a German survey in the early 1980s. These resulted in an inventory of archaeological sites and an archaeological map of some sorts (D. Kurth in Kurth and Rößler-Köhler 1987: 186–94, with maps 1 and 3, plans 6–7). Regrettably the architectural remains recorded for Dayr al-Gabrawi gave rise to widely divergent interpretations and – as a result – considerable confusion (see the authoritative discussion in Grossmann 1991). The site is redeemed by a Latin inscription that – although non-Christian in character – is of great interest for the Christian history of the region. The stone inscription, commonly designated as *CIL* III 22 (= Mommsen 1873: 8, no. 22), has been known since the seventeenth century.[3] According to rumor, the stone was built into the walls of the modern village church as recently as the 1980s, but it has not actually been seen by visitors since the nineteenth century.

The Latin text is dedicatory in nature. It records the construction of the camp (*castra*) of a Roman military division, the First Cohort of the Lusitanians, by the Emperors Diocletian and Maximian in A.D. 288. The establishment of this stronghold was part of the thorough military reorganization of the empire under Diocletian, which for Egypt resulted among other things in the withdrawal of the southern, Nubian frontier to Philae and the building of a chain of fortifications along the River Nile (van Berchem 1952: 59–71). The ruins of a so-called "basilica" on the northeast of the modern village of Dayr al-Gabrawi are most likely remains of this Roman camp (Grossmann 1991: 171–73; Coquin, Martin and Grossmann 1991b: 811–12).

The official inscription of the year 288 combined with the plausible archaeological identification of Roman military architecture on site allows a series of important historical inferences (for which, see primarily Horn 1988: xi–xix; 1992: 128–37). A somewhat later Roman source, the *Notitia dignitatum*, from the end of the fourth century, situates the camp of the First Cohort of the Lusitanians, mentioned in the Latin inscription, near a place called Theraco, a name that is traditionally emended to Hieracon on the authority of another Roman source, the *Antonine Itinerary*.[4] The *Itinerary*, which dates from the time of Diocletian itself, situates the settlement of Hieracon on the east bank of the Nile, south of Antinoopolis, actually at the approximate location of Dayr al-Gabrawi (cf. Bell 1942: 144; Timm 1984–1992: IV, 1950–51).

In the medieval hagiographical literature about the military martyr St. Victor, the son of Romanus ("Victor the General"), a *kastron* ("fortified settlement") of Hierakion, south of Antinoopolis, is mentioned as the place of his martyrdom and death, most notably in the account of his martyrdom that survives in Sahidic Coptic (Budge 1914: 28; Elanskaya 1969: 47–8). In Jürgen Horn's view, the hagiographic tradition concerning St. Victor can be traced back to the late fourth century, even if the Sahidic version of the martyrdom, which mentions Hierakion, is definitely of a much later date (see Horn 1988: lxxi–lxxiii). In any case, a presumably Sahidic martyrdom of St. Victor circulated already in seventh- and eighth-century Thebes (Crum 1921: no. 281), and his cult is attested for the city of Asyut itself in the sixth century (Papaconstantinou 2001: 63, citing a document that mentions a chapel of the saint), which presupposes the local existence of a martyrology.

Therefore, in spite of the late date of the manuscripts, which are from the tenth century, it seems probable that the *kastron* of Hierakion in the Sahidic martyrdom of Victor preserves the memory of the military camp near Hieracon, likewise situated south of Antinoopolis, which was home to the First Cohort of the Lusitanians at the time of Diocletian.[5] The implication is clear: Dayr al-Gabrawi, where the Lusitanians were stationed, may have been the very place where the military saint Victor suffered his martyrdom. The location of Victor's martyrdom and subsequent cult at Dayr al-Gabrawi would seem to be corroborated by the traditional dedication of the local church to the same St. Victor as well as by other hagiographical traditions. Thus, the medieval literature on St. Claudius of Antioch, whose cult was centered in the Asyut region, locates a sanctuary of St. Victor on the east bank, south of Antinoopolis and north of Asyut, precisely where Dayr al-Gabrawi is situated (Papaconstantinou 2001: 66). The wealth of its church attracted the three robbers in the well-known story, told in a homily attributed to Constantine of Asyut, which relates a predatory foray ending with the arrest and conversion of the thieves by St. Claudius himself (Godron 1970: 640–54; see the reconstruction of their itinerary in Drescher 1942: 77, n. 2, and 85).

Supposing that the various assumptions and identifications proposed above prove to be correct, Dayr al-Gabrawi *might* have housed a major sanctuary of a famous martyr from the great persecutions, Victor the son of Romanus, and the site *might* offer an interesting example of the Christian re-use of Roman military architecture. Regrettably, however, too little archaeological work has been done on the various sites of Dayr al-Gabrawi to be firmly positive on any of these points. Tracing the whereabouts of the Latin inscription of emperors Diocletian and Maximian may be a good starting point for any further research.

Dayr al-'Izam: the hermitage of John of Lykopolis?

Dayr al-'Izam, or "the Monastery of the Bones", is situated in the western necropolis of Asyut.[6] The city's extensive cemeteries are famous as the provenance of many of the Middle Kingdom sarcophagi that bear versions of the so-called Coffin Texts, an important collection of Egyptian funerary rituals. The link with these ancient texts sparked a recent revival of interest in the necropolis as a whole. A German project, directed by Jochem Kahl, not only conducts excavations on site, but also intends to document its history up to modern times (see in particular Kahl 2007 and 2013). The project also pays due attention to the period of its Christian re-use, and we may hope to be better informed in the near future than we are now.

The monastic site of Dayr al-'Izam was excavated, or rather plundered, by two local amateur diggers in September 1897. A brief report of their most important finds appeared a few years later (Maspero 1900). A very schematic map (after V. de Bock) was published in 1912 by Somers Clarke in his *Christian Antiquities in the Nile Valley* (Clarke 1912: 179). Later accounts are largely dependent on this scarce information, as the site appears to have suffered destruction later on (Coquin, Martin and Grossmann 1991a). The finds made in 1897 include the sad remains of a medieval monastic library, comprising twenty-four fragments of parchment and paper codices. These texts are still unpublished, but they were briefly analyzed by W.E. Crum shortly after their discovery (Crum 1902: nos. 8080–8103). The find consists of biblical, apocryphal, homiletic and liturgical texts, predominantly in Sahidic. In spite of its piteous state, the collection has its importance, in particular since Crum was able to link some of the fragments to manuscripts kept in the French National Library, presumably originating from the White Monastery near Sohag. The future study of medieval Coptic literature would undoubtedly benefit from a full publication of the find.

Another important find is an amphora inscribed with a long text in Sahidic Coptic from the year 1155/1156 (Crum 1902: no. 8104, plate I). As the inscription is of considerable historical significance it will be translated in full below, at the end of this paper. It identifies the place where the jar was buried and inscribed as the "monastery of Apa John of the Desert", a name that is also found in some other medieval sources (for which see Coquin, Martin and Grossmann 1991a; Timm 1984–1992: II, 830–31). That this "monastery of Apa John of the Desert" at Dayr al-'Izam is the same as a "monastery of Apa John of the rock (or mountain) of Siout (Asyut)", known from earlier sources, is possible but quite uncertain (see Kahle 1954: 24; Timm 1984–1992: II, 829–33). In any case, the Apa John who gave his name to the monastery may well have been – in the opinion of many scholars – none other than John of Lykopolis, a famous monk from the late fourth century, who indeed must have lived at some distance west of Asyut. Hence, Dayr al-'Izam is generally identified as the historical hermitage of St. John of Lykopolis (for instance both Doresse 2000: 422 and Grossmann 2002: 208, n. 20, find this a likely assumption).

Although the identification is not supported by contemporaneous, late antique material from the site of Dayr al-'Izam itself, it served as the basis for an interesting hypothesis, brought forward some twenty years ago by the papyrologist Constantine Zuckerman (1995: 188–94). In his view, Dayr al-'Izam would be the likely provenance of the so-called archive of Apa John, a group of Greek and Coptic papyri from somewhere in Middle Egypt that used to be loosely associated with the Hermopolite nome, as are many other Middle Egyptian papyri. The bilingual "archive" dates from the late fourth century and consists of letters addressing a monk, Apa John, who apparently was well connected with the higher society of his time and who might very well be the sort of influential person John of Lykopolis was.[7] The papyri had appeared on the antiquities market briefly after the amateur diggings at Dayr al-'Izam in 1897 and were soon dispersed over various, mainly British, collections. According to Zuckerman's reconstruction, the amateur excavators of Dayr al-'Izam would have handed some of their finds over to the Antiquities Service, while keeping the more valuable papyri for themselves to bring them piecemeal on the market.

The hypothesis of Zuckerman is very attractive (for a generally positive discussion, see Wipszycka 2009: 83–5). It would at once confirm the connection of John of Lykopolis with Dayr al-'Izam and provide us with the social network of an important holy man from late antiquity, known from various literary sources as a major figure in early Egyptian monasticism. Yet it is also very fragile. The provenance of the dossier of Apa John in Dayr al-'Izam is far from proven and – even if it were – John is and was a very common name. There must have been hundreds of "Apa Johns" in late antique and early medieval Egypt! Hence, Zuckerman's thesis was accepted by some, but received critically by others, for instance by Malcolm Choat, who announced a comprehensive (re-)edition of the archive (see, preliminarily, Choat 2007: in particular 180–83 for a discussion of Zuckerman's thesis). In my opinion, it can at best be accepted as an attractive working hypothesis that can be falsified, not by reflections of a general nature, but only by a prosopographical analysis of the archive of Apa John. Where does the social network of this Apa John fit in topographically: in the Asyut region, or instead more to the north, in the neighborhood of Hermopolis? Another way by which Zuckerman's thesis could be tested would be fieldwork on the spot. Does the medieval "monastery of Apa John of the Desert" at Dayr al-'Izam indeed hark back to the fourth century? For the time being, unequivocal evidence to this effect, be it archaeological or papyrological, appears to be lacking.

Conclusions

The monasteries of Asyut tell a sad story of destruction and neglect. Asyut must have been an important Christian center in late antiquity, as it is today, but hardly anything survives that can be used to document its history in this period. The connections that have been postulated between Dayr al-Gabrawi and St. Victor the General, and between Dayr al-'Izam and St. John of Lykopolis, and which I have briefly discussed here, rest on slender evidence and the arguments in favor of them are extremely fragile. Only a combination of further research in the field *and* in museum and library collections may help to reconstruct a world that now seems largely lost.

Appendix

The jar from Dayr al-'Izam

The Sahidic inscription on the amphora mentioned above is historically interesting on various scores. It shows that in the twelfth century Dayr al-'Izam bore the name of "monastery of Apa John of the Desert", but also that it maintained contacts with the region of Akhmim farther south. In addition, the text presents a rather gloomy picture of the circumstances under which not only Fatimid rule but also many medieval monasteries succumbed. Yet it has never – to the best of my knowledge – been translated into English (for a French translation, see Mallon 1914: 2866). The following translation is made after the text printed by Maspero (1900: 117–18), but takes into account the important corrections published by V. Loret (1903: 104). I furthermore collated Maspero's readings with the published photograph (Crum 1902: pl. I), which shows only part of the text, however. Later re-editions (such as *SB Kopt.* I 299) lack independent value. Crum (1902: no. 8104) provides a detailed description of the jar.

> God have mercy on us, me, *Papa* Basile, and my brother, the deacon Pakire, the sons of the blessed *Papa* Joseph, from Talmarage (al-Maragha) in the province of Shmin (Akhmim), who assumed – beyond our worth – that name of "monk" in this monastery of Apa John of the Desert – may his blessings be with us. Amen.
>
> In the brief and insignificant days that we have consumed, this tiny bit of perfume, which is this myrrh, came in and we stored it for who will come after us. Its amount is 9 quarters, of which there are 12 in an *oipe*.

SNIPPETS FROM THE PAST

(We did this in the year of) the Era of the Martyrs 872 (A.D. 1155/1156), when there was famine in the land of Egypt, in particular the city of Siout (Asyut), and violence from the part of the authorities in charge and unrest from outside, due to the Arabs and the people who had become infuriated with each other. O Lord, help us!

And we believe in the creed that was established in Nicaea by the 318 bishops, which professes the Father and the Son and the Holy Spirit, one single Godhead – to Him be the glory. Amen.

As an afterword, it may be added that the jar and its contents were found sealed and intact, and that since the days of Basile and Pakire no one had apparently made use of the myrrh.

Notes

1 For the pharaonic antiquities of Dayr al-Gabrawi, see now Moreno García 2012.
2 For the toponym, see Horn 1992: 65, with n. 172. The designation "Dayr" does not necessarily imply that it was the site of an ancient monastery, as it may refer to any ancient Christian settlement. For general accounts of Christian Dayr al-Gabrawi, see Coquin, Martin, Grossmann 1991b; Timm 1984–1992: I, 41–4 (under Abnub); IV, 1949–52; Horn 1992: 65–6, 128–37.
3 See Sicard 1982: 14–15, with the notes by M. Martin, about the place where the stone was located in 1716.
4 Cf. the apparatus of the latest edition of the *Notitia*, Neira Faleiro 2005: 254.
5 Note, however, that the Greco-Latin word *kastron*, "fortified place", has a different range of meaning than the Latin *castra*, "army camp"; see for discussion Łajtar 1997: 44–5.
6 For general accounts, see Clarke 1912: 178–79; Coquin, Martin and Grossmann 1991a; Doresse 2000: 418–23, cf. 571–72; Kahl 2007: 99–102; 2013: 126–29.
7 See, in addition to Zuckerman 1995; van Minnen 1994: 80–5; Choat and Gardner 2006; Choat 2007; Gonis 2008; Wipszycka 2009: 83–5.

Bibliography

Bell, J. 1942, *Egypt in the Classical geographers*, Cairo: Government Press.
Budge, E.A. Wallis. 1914, *Coptic martyrdoms etc. in the dialect of Upper Egypt*, London: British Museum.
Choat, M. 2007, "The archive of Apa Johannes: notes on a proposed new edition", in: J. Frösén, T. Purola, and E. Salmenkivi (eds.), *Proceedings of the 24th International Congress of Papyrology, Helsinki, 1–7 August, 2004*, Helsinki: Societas Scientiarum Fennica, I, 175–83.
Choat, M., and I. Gardner. 2006, "*P. Lond. Copt.* I 1123: another letter to Apa Johannes?", *ZPE* 156: 157–64.
Clarke, S. 1912, *Christian antiquities in the Nile Valley: a contribution towards the study of the ancient churches*, Oxford: Clarendon Press.
Coquin, R-G., M. Martin, and P. Grossmann. 1991a, "Dayr al-'Iẓâm (Asyûṭ)", *CoptEnc* 3: 809–10.
Coquin, R-G., M. Martin, and P. Grossmann. 1991b, "Dayr al-Jabrâwî (Asyûṭ)", *CoptEnc* 3: 810–13.

JACQUES VAN DER VLIET

Crum, W.E. 1902, *Coptic monuments*, Cairo: IFAO; reprinted Osnabrück: Otto Zeller Verlag 1975.

Crum, W.E. 1921, *Short texts from Coptic ostraca and papyri*, London/New York: H. Milford.

Davies, N. de Garis. 1902, *The rock tombs of Deir al Gebrâwi*, London: Egypt Exploration Fund.

Delattre, A. 2012, *Banque de données des textes coptes documentaires*, Université libre de Bruxelles, Centre de papyrologie et d'épigraphie grecque (http://dev.ulb.ac.be/philo/bad/copte/base.php).

Doresse, J. 2000, *Les anciens monastères coptes de Moyenne-Égypte (du Gebel-et-Teir à Kôm-Ishgaou) d'après l'archéologie et l'hagiographie*, Yverdon-les-bains: Institut d'archéologie yverdonnoise.

Drescher, J. 1942, "Apa Claudius and the thieves", *BSAC* 8: 63–87.

Elanskaya, A.I. 1969, *The Coptic manuscripts of the State Public Library M.E. Saltykov-Shchedrin*, Leningrad: Nauka (Russian).

Godron, G. 1970, *Textes coptes relatifs à saint Claude d'Antioche*, Turnhout: Brepols.

Gonis, N. 2008, "Further letters from the archive of Apa Ioannes", *BASP* 45: 69–85.

Grossmann, P. 1991, "Spätantike Baudenkmäler im Gebiet von Dair al-Ğabrawî", in: *Tesserae: Festschrift Josef Engemann*, Münster: Aschendorffsche Verlagsbuchhandlung, 170–80.

Grossmann, P. 2002, *Christliche Architektur in Ägypten*, Leiden: Brill.

Horn, J. 1988, *Untersuchungen zu Frömmigkeit und Literatur des christlichen Ägypten: Das Martyrium des Viktor, Sohnes des Romanos (Einleitung in das koptische Literaturwerk/Kommentar zum "Ersten Martyrium")*, Ph.D-dissertation, Georg-August-Universität Göttingen.

Horn, J. 1992, *Studien zu den Märtyrern des nördlichen Oberägypten* II: *Märtyrer und Heilige des XI. bis XIV. oberägyptischen Gaues. Ein Beitrag zur Topographia christiana Ägyptens*, Wiesbaden: Harrassowitz.

Kahl, J. 2007, *Ancient Asyut: the first synthesis after 300 years of research*, Wiesbaden: Harrassowitz.

Kahl, J. 2013, *Die Zeit selbst lag nun tod darnieder. Die Stadt Assiut und ihre Nekropolen nach westlichen Reiseberichten des 17. bis 19. Jahrhunderts: Konstruktion, Destruktion und Rekonstruktion*, Wiesbaden: Harrassowitz.

Kahle, P.E. 1954, *Bala'izah: Coptic texts from Deir el-Bala'izah in Upper Egypt*, London: Oxford University Press.

Kanawati, N. 2005–2013, *Deir el-Gebrawi* I-III, Oxford: Aris & Phillips.

Kurth, D., and U. Rößler-Köhler. (eds.) 1987, *Zur Archäologie des 12. oberägyptischen Gaues: Bericht über zwei Surveys der Jahre 1980 und 1981*, Wiesbaden: Harrassowitz.

Łajtar, A. 1997, "Τὸ κάστρον τῶν Μαύρων τὸ πλησίον Φίλων: Der dritte Adam über *P.Haun.* II 26", *JJP* 27: 43–54.

Loret, V. 1903. "Carnet de notes égyptologiques", *Sphinx* 6: 97–112.

Mallon, A. 1914, "Copte (épigraphie)", in: *DACL* III/2, col. 2819–86.

Maspero, G. 1900, "Les fouilles de Deir el Aizam", *ASAE* 1: 109–19.

Mommsen, Th. 1873, *Inscriptiones Asiae, provinciarum Europae graecarum, Illyrici latinae* I, Berlin: G. Reimerum.

Moreno García, J.C. 2012, "Deir el-Gabrawi", in W. Wendrich (ed.), *UCLA Encyclopedia of Egyptology*, Los Angeles, (http://escholarship.org/uc/item/99j1g8zh#page-6).

Neira Faleiro, C. 2005, *La Notitia dignitatum: nueva edición crítica y comentario histórico*, Madrid: Consejo superior de investigaciones cientícas.

Papaconstantinou, A. 2001, *Le culte des saints en Égypte des Byzantins aux Abbasides: l'apport des inscriptions et des papyrus grecs et coptes*, Paris: Éditions du CNRS.

Sicard, Cl. 1982, *Œuvres* I: *lettres et relations inédites*, ed. M. Martin, Cairo: IFAO.

Timm, S. 1984–1992, *Das christlich-koptische Ägypten in arabischer Zeit*, 6 vols., Wiesbaden: Reichert Verlag.

van Berchem, D. 1952, *L'armée de Dioclétien et la réforme constantinienne*, Paris: P. Geuthner.

van Minnen, P. 1994, "The roots of Egyptian Christianity", *AfP* 40: 71–85.

Wipszycka, E. 2009, *Moines et communautés monastiques en Égypte (IVᵉ–VIIIᵉ siècles)*, Warsaw: University of Warsaw/Raphael Taubenschlag Foundation.

Zuckerman, C. 1995, "The hapless recruit Psois and the mighty anchorite, Apa John", *BASP* 32: 183–94.

15

MONKS AND SCHOLARS IN THE PANOPOLITE NOME
The epigraphic evidence

Sofia Schaten and Jacques van der Vliet

Introduction

During the conference "Perspectives on Panopolis", which took place in Leiden in 1998, Lucia Criscuolo discussed the evidence of the Greek inscriptions, including Christian ones, from the Panopolite nome, the present-day Sohag-Akhmim area. Already in the beginning of her paper, she observed that it would be "impossible to sketch a coherent picture of Panopolis on the basis of its Greek inscriptions".[1] Regrettably, the same judgment applies to the exclusively Christian sources from late antique and medieval times that are the subject of the present contribution. It is not that Christian inscriptions from the region are scarce, rather to the contrary, but the record is discontinuous and often lacks the context that might give it historical significance. Further problems, as we will see, concern the heuristics and the accessibility of parts of the material.

In the following pages, some problems and challenges of the epigraphic evidence for the Christian history of the region will be briefly discussed. Our discussion will be guided by the geographical distribution of the texts. In fact, most of the inscriptions from Christian Panopolis can be traced to one of four main provenances: first, on the east bank, the necropoleis in the vicinity of the town; and, secondly, stretching into the eastern desert, Wadi Bir al-'Ayin; then, on the opposite bank of the Nile, the White Monastery and its surroundings; and, finally, the more modest site of the Red Monastery. Wadi Bir al-'Ayin and its inscriptions have received considerable attention in Klaus Kuhlmann's book on the archaeology of the Akhmim area,[2] whereas the inscriptions in the church of the Red Monastery will be published in a volume that is due to appear under the editorship of Karel C. Innemée in the IFAO series "La peinture murale chez les coptes" (Institut français d'archéologie orientale, Cairo) [forthcoming]. As the reader can be easily referred to the publications mentioned, only two out of the four geographical clusters of epigraphic material will be dealt with below. First, the numerous group of tombstones from late antique Panopolis, intriguing on account of their idiosyncrasy, will be discussed by Sofia Schaten. Then, Jacques van der Vliet will review the

"Rive droite": the town

From the point of view of epigraphy, the town of Panopolis, modern Akhmim – a town that was a well-known production center of linen fabrics, and had a flourishing Greek-style urban life and a rich monastic *hinterland* – appears like a desert region. All the epigraphic witnesses of urban life, such as inscriptions commemorating the foundation or restoration of churches and other public buildings, seem to have disappeared. The only epigraphic material that we do have originates from funerary contexts.[3] The area around Akhmim had extensive cemeteries and, from the early second half of the nineteenth century, the discovery of tombs, mummies and fine mummy clothes made the town famous for its splendid textiles.[4] Cäcilia Fluck [(2008)] discusses these textile finds in detail and gives extensive information about the various areas and cemeteries where these may have come from. The publications about the textile finds that followed the earliest so-called excavations (for which see Fluck [2008]) contain information about several necropoleis but, to the best of my knowledge, do not mention Christian stelae. The bulk of the funerary inscriptions said to be from Akhmim (some 115 stelae in Lefebvre 1907 alone) were found in the second half of the nineteenth and the beginning of the twentieth century, and made their way into various museums.[5] None of these stelae, however, were found during regular excavations, and no doubt some of them have been given an Akhmim provenance in order to make them more interesting for the antiquities market.[6]

In spite of the relatively large numbers that have survived, many Christian stelae from the Akhmim region, as from the rest of Egypt, must have been lost. Thus, the lack of decorated stelae in various styles is conspicuous, even though there are a few stelae with an *anch*-shaped cross depicted in the center of the stone.[7] Their assignment to Akhmim is often uncertain, however.[8] Like other regions and cities that were centers of Greek culture with inhabitants of both ethnic backgrounds (Greek and Egyptian), workshops in Panopolis must have produced elaborate decorated stelae with Hellenistic and, later on, Byzantine characteristics. Variations in usage can be observed even within the same sites. In all probability, the decorated stelae from the Panopolite region were destroyed owing to natural circumstances or re-used in constructions outside the cemeteries.

Stelae with inscriptions only were sometimes set in a niche in the wall of a tomb. Since very few Coptic cemeteries have been found intact, it is difficult to determine whether stelae like those from Akhmim were placed in a niche or erected on top of a grave. The type that is commonly called typical for Akhmim is a roughly triangular stela with an inscription, a model that seems restricted to this particular region. All stelae in this class are from limestone, mostly roughly cut; their sizes may vary, but only a few measure more than 30 × 60 cm. Their shape suggests that they were erected on top of a tomb, as their relative great thickness gives them considerable stability. A second group in the more common rectangular shape but with a typical inscription possibly shares the same local background. Both groups show an individuality in the textual formulae that deserves further discussion.

Inscribed funerary monuments from medieval Christian Egypt followed the tradition of calling upon the visitor to commemorate the deceased, and recitation was thought to create a relationship between the deceased and the living person. The typical stelae from Akhmim bear a Greek inscription that consists of the introductory formula "stela of the deceased N.N.", followed by the verb "he (or she) lived" and the age of the deceased. Although variations are possible, the date of death is usually given in the form: month name, day of the month (in numbers) and number of the indiction year, a way of dating that does not allow the assignment of absolute dates.[9] In a few cases short prayers or acclamations follow, such as "do not be sorrowful, no one is immortal". The use of the word "to die" is avoided here as in other areas of Egypt, where it may be replaced with expressions like "to lay down the body" or "to go to rest". In Panopolis, the word "stela" takes the place of these phrases, which is unique for Egypt. A lot of these stelae are damaged, but the formulaic pattern of the inscriptions allows suggestions for the restoration of the missing parts. Within this stereotypical group, some variation can nevertheless be observed. Thus, a few inscriptions start with formulae like "God is one who helps" or "Oh God, have mercy on the soul" that are also known from stelae in the Hermonthis area.[10] Small crosses or an alpha and omega are the only decorative elements that are sometimes added to the text.

The second, much smaller group of stelae has the more common rectangular shape and also bears a different text (in Greek or in Coptic). Like many inscriptions from Hermonthis, they state that the deceased "ended his life" or "went to his/her rest".[11]

The only provenance given for all these stelae is Akhmim. No more precise indications are ever given. It would be interesting to investigate whether all these pieces may have come from one place or from different cemeteries, and whether they have been picked up in small compounds, belonging to one or more communities of either laymen or clergy. For questions of prosopography, demography and causes of death, an extensive research into the Akhmim stelae, covering names and occupations,[12] as well as age, gender and the month of death, would undoubtedly yield more and more precise results.[13] Also, dating the stelae from Akhmim (like similar stelae from other regions in Egypt) remains a problem, not only on account of the absence of absolute data but also owing to the lack of information about find circumstances. On various grounds, they may be dated to the sixth or early seventh century, however.[14]

"Rive gauche": the White Monastery

Shenoute and his patrons

Very few inscriptions in the White Monastery can be linked to its most famous abbot and the builder of its monumental basilica, St. Shenoute. The commemorative text inscribed on the inner face of the granite lintel above the main southern entrance to the church is a major exception.[15] Its six lines of Greek are dedicated to "the eternal memory" of "the founder" of the building, a high ranking official, the Count (*komes*) Caesarius (Kaisarios), the son of Candidianus (Kandidianos). The inscription is an invaluable document for several reasons. It is, first of all, an independent, non-literary witness to the importance of Shenoute's social network, confirming that he counted his patrons among the political and economic elite of his time. The text formally identifies one of these patrons, the Count Caesarius, who is known also from the writings of Shenoute, as the principal sponsor of the church. Furthermore, the language of the text, Greek, underlines the socio-linguistically significant observation that Coptic came only to be used for public inscriptions well over a century after the death of Shenoute.[16] Finally, although the inscription is not dated, the activity of Caesarius in this part of Egypt can with a high degree of probability be situated in the very middle of the fifth century.[17] This, in turn, yields a reliable date for the building of the monastery's great church.

Further inscriptions that might shed light on the chronology of Shenoute date from a far later period. They belong to a considerable group of Coptic *dipinti* in and around the northern conch of the church sanctuary, which have now for the greater part disappeared.[18] Two of these (Crum A.1 and A.2) contain a concise *curriculum vitae* of Shenoute himself, which provides precise, though not necessarily accurate dates for his life and times. They may date from as late as the late thirteenth or early fourteenth century.[19] Modern scholars tend to be wary of using this information, and they are obviously right, also because of the uncertainties and the lacunae in the published texts. Nevertheless, it may be observed that, while being all but contemporary, they are at least based on intentional computing, whereas contemporary sources rather tend to produce accidental information. Nor can it be excluded that the medieval chronologist had information at his disposal that is not available to the modern biographer. Thus, according to Crum's *dipinto* A.2, Shenoute was born in A.D. 348 or 349,[20] while, according to A.1, the great church of the monastery was consecrated in the 106th year of his life.[21] Combining these data yields the year 454 or 455 for the latter event, which is certainly not far off the mark. This calculation could be based upon a now lost hagiographic or epigraphic source, perhaps even a foundation inscription that, unlike the lintel of Count Caesarius, did not survive to the present.

Also near the sanctuary of the church another late antique dedicatory inscription is found, one much less known than the Caesarius lintel.[22] Regrettably, its relationship with the construction works undertaken by Shenoute remains doubtful. This brief text, again in Greek, is situated several meters above the floor, on the shaft of the northernmost of two marble columns that originally marked the entrance to the triconch of the sanctuary, but are now partly walled in. Although noticed as early as the late seventeenth century, the inscription has never been properly published and the rare authors that mentioned it failed to interpret it correctly.[23] It consists of three lines of beautifully engraved majuscules surmounted by a big Latin cross, also carefully executed in raised relief.[24] The text reads (in translation):

"In fulfilment of a vow by Heliodoros and Kallirhoe and their children".

The inscription, therefore, commemorates a Christian family who erected this column with its sculptured cross as an *ex-voto*.

At first sight, it might seem indisputable that Heliodoros too, like Caesarius, numbered among the wealthy patrons who had assisted Shenoute in financing and erecting the great monastery church. Some objections can be raised, however. First, according to Peter Grossmann, the present columns at the entrance to the sanctuary were erected there in a relatively late stage, in replacement of earlier, bigger columns.[25] Then, paleographically, the text is rather different from the Caesarius lintel, which is certainly contemporary with Shenoute. Finally, and most compellingly, in February 1909 a similar marble column with an exactly identical inscription and decoration was discovered, not in the White Monastery, but built into a house to the southeast of the town of Akhmim.[26] If both columns once belonged to one and the same building, as seems plausible, the find of the second column is not in favor of identifying this building with the White Monastery church. Although various scenarios can be envisaged, it seems more likely that the columns were originally part of another Christian or christianized building, perhaps situated in or near the town of Panopolis, which was demolished or destroyed at an unknown date and plundered for various other building or restauration works. Thus Heliodoros and his family may have contributed only indirectly and unwittingly to the construction of Shenoute's church.

Medieval patrons, artists and scholars

Before the early twentieth-century restoration that stripped the walls of the church almost entirely of their plaster coating, these must have been covered with numerous inscriptions, both painted and engraved.[27] The very few that survive or have been recorded in the past tell not only about leaders and patrons of the monastery in medieval times, long after Shenoute's death, but also about the use and development of the monastery church as sacred space. Obviously, several of these inscriptions belong to the class of dedicatory inscriptions connected with building, restoration or decoration works. The most sensational among them are undoubtedly the bilingual, Armenian and Coptic set of texts that accompany the monumental painting of Christ enthroned, still visible in the central apse of the church. They comprise both legends and prayers for the artist, the Armenian painter Theodore, and the various Armenian and Coptic sponsors of the project that was apparently completed in A.D. 1123/1124.[28] The whole set-up is reminiscent of similar "multicultural" decoration projects in Dayr al-Suryan, in Wadi al-Natrun. Even if the White Monastery has never really become a "Monastery of the Armenians", the scale and the central position of the apse painting clearly indicate the importance of Armenian patronage in the early twelfth century.

Another set of inscriptions in the northeast area of the church concerns reconstruction and decoration works that took place about a century later. Two of these texts mention a major rebuilding that was completed in the year 1259 under the Archimandrite Ioannes.[29] The former is a long *dipinto* that can still be seen today on the front of the brick pier to the north of the central apse; the latter, much briefer, was situated between two niches in the northern conch. They record the erection of "four columns" in order to reconstruct the roofing of two "tabernacles" and their adjacencies, which had become "uncovered" as the result of an earthquake. In view of the position of one of the texts (A.6) and the use of a Coptic term for "canopy" in both, it may be supposed that they commemorate the reshaping of the original, late antique triconch into its present form, with a central dome carried by four brick piers.[30]

Other texts in the same general area commemorate the donation and execution of a decoration program in this part of the church, but the nature of the works is not always entirely clear. Only in one case, a priest and monk Phibamon, who is also styled a scribe and architect, is clearly credited for sponsoring the painting of an archangel, probably St. Michael, still vaguely visible above the door to the northern *pastophorion*.[31] In addition to their obvious interest for the architectural history of the church, these dedicatory inscriptions provide lots of accidental information. Thus Crum's text A.6 gives in passing a brief biography of the Archimandrite Ioannes (ll. 17–23) as well as a hint of the (very negative) popular opinion about the rule of "the Turks", the first Mamluks (ll. 16–17).

A final group of *dipinti* from the same northeast part of the church, the famous library inscriptions, must date from the same general period as the ones quoted above (twelfth-fourteenth century).[32] They have recently been discussed and partly republished by Tito Orlandi,[33] and need only be mentioned briefly here. Written on the walls of the northern *pastophorion*,[34] they list titles of books with their quantities, and in addition contain a number of apotropaic charms written in Arabic in Coptic characters as well as prayers for the scholarly priest who wrote the inscriptions, a certain Klaute (Claudius). It would seem that they combine a shelving system with an inventory of the library that was kept in the room.

In spite of their obvious interest, some cautionary remarks about the documentary value of these inscriptions are due here, some of which have been made already by earlier authors. First, we have very little idea of the disposition of the texts on the walls, which hampers any reconstruction of the library – if that it was – as a physical and functional unit. Secondly, whereas the room in question may have stored a library, as is not unusual for a sacristy,[35] it cannot automatically be equated with *the* library of the monastery. The place where the inscriptions were situated, the northern *pastophorion*, would suggest that the books stored there were primarily those used for liturgical reading. However, even if this room really did contain a more comprehensive library, the inscriptions can only give an indication of what it may have looked like at an unknown, but certainly quite late stage. A library, even a monastic library, is not a static unity, but subject to constant renewal. This is all the more so when it passes through periods of language shift, as it occurred in the White Monastery, which saw two major language shifts in the course of its history: first from Greek to Coptic, and then from Coptic to Arabic. The library inscriptions reflect the final stages of the last of these shifts. Even if they cannot, therefore, provide a reliable guide to *the* White Monastery library, they do remain an important witness to medieval Coptic literary culture, and as such they deserve further study.

Outside of the sanctuary area, medieval painted inscriptions can be found even today at various places where the original plaster still holds – thus, in particular, on the medieval masonry piers, erected within the so-called "southern narthex", which was perhaps originally a chapter-house.[36] As far as they can be deciphered, they seem to be mostly commemorative in character, combining names and brief prayers. Some are published, like the Coptic prayers for workmen copied in the early years of the twentieth century;[37] others appear to have been never recorded. On the south wall of the same room, a more remarkable text, the Great Doxology in Greek, survived against all odds.[38] In some way, this long and formal liturgical text, enclosed in a *tabula ansata*, must have been connected with the function that this part of the "southern narthex" had in medieval times.

Adding pieces to a puzzle

The walls of the basilica are not the only source for the epigraphy of the White Monastery. The extensive ruins that surround it also yielded many inscriptions, which often still await publication. Some of these afford vivid glimpses of past monastic life. Perhaps the most stunning of them is a unique stone lintel that was found in the 1990s by an Egyptian mission, and is as yet unpublished.[39] It gives a graphical representation of what the White Monastery "federation" must have looked like hierarchically a few centuries after Shenoute's death. In polychrome relief, it depicts eight standing monastic dignitaries, represented *en face*, four to the left and four to the right of a central motif, presumably a cross, now missing. The Coptic inscription that frames the relief panels mentions their names and titles, starting from the left with "the great archimandrite", perhaps called Apa Paniskos. As the fourth in the series, a woman appears, undoubtedly the head of the nunnery. The last person portrayed is an architect and deacon, Apa Stephanos. According to the inscription, he was the founder of the monastic building for which the lintel was intended. The monument's lower architrave bears a prayer for the sculptor.

The top of the hierarchy of the White Monastery is again immortalized on a far later and much different monument. The biggest of the impressive ceremonial keys from Sohag that are now in the Coptic Museum in Cairo is inscribed in copper and silver inlay.[40] The inscriptions commemorate the monastic dignitaries who ordered it, among whom was an Archimandrite Iohannes. It has recently been suggested that he might be the same person as the like-named archimandrite who directed the rebuilding of the church sanctuary in the middle of the thirteenth century, according to the mural inscriptions quoted above.[41] The fabrication of such a monumental key would indeed well fit the wave of architectural renewal attested by these inscriptions.

In addition to other textual remains, the Egyptian mission mentioned earlier also appears to have discovered funerary inscriptions, otherwise hardly known for the White Monastery.[42] This would open an as yet unexplored field of research. Until now, the funerary epigraphy from the White Monastery area seemed to be virtually limited to a single Coptic stela, now apparently lost, which belonged to a monk, Pamin, and was published more than a century ago by W. de Bock.[43] The publication of the recent finds could pave the way for a study of local funerary habits and formulae as it is possible for other centers in Upper Egypt.

This brief review may give an indication of the interest and broad variety of the epigraphic material from the White Monastery. It also shows how much work remains to be done. Even within the relatively well-explored area of the great church, many inscriptions wait to be published or even identified. Others are available only in old or substandard publications. Already quite a lot have vanished completely or are doomed to vanish soon. About recent finds made in the vicinity, almost all information is lacking. It must therefore be urgently recommended that, whatever conservation or research projects are considered for the White Monastery or the surrounding area, they should on principle include a systematic and comprehensive survey of the epigraphic material, either lost or surviving. In reconstructing the long but poorly known history of this important center of learning and piety, the evidence of the inscriptions cannot be dispensed with.

Notes

1 Criscuolo 2002: 56.

2 Kuhlmann 1983: 6–9, with pl. 5–15; two previously unknown texts discovered by Kuhlmann were published by Guy Wagner (1982; cf. Lajtar 1993).

3 For further information about funerary stelae, Greek and Coptic, from Christian Egypt and their classification, see Crum 1902 (catalogue of Christian stelae in Greek and Coptic in the Egyptian Museum; most of these are in the Coptic Museum today), Lefebvre 1907 (catalogue of Christian inscriptions in Greek from collections in Egypt and abroad), Zuntz 1932 (an attempt to locate stelae in different areas and places), Krause 1991 (general introduction to the Christian epigraphy of Egypt), Wietheger 1992 (analysis of the inscriptions from the Monastery of St. Jeremiah at Saqqara), *SB Kopt.* I-III (Coptic "Sammelbuch" including many funerary inscriptions), Thomas 2000 (about decorated stelae). Stelae from Achmim were first collected in Lefebvre 1907 and 1911; for recent reviews of this material, see Timm 1984–1992: I, 90, n. 44–6, and Criscuolo 2002, to which may be added: Lajtar and Twardecki 2003: nos. 94–5 (Adam Lajtar), and Gascou 2004.

4 For the archaeological remains in the area, see Kuhlmann 1983; McNally and Dvoržak Schrunk 1993.

5 Most of them are now in the Greco-Roman Museum in Alexandria, a few others in the Coptic Museum, Cairo, and in various collections in Europe and the United States. As

MONKS AND SCHOLARS IN THE PANOPOLITE NOME

yet no special publication has been devoted to them; most of the stelae in Alexandria are known only from Lefebvre 1907, where no pictures are given.

6 Thus, in Lefebvre 1907 and 1911, over fifty pieces are labelled "Achmîm?".

7 Cramer 1955: 20–6, figs. 19–21; Cramer 1957: no. 22.

8 E.g., in the case of the stelae Crum 1902: nos. 8575 (= Coptic Museum inv. no. 8656) and 8603.

9 For the indiction, a fifteen-year cycle introduced for fiscal purposes in the early fourth century, see now Bagnall and Worp 2004.

10 E.g., Lefebvre 1907: nos. 263, 294, 345.

11 Lefebvre 1907: no. 673, cf. Pelegrini 1907: no. 3; Lefebvre 1911: 238–45.

12 Cf. Timm 1984–1992: I, 90, n. 46.

13 Thus, Scheidel 2001 records seasonal mortality in Egypt from ancient to modern times; his "Appendix one" lists months of death after epitaphs from ancient and medieval Egypt and Nubia, including stelae from Achmim quoted after Lefebvre 1907.

14 See Krause 1991: 1293–94.

15 *SB* III 6311; *ed. princeps* with facsimile: Lefebvre 1920a: 470–75; see also Lefebvre 1920b; Monneret de Villard 1923; 1925–26: I, 18–22; Emmel 1998: 94. The information in Timm 1984–1992: II, 608, n. 44–5, is misleading: the lintel is no tombstone and the text is entirely unambiguous.

16 Cf. Bingen 1999: 613–14.

17 As can be inferred from works by Shenoute that mention him; see Emmel 1998: 94, taking up earlier discussions by Monneret de Villard (1923, 1925–26: I, 18–22). A Coptic fragment edited by Johnson (1976: 10, 1 vo, col. a, l. 29) associates Caesarius with the death of Nestorius, which would date his activity again around 445–55, but the text smells of hagiographical embroidery. No other independent sources for Caesarius appear to exist (cf. Martindale 1980: 249–50). [but see now Gascou 2002]. Lefebvre (1920a: 475) dates the inscription paleographically to the first half of the fifth century.

18 Published by W.E. Crum (1904) after copies by the English clergyman W.J. (not W.T.) Oldfield (1857–1934; *Who was who 1929–1940*: 1021).

19 If the commemoration of the artist Merkouri in A.1, ll. 23–5 (cf. Crum 1904: 555; Monneret de Villard 1925–26: I, 28; Coquin 1975: 277–78) indeed belongs to the same text.

20 Crum 1904: 555–56.

21 Crum 1904: 554, ll. 13–16.

22 Van der Vliet, ["*Parerga*: Notes on christian inscriptions from Egypt and Nubia", *ZPE* 164 (2008), 153–55 = Study 16], offers a fuller discussion of this inscription than is possible here.

23 Vansleb (Wansleben 1677: 374) erroneously describes it as an epitaph. Coquin and Martin (1994: 765) think it belonged to a pagan temple, which may be correct for the column, but not for the inscription. They are followed by Criscuolo 2002: 60–1, who tentatively links it to a far older and entirely unrelated monument of a Triphis priest (*SEG* 43, no. 1124). According to Monneret de Villard (1925–26: I, 25, n. 6) it had disappeared; other descriptions of the church do not appear to mention it.

24 It is technically impossible that the cross is a Christian addition, post-dating the inscription. The two belong together, as in the column's exact counterpart quoted below.

25 Grossmann 2002b: 127; cf. 2002a: 533–34 (where he dates the capitals on top of the present columns to the sixth century).

26 Lefebvre 1910: 62–3, no. 815 (*SB* I 1597). This column was transported to Cairo in 1909, but I have been unable to trace it.

27 Cf. Crum 1904: 552.

28 See Kapoïan-Kouymjian 1988: 16–17 (includes photographs and extensive bibliography); also Crum 1904: 556–57.

29 Crum 1904: A.6 and A.7.
30 Cf. C.R. Peers *apud* Crum 1904: 569; Monneret de Villard 1925–26: I, 28–31; Grossmann 2002a: 535–36.
31 Crum 1904: A.10. Monneret de Villard 1925–26: I, 30, tends to make this Phibamon the architect responsible for the restoration works described by the other inscriptions, but this is far from certain.
32 The texts are now completely lost; the copies by W.J. Oldfield as published by Crum 1904: 564–69, nos. B.12–31, are our only source.
33 Orlandi 2002: 213–15; cf. Khosroyev 2003; Takla 2005.
34 See Crum 1904: 552–53; *pace* Orlandi 2002: 211–12.
35 Cf. Crum 1904: 553, n. 4.
36 See Grossmann 2002a: 531–32.
37 Lefebvre 1920a: 485–86, 488, fig. 3660.
38 Lefebvre 1907: no. 237; 1920a: 485, fig. 3658; cf. Leclercq 1921: 2511–12; 1925: 2891–93; Quecke 1970: 276; Van Haelst 1976: no. 773; Tidda 2001: 118–19.
39 I owe my knowledge of this piece to a set of photographs made by the excavators. It presently consists of two blocks of about 36 × 46 cm each; it can be dated to about the seventh-ninth century with the greatest caution only.
40 Inv. no. 5915; see Bénazeth and Boud'hors 2003 (*editio princeps* of the texts and extensive discussion).
41 Bénazeth and Boud'hors 2003: 31 and 36.
42 Known to me only in inexpert transcriptions.
43 De Bock 1901: 69, no. 81 ("une stèle calcaire trouvée près des ruines de la ville d'Athribis").

Bibliography

Bagnall, R.S., and K.A. Worp. 2004, *Chronological systems of Byzantine Egypt: second edition*, Leiden/Boston: Brill.

Bénazeth, D., and A. Boud'hors. 2003, "Les clés de Sohag: somptueux emblèmes d'une austère reclusion", in: Chr. Cannuyer (ed.), *Études coptes VIII: Dixième journée d'études, Lille 14–16 juin 2001*, Lille/Paris: Association francophone de coptologie, 19–36.

Bingen, J. 1999, "L'épigraphie grecque de l'Égypte post-constantinienne", in: *XI Congresso internazionale di epigrafia greca e latina, Roma, 18–24 settembre 1997. Atti II*, Rome: Edizioni Quasar, 613–24.

Coquin, R-G. 1975, "Les inscriptions pariétales des monastères d'Esna: Dayr al-Šuhadâ – Dayr al-Fahûrî", *BIFAO* 75: 241–84.

Coquin, R-G., and M. Martin. 1994, "Dayr Anbâ Shinûdah: history", in: *CoptEnc* 3: 761–66.

Cramer, M. 1955, *Das altägyptische Lebenszeichen im christlichen (koptischen) Ägypten*, third edition, Wiesbaden: Harrassowitz.

Cramer, M. 1957, *Archäologische und epigraphische Klassifikation koptischer Denkmäler des Metropolitan Museum of Art, New York, und des Museum of Fine Arts, Boston, Mass*, Wiesbaden: Harrassowitz.

Criscuolo, L. 2002, "A textual survey of Greek inscriptions from Panopolis and the Panopolite", in: A. Egberts, B.P. Muhs, and J. van der Vliet (eds.), *Perspectives on Panopolis*, Leiden/Boston/Cologne: Brill, 55–69.

Crum, W.E. 1902, *Coptic monuments*, Cairo: IFAO; reprinted Osnabrück: Otto Zeller Verlag 1975.

MONKS AND SCHOLARS IN THE PANOPOLITE NOME

Crum, W.E. 1904, "Inscriptions from Shenoute's monastery", *Journal of Theological Studies* 5: 552–69.

De Bock, W. 1901, *Matériaux pour servir à l'archéologie de l'Égypte chrétienne*, St. Petersburg: E. Thiele.

Emmel, S. 1998, "The historical circumstances of Shenute's sermon *God is blessed*", in: M. Krause and S. Schaten (eds.), *ΘΕΜΕΛΙΑ: Spätantike und koptologische Studien Peter Grossmann zum 65. Geburtstag*, Wiesbaden: Reichert Verlag, 81–96.

Fluck, C. 2008, "Akhmim as a source of textiles", in: G. Gabra and H.N. Takla (eds.), *Christianity and monasticism in Upper Egypt*, vol. 1: *Akhmim and Sohag*, Cairo/New York: American University in Cairo Press, 211–23.

Gascou, J. 2002, "Décision de Caesarius, gouverneur militaire de Thébaïde", in: Vincent Déroche et al. (eds.), *Mélanges Gilbert Dagron*, Paris: Centre de recherche d'histoire et civilisation de Byzance, 269–77.

Gascou, J. 2004, "Une stèle funéraire panopolite du Musée archéologique de Strasbourg", *Cahiers alsaciens d'archéologie, d'art et d'histoire* 47: 7–10.

Grossmann, P. 2002a, *Christliche Architektur in Ägypten*, Leiden: Brill.

Grossmann, P. 2002b, "Die klassischen Wurzeln in Architektur und Dekorsystem der grossen Kirche des Schenuteklosters bei Suhâg", in: A. Egberts, B.P. Muhs, and J. van der Vliet (eds.), *Perspectives on Panopolis*, Leiden/Boston/Cologne: Brill, 115–31.

Johnson, D.W. 1976, "Further fragments of a Coptic history of the Church: Cambridge OR. 1699 R", *Enchoria* 6: 7–17.

Kapoïan-Kouymjian, A. 1988, *L'Égypte vue par des arméniens (XIe-XVIIe siècles)*, Paris: Editions de la Fondation Singer-Polignac.

Khosroyev, A.L. 2003, "Aus der Lektüre der koptischen Mönche in arabischer Zeit", in: W. Beltz (ed.), *Die koptische Kirche in den ersten drei islamischen Jahrhunderten*, Halle: Martin-Luther-Universität, 121–29.

Krause, M. 1991, "Inscriptions", in: *CoptEnc* 4: 1290–99.

Kuhlmann, K.P. 1983, *Materialien zur Archäologie und Geschichte des Raumes von Achmim*, Mainz: Philipp von Zabern.

Łajtar, A. 1993, "Zu einer christlichen Inschrift mit dem Gebet an den heiligen Georgios aus Wadi Bir el-Ain, Ägypten", *ZPE* 98: 243–44.

Łajtar, A., and A. Twardecki. 2003, *Catalogue des inscriptions grecques du Musée national de Varsovie*, Warsaw: University of Warsaw/Raphael Taubenschlag Foundation.

Leclercq, H. 1921, "Égypte", in: *DACL* 4/2: 2401–2571.

Leclercq, H. 1925, "Hymnes", in: *DACL* 6/2: 2826–2928.

Lefebvre, G. 1907, *Recueil des inscriptions grecques-chrétiennes d'Égypte*, Cairo: IFAO; reprinted Chicago: Ares 1978.

Lefebvre, G. 1910, "Égypte chrétienne II", *ASAE* 10: 50–65.

Lefebvre, G. 1911, "Égypte chrétienne IV", *ASAE* 11: 238–50.

Lefebvre, G. 1920a, "Deir-el-Abiad", in: *DACL* 4/1: 459–502.

Lefebvre, G. 1920b, "Inscription grecque du Deir-el-Abiad", *ASAE* 20: 250.

McNally, S., and I. Dvoržak Schrunk. 1993, *Excavations in Akhmîm, Egypt: continuity and change in city life from late antiquity to the present*, Oxford: Tempus Reparatum.

Martindale, J.R. 1980, *The prosopography of the later Roman Empire* II: *A.D. 395–527*, Cambridge: Cambridge University Press.

213

SOFIA SCHATEN AND JACQUES VAN DER VLIET

Monneret de Villard, U. 1923, "La fondazione del Deyr el-Abiad (SB III 6311)", *Aegyptus* 4: 156–62.

Monneret de Villard, U. 1925–1926, *Les couvents près de Sohag (Deyr el-Abiad et Deyr el-Ahmar)*, Milan: Tipografia e libreria pontificia arcivescovile S. Giuseppe.

Orlandi, T. 2002, "The library of the Monastery of Saint Shenute at Atripe", in: A. Egberts, B.P. Muhs, and J. van der Vliet (eds.), *Perspectives on Panopolis*, Leiden/Boston/Cologne: Brill, 211–31.

Pellegrini, A. 1907, "Stele funerarie copte del Museo archeologico di Firenze", *Bessarione* 22 [ser. 3/3]: 20–43.

Quecke, H. 1970, *Untersuchungen zum koptischen Stundengebet*, Louvain: Peeters.

Scheidel, W. 2001, *Death on the Nile: disease and the demography of Roman Egypt*, Leiden: Brill.

Takla, H.N. 2005, "The library of the Monastery of Saint Shenouda the Archimandrite", *Coptica* 4: 43–51.

Thomas, Th. K. 2000, *Late antique Egyptian funerary sculpture: images for this world and the next*, Princeton: Princeton University Press.

Tidda, F. 2001, "Terminologia della luce e battesimo nelle iscrizioni greche cristiane", *Vetera Christianorum* 38: 103–24.

Timm, S. 1984–1992, *Das christlich-koptische Ägypten in arabischer Zeit*, 6 vols., Wiesbaden: Reichert Verlag.

Van der Vliet, J. 2008, "*Parerga*: two notes on Christian inscriptions from Egypt", *ZPE* 164: 153–58 [Study 16].

Van Haelst, J. 1976, *Catalogue des papyrus littéraires juifs et chrétiens*, Paris: Sorbonne.

Vansleb [J.M. Wansleben]. 1677, *Nouvelle relation en forme de journal d'un voyage fait en Egypte . . . en 1672 & 1673*, Paris: Estienne Michallet.

Wagner, G. 1982, "Deux prières chrétiennes du Wadi Bir el-Aïn", *BIFAO* 82: 349–54.

[N.N.], *Who was who 1929–1940*, London 1949: A. & C. Black.

Wietheger, C. 1992, *Das Jeremias-Kloster zu Saqqara unter besonderer Berücksichtigung der Inschriften*, Altenberge: Oros Verlag.

Zuntz, D. 1932, "Koptische Grabstelen: Ihre zeitliche und örtliche Einordnung", *MDAIK* 2: 22–38.

16

PARERGA
Notes on Christian inscriptions
from Egypt and Nubia

Jacques van der Vliet

1. A family of donors from the Panopolite: *SB* I 1597
and White Monastery

The mid-fifth-century basilica of Shenoute's White Monastery near modern Sohag is even in its present mutilated state by far the most impressive late antique monument of Egypt. The entrance to the original sanctuary, which has the form of a triconch, is marked by two marble columns that are now partly walled in. The shaft of the northernmost of these columns bears a Greek inscription that has been known since the seventeenth century but has never been published. This may seem not to be a serious omission, as it is identical to another one, found in *SB* I under no. 1597. In fact, the connection between both inscriptions was only made quite recently, and gave rise to a number of questions and hypotheses. A personal inspection of the White Monastery inscription in February 2006 answered some of these questions, but also raised other ones. It may therefore be worthwhile to have a look at both inscriptions, taking the White Monastery as our point of departure.[1]

The White Monastery inscription (Figure 16.1) is situated several meters above floor level, and therefore I was unable to take measurements. The text consists of three lines of large and carefully engraved late antique majuscules, provided with *apices* throughout. A, Δ and Λ have a horizontal stroke on top; the symmetrical A has a broken bar; the characteristic, very wide ω consists of two detached near-circular halves connected by a horizontal bar; E, by contrast, is relatively narrow and angular. The text is surmounted by a large Latin cross, also carefully executed in raised relief. The column is partly built into a much more recent brick pier, and traces of paint and mortar affect the lefthand part of the inscription. Nevertheless, the text is apparently complete and easily legible, apart from the very first letters of each line.

White Monastery (Sohag),   sixth century?

Ὑπὲρ εὐχῆς Ἡλιοδώρου
καὶ Καλλιρόης καὶ τῶν
τέκνων αὐτῶν

Figure 16.1 Inscription of Heliodoros in the White Monastery (Sohag) (photo: Jacques van der Vliet)

In fulfillment of a vow by Heliodoros and Kallirhoe and their children.

A perfect twin to this inscription was discovered by Gustave Lefebvre in 1909 on another marble column, found in a house to the southeast of the city of Akhmim. Gustave Lefebvre published this piece in 1910,[2] and the text entered the *Sammelbuch* under no. 1597. His edition is accompanied by a brief description and a facsimile made after a rubbing of the text (Figure 16.2). In his summary commentary to the text, Lefebvre stresses its Christian character. Comparison of his information with the inscription as visible in the White Monastery shows that both monuments must be completely identical as regards material, script, text, disposition of the text and decoration.[3] Lefebvre's column was transported to Cairo in 1909, but its present whereabouts are unknown to me.

The White Monastery inscription was first noticed in 1673 by the early orientalist J.M. Wansleben (Vansleb), who describes it as "an epitaph in Greek letters of a man named Heliodoros".[4] Subsequent descriptions of the monastery do not mention it. Even Lefebvre himself, who wrote a long and thorough study of the monastery,[5] apparently overlooked it. U. Monneret de Villard, the author of a standard work on the monasteries of Sohag, knew Wansleben's description, but he believed the inscription had disappeared.[6]

Only quite recently, R.-G. Coquin and M. Martin noticed the continued presence of the inscription on the same spot where it had been seen by Wansleben two

Figure 16.2 Inscription of Heliodoros from Akhmim (after Lefebvre's facsimile)

hundred years earlier.[7] Moreover, Coquin and Martin were the first to connect it with the similar inscription that had been found by Lefebvre in 1909. They did not seem to have any doubts about Wansleben's identification of the text as an epitaph, while at the same time they suggested that the columns were "a pair of columns from a pagan temple". Some years ago, L. Criscuolo, following Coquin and Martin, took the discussion one step further.[8] She proposed to connect the donor of the columns with another Heliodoros, a priest of Triphis, whose *topos*, presumably at Panopolis, is mentioned on a stela published in 1993 in this journal by E. Bernand.[9] She cast doubt on the funerary character of the texts on the columns, and suggested, with due caution, that they might come from the same *topos* as Bernand's stela.

From this rather meager dossier, three questions emerge. Are the inscriptions funerary? Are they pagan or Christian? And can they be connected with the *topos* of a certain Heliodoros who was a local priest of Triphis? Even a brief analysis of both monuments in conjunction can answer these questions. Further questions, however, concerning their original context and their date, are less easily resolved.

The textual formula used on the columns (ὑπὲρ εὐχῆς . . .) is not, as Lefebvre assumed, exclusively Christian. In Egyptian epigraphy it is very rare and, actually, not one of the five other Egyptian occurrences known to me can be considered Christian.[10] Christian examples from Syria and elsewhere in the East abound, however. The present text, too, is unmistakably Christian in nature. On either column, the inscription is intentionally conceived in relation to the sculptured cross above it and, since the cross is in raised relief, it must have been executed prior to or simultaneously with the engraved text immediately below. Clearly, text and relief constitute one single unit. This automatically precludes the identification of the present donor, Heliodoros, with his namesake, the priest of Triphis. The latter's modest sandstone stela bears a relief of the young god Horus on the lotus, and may date from about the first or second century.[11] Also artistically and paleographically it is far distant from the superb marble columns. These columns may once have belonged to a pagan temple, as Coquin and Martin supposed, but in that case their text and decoration are indicative of a later re-use. In view of their similarity, they must have been destined for one and the same Christian or Christianized building, presumably a church or oratory. A funerary function is hardly likely. The textual formula is typically dedicatory in character, and there is nothing distinctly funerary about the text or the monuments themselves.

Apparently, therefore, both inscriptions commemorate a wealthy Christian family who erected these two columns with their sculptured crosses as an *ex-voto*, to adorn a monumental building in or near Panopolis. As one of them is still *in situ*, it is tempting to assume that Heliodoros and his wife had assisted Shenoute in erecting the great basilica of the White Monastery around 450. It is a well-known fact, also reflected in his own writings, that Shenoute counted his patrons among the members of the political and economic elite of his time.[12] That these effectively took pride in sponsoring his building operations is shown by *SB* III 6311. This inscription on a lintel that is still *in situ* above the main southern entrance to the church celebrates the *comes* Caesarius (Kaisarios), son of Candidianus, as its founder (κτίστης).[13] Heliodoros would seem to belong in the same category of rich benefactors.

Against such a reconstruction of the facts, several objections can be raised, however. First, the column published by Lefebvre was not discovered in the neighborhood of the White Monastery, but on the opposite bank of the Nile, near Akhmim. If both columns had once belonged to one and the same building, as seems plausible, the provenance of this second column is not in favor of identifying it with the still standing White Monastery church. Secondly, according to P. Grossmann's reconstruction of the architectural history of the monastery church, the columns at the entrance to the sanctuary have been erected there at a relatively late stage, in replacement of earlier, taller columns.[14] In his view, this work was part of a restoration following a fire that destroyed part of the church, presumably during the Persian occupation of the early seventh century. Grossmann dates the capitals on top of the present columns to the sixth century. If this applies to the columns themselves as well, the inscriptions would post-date Shenoute, who probably died in 465. The mannerisms of the script would not contradict such a date.

Although various scenarios can be envisaged, it seems most likely that the columns inscribed for Heliodoros were originally part, not of the White Monastery complex, but of another building, perhaps a church in or near the city of Panopolis. This building was demolished or destroyed at a date that cannot be determined with certainty, to be re-used in building or restoration works at various other places in the neighborhood. Not only this building, but also its founders or benefactors, the Heliodoros family, cannot be further identified. The name Heliodoros was borne by many late antique Egyptians, but none of those presently known can be confidently identified with the rich donor from Panopolis. The name Kallirhoe, by contrast, is far from banal, but a search in the existing databases produced nobody who can possibly have been the wife of the Christian Heliodoros. Ironically, as pious founders, Heliodoros and Kallirhoe remain foremost connected with Shenoute's church, to the construction of which they probably contributed only unwittingly.

PARERGA

2. The day young Gregory died: Coptic stela Liebieghaus 13
(*SB Kopt*. III 1596)[15]

The date of the Coptic funerary stela of a boy or young man Gregory, presently in the Museum Liebieghaus in Frankfurt am Main, has been disputed for over a decade. The stela formerly belonged to the collections of the renowned Christian epigraphist and archaeologist Carl Maria Kaufmann (1872–1951). Its text, inscribed on a re-used slab of white marble is, for its general inspiration and personal tone, quite close to the poetical stelae of the "Totenklage" type from Middle Egypt.[16] Although its exact provenance is unknown, it may safely be supposed to come from the same region, most likely Antinoe or its hinterland.

Here we are only concerned with its year date, contained in l. 17. At this particular spot, the otherwise well-preserved stone suffered from minor corrosion.[17] In his 1994 *editio princeps* of the stone, W. Brunsch avowed himself unable to read the date in a satisfactory way. He noted rightly, though, that after the words ntïpomпe taï, "in this very year", a year (indiction or Diocletian) is expected. Only, the traces which he had deciphered, half-way correctly as it appears, as ϫүрⲁ, did not, in his opinion, fit such a reading.[18]

A few years later, in this journal, L.S.B. MacCoull tried to correct Brunsch's readings after the mediocre reproduction in the *editio princeps*. In her view, the group following ntïpomпe taï could be read: (ⲁ)ⲡ(ⲟ) ϙ (ⲙⲁⲣⲧⲩⲣⲱⲛ) ү4ⲁ, "Martyrs year 491", correcting Brunsch's rho into a qoppa.[19] The resulting date, which is basically correct, was claimed by the author to be possibly the earliest attestation of a Martyrs year, the earliest known then being 502.[20] Her interpretation, however, demands a highly questionable abbreviation of ⲁⲡⲟ, whereas she added the era name "of the Martyrs" out of the blue.

These aspects of her readings were criticized by R.S. Bagnall and K.A. Worp in the revised 2004 edition of their handbook about the chronology of Byzantine Egypt.[21] Instead, they proposed to read: ⲧⲓⲁ. ⲩⲣⲁ, with ⲁ. ⲩⲣⲁ as a plausible abbreviation for δευτέρα. This interpretation yields a second indiction year, but no absolute date at all. Bagnall and Worp were certainly right in expecting an indiction year after the phrase ntïpomпe taï. This is a common pattern that one finds frequently in stelae from, for example, Antinoe.[22] Their supposed article ⲧⲓ- looks somewhat out of place, however. As a rule such Coptic indiction dates are constructed with the bare Greek numeral in the genitive.[23]

In fact, both the indiction year and the absolute date are there. An examination of the stone itself in June 2006 permitted the following reading of the dating lemma (ll. 14–17):

     ⲛⲧⲁ ⲡⲙⲁⲕⲁⲣ(ⲓⲟⲥ)
15. ⲅⲣⲏⲅⲟⲣⲉ ⲉⲙⲧⲟⲛ ⲛⲙⲟⲃ ⲛ-
     ⲥⲟⲩ ⲙ̄ⲛ̄ⲧⲱⲙⲏⲛ ⲛⲡⲁⲣⲙ\ⲟⲩ/
     ⲛⲧⲓⲣⲟⲙⲡⲉ ⲧⲁⲓ ⲓⲅ ⲭ(ⲣⲟⲛⲱⲛ) ⲩ4ⲁ

14. ⲡⲙⲁⲕⲁⲣⲥ ‖ 16. l. ⲡⲁⲣⲙ`ⲟⲩ´ⲧⲉ ‖ 17. ⲭ': χρόνων

The blessed Gregory went to rest (i.e. died) on the eighteenth of Parmoute in this very year (of the indiction) 13, of the era 491 (i.e. 13 April A.D. 775).

The abbreviation of χρόνος by a single χ (in l. 17) is unusual,[24] and the supposed stroke to mark it is uncertain. Also the *qoppa* looks very much like a *rho*, only slightly weaker in shape perhaps. Nevertheless, the entire group leaves no room for a different interpretation. The Diocletian year 491 actually coincides with a thirteenth indiction year, which confirms the present reading. Since the era name (Diocletian or Martyrs) is not specified, there is no reason to consider this the earliest attestation of a Martyrs year, however.

The newly obtained readings show the stela of young Gregory to be also chronologically close to the "Totenklage" stelae, which usually bear eighth-ninth-century dates.

3. A Bishop Leo in Saï?

In a recent article on Christian Saï, Adam Łajtar offers a useful survey of the written sources presently available for this little-known administrative and ecclesiastical center in northern Makuria.[25] These include some stone epitaphs of local bishops, mostly in Sahidic Coptic.[26] A single one in Greek was published by B. Boyaval in 1972.[27] In his edition, he transcribed the lemma containing the name of its owner (l. 7) as:

τοῦ [ἀ]ββᾶ Λ[έ]ων ἐπισκόπου ΖΑΗ
(God . . . give rest to the soul) of Abba Leo, bishop of Saï.

Łajtar briefly discusses this stela, which he attributes to the eleventh-twelfth century on paleographical grounds, but he does not question Boyaval's reading of the name.[28] Yet, Leo is an "impossible" name for a Monophysite bishop. In the Coptic tradition, this name was firmly associated with the abhorred "Tome of Leo", Leo the Great's *Tomus ad Flavianum* of 449, which had become the embodiment of all the errors imputed to the Council of Chalcedon (451).[29] In the absence of any evidence for a Melkite hierarchy in Saï, the occurrence of such a name looks suspect.[30]

In fact, even from the mediocre photo in the *editio princeps*, it can be seen that the entire phrase has been rendered incorrectly by Boyaval. Instead of τοῦ, the abbreviated group τὸν δοῦλον is to be read, as usually in an incorrect accusative case (for: τοῦ δούλου σου).[31] The title ἀββᾶ is really there, but it is abbreviated as ⲁⲃ̄, with a clearly visible stroke above, which is common enough.[32] It is followed by a small blank to separate it from the following name. This name, accordingly, should count six letters, not four. I propose to read ⲥⲩⲙ[ⲉ]ⲱⲛ, Συμεῶν. The initial

PARERGA

c is of a rather ugly squarish shape, but certainly not a в. The following two letters are hardly doubtful. The name retains the nominative form, instead of the required genitive case, as is quite common in late epitaphs from Nubia.[33] To sum up, a correct rendering of the full lemma in l. 7 of the stela, retaining the case endings of the original, would be:

τὸν δ(οῦλον σου) ἀβ(βᾶ) Συμ[ε]ῶγ ἐπισ‵κ′(όπου) Ζαη
(God . . . give rest to the soul) of your servant Abba Symeon, bishop of Saï.

A Bishop Symeon of Saï is actually known from a juridical document in Old Nubian, which is dated to the reign of a King Basil and the episcopacy of ⲍⲓⲙⲉⲱⲛ, ⲡⲁⲡⲁⲥ of ⲍⲁⲉⲓ.[34] F.L. Griffith, the first editor of the document, tended to identify this King Basil with the like-named king who sent an embassy to Cairo in or around 1089.[35] In that case, the bishop of the document and the Abba Symeon of the Greek epitaph could well be one and the same person. Other Nubian kings named Basil are known, however, and the document offers no further dating criteria. It cannot be excluded that several bishops called Symeon have occupied the see of Saï in the course of its long history, as there were at least two called Iesou.[36] In any case, the name "Leo" can safely be deleted from the list of its incumbents.

Notes

1   I thank the conveners and hosts of the 2006 Sohag Symposium for their hospitality, and P. Grossmann and H.-G. Severin for kindly answering my questions.

2   "Égypte chrétienne II", *ASAE* 10 (1910), 50–65 at 62–3, no. 815.

3   Lefebvre notes that the column is 3.50 m high. From his facsimile, only very minor differences can be discerned: in the White Monastery inscription, l. 3 takes up more space and ends further to the right than in the other one, and the cross seems to be placed a little more to the right.

4   Vansleb [J.M. Wansleben], *Nouvelle relation en forme de journal d'un voyage fait en Egypte . . . en 1672 & 1673*, Paris: Estienne Michallet 1677, 374.

5   Art. "Deir-el-Abiad", in: *DACL* 4/1 (1920), 459–502.

6   *Les couvents près de Sohâg (Deyr el-Abiad et Deyr el-Ahmar)*, Milan 1925–1926, I, 25, n. 6.

7   Art. "Dayr Anbâ Shinûdah: history", in: *CoptEnc* 3, 761–66 at 765.

8   "A textual survey of Greek inscriptions from Panopolis and the Panopolite", in: A. Egberts, B.P. Muhs, and J. van der Vliet (eds.), *Perspectives on Panopolis*, Leiden/Boston/Cologne: Brill 2002, 55–69 at 60–1.

9   "Un prêtre de la déesse Triphis", *ZPE* 96 (1993), 64–6; cf. *SEG* 43, 1124.

10   *SB* XII 11.100 may date from the fourth-fifth century, but appears to be Jewish; cf. K. Herbert in *I. Brooklyn* 32; W. Horbury and D. Noy, *Jewish inscriptions of Graeco-Roman Egypt*, Cambridge: Cambridge University Press 1992, no. 134.

11   See note 9 above.

12   See e.g. J. Leipoldt, *Schenute von Atripe*, Leipzig: J.C. Hinrichs 1903, 162–65; J. Hahn, "Hoher Besuch im Weissen Kloster: Flavianus, praeses Thebaidis, bei Schenute von Atripe", *ZPE* 87 (1991) 248–52.

13 *Ed. princeps* with facsimile: Lefebvre, "Deir-el-Abiad", 470–75; for further discussion, see Monneret de Villard, *Les couvents près de Sohâg* I, 18–22; S. Emmel, "The historical circumstances of Shenute's sermon *God is blessed*", in: M. Krause and S. Schaten (eds.), *ΘΕΜΕΛΙΑ: Spätantike und koptologische Studien Peter Grossmann zum 65. Geburtstag*, Wiesbaden: Reichert Verlag, 81–96 at 94.

14 P. Grossmann, "Die klassischen Wurzeln in Architektur und Dekorsystem der grossen Kirche des Schenuteklosters bei Suhâg", in: A. Egberts, B.P. Muhs, and J. van der Vliet (eds.), *Perspectives on Panopolis*, Leiden/Boston/Cologne: Brill 2002, 115–31 at 127; idem, *Christliche Architektur in Ägypten*, Leiden: Brill 2002, 533–34.

15 I am much obliged to the staff of the Museum Liebieghaus, in particular P.C. Bol and G. Kaminski, for their hospitality, to O.E. Kaper, for an earlier collation of the text, and to K.A. Worp, for discussion. A full publication of the inscriptions from the Kaufmann collection is a desideratum.

16 Studied by M. Cramer, *Die Totenklage bei den Kopten*, Vienna/Leipzig: Hölder-Pichler-Tempsky 1941, and H.D. Schneider, "The lamentation of Eulogia: a Coptic dirge in the Leiden Museum of antiquities", *Oudheidkundige mededelingen* 50 (1969), 1–7.

17 For a good photo, see H. Beck, and P.C. Bol (eds.), *Liebieghaus: Museum alter Plastik, Frankfurt am Main*, Frankfurt am Main: Liebieghaus 1978, fascicle "Ägypten", Blatt 3.

18 W. Brunsch, "Koptische Grabinschriften aus der Sammlung des Liebieghauses in Frankfurt am Main", in: S. Giversen, M. Krause, and P. Nagel (eds.), *Coptology: past, present, and future. Studies in honour of Rodolphe Kasser*, Louvain: Peeters 1994, 1–6 at 6, on l. 17. He apparently did not read the group between ⲧⲁⲓ and his ⲭⲩⲣⲁ. Brunsch's text is reprinted with some corrections in *SB Kopt.* III 1596.

19 "Lesefrüchte", *ZPE* 123 (1998), 204–06 at 204, no. 1.

20 See R.S. Bagnall and K.A. Worp, *Chronological systems of Byzantine Egypt: second edition*, Leiden: Brill 2004, 82.

21 *Chronological systems*, 87.

22 See H. Munier, "Stèles chrétiennes d'Antinoé", *Aegyptus* 29 (1948), 126–36 at 129–34, for several representative examples (republished as *SB Kopt.* I 1069–77).

23 As e.g. Munier, "Stèles chrétiennes", 131, no. 4, ll. 9–11: ⲛⲧⲓⲣⲟⲙⲡⲉ ⲧⲁⲓ ⲁⲉⲩⲧⲉⲣⲁⲥ (*SB Kopt.* I 1072).

24 Note that the single example quoted by M. Avi-Yonah, "Abbreviations in Greek inscriptions", reprinted in: A.N. Oikonomides, *Abbreviations in Greek*, Chicago: Ares 1974, 110, is misleading (the ⲭ borders a lacuna). The ⲭⲣ ligature read by MacCoull is certainly not found on the stone.

25 "Christian Saï in written records (inscriptions and manuscripts)", *JJP* 36 (2006), 91–104.

26 *I. Cambridge Egypt* 155; *I. Khartoum Copt.* 27 (cf. Łajtar, "Christian Saï", 92–3, nos. 1 and 2). A third one (unpublished) was found in 2002; see N. Grimal and E. Adly, "Fouilles et travaux en Égypte et au Soudan, 2002–2003", *Orientalia* 73 (2004), 1–149 at 130–31, fig. 41.

27 "Une stèle grecque inédite de l'île de Saï", *REG* 24 (1972), 20–4. According to Łajtar, the stone is now kept in Lille.

28 Łajtar, "Christian Saï", 93, no. 3.

29 See e.g. M. Cramer and H. Bacht, "Die antichalkedonische Aspekt im historisch-biographischen Schrifttum der koptischen Monophysiten (6.-7. Jahrhundert): Ein Beitrag zur Geschichte der Entstehung der monophysitischen Kirche Ägyptens", in: A. Grillmeier, and H. Bacht (eds.), *Das Konzil von Chalkedon: Geschichte und Gegenwart*, Bd. II: *Entscheidung um Chalkedon*, Würzburg: Echter Verlag 1953, 315–38 at 319–22; D.W. Johnson, "Anti-Chalcedonian polemics in Coptic texts", in: B.A. Pearson and J.E. Goehring (eds.), *The roots of Egyptian Christianity*, Philadelphia: Fortress Press 1986, 216–34.

PARERGA

30 The meager sources for a Melkite hierarchy in Nubia are reviewed by U. Monneret de Villard, "Note Nubiane: 1. Articula (Plin., *N.H.*, VI, 184). 2. La Chiesa melkita di Nubia", *Aegyptus* 12 (1932), 305–16 at 309–16; idem, *Storia della Nubia cristiana*, Rome: Pontificium Institutum Orientalium Studiorum 1938; reprinted 1962, 158–60. Theories about a Melkite hierarchy in Faras are discussed critically by M. Krause, "Bischof Johannes III von Faras und seine beiden Nachfolger: Noch einmal zum Problem eines Konfessionswechsels in Faras", in: *Études nubiennes: colloque de Chantilly, 2–6 juillet 1975*, Cairo: IFAO 1978, 153–64; see also idem, "Zur Kirchengeschichte Nubiens", in: T. Hägg (ed.), *Nubian culture: past and present*, Stockholm: Almqvist & Wiksell 1987, 293–308.

31 Discussed by A. Łajtar, in *I. Khartoum Greek* 24, ad 6. Note also that the abbreviation of the title ἐπίσκοπος, with a raised κ, was not properly recognized by Boyaval.

32 Further examples from Nubia: *I. Khartoum Copt.* 208, s.v. (Coptic); *I. Khartoum Greek* 231, s.v. (Greek).

33 See A. Łajtar, in *I. Khartoum Greek*, xx and 258.

34 F.L. Griffith, "Christian documents from Nubia", *PBA* 14 (1928), 117–46 at 128–34; E. Zyhlarz, "Neue Sprachdenkmäler des Altnubischen", in: *Studies Griffith*, London: Egypt Exploration Society 1932, 187–95 at 187–90, no. 1; cf. Łajtar, "Christian Saï", 94, no. 5.

35 "Christian documents", 129, 133; single source: *History of the Patriarchs of the Egyptian Church*, ed. A.S. Atiya et al., Cairo: Société d'archéologie copte 1959, II/3, 222 (Arabic), 349–50 (English); cf. G. Vantini, *Oriental sources concerning Nubia*, Heidelberg: Heidelberger Akademie der Wissenschaften /Warsaw: Polish Academy of Sciences 1975, 217.

36 Łajtar, "Christian Saï", 93–4, nos. 2 and 4.

17

EPIGRAPHY AND HISTORY IN THE THEBAN REGION

Jacques van der Vliet

Introduction

Epigraphy, the study of inscriptions, is usually considered a dull subject. It is true that the information provided by individual inscriptions may often not seem very exciting, as they provide mainly names and rarely dates. Also their literary value is hardly ever great, a majority consisting of brief prayers. Yet there are a number of reasons why, in my opinion, inscriptions do have considerable value as a historical source.

Taken in isolation, inscriptions may not be very informative, but the picture becomes different when they are part of a broader dossier, including archaeo-logical, art-historical and papyrological material. Reviewing the various bits of documentation assembled during his excavations in the hermitages of Esna, Serge Sauneron observed: "when they remain isolated, they are nothing but the frag-ments of an inventory-list. Their interest is *in the relations* that connect them mutually, so to say *in their syntax*" (Sauneron 1972: I, xix, emphasis in original). Within the context of larger units of documentation, inscriptions acquire meaning on a higher level, that of the overall interpretation of a given site and its history. The particular interest of inscriptions on this higher level of interpretation derives from their very nature. Inscriptions, carved in stone monuments or painted upon walls, were never meant to travel, as were letters, for example, or works of lit-erature, which may cross linguistic, political or even cultural borders. They are linked with space both in its physical and its ideological dimensions. Inscriptions, therefore, derive their value from a double insertion in greater contexts. Their significance becomes apparent when they are studied as part of a broader group of source material, in what Sauneron calls their "syntax", and within a wider spatial and ideological setting, in what I would call "landscape".

The epigraphic material from the region between Nag Hammadi and Esna illus-trates these points in an exemplary way. As the region is vast and diverse, and offers a great amount and variety of material, this presentation will focus on the Theban area, which makes up its core both geographically and in terms of the density and distribution of the epigraphic material, and on the important Christian monuments of Esna, further south.

JACQUES VAN DER VLIET

The Theban temples: a dialogue with the past

One of the Sahidic *Lives* of Pachomius tells us how the young saint, although being not particularly strong physically, was recruited for the army. From his native town, Esna, he was shipped northwards with other recruits and arrived in "Ne, the ancient capital" (Lefort 1933: 212). The latter qualification undoubtedly refers to the biblical No-Ammon, to what we call Thebes today (cf. Vycichl 1983: 137). It is one of the very few passages in Christian sources from Egypt to show that there is something special about the place. Yet, even today, Luxor and its vicinity are dominated by the monuments of its pharaonic past, in particular by its impressive temple complexes.

The Christian re-use of pagan temples has often been discussed in recent years (for example, Dijkstra 2008; Hahn, Emmel and Gotter 2008).Whereas hagiographic and historiographic literature preserves a record of devastation and iconoclasm, temples being violently turned into churches as the symbolic expression of the triumph of Christianity, the archaeological and epigraphic sources show a much more diverse and moderate picture. More realistic views of temple conversion draw the attention to both the chronology and the variety of later re-use. Often there had been a lapse of time between the abandonment of a temple as a traditional cult center and its re-use as a church. Moreover, temples could simply lie neglected forever, be used as quarries, or serve as the foundation for an entire new settlement – a village, a town or a monastery – in addition to actual conversion into a church. Furthermore, the process of conversion itself is not so easy as it would seem at first sight. Of course, a deserted temple offers a cheap and durable stone structure, but not one that is per se suitable for the celebration of liturgy. In fact, its decoration, its orientation and the arrangement of its rooms may offer strong impediments.

Regrettably, most of the archaeological material for the study of temple conversion in the Theban area is lost. In the nineteenth and twentieth centuries, pharaonic temples were stripped bare of their later overbuildings by zealous Egyptologists who cared very little about documentation. As a result, hardly anything now remains except for the rare traces of inscriptions, paintings, sockets and niches, which demand careful deciphering and interpretation. A fine piece of such detection work was done for the temples of Karnak by René-Georges Coquin (1972), who was able to build upon earlier work by Michel Jullien (1902) as well as Henri Munier and Maurice Pillet (1929). Especially noteworthy is his careful reconstruction of the church in the Akh-Menou, the Festival Hall of Thutmosis III. Among the few inscriptions that have come to light there, one is particularly intriguing. It is a list of archimandrites, which probably served as an *aide-mémoire* for deacons during the liturgical commemoration of the deceased leaders of the monastic community that used the church. It assigns a prominent place to Shenoute and Besa, presumably the famous abbots of the White Monastery near Sohag, and thereby suggests a link between the local community at Karnak and the better-known one at Sohag (Coquin 1972: 173–76).

The Theban West Bank also counts several major temples that have been re-used in Christian times. But the nature and history of their secondary use varies widely. Thus the massive mortuary temple of Ramses III, at Medinet Habu, had already in pre-Christian times become the core of a large village or small town, Jeme. Thanks to its intensive re-use as a habitation site, the temple still stands relatively unharmed today. Civic life in ancient Jeme is best known from non-epigraphic sources, in particular the wealth of legal and other documents found there (Wilfong 1989, 2002). Yet Christian inscriptions – sometimes quite important ones – can be found everywhere (recorded in Edgerton 1937: pls. 92–102, nos. 342–405). A fascinating one can still be clearly seen in a conspicuous place on the northern jamb of the main entrance gate to the temple. A *lashane*, a village headman, called Peter boasts of having bought a scale for the village and set it up in its butchery (Edgerton 1937: pl. 93, no. 343; *SB Kopt.* III 1538). The conspicuous position of the text must reflect Peter's pride in this instance of civic munificence.

An entirely different situation prevails in temples that became centers of monastic communities. A relatively modest example is the small Ptolemaic temple of Hathor at Dayr al-Medina. Thanks to Chantal Heurtel we now possess a complete edition of its Greek and Coptic inscriptions (Heurtel 2004). These show that at a certain moment the temple must have been converted into a church, apparently dedicated to a holy martyr, St. Isidore.

The inscriptions themselves are varied in nature. Most are short prayers, often commemorative in nature, such as a list of dates for the commemoration of the church's deceased clergy (Heurtel 2004: 28–30). Among others, an interesting prescription (*logos*) that provides measurements for the fabrication of tunics (*lebiton*) and shirts (*thalis*) may be indicative of the presence of a dressmaker's workshop in the neighborhood (Heurtel 2004: 20–3). Some of the persons mentioned are known from elsewhere, and the texts can therefore be dated to the seventh or eighth century, like most of the written material from Western Thebes. This suggests a considerable lapse of time between the end of the pagan cult in the temple and its conversion into a church.

The same phenomenon can be observed in a far more important pharaonic monument, at just a short distance from Dayr al-Medina, the mortuary temple of Queen Hatshepsut at Dayr al-Bahari. In Ptolemaic and early Roman times, it housed a thriving healing shrine patronized by the "pagan saints" Amenhotep, son of Hapu, and Imhotep (Laskowska-Kusztal 1984; Łajtar 2006: 3–94). In the third and fourth centuries, part of the temple was used as a cemetery (Łajtar 2006: 94), and another part had come to serve for the social and ritual gatherings of the members of the corporation of ironworkers from nearby Hermonthis (Łajtar 2006: 95–103). Adam Łajtar's monumental edition of the pre-Christian Greek inscriptions in the temple affords a fascinating glimpse of this final stage of pagan worship at the spot. After the middle of the fourth century its traces disappear, and only at a quite late stage, toward the end of the sixth century, did the temple become the core of an impressive monastery, named after the holy martyr Phoibammon. This dominated the area until its remains were demolished by nineteenth-century archaeologists (Godlewski 1986). Again a great gap can be observed between the end of the traditional cults at the spot and the Christian re-use of the temple as a monastery.

The Christian inscriptions from the temple have been assembled by Włodzimierz Godlewski in his 1986 monograph on the Monastery of Saint Phoibammon. In addition to a great number of commemorative graffiti and brief prayers, these include moralizing texts, such as this rather cryptic one: "Do not approach a child; you are yourself a child" (Godlewski 1986: 149, no. 26, my readings). Also noteworthy is a Coptic sapiential text, noticed earlier by W.E. Crum (Winlock and Crum 1926: 12–13). It is a *dipinto* written near the entrance of what had originally been the Chapel of the Nocturnal Sun, and consists of a number of epigrammatic statements about sins and virtues, such as "wisdom is a grace, virginity a commendation" (Godlewski 1986: 150–51, no. 27). These maxims, which may derive from some anthology, were apparently popular all over Egypt, for they are found, with slight variations, in other contexts, for example in the Bohairic *dipinti* at the Monastery of the Syrians (Dayr al-Suryan) (cf. Van der Vliet 2004: 196–97 [Study 5]) and Kellia (Bridel 1999: 314–15, no. 156). Such moralizing texts encode a certain type of behavior and were undoubtedly meant to be read and reread time and again. They and others are an important witness to the re-use and transformation of the originally pagan sacred space over time. Overlaying the walls of Hatshepsut's funerary temple like a palimpsest, they both reflect and regulate the piety of late antique monasticism.

Mummies, monks and pilgrims: a necropolis for the living

The mortuary temples mentioned above are already part of the world of the West, the great pharaonic necropolis that stretches from north to south over many kilometers. Late antiquity saw the transformation of the valleys of the Theban Mountain from a city of the dead into a tomb for the living, serving as the dwelling place for large communities of ascetics. These barren wadis are the setting of perhaps the nicest of all stories about a monk having "some words with a mummy", the one showing Pesynthios of Koptos discovering a cache of ancient mummies "in the Mountain of Jeme" (*Encomium of Pesynthios*, Bohairic; Amélineau 1887: 141–51). The broader subject of the Christian transformation of the Theban Necropolis is covered by Elisabeth O'Connell [(2010)]. Here only the testimony of the inscriptions will be discussed.

The monastic re-use of the pharaonic necropolis covers both extensive monasteries, like that of Epiphanius, and much smaller units, which may occupy one single tomb only. It was marked by an intensive scribal activity, resulting among other things in thousands of more or less informal inscriptions and graffiti that can be found both within the ancient tombs and outside of them, on the rocks of the Theban Mountain. Among longer, more formal inscriptions, those from the tomb of Daga, supposedly the core of the Monastery of Epiphanius, are justly the most famous. They include an entire set of dogmatic texts of a clear Monophysite

EPIGRAPHY AND HISTORY IN THE THEBAN REGION

tendency, among them lengthy quotes from Patriarchs Cyril and Damian of Alexandria and Severus of Antioch in Sahidic Coptic. For the study of this multilingual ensemble, where also Greek and Syriac are represented, and its context, the two magnificent volumes of *The Monastery of Epiphanius at Thebes* (Winlock and Crum 1926; Crum and Evelyn White 1926) remain the standard reference. Since the publication of these volumes, the inscriptions have retained their importance as witnesses, for example, to the textual tradition of the writings of Severus of Antioch (see most recently Lucchesi 2007) or Athanasius' *Letter to the monks* (de Jerphanion 1930). In addition to their philological interest, they illustrate a phenomenon that could also be observed in the Monastery of Saint Phoibammon. Apparently, the walls of monastic habitation sites were privileged bearers of texts that deserved to be read aloud or even learned by heart by the monks who lived within them.

Compared to these well-executed literary texts, the often quite insignificant graffiti left by simple passersby as well as real pilgrims may seem to belong to an entirely different class of texts, which is true to a degree. Yet these less sophisticated inscriptions were also undoubtedly meant to be read aloud in order to acquire their full commemorative function. Large numbers of them survive from late antiquity, both in Greek and in Coptic, many of them hidden in unwieldy collections such as the *Graffiti de la Montagne thébaine* (Černý et al. 1969–1983) or the *I. Syringes* (Baillet 1920–1926), not to mention purely Egyptological publications. Only recently a project was announced that goes beyond mere data collecting by inserting this epigraphic material into the larger complex of textual and material remains from Western Thebes. It is undertaken by a Franco-Belgian team and aims at studying the distribution of Christian settlements and inscriptions in the southern part of the Theban Mountain, from the Valley of the Queens southward (see preliminarily Delattre, Lecuyot, and Thirard 2008). The study of the inscriptions will thereby take its place in a comprehensive survey of the archaeological and architectural remains, which will reveal not merely patterns of settlement, but also patterns of piety and commemoration. For to know where people traveled and for what reasons is no less important than to know where and how they lived.

This is of course best illustrated by pilgrimage sites, of which there were several in Western Thebes. A well-known example is the so-called Wadi of the Spanish Pilgrims, which will be included in the Franco-Belgian survey (Delattre, Lecuyot, and Thirard 2008: 123, 127–28). Although the pilgrims after whom the wadi is named did not actually come all the way from Spain, but from a village called Spania in Middle Egypt, near Kbehs/Aqfahs (Padró 1999–2000), their signatures on this somewhat forbidding spot do show its religious importance in late antiquity. Recent publications by Alain Delattre have already yielded important results, both by tracing more pilgrims (Delattre, forthcoming) and identifying local cults (Delattre 2008). The integrative approach advocated by the Franco-Belgian team, together with the broader resurgence of interest in the Christian textual and material remains from the area since the late 1980s, will gradually reveal an entirely

229

different vision of late antique Western Thebes, one where the living easily outnumber the dead.

Esna: on the road to Nubia

Heading southward to Esna, the home of St. Pachomius, we are leaving the Theban area proper and approach the Nubian frontier. The ancient town of Esna is largely buried beneath the modern one. Yet most of its surviving Christian monuments have been studied and published with great love and care. This is mainly due to one person, the Egyptologist Serge Sauneron. He rightly considered it his task, not merely to publish Esna's famous temple, but to collect everything related to the history of Esna through the ages.

With this in mind, Sauneron also undertook a corpus of the known Christian funerary stelae from Esna, completed after his death by René-Georges Coquin (Sauneron and Coquin 1980). The collection distinguishes among various types of sculptured decoration and textual formulae. Most of them are closely related to those known from other centers in the region, such as Hermonthis/Armant and Edfu. Others are more peculiar, however. Thus, several funerary inscriptions, in Greek and in Coptic, from various periods, bear the invocation "God of the spirits and Lord of all flesh", with its echo of Numbers 16:22 and 27:16 (Sauneron and Coquin 1980: nos. 20, 29, 57, 80; Coquin 1975: 257–60, 265–66). In medieval Nubia this formula, mainly in Greek, was exceedingly common as the opening of the prayer for the deceased, while it was far less so in Egypt (Roquet 1977: 166–71; *I. Khartoum Copt.*, p. 141). The relatively numerous Esna prayers of this type may be indicative of cultural contacts with neighboring Nubia in the medieval period, as are several pieces from the Esna-Edfu lot of codices now in the British Library (Layton 1987: xxvi–xxx).

None of the funerary stelae assembled by Sauneron and Coquin were found *in situ*. Fortunately the situation is much different for both other ensembles of Christian inscriptions from Esna, the mural *dipinti* in the two medieval monasteries west of Esna and those from the late antique hermitages excavated by Sauneron and his team.[1] The underground hermitages, which may date from the sixth to seventh centuries, are rich in inscriptions providing various kinds of information, such as the names of former inhabitants and the saints venerated by them. Yet the main interest of this ensemble is not so much in onomastics, but rather – as Sauneron and Coquin, the principal editor of the inscriptions, were quick to realize – in the forms of piety they attest. One striking example may be quoted. A well-known type of funerary inscription is characterized by a list-like enumeration of saints preceding the name or names of the deceased. It was popular in monastic sites in particular in Middle Egypt. They are often referred to as the "litany type" of funerary inscriptions (Wietheger 1992: 210–19; Papaconstantinou 2001: 22, 387–402). Usually these are inscribed on stone slabs, commemorating one or more deceased persons, which have hardly ever been found in their original context. In the Esna hermitages, inscriptions of the litany type were found *in situ*, as *dipinti* below the central niches of the oratories. As their position shows, they were really what their conventional name suggests, a kind of litany meant to be performed orally as part of the monks' regular praying practice. It also connects them to the other examples of monastic mural inscriptions, meant for recitation and perhaps even learning by heart, discussed above (Monastery of Saint Phoibammon, Monastery of Epiphanius).

The medieval churches of al-Shuhada Monastery (Dayr al-Shuhada, or Monastery of the Martyrs) and Dayr al-Fakhuri (the Monastery of the Potter) are much different in nature. They provide characteristic examples of decorated monastic churches, in which architecture, wall paintings and texts together shape the liturgical space. As such, they can be compared to the more impressive ensembles in, for example, the Monastery of the Archangel Gabriel (Dayr al-Malak) at Naqlun (Fayoum, eleventh century) or the Monastery of St. Antony (Dayr Anba Antunius) near the Red Sea (thirteenth century). As in these other monasteries, the texts painted on the plastered wall include a rich variety of genres: legends that are part of the wall paintings, prayers for founders and benefactors, commemorative texts, visitors' inscriptions (Coquin 1975; Dekker and van der Vliet 2008). They provide twelfth- to fourteenth-century dates and presumably also the medieval name of al-Shuhada Monastery, which was named after Isaac the Anchorite (Coquin 1975: 247–48 [Inscr. G], but cf. p. 249, and Khater 1981: 23). It is to be regretted that the inscriptions and the wall paintings, which form a closely knitted whole in both churches, were published separately (Coquin 1975; Leroy 1975). Apart from having to handle two incomplete publications, art historians may now easily miss, for example, the fact that Dayr al-Fakhuri has a rare representation of St. John of the Golden Gospel (Coquin 1975: 275; unidentified in Leroy 1975: 28, but cf. 63). The interest of both rapidly dilapidating churches for the medieval history and culture of the Esna region is great and certainly warrants a serious effort of consolidation and restoration.

Conclusions

As these few examples show, inscriptions provide invaluable clues to the life of the civic and monastic communities that inhabited the Theban area and its ancient monuments in late antiquity. They bear witness to the Christian transformation of the landscape, attesting forms and patterns of cult and piety in addition to more wordly activities, like weighing meat or producing garments. The inscriptions from pharaonic monuments, moreover, reveal a remarkable chronological gap between the demise of the traditional cult in the area, which had occurred by the end of the fourth century even at Dayr al-Bahari, and their intensive Christian re-use around the seventh and eighth centuries, a phenomenon that remains to be explained. The inscribed monuments of Christian Esna, while offering much less of a coherent corpus, attest a similar interaction between space and written text. Finally, the important new project of the Franco-Belgian mission carries the promise of a more contextual study of the rich epigraphic material, adding exciting new pages to the Christian history of the Theban region.

Note

1 Sauneron 1972, with major contributions by R.-G. Coquin; the inscriptions are in vol. 2, commentaries in vol. 4; supplementary: Krause 1990.

Bibliography

Amélineau, E. 1887, *Étude sur le christianisme en Égypte au septième siècle*, Paris: E. Leroux; reprint from *Mémoires de l'Institut d'Égypte* 2 (1889): 261–424.

Baillet, J. 1920–1926, *Inscriptions grecques et latines des tombeaux des rois ou Syringes à Thèbes*, Cairo: IFAO.

Bridel, Ph. (ed.) 1999, *Explorations aux Qouçoûr el-Izeila lors des campagnes 1981, 1982, 1984, 1985, 1986, 1989 et 1990*, III, Louvain: Peeters.

Černý, J. et al. 1969–1983, *Graffiti de la montagne thébaine*, Cairo: Centre de documentation et d'études sur l'ancienne Égypte.

Coquin, R-G. 1972, "La christianisation des temples de Karnak", *BIFAO* 72: 169–78.

Coquin, R-G. 1975, "Les inscriptions pariétales des monastères d'Esna: Dayr al-Šuhadâ' – Dayr al-Fahûrî", *BIFAO* 75: 241–84.

Crum, W.E., and H.G. Evelyn White. 1926, *The monastery of Epiphanius at Thebes* II, New York: Metropolitan Museum of Art.

de Jerphanion, G. 1930, "La vraie teneur d'un texte de saint Athanase rétablie par l'épigraphie: l'*Epistula ad monachos*", *Recherches de science religieuse* 20: 529–44; reprinted in *La voix des monuments: études d'archéologie*, Rome: Pontificio istituto orientale/ Paris: Éditions d'art et d'histoire 1938.

Dekker, R., and J. van der Vliet. 2008, "From Naqada to Esna: a late Coptic inscription at Deir Mari Girgis (Naqada)", *ECA* 5: 37–42 [Study 18].

Delattre, A. 2008, "Inscriptions grecques et coptes de la montagne thébaine relatives au culte de saint Ammônios", in: A. Delattre and P. Heilporn (eds.), *"Et maintenant ce ne sont plus que des villages . . .": Thèbes et sa région aux époques hellénistique, romaine et byzantine*, Brussels: Association égyptologique Reine Elisabeth, 183–88.

Delattre, A. forthcoming, "L'itinéraire thébain d'un prêtre copte".

Delattre, A., G. Lecuyot, and C. Thirard. 2008, "L'occupation chrétienne de la Montagne thébaine: première approche", in: A. Boud'hors and C. Louis (eds.), *Études coptes X: Douzième journée d'études (Lyon, 19–21 mai 2005)*, Paris: de Boccard, 123–35.

Dijkstra, J.H.F. 2008, *Philae and the end of ancient Egyptian religion: a regional study of religious transformation (298–642 CE)*, Louvain/Paris/Dudley, MA: Peeters.

Edgerton, W.F. 1937, *Medinet Habu graffiti: facsimiles*, Chicago: University of Chicago.

Godlewski, W. 1986, *Le monastère de St Phoibammon*, Warsaw: PWN–Éditions scientifiques de Pologne.

Hahn, J., S. Emmel and U. Gotter. 2008, *From temple to church: destruction and renewal of local cultic topography in late antiquity*, Leiden/Boston: Brill.

Heurtel, Ch. 2004, *Les inscriptions coptes et grecques du temple d'Hathor à Deir al-Médîna*, Cairo: IFAO.

Jullien, M. 1902, "Le culte chrétien dans les temples de l'antique Égypte", *Études* 92: 237–53.

Khater, A. (ed.) 1981, *Martyre des citoyens d'Esna, de leur évêque Amon et de leur persécuteur Arien: homélie par Paul, évêque d'Assiout, d'Abou Tîg et de Manfalout*, Cairo: Centre franciscain d'études orientales chrétiennes.

Krause, M. 1990, "Verlorene Inschriften und Beischriften der Eremitensiedlungen 3 und 4 bei Esna", in: *Festschrift Jürgen von Beckerath*, Heidelberg: Gerstenberg, 147–70.

Łajtar, A. 2006, *Deir el-Bahari in the Hellenistic and Roman periods: a study of an Egyptian temple based on Greek sources*, Warsaw: University of Warsaw/Raphael Taubenschlag Foundation.

Laskowska-Kusztal, E. 1984, *Le sanctuaire ptolémaïque de Deir el-Bahari*, Warsaw: PWN–Éditions scientifiques de Pologne.

Layton, B. 1987, *Catalogue of Coptic literary manuscripts in the British Library acquired since the year 1906*, London: British Library.

Lefort, L. Th. 1933, *S. Pachomii vitae sahidice scriptae*, Paris: Imprimerie nationale.

Leroy, J. 1975, *Les peintures des couvents du désert d'Esna*, Cairo: IFAO.

Lucchesi, E. 2007, "L'homélie cathédrale II de Sévère d'Antioche en copte", *Analecta Bollandiana* 125: 7–16.

Munier, H., and M. Pillet. 1929, "Les édifices chrétiens de Karnak", *Revue de l'Égypte ancienne* 2: 58–88.

O'Connell, E.R. 2010, "Excavating Christian Western Thebes: a history", in: G. Gabra and H.N. Takla (eds.), *Christianity and monasticism in Upper Egypt*, vol. 2: *Nag Hammadi – Esna*. Cairo/New York: American University in Cairo Press, 253–70.

Padró, J. 1999–2000, "Espania en Egipto", *Aula Orientalis* 17–18 (*Homenaje G. del Olmo Lete*): 483–92.

Papaconstantinou, A. 2001, *Le culte des saints en Égypte des Byzantins aux Abbasides: l'apport des inscriptions et des papyrus grecs et coptes*, Paris: Éditions du CNRS.

Roquet, G. 1977, "Inscriptions bohaïriques de Dayr Abû Maqâr", *BIFAO* 77: 163–79.

Sauneron, S. 1972, *Les ermitages chrétiens du désert d'Esna*, Cairo: IFAO.

Sauneron, S., and R-G. Coquin. 1980, "Catalogue provisoire des stèles funéraires coptes d'Esna", in: *Le livre du centenaire de l'Institut français d'archéologie orientale du Caire: 1880–1980*, Cairo: IFAO, 239–77.

Van der Vliet, J. 2004, "History through inscriptions: Coptic epigraphy in the Wadi al-Natrun", *Coptica* 3: 187–207; republished in: M.S.A. Mikhail and M. Moussa (eds.), *Christianity and monasticism in the Wadi al-Natrun: essays from the 2002 international symposium*, Cairo/New York: American University in Cairo Press 2009, 329–49 [Study 5].

Vycichl, W. 1983, *Dictionnaire étymologique de la langue copte*, Louvain: Peeters.

Wietheger, C. 1992, *Das Jeremias-Kloster zu Saqqara unter besonderer Berücksichtigung der Inschriften*, Altenberge: Oros.

Wilfong, T.G. 1989, "Western Thebes in the seventh and eighth centuries: a bibliographic survey of Jême and its surroundings", *BASP* 26: 89–145.

Wilfong, T.G. 2002, *Women of Jeme: lives in a Coptic town in late antique Egypt*, Ann Arbor: University of Michigan Press.

Winlock, H.E., and W.E. Crum. 1926, *The monastery of Epiphanius at Thebes* I, New York: Metropolitan Museum of Art.

18

FROM NAQADA TO ESNA
A late Coptic inscription at Dayr Mari Girgis (Naqada)[1]

Renate Dekker and Jacques van der Vliet

Introduction

On 6 February 2008, just before the start of the international symposium "Christianity and monasticism in Upper Egypt: The region from Nag Hammadi to Esna" at Naqada, the present authors had the opportunity to visit some of the monasteries between Naqada and Qamula.[2] Dayr Mari Girgis, the Monastery of St. George, in particular attracted our attention because of the previously unknown Coptic inscription that we publish here. The history of Dayr Mari Girgis, which is also known as Dayr al-Majma' (the Monastery of the Synod), dates back well before the twelfth century, as the remains of one of its four churches suggest. The church of St. John is one of the two medieval structures that remain in this area, the other one being the main church of Dayr Mari Buqtur (the Monastery of St. Victor).[3] The present Church of St. George at Dayr Mari Girgis is a modern building, however, and replaces the original church, which was leveled in the 1920s.[4] The inscription is found on an isolated stone slab, standing against the inner west wall of this church, to the left of the entrance. Its exact provenance is unknown, but it may have been discovered during undocumented excavations that were carried out on the site by the Antiquities Service in the later part of the twentieth century.

A. Dayr Mari Girgis

The text is inscribed on a slab of hard sandstone, its greatest measurements being 51 × 75 × 11.5 cm (Figure 18.1). The upper part of the stone seems complete; the lower end is broken away. On the lefthand side, the epigraphic field is marked off from a broad blank margin by a deeply engraved vertical line; on the righthand side, no such margin is found, but a protruding tenon suggests that this side of the stone was set into a wall. The back side is only very roughly dressed.

Ten lines of continuous text survive. About three lines must have preceded the present text, but these have entirely disappeared except for some vague traces. Surface wear and candle grease affect the legibility of the remainder of the inscription.

The text is written in extremely inelegant incised uncials of a medieval type, showing an angular U-like *mu*; the lower end of the *rho* has an upward stroke to

Figure 18.1 Stone inscription at Dayr Mari Girgis (Naqada)
(photo: R. Dekker)

the right. Numerals are irregularly marked off by high points. The script of our l. 1 is considerably smaller than the following lines, and this line may have been inscribed by a different hand. The height of the letters is ca. 1.5–2 cm in l. 1, and 4.5–5 cm in l. 10.

The language is Sahidic Coptic with some irregularities.

Dayr Mari Girgis (Naqada),   A.D. 1316/1317

[about three lines precede]
ϩⲛ ⲧⲉⲣⲟ[ⲙⲡⲉ] ⲛ̄[ϣⲟⲣⲡ ⲉⲡ]ϣⲟ ⲗ̄ⲃ̄
ⲁϥⲃⲱⲕ ⲛ[ϭ]ⲓ ⲁⲙⲁⲃⲣⲁⲧ
ϩⲙ ⲡⲉϩⲟⲟⲩ ⲙ̄ⲡⲥⲁⲃⲃⲁⲧⲟⲛ
4. ⲕ̄ · ϩⲛ ⲙⲉⲥⲣⲁ ⲁⲩⲱ {ϩⲛ}
ϩⲙ ⲡⲟⲟϩ · ⲓ̄ · ⲁⲩⲱ ϩⲛ
ⲧⲙⲉϩⲥⲛ̄ⲧ`ⲉ´ ⲛ̄ⲣⲟⲙⲡ`ⲉ´
ⲉⲡⲁ/·ⲗ̄ⲅ̄ · ⲁϥⲃⲱⲕ <ⲛϭⲓ> · ⲁⲙⲁⲃ-
8. ⲣⲁⲧ ϩⲙ ⲡⲉϩⲟⲟⲩ ⲛ̄ⲧⲕⲩⲣⲓⲁⲕⲏ
ⲕ̄ⲅ̄ ⲙⲛ̄ ⲉⲡⲏⲡ ⲁⲩⲱ
ϩⲙ] ⲡⲟⲟϩ ⲍ̄ vac ϩⲛ ⲧⲁⲓⲣⲟⲙ-
[ⲡⲉ ⲛⲟⲩⲱⲧ - - -]

[1. ϣⲟ ⲗ̄ⲃ̄: alternative for ⲁⲗ̄ⲃ̄ || 7. ⲁ/·ⲗ̄ⲅ̄: l. ⲁⲗ̄ⲅ̄] || 9. ⲙⲛ̄, l. ϩⲛ ||10. ⲧⲁⲓ-: l. ⲧⲉⲓ-.

[. . .] in the first year, in 1032.
(A)l-Mabrat went on Saturday 20 of Mesra (Mesore), and of the moon (day) 10, and in the second year, in 1033.

236

(A)l-Mabrat went on Sunday 23 of Epêp, and of the moon (day) 7, in this [same] year [. . .]

The inscription commemorates the "going" (вⲱк) of a man called al-Mabrat on various precisely recorded dates. Two successive years are mentioned, A.M. 1032 and 1033, which correspond to A.D. 1315/1316 and 1316/1317 respectively. Line 1, mentioning the year A.M. 1032, shows a smaller script and possibly a different hand. Presumably, therefore, the remainder of the surviving text (ll. 2–10) was written one year later (in A.M. 1033). Mesore 20 (13 August) is a Saturday in A.D. 1317, and 23 Epêp (17 July) a Sunday in that very same year. Since the latter date is the earlier one, this must have been added as an afterthought, which is confirmed by the presence of the demonstrative ⲧⲁⲓ- in l. 10, presumably meaning "this (very same) year", i.e. A.M. 1033.

The entries for all three dates, as far as they are preserved, show an identical structure: weekday, day of the month, moon-day, year A.M. The years A.M. are, moreover, designated as the first and the second year respectively (in ll. 1 and 6). The way in which these dates are constructed shows a clear lack of familiarity with the traditional scribal habits of Sahidic. In particular, the odd preposition ⲙⲛ̄ in l. 9 (instead of ϩⲛ in the parallel phrase in l. 4) may have been chosen under the influence of the Arabic *min*.[5] The month name Mesore is also written in its Arabic form, Mesra, in l. 4. The habit of dating by moon-days, which is attested in late Greek and Coptic inscriptions from Nubia and the First Cataract area,[6] undoubtedly reflects a similar influence. The demonstrative ⲧⲁⲓ- in l. 10 must be a Bohairicism. On the whole, the text bears witness to a very late stage of Sahidic, in which it was apparently not in use for writing purposes anymore.

The main questions raised by the inscription are, who al-Mabrat was and why the trouble was taken to commemorate his "going" on at least three different days in two successive years in a stone inscription. The stone itself clearly does not answer these questions. The Arabic name Mabrat and its variants occur in some other Coptic texts,[7] but in only one instance possibly the same person was intended. This is another inscription from the same general region, a *dipinto* from the Church of Dayr al-Fakhuri, near Esna, which we republish below, with some minor modifications, after René-Georges Coquin's excellent first edition of the text.

B. Dayr al-Fakhuri

Dipinto on the western face of the southeastern pier of the central domed room of the monastery church, at about eye level (Figure 18.2). Remains of eight lines of text painted on plaster in black ink. The lower part of the text, beyond l. 9, is broken away. L. 1 must have consisted of monograms only; Coquin's l. 2 is non-existent, as is shown by the decorative use of raised letters in l. 3, which is actually the first line of the text proper. We nevertheless retained his line numbering to avoid confusion.

Figure 18.2 Dipinto in the church of Dayr al-Fakhuri (Esna)
(photo: R. Dekker)

The text appears to be written by a single practiced hand in a clear late uncial. Both a decorative four-stroke *mu* (in l. 3 only) and a three-stroke U-like *mu* (remainder of the text) are used. A staurogram appears to precede the text of l. 3 (not read by Coquin). A long line separates ll. 6 and 7. The date in the beginning of l. 7 is written in much bigger script, occupying also the beginning of l. 8.

The inscription is bilingual, in Greek and Sahidic Coptic. In order to bring out the use of different codes, we render those parts of the text that were apparently meant to be Greek in minuscules.

Bibl.: Coquin 1975: 277–79, inscr. P, pl. XLIX; *SB Kopt.* I 348 (M.R.M. Hasitzka); cf. Leroy 1975, pl. 67.

Dayr al-Fakhuri (Esna),   A.D. 1316

    [ι(ηcoυ)c +] x[(ριcτo)c]
    vac.
    ☧ δούλου ἐλαχ(ίστου) κ(αὶ) εὐτηλοῦ ΜΕΡΚΟΎΡΙ ΖΩΚ-
4. ΡΑΦΟC Υ(ΙΟ)Υ ΠΔΙΑΚΟΝ ΑΝΑΤΟΛΕ [ΜΟΝ]ΑΧ-
    ΟC Υ(ΙΟ)Υ ΠΜΟΥΝΑCΤΗΡΙΟΝ ΜΠΝΙΩΤ [Π]ΡΟΦ(Η)ˋΤˊ(ΗC)

FROM NAQADA TO ESNA

[ⲁ]ⲡⲁ ϣⲉⲛⲟⲩⲧⲉ ⲡⲛⲟⲩⲧⲉ ⲛⲁ ⲛⲁϥ Ἀμήν κ(αὶ) (ἀμήν).

x`ρ´(ⲟⲛⲱⲛ) ⲙ(ⲁ)`ρ´(ⲧⲩⲣⲱⲛ) ⲁ̅ⲗ̅ⲃ̅ ⲁⲡⲁ ⲗⲙⲁⲃⲣⲁⲧ ⲭⲟⲕ [ⲉⲃ]ⲟⲗ [- - -]
8. ⲙ̄ⲡⲉ̣ⲃⲟⲧ ⲙⲟⲥⲣⲏ ⲥ[- - -]
   [- - -] ⲁ̣ⲡⲁ ⲗⲙⲁⲃⲣⲁⲧ ⲭⲟ̣[ⲕ ⲉⲃⲟⲗ - - -]

1. Coquin [ⲓⲥ ⲡⲉ]ⲭ[ⲥ] ‖ 3. ₱: uncertain, Coquin om.; ⲗⲟ`ⲩ´ⲗⲟ`ⲩ´ ⲉⲗⲁ`ⲭ´ κ; εὐτηλοῦ: 1. εὐτελοῦ, Hasitzka ⲉⲩⲧⲏⲗⲟⲩ`ⲥ´; ⲟ`ⲩ´ (ter): ligature ‖ 3–4. ⲍⲱⲕⲣⲁⲫⲟⲥ: Coquin ⲍⲱⲣⲣⲁⲫⲟⲥ ‖ 5. [ⲡ]ⲣⲟⲫ(ⲏ)`ⲧ´(ⲏⲥ): no room for [ⲙⲡ] or [ⲡⲉⲡ] ‖ 6. In fine: κ, ⳉ[ⲑ] ‖ 7. Two monograms; ⲭⲟⲕ: Coquin ⲭⲱⲕ ‖ 8. First part occupied by date in l. 7; ⲙ̄ⲡⲉⲃⲟⲧ: Coquin ⲙ̄ⲡⲃⲟⲧ; ⲥ[: Coquin om., fortasse ⲥ[ⲟⲩ ‖ 9.] ⲁ̣ⲡⲁ: Coquin ⲁ] ⲡⲁ; ⲭⲟⲕ: Coquin ⲭ[.

[Greek] [Jesus +] Christ.
+ (Memento of) (the) most humble and insignificant servant [Coptic] Merkouri, painter, son (of) the deacon Anatole, monk (and) son (of) the monastery of our father, (the) prophet Apa Shenoute. O God, have mercy on him. [Greek] Amen and Amen.
[Coptic] Anno Martyrum 1032. Apa (a)l-Mabrat finished [- - -] in the month Mosrê (Mesore), [day? - - -]Apa (a)l-Mabrat finished (?) [

Coquin treated the very fragmentary text reproduced above as one inscription (labeled P by him). This is, indeed, what the style of writing and the general appearance of the text seem to suggest. Yet, the long line separating ll. 6 and 7 and the large date that opens l. 7 clearly divide the text into two parts, even though it is likely that these were written at the same time and by the same hand.

The upper part of the *dipinto* (our ll. 3–6) is a bilingual Greek-Coptic prayer for a painter Merkouri.[8] He was the son of a deacon Anatole and a monk of the monastery of St. Shenoute, the famous White Monastery near Sohag. Coquin convincingly argued that he is none other than the painter Merkouri who left more or less similar *dipinti* in the Monastery of Anba Hadra near Aswan in A.M. 1034 (A.D. 1317/1318), in the Red Monastery near Sohag on 11 Mesore 1017 A.M. (4 August 1301 A.D.) as well as in A.M. 1038 (A.D. 1321/1322), and in the church of the White Monastery itself (undated).[9] His professional quality as a painter and a number of similarities with the inscriptions left by Merkouri elsewhere, in particular in the Red Monastery, suggest that he wrote all of these *dipinti* himself.

The exclusively Coptic lower part of the text (ll. 6–9), by contrast, twice mentions a certain Apa (a)l-Mabrat as well as a date in Mesore A.M. 1032, that is between 25 July and 23 August 1316. At first sight, the use of the verb ⲭⲱⲕ ⲉⲃⲟⲗ in l. 1 could suggest that this inscription is commemorative in character. The verb is occasionally used in the sense of "to reach perfection" as a euphemism for "to die".[10] Both the name (a)l-Mabrat and the verb apparently re-occur in l. 9, however, which is less likely to happen in a funerary inscription. Instead, the identical

239

spelling of the name and the similar way of repeating statements about the same person recall the contemporary inscription from Naqada. It seems highly plausible that both inscriptions mention the same person, who left Naqada on at least three dates in 1316 and 1317, and who "finished" or "completed" something in Esna on various occasions in 1316.

Regrettably, although the Esna inscription does provide some additional information about al-Mabrat, the text is very incomplete and, therefore, not as helpful as one might hope. In Esna, al-Mabrat receives the honorific title Apa, which suggests that he could have been a monk, but this is by no means certain.[11] The close way in which he is associated with the painter Merkouri, however, may prove to be more significant. It is conceivable that he was an itinerant artist, like Merkouri, and "completed" some painting or restoration work initiated by the other or undertaken in conjunction by both men. In that case, the Naqada inscription might record the occasions on which he was called away from home to work elsewhere. Yet, precisely this inscription suggests that he may have been a personality of more consequence than a mere artist. In the fourteenth century, the Naqada region appears to have been less of a backwater than it might seem. Thus, some fifty years later, in 1372, Bishop Timotheos of Faras and Ibrim in Nubia was enthroned in the nearby monastery of St. Victor (Dayr Mari Buqtur), in the presence of none other than the well-known scholar and bishop Athanasius of Qus.[12] Whatever role Apa al-Mabrat may have played in Naqada and Esna in the early years of the century is likely to remain unknown, unless additional sources add new pieces to the puzzle.

Notes

1   We thank His Grace Bishop Biemen, Bishop of Naqada and Qus, and his staff, in particular Sister Salwa, as well as the sisters at Dayr Mari Girgis, for their assistance and hospitality.

2   See Clarke 1912: 121–40; Johann Georg 1914: 56–8; idem 1930: 47–8; Timm 1984–1992: IV, 1727–32; Samuel al Syriani and Badii Habib 1990: 50–7, nos. 54–61; Grossmann 2004.

3   For the church of St. John, see Grossmann 2004: 26, fig. 2, pls. III-IV (the dome shown there has now collapsed); cf. Clarke 1912: 137–39, pls. XXXIX-XL; Grossmann 1982: 22–5, fig. 8; idem 1991: 820–21.

4   See Grossmann 1991: 820; cf. Samuel al Syriani and Badii Habib 1990: 53–4, no. 58; Grossmann 2004: 26–7.

5   As was suggested to us by Dr. Samuel Moawad.

6   Nubia: Bagnall and Worp 1986; cf. Bagnall and Worp 2004: 313–14; Qubbat al-Hawa (Aswan): *SB Kopt.* II 1060; Edel 2008, 514–17 (QH 34f). Earlier Coptic examples of dating by the moon appear to belong to the domain of magic and divination; e.g. P.Cairo 45060 (ed. Kropp 1930–1931, I, 50–4).

7   The name occurs as ⲙⲟⲩⲃⲣⲁⲧ and ⲙⲉϥⲣⲁⲧ in various unpublished eleventh-century documents from Naqlun, including the account book British Library Or. 13885; as ⲁⲗⲙⲱⲃⲣⲁⲧ in a letter from Ashmunayn: Crum 1905· no. 1155, ro., ll. 2–3. Cf. Heuser 1929: 116.

8   For a similar late bilingual *dipinto*, see Dekker 2008: 32–6, no. 3 (Aswan); in general, for the learned interest in Greek in this period and region, see Sidarus 2000.

FROM NAQADA TO ESNA

9   See Coquin 1975: 277–78, where all references are given; the Red Monastery *dipinti* of A.M. 1038 are as yet unpublished.
10 For its background, see Van der Vliet 1988.
11 See Derda and Wipszycka 1994; Gonis 2001.
12 See Plumley 1975; cf. Brakmann 2006: 327–28; Coquin 1977, in particular 145–46; Sidarus 2000: 293–94; Timm 1984–1992: II, 723–26.

Bibliography

Bagnall, R.S., and K.A. Worp. 1986, "Dating by the moon in Nubian inscriptions", *CdÉ* 61: 347–57.

Bagnall, R.S., and K.A. Worp. 2004, *Chronological systems of Byzantine Egypt: second edition*, Leiden/Boston: Brill.

Brakmann, H. 2006, "*Defunctus adhuc loquitur:* Gottesdienst und Gebetsliteratur der untergegangenen Kirche in Nubien", *Archiv für Liturgiewissenschaft* 48: 283–333.

Clarke, S. 1912, *Christian antiquities in the Nile Valley: a contribution towards the study of the ancient Churches*, Oxford: Clarendon Press.

Coquin, R-G. 1975, "Les inscriptions pariétales des monastères d'Esna: Dayr al-Šuhadâ' – Dayr al-Fahûrî", *BIFAO* 75: 241–84.

Coquin, R-G. 1977, "A propos des rouleaux coptes-arabes de l'évêque Timothée", *BiOr* 34: 142–47.

Crum, W.E. 1905, *Catalogue of the Coptic manuscripts in the British Museum*, London: British Museum.

Dekker, R.E.L. 2008, "'New' discoveries at Dayr Qubbat al-Hawâ, Aswan: architecture, wall paintings and dates", *ECA* 5: 19–36.

Derda, T., and E. Wipszycka. 1994, "L'emploi des titres abba, apa et papas dans l'Egypte byzantine", *JJP* 24: 23–56.

Edel, E. 2008, *Die Felsgräbernekropole der Qubbet el-Hawa bei Assuan* I/1–3, eds. K.J. Seyfried and G. Vieler, Munich: F. Schöningh.

Gonis, N. 2001, "Abû and apa. Arab onomastics in Egyptian context", *JJP* 31: 47–9.

Grossmann, P. 1982, *Mittelalterliche Langhauskuppelkirchen und verwandte Typen in Oberägypten: eine Studie zum mittelalterlichen Kirchenbau in Ägypten*, Glückstadt: J.J. Augustin.

Grossmann, P. 1991, "Dayr al-Majma': architecture", in: *CoptEnc* 3: 820–22.

Grossmann, P. 2004, "A journey to several monasteries between Naqada and Qamula in Upper Egypt", *BSAC* 43: 25–32.

Heuser, G. 1929, *Die Personennamen der Kopten*, Leipzig: J.C. Hinrichs.

Johann Georg, Herzog zu Sachsen. 1914, *Streifzüge durch die Kirchen und Klöster Ägyptens*, Leipzig/Berlin: B.G. Teubner Verlag.

Johann Georg, Herzog zu Sachsen. 1930, *Neue Streifzüge durch die Kirchen und Klöster Ägyptens*, Leipzig/Berlin: B.G. Teubner Verlag.

Kropp, A.M. 1930–1931, *Ausgewählte koptische Zaubertexte*, Brussels: Fondation égyptologique Reine Elisabeth.

Leroy, J. 1975, *Les peintures des couvents du désert d'Esna*, Cairo: IFAO.

Plumley, J.M. 1975, *The scrolls of Bishop Timotheos: two documents from medieval Nubia*, London: Egypt Exploration Society.

Samuel al, Syriani, and Badii Habib 1990, *Guide to ancient Coptic churches & monasteries in Upper Egypt*, Cairo: Institute of Coptic Studies, Department of Coptic Architecture.

Sidarus, A. 2000, "La tradition sahidique de philologie gréco-copto-arabe (manuscrits des XIIIᵉ–XVᵉ siècles)", in: N. Bosson (ed.), *Études coptes VII: Neuvième journée d'études, Montpellier 3–4 juin 1999*, Louvain/Paris: Peeters.

Timm, S. 1984–1992, *Das christlich-koptische Ägypten in arabischer Zeit*, 6 vols., Wiesbaden: Reichert Verlag.

Van der Vliet, J. 1988, "A note on ϫⲱⲕ ⲉⲃⲟⲗ, 'to die'", *Enchoria* 16: 89–93.

19

"IN YEAR ONE OF KING ZACHARI"
Evidence of a new Nubian king from the Monastery of St. Simeon at Aswan[1]

Jitse H.F. Dijkstra and Jacques van der Vliet

Introduction

Among the landmarks of Christian Aswan are the monumental remains of a monastery, commonly called the Monastery of St. Simeon. Situated on the west bank of the Nile, opposite the town, the site is familiar to all visitors of the First Cataract region. It is less well known that this monastery, which actually bore the name of a fourth-century bishop of Aswan, Hadra,[2] is a source of inscriptions of considerable interest for the medieval history of both Egypt and Nubia. Regrettably, the fate of these inscriptions in modern scholarship is not an entirely happy one.[3] Whereas a large ensemble of Coptic funerary stelae from the monastery grounds was competently published by H. Munier in the early 1930s,[4] the equally numerous *dipinti* and graffiti on its walls can only be gleaned from disparate publications dating back to the late nineteenth and early twentieth centuries.[5] Most of these publications, moreover, can hardly be called satisfactory, even when judged by the scholarly standards of their time. In the 1920s, the monastery was fully excavated and recorded by U. Monneret de Villard. Unfortunately, only the first volume of his final report, which was intended to include the inscriptions, was published.[6] Although this volume does occasionally quote epigraphic material,[7] no comprehensive discussion of the available sources ever appeared. The history of the monastery and its role in the life of the First Cataract region largely remains to be written.

The aim of the present contribution is to provide a re-edition and first translation of an almost completely ignored Coptic *dipinto* on one of the monastery walls and to briefly discuss its historical context. The inscription was first published in 1894 by U. Bouriant in J. de Morgan's ambitious catalogue of ancient Egyptian monuments and inscriptions, which reproduces both a printed text and a handcopy.[8] Its neglect by subsequent scholarship is readily explained by the quality of this *editio princeps*. The text as presented by Bouriant is incomplete and barely understandable and no translation or commentary are provided. Some of the information which it contains was quoted, though, by Monneret de Villard, in the first volume of his unfinished study of the monastery,[9] by Munier, in his edition of the Coptic funerary stelae from Aswan,[10] and by W.E. Crum, in his great dictionary.[11] Fortunately, a large part of the inscription still survived *in situ* in the autumn of 2002, when one of the present authors was able to record it.[12] The following re-edition is based partly on the information assembled on that occasion, and partly on the handcopy reproduced in the *editio princeps*, which proved particularly useful for those elements of the text that can no longer be confidently read on the original.

The inscription is a commemorative *dipinto*, written in red-brown paint upon the plastered wall. It is situated on the south wall of the northern aisle of the church on the lower level of the monastic complex,[13] at a height of 1.90 m above floor level. It thereby remained just beyond the reach of the majority of later visitors who incised their numerous graffiti in the lower part of the wall. Somewhat to its left, the remains of another extensive Coptic *dipinto* published by Bouriant can still be found.[14] The size of the epigraphic field is 22 × 39 cm and the average height of the letters about 1.5 cm (with actual variation from 0.5–3.5 cm). The state of preservation is mediocre, but not bad in comparison with the other texts on the wall. Most of these have suffered greatly from wear, dirt, later scratchings, clumsy modern repairs and the greasy hands of tourists. At several places the plaster has become detached from the wall or threatens to do so. Although good fortune has preserved our inscription until now, its existence is menaced by a large crack on its righthand side. Since the time of Bouriant's copy, already the very ends of ll. 1 and 2 have disappeared completely and ll. 4 and 6 have much deteriorated. Unless further action is undertaken, the inscription will soon be lost forever.

The text consists of six lines of badly ruled Coptic uncials in a swift and practiced but rather careless hand. The scribe has a tendency to lapse into a more cursive style of writing (thus he uses two types of π, e.g. in l. 5, the more cursive one taking a Λ-like form)[15] and there are a few ligatures (in ll. 2 and, perhaps, 6). Only sporadically superlinear strokes occur (as far as can be seen, only on the first letters of ll. 3 and 6; in both cases, strongly curved). The first ρ in l. 2 and the ρ of l. 6 are written with a long downward tail, as are occasionally other letters (e.g. ⲁ in l. 5). The text opens with a simple Greek cross and concludes with two oblique strokes. There are no traces of a frame or additional decorative elements. The language is Sahidic Coptic.

Aswan, 19 April   A.D. 962

+ ⲡⲉϩⲟⲟⲩ ⲛⲧⲁϥⲙⲧⲟⲛ ⲙⲙ[ⲟϥ]
ⲛϭⲓ ⲡⲉⲛⲥⲟⲛ ⲡⲉⲧⲣⲟ ϩⲓ ⲧⲣⲟⲙⲡⲉ [ⲁ̄]
ⲛ̄ⲍⲁⲭⲁⲣⲓ ⲡⲣⲣⲟ ⲉϥϩⲛ ⲡⲙⲁⲛϣⲱ(ⲡⲉ)
4.   ⲛⲁⲡⲁ ϩⲁⲧⲣⲉ ⲛϭⲟⲩⲁⲛ̣ ⲁⲡⲟ ⲇⲓⲟⲕ(ⲗⲏⲧⲓⲁⲛⲟⲩ) ⲭⲟ̄ⲏ̄
ⲡⲁⲣⲙⲟⲩⲧⲉ ⲥⲟⲩ ⲕⲁ ⲡϫⲟⲉⲓⲥ
ⲣ̄ ⲟⲩⲛⲁ ⲙⲛ ⲧⲉϥⲯⲩⲭⲏ //

"IN YEAR ONE OF KING ZACHARI"

Figure 19.1 Monastery of St. Simeon, Aswan: general view of the wall with the Coptic *dipinto* of Peter in the center
(photo: Kathryn E. Piquette 2017)

1. Last two letters and most of a third letter have disappeared since *ed. princ.* According to *ed. princ.* there was no trace of the optional resumptive pronoun (ⲛϩⲏⲧϥ).
2. ⲡⲉⲛⲥⲟⲛ: ⲡⲉ ligatured. ⲡⲉⲧⲣⲟ: ⲉⲧ ligatured. ⲣⲟⲙⲡⲉ̣: ⲣ corrected out of ⲋ. ⲁ̄: according to the handcopy in *ed. princ.*, the numeral seems to have been written slightly below the line; it has since disappeared.
3. ⲡⲣⲣⲟ: *ed. princ.* ⲟⲣⲣⲟˋⲥ́, but the supposed ⲥ above the line is in fact part of the long tale of the ⲣ in ⲡⲉⲧⲣⲟ (l. 2).
4. ⲛⲁⲡⲁ up to ⲁⲡⲟ: text now faded and lacunary. ⲛⲁⲡⲁ ϩⲁⲧⲣⲉ: *ed. princ.* ⲛⲁⲙ ϩⲁⲧⲣⲉ, but already Crum read ⲛⲁⲡⲁ; indeed, vague traces, showing a "cursive" ⲡ, are still visible. ⲛ̇ⲥ̣ⲟ̣ⲩⲁⲛ: ⲥ, ⲟ and ⲛ barely readable. x̄oⲏ̄: with Monneret de Villard and Munier; *ed. princ.* x̄oⲕ̄.
5. ⲥⲟⲩ ⲕⲁ: now much faded. ⲥⲟⲩ was written quite small; no traces of a stroke over ⲕⲁ. *Ed. princ.* was unable to decipher these words (although transcribed in the handcopy, they are omitted from the printed text). ⲡϫⲟⲉⲓⲥ: well readable but entirely omitted by *ed. princ.*
6. p̄ up to ⲧⲉϥⲯⲩⲭⲏ: this entire line is much faded and partially damaged by an Arabic graffito below; owing to its formulary character, however, readings are practically certain. *Ed. princ.* offers no coherent text. The element between p̄ and ⲟⲩⲛⲁ is part of the tail of ⲣ in l. 5. ⲧ̣ⲉ̣ϥⲯⲩⲭⲏ: ⲧⲉ possibly ligatured.

245

+ The day on which our brother Petro went to rest, in year 1 of King Zachari, while (living) in the Monastery of Apa Hatre (Hadra) in Aswan: (in the year) since Diocletian 678, (in the month of) Parmoute, day 24.
Lord, have mercy on his soul.

The inscription commemorates the death of a member of the monastic community ("our brother"), called Peter (Petro, apparently a variant of Petros).[16] The event is dated both to the first year of a King Zachari and to the Diocletian era. The latter date corresponds to 19 April A.D. 962. The text exhibits a bipartite pattern which is extremely common in commemorative inscriptions and epitaphs throughout Upper Egypt and Lower Nubia. In fact, the same textual structure can be found on many of the funerary stelae from the monastery grounds itself.[17] The first part (ll. 1–5) primarily serves to communicate the name of the deceased and the date of his passing away, which are essentials to his commemoration. The fact of death is designated by the common euphemism "to go to rest: ⲙⲧⲟⲛ ⲙⲙⲟϥ" (l. 1). The second part consists of a brief prayer, asking for divine mercy (ll. 5–6).

The scribe is competent and writes a correct Sahidic with few errors or peculiarities. In ⲧⲣⲟⲙⲡⲉ, in l. 2, the false start with ϭ- instead of ⲣ- my explain the use of the article ⲧ- (instead of ⲧⲉ-); the preposition ϩⲓ denoting time, instead of ϩⲛ, is less common but by no means irregular.[18] In l. 3, the expected assimilation in ϩⲛ is not realized, as quite often. In l. 6, the imperative ⲡ̄ ⲟⲩⲛⲁ instead of the habitual

Figure 19.2 Monastery of St. Simeon, Aswan: the *dipinto* of Peter
(photo: Kathryn E. Piquette 2016)

ⲁⲣⲓ ⲟⲩⲛⲁ may be noted. One might feel tempted to correct this into a Third Future and emend either <ⲉⲣⲉ> ⲡⲭⲟⲉⲓⲥ ⲣ̄ ⲟⲩⲛⲁ or ⲡⲭⲟⲉⲓⲥ <ⲉⲕⲉ> ⲣ̄ ⲟⲩⲛⲁ, but the text as it stands, though uncommon, is not incorrect.[19] At the end of l. 3, ⲡⲙⲁⲛϣⲱ for ⲡⲙⲁⲛϣⲱⲡⲉ may be an abbreviation rather than a casual error, although no traces of a stroke are visible.[20] A certain sense of informality is conveyed primarily by the absence of proper ruling.

Historical context

This precisely dated *dipinto* commemorating the monk Peter offers two major points of interest. As Monneret de Villard was the first to observe, it is one of the few ancient sources that have preserved the true name of the so-called Monastery of St. Simeon (ll. 3–4).[21] Apart from the topographical handbook formerly attributed to Abu Salih and now to Abu 'l-Makarim[22] and other, more recent sources,[23] epigraphic attestations of its proper designation as Monastery of St. Hadra are scarce. The only other unquestionable example that has been published is the Coptic funerary stela of a superior of the monastery, which is there designated as ⲡⲙⲟⲛⲁⲥⲧⲏⲣⲓⲟⲛ ⲙⲡⲉⲛⲉⲓⲱⲧ ⲁⲡⲁ ϩⲁⲧⲣⲉ, "the Monastery of Our Father Apa Hatre (Hadra)".[24] Whether a lacunary *dipinto* copied by J. Clédat, which appears to mention [ⲡⲉ]ⲛⲉⲓⲱⲧ ⲉⲧ[ⲟⲩⲁⲁⲃ] ⲁⲡⲁ ϩⲓⲧⲣⲉ,[25] also refers to the monastery cannot be decided on the basis of the edited text. In the present *dipinto*, instead of ⲙⲟⲛⲁⲥⲧⲏⲣⲓⲟⲛ, the term ⲙⲁⲛϣⲱⲡⲉ, literally "dwelling, residence", is used (l. 3), undoubtedly with exactly the same sense of "monastery".[26] The topographical specification "in Aswan (ⲥⲟⲩⲁⲛ)", if not added mechanically, could be meant to distinguish the monastery from other ones dedicated to the same saint or a namesake.[27]

The text also contains hitherto unnoticed information. Thus, it dates Peter's death to the first year of a king named Zachari, a variant form of Zacharias (ll. 2–3).[28] Who was this Zacharias? As he cannot have been one of Egypt's contemporary Muslim rulers, he is likely to have been a Nubian king. In fact, among the kings of ninth- and tenth-century Makuria, precisely George and Zacharias were favorite royal names. According to the extremely lacunary list of Makurian kings for this era, at least four of them bore one of these names.[29] However, the present king Zacharias, whose first regnal year either fell in 961/962 or in 962/963, is absent from existing reviews of Nubian medieval history and new to the list. His reign should be intercalated between that of two contemporary rulers, Zacharias "II", who acceded to the throne in 915 and was still in authority in 930,[30] and George "II", who is known to have been reigning in the year 969.[31] The relationship between the new Zacharias "III"[32] and the latter king, who may have been his immediate successor, can perhaps be further specified.

In two medieval Nubian churches, at Faras and Sonqi Tino, wall paintings have been discovered which represent a King George, protected, in a typically Nubian way, by Christ (Sonqi Tino) and the Virgin Mary (Faras).[33] The generally accepted chronology of the Faras painting identifies this King George with George "II"

247

who reigned in 969. In both churches, the picture is accompanied by a Greek legend which names the king in question, according to the Faras version, as "King Geôrgiou (George) . . . son of King Zacharias".[34] The latter name is abbreviated as ᴢᴀˋх́ρι, which is reminiscent of the spelling ᴢᴀхᴀρι in the Aswan inscription of Peter. If the identification of the painting's subject is correct, we may quite plausibly recognize in the new King Zacharias "III" from Aswan the father of King George "II", who must have succeeded him after a short reign of seven or eight years at most.[35]

Thanks to the inscription of Peter something more can be said about this reign, which may in fact have been a quite remarkable one. On account of the precise date provided by the inscription, the name of King Zacharias "III" can almost certainly be connected with the campaign which took the Nubian troops in 963, only a few months after Peter's death in Aswan, as far north as Akhmim.[36] In an era that saw "the climax of Christian Nubian political power",[37] around the tenth century, this campaign stands out as a major military feat.

Against the background of Nubian intervention in Egypt, the fact that, in the Aswan inscription of Peter, the name of a Nubian king is used without any further comment as a dating reference, acquires an historical significance of its own. Nubia's political and military presence in Egypt during this period is mainly documented by usually later, Arabic sources. These create an impression of haphazard military actions, of raids and plunder without much political consequences.[38] The contemporary Coptic inscription presented here suggests a different picture. It follows the traditional Nubian custom of dating after the regnal years of indigenous kings, a custom that was inherited from the Byzantine world and normally limited to official documents.[39] It appears therefore that, at a time when Cairo was nominally governed by the Ikhshidids, in southern Egypt Nubian rule had taken more permanent forms and gained wide acceptance.[40] However, Peter's *dipinto* is a private, almost informal commemorative text for a modest monk. Here, adopting official Nubian usage probably means more than the mechanical reflection of the contemporary political and military situation. Far from the battlefields, in a monastery church, dating after a Nubian king may also have been the expression of religious loyalties and aspirations alive among Christian Egyptians who looked southward for the political means to realize their dreams.[41] In the eyes of some of them, year one of the military successful King Zacharias "III" must have looked like the beginning of a new era.

Notes

1 We wish to thank Dr. B.P. Muhs, Leiden, for his advice on the English of this paper.
2 For this saint, Gawdat Gabra, "Hatre (Hîdra), Heiliger und Bischof von Aswan im 4. Jahrhundert", *MDAIK* 44 (1988), 91–4; R-G. Coquin, "Hadrâ of Aswan (Saint)", in: *CoptEnc* 4, 1190; A. Papaconstantinou, *Le culte des saints en Égypte des Byzantins aux Abbasides: l'apport des inscriptions et des papyrus grecs et coptes*, Paris: Éditions du CNRS 2001, 224. See further below.

"IN YEAR ONE OF KING ZACHARI"

3   A somewhat muddled and incomplete but nevertheless useful account of ancient sources and modern research on the monastery can be found in S. Timm, *Das christlich-koptische Ägypten in arabischer Zeit*, Wiesbaden: Reichert Verlag 1984–1992, II, 664–67, s.v. "Dêr Anbâ Hadrâ".

4   H. Munier, "Les stèles coptes du Monastère de Saint-Siméon à Assouan", *Aegyptus* 11 (1930–1931), 257–300; 433–84. All have been republished as *SB Kopt.* I 498–675.

5   In particular, U. Bouriant, "Notes de voyage, § 19: Le Déir Amba-Samâan, en face d'Assouân", *RdTr* 15 (1893), 179–80; the same, in J. de Morgan et al., *Catalogue des monuments et inscriptions de l'Égypte antique* I, Vienna: Adolphe Holzhausen 1894, 136–40; cf. U. Bouriant, and A.F. Ventre, "Sur trois tables horaires coptes", *Mémoires de l'Institut Égyptien* 3/7 (1898), 575–604; J. Clédat, "Les inscriptions de Saint-Siméon", *RdTr* 37 (1915), 41–57. Some Greek and Old Nubian texts were (re-) edited by F.L. Griffith, *The Nubian texts of the Christian period*, Berlin: Verlag der Königl. Akademie der Wissenschaften 1913, 57–8, and "Christian documents from Nubia", *PBA* 14 (1928), 117–47 at 134–45.

6   U. Monneret de Villard, *Il monastero di S. Simeone presso Aswân* I: *Descrizione archeologica*, Milan: Tipografia e libreria pontificia arcivescovile S. Giuseppe 1927. The same author devoted several preliminary publications to the subject.

7   Particularly in chapters IX (on the internal organization of the monastery, with a brief discussion of the mural inscriptions on p. 140) and X (on Christianity in Aswan and the history of the monastery).

8   De Morgan, *Catalogue*, 136 (printed text under no. 4) and p. 140 (handcopy under no. 1); for Bouriant as responsible for the edition of the mural inscriptions, see p. XII. Clédat's later collection does not give the text. Possibly, it can be identified with Inscriptions, 44, "paroi sud d", no. 2, which he describes as: "Inscription d'au moins six lignes. Elle était écrite en rouge. Je n'ai pu en prendre copie par suite de son mauvais état".

9   Monneret de Villard, *Monastero*, 9.

10 Munier, "Stèles coptes", 479, under no. 176.

11 *A Coptic dictionary*, Oxford: Oxford University Press 1939, 580b, s.v. ϭⲱⲡⲉ (ⲙⲁ ⲛϭⲱⲡⲉ).

12 J.H.F. Dijkstra wishes to thank H. Thaler and B. Tratsaert for their kind assistance in recording the inscription.

13 Wall d in room II on the plan in Clédat, "Inscriptions", 54.

14 De Morgan, *Catalogue*, 140, no. 7. An even less reliable text is given by Clédat, "Inscriptions", 44, under "paroi sud d", no. 1.

15 Cf. V. Stegemann, *Koptische Paläographie*, Heidelberg: im Selbstverlag 1936, Tafelband, pls. 13 and 16 (examples from late eighth and ninth century documentary texts).

16 Cf. e.g. S. Jakobielski, *A history of the bishopric of Pachoras on the basis of Coptic inscriptions*, Warsaw: PWN–Éditions scientifiques de Pologne 1972, 199.

17 An identical opening formula (but with a different prayer) in Munier, "Stèles coptes", no. 114 (A.D. 872; *SB Kopt.* I 612). The same prayer for mercy, always directly following the date, in e.g. ibidem, nos. 54 (A.D. 718; *SB Kopt.* I 552) and 11 (A.D. 855; *SB Kopt.* I 609). The opening lines correspond to the third Coptic formulary in Krause's classification of the stelae from Sakinya (Toshka West), see M. Krause, "Die Formulare der christlichen Grabsteine Nubiens", in K. Michałowski (ed.), *Nubia: Récentes recherches*, Warsaw: Musée national 1975, 76–82.

18 See Crum, *Coptic dictionary*, 644b-45a; for some examples in dating formulae, see W.E. Crum, *Catalogue of the Coptic manuscripts in the collection of the John Rylands Library*, Manchester: University Press 1909, no. 464, l. 1 (tax-receipts, A.D. 1006/1007); H. Munier, "Stèles chrétiennes d'Antinoé", *Aegyptus* 29 (1949), 126–36 at 129, no. 1, ll. 7–8 (epitaph). Cf. J. van der Vliet, *I. Khartoum Copt.* 2, l. 5, with note.

19 Cf. L. Stern, *Koptische Grammatik*, Leipzig: T.O. Weigel 1880, 225.

20 The abbreviation ⲙⲁⲛϣⲱ(ⲡⲉ) occurs in the tenth-century lintel Berlin 9898, l. 7. See the re-edition by J. van der Vliet, "Monumenta fayumica", *Enchoria* 28 (2002–2003), 143–46 [Study 7].

21 Monneret de Villard, *Monastero*, 9.

22 For a thorough discussion of the (single) manuscript and the question of authorship, see U. Zanetti, "Abû l-Makârim and Abû Sâlih", *BSAC* 34 (1995), 85–138.

23 For which see Gawdat Gabra, "Hatre".

24 Munier, "Stèles coptes", 479–80, no. 176 (= *SB Kopt*. I 674), ll. 12–14. Cf. Monneret de Villard, *Monastero*, 9 and 135.

25 Clédat, "Inscriptions", 56 ("étage supérieur"), l. 3 (our tentative reconstruction). The text may simply invoke the prayers of the Saint. The form ⲉⲓⲧⲣⲉ, if read correctly, would reflect the pronunciation Hîdra, which is still current in Aswan. Cf. Gawdat Gabra, "Hatre", 92 and 94, n. 40.

26 Thus already Crum, *Coptic dictionary*, 580b.

27 One was situated on nearby Elephantine Island according to "Abu Salih", vol. 101b (Evetts), but cf. Timm, *Ägypten* II, 665–66; III, 1046, n. 13; Gawdat Gabra, "Hatre", 94. For another one, in the Coptite nome, see Papaconstantinou, *Le culte des saints*, 224.

28 Similar truncated forms of the name are quite common in Nubia. Cf. Jakobielski, *Bishopric of Pachoras*, 197, n. 21; for the present one, G.M. Browne, *Old Nubian texts from Qasr Ibrîm* III, London: Egypt Exploration Society 1991, 112 s.v.

29 For the latest update of this "list", see D.A. Welsby, *The medieval kingdoms of Nubia: Pagans, Christians and Muslims along the Middle Nile*, London: British Museum Press 2002, 259–61. Cf. U. Monneret de Villard, *Storia della Nubia cristiana*, Rome: Pontificium Institutum Orientalium Studiorum 1938, 223. The fragmentary list of names from Faras, which may have been a dynastic list, mentions three persons with the name of Zacharias; see Jakobielski, *Bishopric of Pachoras*, 195–98. If kings, they could well belong to this period; see S.C. Munro-Hay, "Kings and kingdoms of ancient Nubia", *Rassegna di Studi Etiopici* 29 (1982–1983), 97–137 at 110–11.

30 According to *I. Khartoum Copt*. 2.

31 See Monneret de Villard, *Storia*, 123.

32 King Zacharias "II" is called Zacharias "III" by most earlier authors, including Monneret de Villard, *Storia*, 223, and G. Vantini, *Christianity in the Sudan*, Bologna 1981, 113, because they count a crown prince of the same name, the son of King Mercurius (early eighth century), as the first Zacharias. In addition, a King Kubra, son of Surûr, may have reigned around 943, but apart from a single reference in an Arabic source, nothing is known about him. See Monneret de Villard, *Storia*, 120–21, and G. Vantini, *Oriental sources concerning Nubia*, Heidelberg: Heidelberger Akademie der Wissenschaften/Warsaw: Polish Academy of Sciences 1975, 130.

33 Faras: K. Michałowski, *Faras: Die Wandbilder in den Sammlungen des Nationalmuseums zu Warschau*, Warschau: Wydawnictwo Artystyczno-Graficzne /Dresden: VEB Verlag der Kunst 1974, no. 34. Cf. Jakobielski, *Bishopric of Pachoras*, 127–29. Sonqi Tino: Vantini, *Christianity*, 142–43, pl. XVI. Cf. S. Donadoni, "Les fouilles à l'église de Sonqi Tino", in: E. Dinkler (ed.), *Kunst und Geschichte Nubiens in christlicher Zeit*, Recklinghausen: Verlag Aurel Bongers 1970, 209–18 at 214–15.

34 Edited by S. Jakobielski, in: Michałowski, *Faras: Die Wandbilder*, 304–05, no. 18.

35 In any case, the speculative reconstruction of the royal succession for this period by Munro-Hay, "Kings and kingdoms", 107–11, can now be proven wrong.

36 See Monneret de Villard, *Storia*, 122, with reference to the Arabic sources, which date the event to A.H. 352 (= A.D. 963), not 962. Cf. Vantini, *Oriental Sources*, 735.

37 W.Y. Adams, *Nubia: corridor to Africa*, London: Penguin 1977, 455.

"IN YEAR ONE OF KING ZACHARI"

38 See e.g. the accounts in Vantini, *Christianity*, 116, and Welsby, *Medieval kingdoms*, 73–5.

39 For the Byzantine usage, see R.S. Bagnall, and K.A. Worp, *Regnal Formulas in Byzantine Egypt*, Missoula, Mont.: Scholars Press 1979; D. Feissel, "La réforme chronologique de 537 et son application dans l'épigraphie grecque: Années de règne et dates consulaires de Justinien à Héraclius", *Ktema* 18 (1993), 171–88. For its use in Nubian legal documents and foundation inscriptions, see the comments by A. Łajtar, *I. Varsovie* 101, and J. van der Vliet, *I. Khartoum Copt.* 1 and 2. In Nubia, it never occurs in the private funerary domain.

40 This was already convincingly argued from other sources by Monneret de Villard, *Storia*, 122–25.

41 Cf. Adams, *Nubia*, 455, and, in particular, J. van Lent, *Koptische apocalypsen uit de tijd na de Arabische verovering van Egypte*, Leiden: Oosters genootschap in Nederland 2001, 36–40.

20

CONTESTED FRONTIERS
Southern Egypt and Northern Nubia, A.D. 300–1500. The evidence of the inscriptions

Jacques van der Vliet

Introduction: frontiers and beyond

Aswan traditionally represents Egypt's southern frontier. This was already the case in pharaonic times, if not in reality, at least symbolically.[1] The very notion of a frontier implies the notion of discontinuity. Frontiers may mark and even create differences. On this side of the border things are normal, but beyond things are different – or that is what we expect. Yet frontiers are also zones of passage: places of contact where we get to experience the differences, negative or positive, and where we are forced to react accordingly. Frontiers, which may seem static and eternal at first sight, can acquire a dynamic aspect and come to foster processes of change, for example, linguistic change or cultural change.

Such processes of change are not to be conceived simply as the mechanical outcome of contact. Instead they depend on a wide variety of stimuli, going all the way from a liberal exchange to a strict refusal of contact, and from borrowing to the assertion or even creation of distinct identities on either side of the boundary line. Political frontiers, for example, draw seemingly abrupt and clear boundary lines, but at the same time create networks to maintain and – at once – transgress these lines, facilitating specific modes of passage that may become institutionalized in time.

Cultural, linguistic and religious boundaries are less easy to grasp, even though they may partly or wholly coincide with political ones. They are, moreover, constantly open to redefinition and may reflect not only changing political and economical circumstances, but also react to shifts within the social networks that are the bearers of religious or linguistic identity. Frontiers are not merely the end of something and the beginning of something else, but zones of contact where identities are defined and redefined within a given spatial setting.

The Aswan region, and in particular Elephantine Island at the northern end of the cataract zone, have for many centuries been the southern frontier of Egypt. Here, Egypt ends and Africa begins. For many centuries, though not always, this has been a political frontier, dividing two nations – hence its great military importance during some periods of history. Its status as a linguistic and cultural boundary zone is much less clearly defined. For example, not only in remote antiquity but even today Nubian is spoken also north of the cataract. In ancient and medieval times, Egyptian – in all its historical forms, from hieroglyphic to Coptic – was used as a written language along the whole of the Middle Nile, as is Arabic today. The same applies to its role as a religious boundary. For several centuries, Aswan marked a frontier between Christianity and traditional religion, and later between Islam and Christianity, but at the same time both religions were found north and south of the cataract region.

Actually, we are not dealing with one static frontier but with various contested frontiers: a zone where political, linguistic and religious identities are negotiated and articulated. Part of this ongoing process is reflected in inscriptions. Inscriptions are, in fact, privileged means to express these shifting identities. As written text, inscriptions are linguistic utterances, vehicles of the forms and values of a chosen language. They are also public utterances. As a rule they were visibly exposed and primarily meant to be seen, even by those who could not read them. Finally, they are strongly formalized. They use the specific codes of the social group that produced them and – inversely – these codes serve to define and reproduce these very groups. Better than any other genre of documents, therefore, inscriptions reflect the dynamics of a frontier zone. They mark the theater where the various social groups act out their identity and assert their authority.

The remainder of this essay will be devoted to a chronological discussion of three groups of selected inscriptions that cover a thousand years of Christianity in the broader Aswan region. Although occasionally reference will be made to Meroitic and Arabic inscriptions, the bulk of these texts are in Greek and Sahidic Coptic, representing the written codes of the elite groups that produced them. The principal vernacular of the region, Nubian, is practically absent from the written record. The first group of inscriptions illustrates the spread of Christianity and the assertion of Christian institutions on both sides of the cataract in late antiquity. The second group shows the cataract region as a frontier where the world of Islam confronted Christian Africa, represented politically by the Kingdom of Makuria. Finally, some later medieval inscriptions are witnesses to the cracking of the "Nubian dam" that had been constructed in the seventh century.

Traditional religion and the rise of Christianity (298–641)

Kalabsha and Tafa: political turmoil and religious change

The impressive temple of Kalabsha (Talmis), one of the religious centers of Roman Nubia, preserves a series of inscriptions that reflect on a local scale some of the epochal historical events of late antiquity. Most but not all of these are in Greek and thus bear witness to the strong Hellenization of northern Nubia (the Dodekaschoinos) in this period. One of them, inscribed in Greek in the *pronaos* of temple, dates from the middle of the third century (248–49). It is a rendering in stone of a decree by "the *strategos* of Omboi and Elephantine", a certain Besarion surnamed Ammonios, to the effect that "all pigs be driven out of the temple of the village of Talmis . . . so that the holy rites may take place in the customary way" (Eide et al. 1998: no. 248). Besarion, a military man, had issued this decree on behalf of the high priest of all Egypt, Myron, who resided at Alexandria.

The priests of the temple of Kalabsha, a renowned cult center of the local god Mandulis, must have been offended by the amount of pigs (unclean animals for them, but not for the local villagers who owned the pigs) walking around in the temple area. They were now authorized to chase them from the temple and publicized this authorization on one of the temple walls. The text shows that the region of Kalabsha resorted to the military authority of the *strategos* of Omboi (Kom Ombo) and the spiritual authority of the high priest of Alexandria, two Roman officials. Kalabsha was part of the Empire and northern Nubia was firmly embedded in the religious and political institutions of Roman Egypt.

Another inscription, still to be seen in the same temple, is the long inscription of Kharamadoye (Eide et al. 1998: no. 300). Its precise date is unknown but it must belong to the fourth or even the early fifth century. Significantly, it is not in Greek but in Meroitic. As the Meroitic language is still imperfectly understood, its contents remain largely obscure, but Kharamadoye was apparently a king, a *qore* in Meroitic. The inscription reflects a well-known fact: in 298 Diocletian had withdrawn the southern frontier of Egypt to its traditional place, the Aswan region, more precisely, Philae. The Dodekaschoinos, the Roman buffer zone in northernmost Nubia, had become part of the Meroitic empire. It was ruled by governors and, from some time after the fall of the Meroitic empire, around the middle of the fourth century, by apparently independent kings who were buried near Qustul and Ballana, in the neighborhood of Abu Simbel (see Török 2009: 515–30).

As far as the political situation in Lower Nubia after the withdrawal of Roman control and the subsequent fall of Meroe can be judged, it was characterized by great unrest, usually associated with the rivalry between two competing ethnic groups, the Noubades and the Blemmyes. Greek graffiti at Kalabsha mention obscure kings who have been interpreted as Blemmye chieftains, acting as patrons of local cults and their priests (Eide et al. 1998: nos. 310–11). They suggest that at least some of the Blemmyes had settled in the Nile Valley and had become integrated into the Hellenized framework of Lower Nubian society that had been part of Roman Egypt for over three centuries.

This process of acculturation is illustrated even more clearly by a Greek inscription in the temple of Tafa (Taphis) that belongs to the same general period, the late fourth and fifth centuries (Eide et al. 1998: no. 312). It records that a Blemmye *klinarchos* (president) of a *synodos* (a pagan cultic society) had restored the temple hall, called – with an Egyptian word – a *chant*. The persons mentioned bear Blemmye names with an admixture of Egyptian elements. A longer and even more difficult Greek inscription from the temple of Kalabsha, which mentions a fifth-century *phylarchos* (tribal chief) Phonen, also shows the continuity of both traditional Egypto-Nubian cults and traditional Hellenistic-Egyptian social institutions under Blemmye rule (Eide et al. 1998: no. 313).

Blemmye rule in post-Roman Nubia may have been ephemeral only. The Blemmyes were ousted very soon by the Noubades, who probably were the native inhabitants of the Middle Nile Valley, inheriting Meroitic authority in this part of Nubia. The best-known witness to the Nobadian take-over is the triumphal inscription of Silko on the west wall of the forecourt of the temple of Kalabasha (Eide et al. 1998: no. 317). Here Silko, "king of the Noubades and all the Ethiopians", proclaims his victory over the Blemmyes. The inscription is no longer in Meroitic, but in Greek. Reading this rather pompous text reveals a striking phenomenon. It contains phrases such as "God granted me victory" and, referring to the king's prowess, "I am a lion in the lower regions and a bear in the upper regions". Bears are foreign to both Nubia and Egypt and, in fact, this is the language of the Greek Bible.

Whether or not Silko was a Christian or whether he merely employed a Christian or Jewish scribe are moot points that are hardly important. Silko's monumental inscription announces a profound change of cultural and religious paradigm. The language of the Greco-Egyptian cult societies was replaced by the language of the Bible. It illustrates how finally Christianity came to Lower Nubia: not through the efforts of heroic missionaries, but by a gradual process of acculturation in which religion may not even have been a central concern. The new rulers of post-Roman Nubia adopted cultural models from the Empire in order to strengthen their position within Nubian society. The symbolic power afforded by these cultural alliances was no less real than that provided by their political alliances.

The final outcome of this process within the domain of religion can be observed again in Kalabsha. Two Coptic inscriptions incised by a single hand on the temple front, on the lefthand pylon, read: "I, Paul, the priest, prayed here for the first time", and "I, Paul, the priest, set up the Cross here for the first time" (Richter 2002: 162–63). These two very similar texts signal the conversion of the Kalabsha temple into a church, presumably in the middle or the second half of the sixth century. The script and the nature of the texts strongly suggest that they are contemporaneous with the better-known inscription that commemorates the conversion of the temple of Dendur during the episcopacy of Bishop Theodore of Philae (ca. 525 – after 577). They are written in Sahidic Coptic, but show dialectal forms that are attested in particular in southernmost Egypt (Roquet 1978). The priest Paul was definitely a local man.

Overviewing the entire group of inscriptions from Kalabsha, two remarks must be made. First, the inscriptions from Kalabsha are important witnesses to the major historical events of the period: the integration of northern Nubia into the administrative, religious and military structures of Roman Egypt; its return to Meroitic authority under the reign of Diocletian; the political instability of the region concomitant with the long-term persistence of local social and religious structures; and finally the gradual adoption of new cultural models, among them a new religion, Christianity. Secondly, in spite of the occasional appearance of a Meroitic king or an Alexandrian high priest, and the decisive intervention of the emperor Diocletian, they reflect primarily local issues: from the Nubians' fondness for pig raising to the priest Paul's conversion of the temple into a church.

Bishop Theodore and Philae as a symbol of religious change

Theodore of Philae, bishop for many years in the sixth century, is known from literary sources as well as from a considerable number of inscriptions. One of these, already mentioned, commemorates in Sahidic Coptic the conversion of the temple of Dendur into a church (Eide et al. 1998: no. 330; Dijkstra 2008: 299–302). It begins:

> By the will of God and the order of King Eirpanome and the eager student of the word of God, Joseph, the exarch of Talmis (Kalabsha), and as we received the Cross from Theodore the bishop of Philae, I, Abraham, the most humble priest, set up the Cross on the day on which the foundations of this church were laid (and a date follows, for which, see Ochała 2011).

As in Kalabsha, a priest inaugurated the new church and exactly the same formula is used to express this event in both places ("I, so-and-so, set up the Cross"). The Dendur inscription, however, in addition spells out the patronage of the political and religious authorities of the moment: a Nobadian king with a clearly indigenous name, Eirpanome; a local military commander, Joseph, based at Kalabsha, whose piety is particularly praised; Theodore, the bishop of Philae; and various civil officials. Apparently there was not yet a Nubian hierarchy of bishops in place. The temples of Kalabsha and Dendur were converted by priests, in Dendur and plausibly also in Kalabsha, under the supervision of the bishop of Philae, the frontier town between Byzantine Egypt and the Nobadian kingdom. Theodore acted as the patron of the official Christianization of Northern Nubia, together with a Nubian king, Eirpanome, about whom nothing is known beyond his name, but who was most likely a Christian too.

Philae, the see of Theodore, was both a frontier town and a religious center. As an ancient center of pilgrimage it had attracted for centuries pilgrims from south and north who came to visit the shrine of Isis and other gods of the traditional Egyptian religion (Rutherford 1998). From the north, these were Greeks and Romans in addition to Egyptians, and from the south, Meroites, sometimes from Meroe itself, altogether an immense area. These pilgrims left a variety of inscriptions, mostly graffiti, in various languages: Greek, Latin, Demotic Egyptian and Meroitic (Dijkstra 2008: 175–92). The Demotic scribal tradition in Philae, closely associated with the local priesthood, petered out around the middle of the fifth century; the last pagan inscription in Greek on Philae dates from about the same time (456–67). The fate of the traditional cults after this period is uncertain. Yet it took over seventy years before one of the generals of the emperor Justinian, the famous Narses, closed down the temples of Philae, around 535–37.

The latter date is usually taken to be the end of the traditional cults within Egypt, followed by a massive conversion of Philae's temples into churches. The evidence of the various Greek inscriptions left by Bishop Theodore in the great temple of Isis is taken as the expression of a decisive victory over a deeply rooted paganism, most eloquently summarized in one of them: "The Cross has prevailed (and) it always prevails" (Bernand 1969: no. 201). This picture has rightly been modified in recent publications, in particular by Jitse Dijkstra's monograph *Philae and the end of ancient Egyptian religion* (2008). Dijkstra correctly observed that the pagan cults in Philae had been moribund for a considerable time before Justinian and that Christianity was installed on the island far earlier than hitherto believed. In fact, bishops of Philae are known from the middle of the fourth century onwards. The Christian communities already had churches of their own and hardly needed the ancient temples for this purpose. The transition was a process at once more gradual and more complex than has usually been supposed, and much less a single dramatic event, culminating in the conversion of the great temple of Isis into a church devoted to St. Stephen.

Yet the evidence of the inscriptions remains quite impressive. Not only Theodore but also an earlier bishop, Daniel, active around the middle of the fifth century, left a considerable number of inscriptions, all in Greek (Daniel: Bernand 1969: nos. 194–95; Theodore: ibidem, nos. 200–04, 216; name of bishop lost: ibidem, nos. 220–21, 227). Whereas some of these inscriptions are connected with the conversion of the temple of Isis into the church of St. Stephen (Bernand 1969: nos. 200–04), most others record episcopal involvement in civil building activity on Philae. Actually, from no other place in Egypt do we have so many inscriptions left by bishops. In order to understand this considerable epigraphic activity we have to keep in mind that Philae was a place of symbolic value on either side of the political frontier and had religious as well as military importance. It was moreover dominated by the impressive temple buildings of traditional religion, themselves heavily inscribed with monumental texts in various languages. Understood within this context, the inscriptions left by bishops Daniel, Theodore and possibly others articulated the new Christian identity of the place and were a means of appropriating the symbolically important landscape of Philae.

In particular, the role of Theodore must have been of crucial importance. From various sources, including historical and church-historical ones, we know that he took a prominent part in the ecclesiastical life of his time, far beyond the boundaries of his diocese. He had the advantage of a long episcopacy, lasting over fifty years, and was clearly held in high regard by military and civil authorities in both Egypt and Nubia. Also, presumably in his time, the early history of the diocese of Philae was codified in a wonderful text that survives in Sahidic Coptic and is now known as the *Life of Aaron*. This collection of edifying stories relates how even the very first bishops of Philae had successfully combated traditional religion, encouraged monasticism, and showed pastoral care for the pagan Nubians.[2] This is precisely what Theodore did after them and what the inscriptions from both Nubia and Philae itself underscore (Dijkstra 2008: 324–33).

Christianity and Islam: a precarious balance (639–1173)

War, more war and peace: the years up to the baqt *(652)*

Very little is known about the political and military situation in Aswan following the Arab invasion of Egypt in the end of 639 (Bruning, forthcoming). Whereas Alexandria surrendered to the conquerors in November 641, they had to face unprecedented resistance at Egypt's southern frontier. According to a later literary source, Aswan was conquered by the Arabs only in 651/652 (Ibn Hawqal, quoted in Vantini 1975: 153), apparently following an earlier occupation. South of Aswan, the earlier Nobadian kingdom had shortly before become integrated into the powerful Christian state of Makuria, which had its center in the Dongola reach, between the Third and Fourth Nile Cataracts. In a historically important battle near Dongola, the Makurian infantry succeeded in decisively checking the Arab advance in this part of Africa, putting up what has been called "the Nubian dam" (Ayalon 1989). For this liminal region, a long period of turmoil came to an end only when in the summer of 652 the famous *baqt*, a bilateral peace treaty between Muslim-ruled Egypt and the Makurian kingdom, was concluded.

A beautiful inscription discovered in Dongola, the Makurian capital, in 2004 can perhaps be linked to these events (Jakobielski and van der Vliet 2011). It contains the long eulogy of a "former bishop" of Aswan, Joseph, who died in Dongola in 668, some sixteen years after the definitive reconquest of Aswan by the Arabs. The text is bilingual. Whereas Greek is used for the formulaic parts, the Sahidic Coptic body of the text contains a prose poem that is unique in the epigraphy of the Nile Valley on account of its rhetorical structure and panegyrical style. The inscription is of special interest, moreover, as it is the earliest precisely dated document found in Christian Dongola so far, and the earliest dated Coptic funerary inscription from Nubia. Since the focus of the text is on the bishop's spiritual and pastoral merits, it does not go into biographical or historical details. Thus it does not explain why this bishop of Aswan had come to end his days in Dongola, even though it emphasizes his status as a foreigner in the Makurian capital. Indeed, however highly Joseph may have been esteemed in his new homeland, where apparently he had lived and worked for some time, he did not occupy any other formal position than that of a "former bishop" of Aswan. This picture suggests that Joseph may have been a refugee who had fled the ongoing violence in Aswan or, perhaps more likely, an exile who as a public personality had sided with the Christian Makurians and was therefore unable to return to his see after 652. A more precise answer can only be expected from the discovery of further sources on Aswan's fate in the eventful years between 639 and 652.

JACQUES VAN DER VLIET

Building a frontier: stronghold Aswan and the "Castle of Philae"

As a result of more than ten years of struggle in and around the Aswan region, Aswan had developed into a major garrison town and administrative center of the new Islamic regime, second only to the new capital of Fustat. From the following centuries, hundreds of Muslim funerary stelae in Arabic have been preserved, a number that is again equaled only in Fustat (Su'ad Mahir 1977; Abd al-Tawab, Abd al-Rahman and Ory 1977–86). Aswan's newly gained prominence offers an example of a contested political frontier favoring the growth of new centers of power and new identities. A somewhat forgotten Coptic inscription, dated to 25 February of the year 693, explicitly attests to activity connected with the improvement of infrastructure around Aswan. It formally thanks "the wholly praiseworthy Amir Abu'l-Azz" for improving the road between Aswan and Kom Ombo, "so that men and animals travel on the road comfortably" (Mallon 1911–12: 132*–34*). Earlier Greek inscriptions from the sixth and seventh centuries attest to a similar concern for local infrastructure and record the involvement of the high military, civil and ecclesiastical officials representing the Christian Empire.[3] Now – merely forty years after the pacification of the region – a Coptic inscription shows a Muslim official with an Arabic name to be in charge of the work.

On the Nubian side of the frontier, too, its newly acquired importance became apparent in intensified political and infrastructural activity. In the reign of the Makurian king Merkourios, around 700, whose piety earned him the surname of "the New Constantine" (Evetts 1910: 140), new churches were inaugurated in the northern Nubian towns of Faras and Tafa (Taphis), foundations commemorated by inscriptions in Greek and Coptic (Faras: *I. Varsovie* 101 = *I. Khartoum Copt.* 1; Tafa: Maspero 1910). Perhaps around the same time or somewhat later, a new office was created, that of eparch of Nobadia. Residing in Qasr Ibrim, the capital of the former kingdom of Nobadia, this was a royal deputy, whose main function consisted in overseeing and regulating the diplomatic and commercial contacts with Islamic Egypt (Godlewski [2013]).

A dossier from the middle of the eighth century, consisting of one Arabic letter and several Coptic ones and found at Qasr Ibrim, gives us a fascinating glimpse of these contacts (Arabic letter: Hinds and Sakkout 1981; Coptic letters: publication forthcoming by J.L. Hagen, Leiden). They involved not only the eparch of Nobadia, but also various other agents, among them a permanent representative of the Nubian kingdom at the office of the Muslim governor in Aswan. More importantly, these documents show that the contacts between the two parties were not limited to matters pertaining to the *baqt* and its conditions. The frontier at Aswan was not a closed one and allowed a variety of contacts, from military alliances to conflicts over commercial interests.

Meanwhile the actual frontier remained where it had been in the Byzantine period, at Philae or – more precisely – a set of military strongholds on the east bank near Philae (Gascoigne and Rose 2010). Until well into medieval times, the Makurian kings claimed that their authority extended over all the Nubians from

Tilimauara, an unknown locality in the south, "up to the *kastron* (castle) of Pilak (Philae)" in the north (Crum 1905: no. 449; cf. Plumley 1981). Corresponding to "the *kastron* of Pilak", we find on the Nubian side "the *kastron* of the Moors". It is known from a late antique Greek papyrus and was tentatively identified by Adam Łajtar with an ancient fortress-like building on the east bank, opposite Bigeh, about 1.5 km south of Philae. According to medieval sources, this was the northernmost stronghold of the Nubian kings (Łajtar 1997).

An interesting Greek inscription from the east bank near Philae probably dates from the Byzantine period, and underscores the liminal character of this frontier zone. It records a foundation, perhaps by an *actuarius* (administrator) of the *kastron* of Philae and contains this prayer: "Lord God protect me, the master of this house, and those who live in it and deliver us from the craftiness of the devil" (Bernand 1989: no. 239). The exact relationship of the inscription with the *kastron* of Philae remains to be established, but it was clearly meant to avert one of the inherent dangers of any liminal zone: attacks not only by visible foes, but also by invisible ones.

An ecclesiastical stronghold: the Monastery of St. Hatre

With the growth of Islamic Aswan, it seems as if the main center of Christian life in Aswan shifts to the West Bank. The impressive site of St. Hatre becomes our main source of documents for the post-conquest period (Monneret de Villard 1927). These include numerous inscriptions, in particular the *dipinti* and graffiti in the church and other monastic buildings, and the hundreds of Coptic epitaphs that have been published by Henri Munier (1930–1931; *SB Kopt.* I 498–675). The monastery was apparently a center of supra-regional importance. Several of the monks were Nubians (*noubas*), whereas others came from far away places such as Pemje (Oxyrhynchos) in Middle Egypt. From a certain period onwards the monastery may have served as a residence for bishops. Two epitaphs mention bishops and a Coptic dedicatory inscription from the monastery records a pious foundation by Abraham, bishop of Aswan and Elephantine, bequeathing half of its income for "the poor of Aswan [and Elephantine]" (De Morgan 1894: 139, n. 1; cf. Crum 1902: no. 8322).

The presence of Nubians among the inhabitants of the Monastery of St. Hatre raises the question of the contacts between the respective Christian communities on either side of the frontier after the Arab conquest. In spite of a clearly drawn frontier, demarcated by fortresses, separating a Muslim state in the north and a Christian one in the south, contacts were apparently varied and lively. The Coptic tombstones of Nubians in St. Hatre are balanced by Arabic tombstones of Muslims living in Lower Nubia, mostly from Fatimid times. One of these even uses the Egyptian (Coptic) calendar that was current in medieval Nubia (Nigm ed Din Mohammed Sherif 1964: 249).

During the entire Fatimid period, the Makurian kingdom held a strong military and political position in the region. At various periods, Christians in medieval

Egypt looked to Nubia for patronage in times of conflict with the Muslim authorities. Such patronage is attested for the Aswan region as well. A commemorative *dipinto* in Coptic on one of the walls of the monastery church of St. Hatre dates the death of a monk, Peter, to 19 April 962 "in year 1 of King Zachari (Zacharias)", one of several Makurian kings of this name (Dijkstra and van der Vliet 2003). Did King Zacharias have real authority over the Aswan region or is this merely the nostalgia of a monk who dreamt of living under a Christian king? In any case, such a dream must have seemed at least reasonably realistic at the time.

The years of decline (1173–1500)

The fall of Qasr Ibrim and its echoes

The tenth and eleventh centuries were the period of the greatest political power and prosperity of the Makurian kingdom. Its decline is usually taken to set in with the conquest of Qasr Ibrim by Shams al-Dawla, the brother of Salah al-Din (Saladin), in 1173. The impact of this event was considerable and its echoes can be found in both Christian and Muslim authors, but also in a contemporary inscription from Aswan. In the ancient tombs hewn in the mountain of Qubbat al-Hawa, on the west bank, a church and monastic dwellings had been installed. One of the hermitages contains a Coptic *dipinto*, probably written by a monk, who, from his elevated position, had seen the armies march by. The text runs:

> On this very day, the 22nd of Tobe, the 1st of the moon, (year) 889 of the Era of the Martyrs (17 January 1173). It happened during the reign of the Turks over the entire land of Egypt, while our father Amba Markos was archbishop of the city of Alexandria and in the days also of Theodore, bishop of the city of Aswan, that the Turks [came south?]. They went up to Prim (Qasr Ibrim) and captured it on the 7th of Tobe. They seized everybody (?) who was inside and came (back) and sold them together with. . . .
> (T.S. Richter in Edel 2008: 515)

This is a rather matter-of-fact and neat eyewitness account of what was apparently quite a shocking event: Shams al-Dawla ("the Turks") returning with his booty to Aswan after a campaign of a few weeks only. The fall of Qasr Ibrim must have dealt a fatal blow to any Christian hopes for support from the south to counter the effects of Muslim rule.

The inscription was clearly written by someone who felt himself to be an Egyptian Christian, but Nubian influence is perhaps discernible in the use of a moon date, which was widespread in contemporaneous inscriptions from Nubia. Quite likely, medieval Christians from the Aswan region looked to Christian Nubia not only for patronage, but also as a cultural model, at least in some respects.

Other inscriptions from the same area likewise suggest that Nubian scribal habits were occasionally followed. This is the case of the commemorative *dipinto* dated to year one of King Zacharias at St. Hatre, quoted above. At Qubbat al-Hawa another instance can be found in a bilingual inscription recently published by Renate Dekker. The dedicatory inscription of the Church of St. Severus of 11 March 1180 opens with the same kind of garbled Greek formulae that are found in Nubian graffiti hundreds of kilometers farther to the south. It then continues in Coptic to record the consecration of the church and its baptismal font by a bishop of Aswan, Severos, and the sponsoring of this event by a certain David from the city of Hermonthis, in the Theban region (Dekker 2008: 32–4). In view of the date (only a few years after the fall of Qasr Ibrim) one may wonder whether there was any connection between the military expedition of Shams al-Dawla and the renovation of this church, which may have been damaged by marauding troops.

King Kudanbes and his retinue

If the neat account of the return of the victorious Shams al-Dawla to Aswan may be called a pathetic document, this is all the more true for the last inscription to be discussed here. It is again situated in the church of St. Hatre. In its latter days, the monastery apparently functioned as a kind of caravanserai giving shelter to travelers and pilgrims, including Muslims from abroad making the Hajj, who left ample traces in the form of both painted and engraved inscriptions in various languages. Thus, in April 1322, a certain Kartolaos left a long inscription in a highly obscure form of Nubian Greek, near the sanctuary of the monastery church, "on the front of the left jamb of the apse" (Griffith 1928: 134–45).

Following an invocation of the Archangel Michael, it enumerates what is apparently the retinue of a Nubian king called Koudanpes (Kudanbes). The latter must have visited the monastery church in great pomp, accompanied by a whole series of bishops, priests and civil dignitaries, with impressive Greek and Latin titles. The scribe expresses his awe upon seeing so many state and church celebrities, but – as far as we can understand the text – he does not state the occasion of the monarch's visit to the monastery.

Fortunately, the historian al-Maqrizi comes to our help here and reveals the sad reality behind Kartolaos' pompous inscription. The events date to the third period of the reign of the Mamluk sultan al-Nasir. In 1315, the latter sent an army to Dongola, the capital of Makuria, to install a puppet king on the Nubian throne. King Kudanbes, a usurper who had previously murdered his own brother, fled together with another brother, Abraham, but both were captured and imprisoned in Cairo in July 1317. Meanwhile another usurper, Kanz al-Dawla, attacked al-Nasir's puppet king and seized the Makurian throne for himself. Then the sultan released Kudanbes' brother Abraham and promised him to release Kudanbes as well if he were to defeat Kanz ad-Dawla. Abraham succeeded and Kanz al-Dawla surrendered to Abraham, but the latter died three days later and the Nubians re-installed Kanz al-Dawla. Thereupon, in 1323, the sultan sent an army to reinstate Kudanbes. The army reached its aim and chased Kanz al-Dawla but as soon as it returned to Egypt, the latter returned and deposed Kudanbes.

As the date of the inscription shows, Kudanbes' visit to the monastery of St. Hatre in Aswan must have marked a stage on his way from his Cairo prison to be reinstated briefly as a king at Dongola in 1323. In spite of all the glorious titles of his retinue, Kudanbes was no more than a puppet of the sultan and, so it seems, one in a series of otherwise inglorious usurpers. Kartolaos' long inscription looks at the facts from an entirely different angle than the Egyptian historian. As a faithful courtier he calls Kudanbes "a king worthy of three hundred years"! As it appeared, his reign was only a prelude to the final disintegration of the Makurian kingdom and the gradual Islamization of its territory in the following centuries.

Conclusions

From a great distance, the past seems orderly. Clear-cut periods are demarcated by successive wars, revolutions, councils and dynasties. Conflicting nations and religions face each other as solid blocks. Real history is different and far more confusing. It is the history written by normal people who, in spite of everything, succeed in surviving disasters and crossing borders. They sometimes leave behind simple documents such as a tombstone or a letter on papyrus. Only rarely do these enable us to discern some of their motives and ambitions. The region of Aswan, situated on the borderline between Egypt and Africa, is one of those rare centers where we can observe not merely the clash of empires and civilizations, but also the ways in which societies reacted to these events and gave shape to their existence. A very small part of all this is reflected in inscriptions, but for the history of Christianity on either side of Egypt's southern frontier, this is by no means the least important part.

Notes

1 For a recent discussion of this frontier in antiquity, see Török 2009: especially 7–22.
2 A rather inadequate edition of the text: Budge 1915: 432–95; a new edition is being prepared by Dijkstra and van der Vliet (forthcoming).
3 For example, at Kom Ombo (?): Gascou 1994; Aswan: Bernand 1989: nos. 235–37; Philae: Bernand 1969: nos. 194–95, 216–28; probably also Bernand 1989: nos. 239, discussed below.

Bibliography

Abd al-Tawab, M., 'Abd al-Rahman and S. Ory, 1977–1986, *Stèles islamiques de la nécropole d'Assouan*, Cairo: IFAO.
Ayalon, D. 1989, "The Nubian dam", *Jerusalem studies in Arabic and Islam* 12: 372–90; reprinted in *Islam and the abode of war: military slaves and Islamic adversaries*, Aldershot: Variorum 1994.
Bernand, A. 1989, *De Thèbes à Syène*, Paris: Éditions du CNRS.
Bernand, E. 1969, *Les inscriptions grecques et latines de Philae* II: *Haut et Bas Empire*, Paris: Éditions du CNRS.

Bruning, J., forthcoming, "Egyptian control over the Egyptian-Nubian frontier between A.D. 600 and 750", in: A. Delattre, M.A.L. Légendre, and P.M. Sijpesteijn (eds.), [*The late Roman and early Islamic Mediterranean and Near East: authority and control in the countryside*, Princeton: The Darwin Press].

Budge, E.A. Wallis 1915, *Miscellaneous Coptic texts in the dialect of Upper Egypt*, London: British Museum.

Crum, W.E. 1902, *Coptic monuments*, Cairo: IFAO; reprinted Osnabrück: Otto Zeller Verlag 1975.

Crum, W.E. 1905, *Catalogue of the Coptic manuscripts in the British Museum*, London: British Museum.

Dekker, R. 2008, " 'New' discoveries at Dayr Qubbat al-Hawâ, Aswan: architecture, wall paintings and dates", *ECA* 5: 19–36.

De Morgan, J., et al. 1894, *Catalogue des monuments et inscriptions de l'Égypte antique* I, Vienna: Adolphe Holzhausen [Coptic texts edited by U. Bouriant].

Dijkstra, J.H.F. 2008, *Philae and the end of ancient Egyptian religion: a regional study of religious transformation (298–642 CE)*, Louvain/Paris/Dudley, MA: Peeters.

Dijkstra, J.H.F., and J. van der Vliet. 2003, "In year one of King Zachari": evidence of a new Nubian king from the Monastery of St. Simeon at Aswan", *BSF* 8: 31–9 [Study 19].

[Dijkstra, J.H.F, and J. van der Vliet. (eds.), forthcoming, *The life of Aaron: a critical edition, translation and commentary*, Leiden.]

Edel, E., et al. 2008, *Die Felsgräbernekropole der Qubbet el-Hawa bei Assuan*, Part I/1–3, München: F. Schöningh [Coptic texts edited by T.S. Richter].

Eide, T., T. Hägg, R.H. Pierce, and L. Török (eds.) 1998, *Fontes historiae Nubiorum: textual sources for the history of the Middle Nile Region between the eighth century BC and the sixth century AD*, vol. 3, Bergen: University of Bergen [*FHN* III].

Evetts, B. (ed.) 1910, *History of the Patriarchs of the Coptic Church of Alexandria* III: *Agathon to Michael I (766)*, Paris: Firmin-Didot; reprinted Turnhout: Brepols 1947.

Gascoigne, A.L., and P.J. Rose. 2010, "Fortification, settlement and ethnicity in Southern Egypt, in: P. Matthiae et al. (eds.), *Proceedings of the 6th International Congress on the Archaeology of the Ancient Near East, May, 5th–10th 2009, "Sapienza" – Università di Roma*, Wiesbaden: Harrassowitz, III, 45–54.

Gascou, J. 1994, "Deux inscriptions byzantines de Haute-Égypte (réédition de *I. Thèbes – Syène* 169 r° et v°)", *Travaux et mémoires* 12: 323–42.

Godlewski, W. 2013, "A short essay on the history of Nobadia from Roman to Mamluk times", in: J.L. Hagen and J. van der Vliet (eds.), *Qasr Ibrim, between Egypt and Africa: a case study in cultural exchange (NINO Symposium, Leiden, 11–12 December 2009)*, Leiden: NINO, 123–33.

Griffith, F.L. 1928, "Christian documents from Nubia", *PBA* 14: 117–46.

Hinds, M., and H. Sakkout. 1981, "A letter from the governor of Egypt to the king of Nubia and Muqurra concerning Egyptian-Nubian relations in 141/758", in: Wadâd al-Qâdî (ed.), *Studia arabica et islamica: Festschrift for Ihsân 'Abbâs on his sixtieth birthday*, Beirut: American University in Beirut Press, 209–29; reprinted in: M. Hinds. 1996, *Studies in early Islamic history*, Princeton, NJ: Darwin Press, 160–87.

Jakobielski, S., and J. van der Vliet. 2011, "From Aswan to Dongola: the epitaph of Bishop Joseph (died A.D. 668)", in: A. Łajtar and J. van der Vliet (eds.), *Nubian voices: studies in Christian Nubian culture*, Warsaw: University of Warsaw/Raphael Taubenschlag Foundation, 15–35.

Lajtar, A. 1997, "Τὸ κάστρον τῶν Μαύρων τὸ πλησίον Φίλων: Der dritte Adam über *P.Haun.* II 26", *JJP* 27: 43–54.

Mallon, A. 1911–12, "Coptica", *Mélanges de la Faculté orientale (Université Saint-Joseph, Beyrouth)* 5: 121*–34*.

Maspero, J. 1910, "Le roi Mercure à Tâfah", *ASAE* 10: 17–20.

Monneret de Villard, U. 1927, *Il monastero di S. Simeone presso Aswân* I: *descrizione archeologica*, Milano: Tipografia e libreria pontificia arcivescovile S. Giuseppe.

Munier, H. 1930–1931, "Les stèles coptes du Monastère de Saint-Siméon à Assouan", *Aegyptus* 11: 257–300, 433–84.

Nigm ed Din Mohammed Sherif 1964, "The Arabic inscriptions from Meinarti", *Kush* 12: 249–50.

Ochała, G. 2011, "The date of the foundation inscription from Dendur revisited", *BASP* 48: 217–24.

Plumley, J.M. 1981, "A Coptic precursor of a medieval Nubian protocol", *Sudan Texts Bulletin* 3: 5–8.

Richter, S.G. 2002, *Studien zur Christianisierung Nubiens*, Wiesbaden: Reichert Verlag.

Roquet, G. 1978, "Le morphème *(E)TAH-* et les graffites coptes de Kalabcha", *BIFAO* 78: 533–38.

Rutherford, I. 1998, "Island of the extremity: space, language and power in the pilgrimage traditions of Philae", in: D. Frankfurter (ed.), *Pilgrimage and holy space in late antique Egypt*, Leiden: Brill, 229–56.

Su'ad Mahir Muhammad. 1977, *Madinat Aswan wa-aṭaruha fi al-'asr al-islami* (The city of Aswan and its antiquities in the Islamic period), Cairo: Central Book Organization.

Török, L. 2009, *Between two worlds: the frontier region between ancient Nubia and Egypt 3700 BC-AD 500*, Leiden/Boston: Brill.

Vantini, G. 1975, *Oriental sources concerning Nubia*, Heidelberg: Heidelberger Akademie der Wissenschaften/Warsaw: Polish Academy of Sciences.

Part 3

Nubia

21

COPTIC AS A NUBIAN LITERARY LANGUAGE
Four theses for discussion

Jacques van der Vliet

Introduction

Most current interpretations of the emergence, role and death of Coptic in medieval Nubia stand in need of revision.[1] They are usually based upon impressions gained from only part of the available material and lack the support of appropriate sociolinguistic models. In an attempt to stimulate the debate about multilingualism in Christian Nubia, the present paper proposes four statements for discussion, preceded by a brief survey of the surviving Coptic textual material from Nubia, which is our main source of information. The following remarks do not claim to be in any way final. Not only are new finds continuously being made and unknown texts published, also a number of central issues, for example in the domain of Nubian-Coptic language contact, have not been studied adequately yet.

Some basic facts about Nubian Coptic

Medieval Nubia was a multilingual society. Whereas one or more forms of Nubian must have been predominant within the domain of oral communication, no less than four written languages were in use, though not always simultaneously: Greek, Sahidic Coptic, Old Nubian and Arabic. Sahidic is the variety of written Coptic that had become standard in Upper Egypt from the fourth century onwards. Spreading southward at an early stage, it was written and read in most of Christian Nubia, not merely in liminal regions or commercial centers. It is only weakly attested in the southernmost kingdom of Nubia, Alwa, but then any written record from that area is scarce.[2] In Makuria, it is strongly attested by inscriptions from monastic centers like Mushu and Ghazali as well as, more marginally, from its capital city, Old Dongola.[3] Coptic sources abound in Nobadia, the northernmost part of Nubia. These are not merely inscriptions, but include documentary, literary, liturgical and even school texts, with clear centers of literary activity in and around Faras and Qasr Ibrim.[4]

Chronologically, Sahidic Coptic is attested in Nubia from the fifth century (the presumed date of the so-called Tantani letters, from Qasr Ibrim, *FHN* III 320–22)

269

to the late twelfth century (the date of a letter from the chancellery of the Nubian King Moses George, also from Qasr Ibrim).[5] It thereby covers almost the entire time span in which Sahidic Coptic was used in Egypt, which was roughly from the fourth to the eleventh century. The letters addressed to the local chieftain Tantani are witnesses to the relatively early use of Coptic in Nubia, as is the mid-sixth-century building inscription from the temple of Dendur in northern Nubia, mentioning King Eirpanome.[6] The latter is even the earliest datable building inscription in the Coptic language. Similar though somewhat earlier inscriptions from the White Monastery of Shenoute, near Sohag, which is generally considered an early center of Coptic literature, are still in Greek (Schaten and van der Vliet 2008: 134–35). The letter of King Moses George, on the other hand, is an extremely late example of an official document in Sahidic, dating from a period when Arabic and Bohairic Coptic had replaced Sahidic for almost all practical purposes in Egypt.

Coptic sources from Nubia, therefore, show a remarkably wide geographical and chronological scope. Also their sheer number is impressive, when we take into account that so much less written material survived in Nubia than in Egypt. The literary and documentary texts from Qasr Ibrim, for example, or the tomb-stones from Sakinya (Toshka West) and Ghazali make up sizeable corpora even according to Egyptian standards.[7] In terms of quality, the literary texts from Nubia in particular are a major addition to Coptic literature at large. Thus, the principal Coptic manuscripts of some well-known apocryphal texts have come to us from a Nubian environment. Examples are the *Book of the Resurrection of our Lord Jesus Christ*, by Pseudo-Bartholomew (*CANT*, no. 80)[8] or the first Coptic witness of the *Letter from heaven on the observation of the Sunday* (*CANT*, no. 311),[9] to mention but two. Sahidic Coptic, we may conclude, enjoyed particular favor in Christian Nubia.

1. Coptic in Nubia is not imported but adopted

As has been observed above, all our sources are witnesses of Nubian Coptic as a written language only. What the status and role of spoken Coptic in Nubia may have been, we do not know. In any case, it is hardly likely that people who bore "exotic" names like Eirpanome, Eiñitta or Tirsakouni spoke Egyptian as their native tongue.[10] It is important to realize that the mother tongue in which one speaks is, in a sense, an accident of nature. The language a society uses as a written language is not. It represents, on the contrary, an act of cultural significance. In particular, when a society begins to use a language that is not its own as a written language, it follows a certain model. A foreign language is either adopted by a society, as for example Latin in the medieval Germanic West, or imposed upon it, as Arabic in early medieval Egypt.

Since Nubian society was politically independent and in other respects had retained much of its native culture, we may suppose that Coptic was adopted by choice, not imposed by force. It is one of the literary languages which late antique and early medieval Nubians selected for encoding high-status written messages, as for example monumental inscriptions and literary codices. Now why are languages adopted in such a way, if they are not forced upon a society? This depends on what the sociologist Pierre Bourdieu describes as their market value, that is, the value they acquire in the exchange of social contacts and in the formation of social status (Bourdieu 2001). Languages that lack or lose their cultural market value are destined to die. Coptic Egyptian came to a rather sudden development following the birth and spread of the monastic movement, particularly in Upper Egypt in the third-fourth centuries. In the deeply Hellenized society of late antique Egypt, Coptic could obtain considerable status thanks to its close association with monastic leaders like Pachomius and his successors (Torallas Tovar 2005: 82–97).

The spread of Coptic in Nubia can therefore be best considered as a corollary of the conscious adoption of high-status cultural models from late antique Egypt, Christianity itself in the first place with monasticism in its wake and the radically new lifestyle which it implied (cf. now Dijkstra 2008: 293–94). I would again like to contradict emphatically the obsolete view that Coptic in Nubia was a language of immigrants. Certainly, there must have been refugees from early Islamic Egypt seeking shelter in Nubia, and we do in fact know one of them, Bishop Joseph, a former bishop of Aswan, who died and was buried in Old Dongola in A.D. 668.[11] Yet neither immigration nor commercial contacts can account for the wealth, variety and diffusion of Coptic in Nubia over a period of many centuries. Obviously, Coptic was chosen to function, in addition to Greek, as the privileged means of communication of the new political and ecclesiastical elite that rose to power in Nobadia in the fifth and sixth centuries and of which Tantani and Eirpanome must have been representative examples.

Furthermore, even though language contact phenomena in Nubian Coptic still await thorough study, substrate influence can be observed in a number of cases. Nubian Sahidic, not surprisingly, shares certain characteristics with the Sahidic of southernmost Egypt, in particular the Theban and the Esna-Edfu regions. In funerary inscriptions, for example, the doubling of initial vowels and the non-Sahidic /o/-/a/ shift can be observed with some regularity.[12] Yet also typical Nubian features can be pointed out, which show how much Coptic had become at home in Nubia. Indubitable linguistic interference from a Nubian substrate is apparent, for example, in the frequently occurring errors in gender selection, found in both pronouns and articles,[13] but also in instances of calquing, the use of loan translations, and code-switching, the change of language within one context.[14]

2. Coptic cannot be opposed to Greek

In the funerary epigraphy of Christian Nubia, Greek and Coptic are found side by side. The Makurian monastic site of Ghazali, with its wealth of inscriptions, provides a good example. There, a substantial minority of surviving tombstones were inscribed in Greek, a clear majority in Coptic.[15] The episcopal epitaphs from Faras, in Nobadia, show a similarly inconsistent picture. Thus, Bishop Petros (I), who died in A.D. 999, had his tombstone inscribed in Sahidic Coptic, whereas his immediate successor Ioannou had one made in Greek only six years later (1005).[16] In the past, such differences have been variously explained, for example as reflecting the opposition between a "Monophysite" and a "Melkite" ecclesiastical orientation, which is hardly likely.[17] In fact, these differences become at once more natural and more difficult to account for when it is realized that the Nubian situation is paralleled in Egypt.

Late antique Egypt was strongly and profoundly Hellenized. Demotic Egyptian had gone out of everyday use by the first century A.D., to be replaced by Greek (see e.g. Muhs 2005). For some centuries, the strange situation prevailed in which native Egyptian was largely absent from the written record (cf. Bingen 2005: 86). When it reappears on the scene as Coptic, it is not as a marker of ethnic or cultural distinctness. Rather to the contrary, Coptic Egyptian is born out of the wish to create an equivalent counterpart to Greek within the cultural context of early Christian monasticism. To which degree it is was shaped by a desire for imitation and emulation can be seen from both the visual and the grammatical characteristics of early Coptic.[18] In other words, the rise of Coptic as the second language of late antique Egypt was the outcome of a process of assimilation rather than differentiation. Contrary to what was often assumed in the past, the use of either language is not the expression of unavowed national, religious or even cultural distinctions, but of its value rating in a particular social context, on a particular cultural market.

That no cultural opposition underlay the choice of one or another language in Nubia can be demonstrated by two easily verifiable external criteria. First, there is the lack of visual distinction. Both languages, written Greek and written Coptic, were used in stone inscriptions that were meant to be exhibited publicly and read aloud (epitaphs, building inscriptions). Yet there is hardly ever any marked visual difference to be observed between them. Modern scholars who are dealing with fragmentary inscriptions need to know Coptic and Greek quite well in order to determine which language is used on a particular fragment, and thus occasionally Greek texts are taken for Coptic or the other way round.[19] Secondly, these same texts show an unparalleled degree of code-switching. Thus, the Coptic stela of the Deacon Petrou from Debeira West incorporates a lengthy doxology in Greek (*I. Khartoum Copt.* 17), whereas the predominantly Greek stela of a high Nobadian official Staurophoros switches for the lemma that states the death of its owner from Greek to Coptic.[20] These are only two examples out of many. Which precise criteria underlie the choice of either code in a given text is a question that demands further investigation. They may reflect varying scribal practices or preferences, but not conflicting cultural or religious standpoints.

3. Local variations in usage and distribution must be explained in local terms

One of the most striking characteristics of the funerary epigraphy of Christian Egypt, either Greek or Coptic, is undoubtedly the amount of regional and even local variation (see Van der Vliet 2006b [Study 1]). To a lesser degree, the same phenomenon can be observed in Nubia, where certain formulae seem to be typical of certain regions and certain periods. To quote but one example, the Coptic funerary prayer that asks the Archangel Michael to protect the bones of the deceased appears to be limited to the Faras-Meinarti region in the early eleventh century (*I. Khartoum Copt.* 17, ll. 13–14, with commentary). It is unknown from Egypt, and neither is it found in Greek epitaphs from Nubia or elsewhere. As in Egypt, we appear to have a situation where regional centers play an important role. Local scribal schools have preferences of their own that enjoy, for reasons difficult to explain, strictly local favor.

Another instance of regional variation in Nubian epigraphy can be observed in the proportional use of Greek versus Coptic. For example, much more Greek than Coptic found in Old Dongola cannot be taken as significant for all Makuria, since further to the south, in Ghazali, the proportions are in favor of Coptic (see above). Perhaps a sociological explanation applies here, Old Dongola being the residence of the royal court, whereas Ghazali was a monastic site. In any case, whatever explanation is correct, it resulted in local variation in language preference for epigraphic purposes.

A somewhat similar remark can be made about the bishops' stelae from Qasr Ibrim. Until now no episcopal epitaphs in Coptic have come to light at Qasr Ibrim, whereas a whole series in Greek has been discovered (*I. Qasr Ibrim* 18–26). This situation may of course be merely accidental, due to the chance of survival, but it contrasts oddly with that at Faras and Saï further to the south where bishops frequently had their funerary monuments inscribed in Coptic.[21] Again it is difficult to say what motivated this apparent preference for Greek epitaphs for the bishops of Qasr Ibrim, whereas this town was at the same time a major center of literary activity in Coptic (see Hagen, forthcoming). It is nevertheless certain that, in spite of the difficulties which this entails, these and similar variations in the geographical distribution of Coptic should be interpreted in terms of either sociological variables or local scribal traditions, or most likely both. To formulate my point negatively, language selection should not be explained in terms of the national or religious identity of the individual deceased or of geography (for example, proximity to Egypt), but in the parameters of a local cultural market in the sense of Bourdieu.

4. Sahidic Coptic dies simultaneously in Egypt and in Nubia

The use of all four literary languages of medieval Nubia is subject to change over time, and Sahidic Coptic is no exception. If the building inscription from Dendur and the Tantani letters were early instances of dedications and official letters in Sahidic Coptic by all standards, the twelfth-century letter of King Moses George is a very late example of an official letter in Sahidic. The latest dated Nubian epitaph known in Coptic is the stela of Bishop Georgios of Faras, of A.D. 1097 (*I. Khartoum Copt.* 5). By that time Sahidic Coptic had largely disappeared from the public sphere in Egypt as well. The number of Coptic documents from, for example, the Fayoum (letters, accounts, etc.) declines steeply in the course of the eleventh century.[22] Arabic took over as the language for all purposes of everyday communication, whereas for ecclesiastical purposes Bohairic, the language of the patriarchal see, soon became the standard for all of Christian Egypt. The diminishing status of Sahidic Coptic in Egypt, which led to its gradual demise, must have influenced its status in Nubia as well.

In the late Christian period of Nubian history, following Shams al-Dawla's sack of Qasr Ibrim in January 1173, Sahidic Coptic was about as absent from the written record in Nubia as it was in Egypt. In a late pilgrimage sanctuary such as the one discovered in Banganarti, Coptic is only marginally attested (Łajtar 2008). Greek, however, retained its symbolic value, not only and perhaps not even primarily as a liturgical language, but also as a reference to the remote but culturally significant center of the Christian world, Byzantium. Actually, a similar reorientation towards the Byzantine world and the Greek language can be observed among Christian intellectuals of later medieval Egypt (see, for example, Sidarus 2000). For the contacts with Egypt, Arabic took over the role of Coptic.[23] In Qasr Ibrim, for which we have the most and the most varied documentation, the decline of Sahidic is mirrored by the rise of Old Nubian as a documentary language, used for letters, contracts and similar documents, a role that hitherto was fulfilled by Coptic (see Hagen [2010]). Graphically, and in some respects perhaps even grammatically,[24] Old Nubian was modeled on Coptic, just as in a far earlier period Sahidic Coptic had been modeled on Greek. The Old Nubian script is a mere variety of the sloping late Sahidic uncials that are found in so many tenth-eleventh-century documents and semi-literary texts from the whole of Upper and Middle Egypt, including the Fayoum.[25] In the fifth century, when the Tantani letters were addressed to the court of the Nobadian chieftains at Qasr Ibrim, Sahidic Coptic was modern and fashionable in both Egypt and Nubia. By the end of the eleventh century it had become obsolete and nearly defunct in both countries.

Finally, I would like to emphasize that the four theses presented above are meant to provoke discussion. By their very nature they are unable to do full justice to the importance of Coptic as a medium of literary communication in Nubia in the long period between the fifth and the twelfth centuries A.D. Yet there can be no doubt that Coptic, together with Greek, decisively helped to shape a Christian Nubian culture that was oriented towards the Mediterranean world while remaining part of "Monophysite" Northeastern Africa. The full publication of the important textual finds from, in particular, Qasr Ibrim and Old Dongola will undoubtedly further fill in and refine the picture of the Coptic contribution to medieval Nubian culture.

Notes

1 Two fairly recent examples, each meritorious in its own right, are Adams 1996: 219–24, and Welsby 2002: 236–41; valuable earlier contributions, like Jakobielski 1970: 31–2, and Zaborski 1986: 406–09, do not reflect the present state of documentation.

COPTIC AS A NUBIAN LITERARY LANGUAGE

2 The graffito of a certain Katenantin (*sic!*), published by S. Jakobielski in Welsby and Daniels 1991: 276, shows that Coptic was at least known in Soba.

3 Mushu: *I. Khartoum Copt.* 28–42; Ghazali: *I. Khartoum Copt.* 43–116; some important Coptic inscriptions from Old Dongola still await publication, most notably the texts from the burial vault of Bishop Georgios (died 1113); see provisionally Jakobielski 2001: 164–65; Müller 2001; for the bishop himself and his other monuments, see Łajtar 2002.

4 School texts from Faras: Jakobielski 1983: 136, Group D; Debeira: Shinnie and Shinnie 1978: 98, no. 195; Meinarti: Browne 2002: 8–9 (there wrongly as Old Nubian).

5 Unpublished; see provisionally Adams 1996: 227–29. The late fourteenth-century scrolls of Bishop Timotheos (published in Plumley 1975) are, as far as their Coptic part is concerned, in Bohairic Coptic, a different language variety. Since they were produced in Egypt with a ceremonial purpose only, they can hardly be considered a witness to the role of Coptic in Nubia.

6 For the inscription and its date, see Richter 2002: 164–72; cf. Dijkstra 2008: 299–302.

7 Qasr Ibrim: Hagen forthcoming; Sakinya: Mina 1942; Ghazali: *I.Khartoum Copt.* 43–116.

8 See Westerhoff 1999: 15–16, 226–27.

9 A codex from Qasr Ibrim; publication forthcoming by J.L. Hagen, Leiden.

10 Eirpanome: the king mentioned in the sixth-century Coptic inscription at Dendur, quoted above (note 6); Eiñitta: the scribe of the twelfth-century letter of King Moses George, also mentioned above; Tirsakouni: a monk, the owner of a Coptic tombstone from the Makurian monastery of Ghazali (*I. Khartoum Copt.* 67).

11 For the publication of his bilingual (Greek-Coptic) tombstone, see Jakobielski and van der Vliet 2011 [Study 29].

12 See *I. Khartoum Copt.* 3, l. 18, with n. 113 (doubling of initial vowels); Van der Vliet 2006a: 221, n. 21 (/o/-/a/ shift). The latter phenomenon also in Coptic texts from the Aswan region: Satzinger 1980: 85–7.

13 Thus already Mina 1942: VII; see also the examples listed in *I. Khartoum Copt.*, p. 209.

14 Calquing: Satzinger 2004: 534–35 (on *I. Khartoum Copt.* 23; *SB Kopt.* III 1646, tenth century); code-switching Coptic-Old Nubian: *I. Khartoum Copt.* 19, ll. 6–7 (A.D. 1046).

15 The bulk of these often fragmentary inscriptions is now in Khartoum, see *I. Khartoum Copt.* 43–116; *I. Khartoum Greek* 31–55; smaller collections are in Berlin and in the Merowe museum, cf. *I. Khartoum Copt.*, pp. 104–05.

16 Petros: Jakobielski 1972: 135–39; Ioannou (Ioannes): *I. Khartoum Greek* 2.

17 For a still valid critical discussion of theories about 'Melkite' bishops in Nubia, see Krause 1978, and similar publications by the same author.

18 For the impact of Greek on Coptic, see Reintges 2004.

19 For instance, the fragment *I. Khartoum Greek* 14, from Saï, was initially described as Coptic by its excavators.

20 Published by T. Hägg, in Säve-Söderbergh 1981: 56–9; for the owner, cf. Łajtar and van der Vliet 1998: 50, no. 11 [= Study 28].

21 Faras: Jakobielski 1972; Saï: Łajtar 2006: 92–4; Grimal and Adly 2004: 130–31, fig. 41.

22 Some of the eleventh-century finds from the monastery of Naqlun in that province are among the latest examples of Coptic documentary texts, and show a remarkable degree of Coptic-Arabic code-mixing; see provisionally Van der Vliet 2005.

23 Thus already Zaborski 1986: 409; cf. Adams 1996: 223.

24 See, for a tentative case, Browne 2004: 25–6.

25 *Pace* Browne 2002: 1, who sees a "fundamental difference" between Coptic and Old Nubian scripts; he unwittingly invalidates his own argument by publishing, ibidem: 8–9, a Coptic school text from Meinarti (see above) as an example of the Old Nubian alphabet.

Bibliography

Adams, W.Y. 1996, *Qaṣr Ibrîm: the late mediaeval period*, London : Egypt Exploration Society.

Bingen, J. 2005, "L'épigraphie grecque de l'Égypte post-constantinienne", in: *Pages d'épigraphie grecque* II: *Égypte (1983–2002)*, Brussels: Epigraphica Bruxellensia, 85–99 (Earlier in *XI Congresso internazionale di epigrafia greca e latina, Roma, 18–24 settembre 1997. Atti* II, Rome: Edizioni Quasar 1999, 613–24.)

Bourdieu, P. 2001, *Langage et pouvoir symbolique*, Paris: Seuil.

Browne, G.M. 2002, *Old Nubian grammar*, Munich: Lincom.

Browne, G.M. 2004, "Marginalia Jernstedtiana", in: M. Immerzeel and J. van der Vliet (eds.), *Coptic studies on the threshold of a new millennium: proceedings of the Seventh International Congress of Coptic Studies, Leiden, 27 August-2 September 2000*, Louvain/Paris/Dudley, MA: Peeters 2004, I, 15–26.

Dijkstra, J.H.F. 2008, *Philae and the end of ancient Egyptian religion: a regional study of religious transformation (298–642 CE)*, Louvain: Peeters.

Grimal, N., and E. Adly. 2004, "Fouilles et travaux en Égypte et au Soudan, 2002–2003", *Orientalia* 73: 1–149.

Hagen, J.L. 2010, " 'A city that is set on an hill cannot be hid': progress report on the Coptic manuscripts from Qasr Ibrim", in: W. Godlewski and A. Łajtar (eds.), *Between the Cataracts: proceedings of the 11th conference for Nubian studies, Warsaw University, 27 August–2 September 2006*, Warsaw: Warsaw University Press, II: *session papers*, fasc. 2, 719–26.

Hagen, J.L. (forthcoming), *The Coptic "Library" of Qasr Ibrim and the churches and saints of Christian Nubia*, Ph.D dissertation, Leiden University.

Jakobielski, S. 1970, "Some remarks on Faras inscriptions", in: E. Dinkler (ed.), *Kunst und Geschichte Nubiens in christlicher Zeit*, Recklinghausen: Verlag Aurel Bongers, 29–40.

Jakobielski, S. 1972, *A history of the bishopric of Pachoras on the basis of Coptic inscriptions*, Warsaw: PWN–Éditions scientifiques de Pologne.

Jakobielski, S. 1983, "Coptic graffiti from Faras", *ÉtudTrav* 13: 133–37.

Jakobielski, S. 2001, "Das Kloster der heiligen Dreifaltigkeit: Bauphasen des nordwestlichen Anbaus", in: S. Jakobielski and P.O. Scholz (eds.), *Dongola-Studien: 35 Jahre polnischer Forschungen im Zentrum des makuritischen Reiches*, Warsaw: ZAŚ PAN, 141–68.

Jakobielski, S., and J. van der Vliet. 2011, "From Aswan to Dongola: the epitaph of Bishop Joseph (died A.D. 668)", in: A. Łajtar and J. van der Vliet (eds.), *Nubian voices: studies in Christian Nubian culture*, Warsaw: University of Warsaw/Raphael Taubenschlag Foundation, 15–35 [Study 29].

Krause, M. 1978, "Bischof Johannes III von Faras und seine beiden Nachfolger: Noch einmal zum Problem eines Konfessionswechsels in Faras", in: *Études nubiennes: colloque de Chantilly, 2–6 juillet 1975*, Cairo: IFAO, 153–64.

Łajtar, A. 2002, "Georgios, archbishop of Dongola (+ 1113) and his epitaph", in Εὐεργεσίας χάριν: *studies presented to B. Bravo and E. Wipszycka*, Warsaw: University of Warsaw/ Raphael Taubenschlag Foundation, 159–92.

COPTIC AS A NUBIAN LITERARY LANGUAGE

Łajtar, A. 2006, "Christian Saï in written records (inscriptions and manuscripts)", *JJP* 36: 91–104.

Łajtar, A. 2008, "Late Christian Nubia through visitors' inscriptions from the Upper Church at Banganarti", in: Godlewski and Łajtar, *Between the Cataracts*, Warsaw: Warsaw University Press, I: *main papers*, 321–31.

Łajtar, A., and J. van der Vliet. 1998, "Rich ladies of Meinarti and their churches: with an appended list of sources from Christian Nubia containing the expression 'having the church of so-and-so'", *JJP* 28: 35–53 [Study 28].

Mina, T. 1942, *Inscriptions coptes et grecques de Nubie*, Cairo: Société d'archéologie copte.

Müller, C.D.G. 2001, "Schutzinschriften einer Grablege in Alt-Dongola: Zu nubischen Geheimwissenschaften", in: Jakobielski and Scholz, *Dongola-Studien*, Warsaw, 321–26.

Muhs, B. 2005, "The grapheion and the disappearance of Demotic contracts in early Roman Tebtynis and Soknopaiu Nesos", in: S. Lippert and M. Schentuleit (eds.), *Tebtynis und Soknopaiu Nesos: Leben im römerzeitlichen Fajum*, Wiesbaden: Harrassowitz, 93–104.

Plumley, J.M. 1975, *The scrolls of Bishop Timotheos: two documents from medieval Nubia*, London: Egypt Exploration Society.

Reintges, Chr. 2004, "Coptic Egyptian as a bilingual language variety", in: P. Bádenas de la Peña et al. (eds.), *Lenguas en contacto: el testimonio escrito*, Madrid: Consejo Superior de Investigaciones Científicas, 69–86.

Satzinger, H. 1980, "Sudan-Ägyptisch und Elephantine-Koptisch", *Bulletin de la Société d'égyptologie, Genève* 4 (*Mélanges W. Vycichl*): 83–7.

Richter, S.G. 2002, *Studien zur Christianisierung Nubiens*, Wiesbaden: Reichert Verlag.

Satzinger, H. 2004, "Some peculiarities of Greek and Coptic epigraphy from Nubia", Immerzeel and van der Vliet (eds.), *Coptic studies on the threshold of a new millennium*, I, 529–35.

Säve-Söderbergh, T. (ed.) 1981, *Late Nubian cemeteries*, Solna: Esselte Studium.

Schaten, S., and J. van der Vliet. 2008, "Monks and scholars in the Panopolite nome: the epigraphic evidence", in: G. Gabra and H.N. Takla (eds.), *Christianity and monasticism in Upper Egypt* I: *Akhmim and Sohag*, Cairo/New York: American University in Cairo Press, 131–42 [Study 15].

Shinnie, P.L., and M. Shinnie. 1978, *Debeira West: a mediaeval Nubian town*, Warminster: Aris & Phillips.

Sidarus, A. 2000, "La tradition sahidique de philologie gréco-copto-arabe (manuscrits des XIIIᵉ–XVᵉ siècles)", in: N. Bosson (ed.), *Études coptes VII: Neuvième journée d'études, Montpellier 3–4 juin 1999*, Louvain/Paris: Peeters, 265–304.

Torallas Tovar, S. 2005, *Identidad lingüística e identidad religiosa en el Egipto grecorromano*, Barcelona: Reial Acadèmia de Bones Lletres.

Van der Vliet, J. 2005, "Naqlun: preliminary observations on the Coptic texts found during seasons 2003 and 2004", *PAM* 16: 191–94.

Van der Vliet, J. 2006a, "Two Coptic epitaphs from Qasr Ibrim", *JEA* 92: 217–23 [Study 25].

Van der Vliet, J. 2006b, "L'épigraphie chrétienne d'Egypte et de Nubie: bilan et perspectives", in: A. Boud'hors and D. Vaillancourt (eds.), *Huitième congrès international d'études coptes (Paris 2004)* I: *bilans et perspectives 2000–2004*, Paris: de Boccard, 303–20 [Study 1].

Welsby, D.A. 2002, *The medieval kingdoms of Nubia: Pagans, Christians and Muslims along the Middle Nile*, London: British Museum Press.

Welsby, D.A., and C.M. Daniels. 1991, *Soba: archaeological research at a medieval capital on the Blue Nile*, London: The British Institute in Eastern Africa.

Westerhoff, M. 1999, *Auferstehung und Jenseits im koptischen "Buch der Auferstehung Jesus Christi, unseres Herrn"*, Wiesbaden: Harrassowitz.

Zaborski, A. 1986, "Marginal notes on medieval Nubia", in: M. Krause (ed.), *Nubische Studien*, Mainz: Philipp von Zabern, 403–12.

GLEANINGS FROM CHRISTIAN NORTHERN NUBIA

Jacques van der Vliet

Introduction

When Diocletian withdrew the Roman troops from the Dodekaschoinos in 298, a particularly turbulent episode in the history of the northernmost part of Nubia began. To the modern historian, working with a limited set of disparate data, it may seem as if this period came to an end only around 700, with the reign of the Makurian king Merkourios, who welded the "corridor to Africa" into a powerful political unit. Yet, in the course of the 400 intervening years, the Dodekaschoinos had become part of an independent Christian kingdom, Nobadia, the political capital of which may have been situated initially at Primis, the present-day Qasr Ibrim, and then at Pakhoras (Faras).[1] In these centuries of transition, the ancient urban centers of the area, like Taphis (Tafa) and Talmis (Kalabsha), appear to have retained their importance. The following notes discuss some Greek and Coptic inscriptions that were part of this historical and geographical setting. Although primarily concerned with textual matters, they may also shed light on the culture and institutions of the early Christian Dodekaschoinos.

1. Aroumi from Taphis, ⲁⲣⲁ (*Tibiletti Bruno*, no. 49)

The extensive Christian necropolis of Tafa-Ginari, which was excavated in the early twentieth century by C.M. Firth, yielded an important find of Greek stelae, carelessly edited by their excavator.[2] The find was dispersed afterwards and only a few of the stelae could be traced and re-edited in recent years.[3] One of these was made for a woman called Aroumi, whose apparently non-Greek name receives the addition ⲁⲣⲁ, for which, to the best of my knowledge, no satisfactory explanation has ever been provided.[4] Maria Grazia Tibiletti Bruno in the apparatus of her re-edition of the epitaph simply states: "ⲁⲣⲁ = ἐκοιμήθη", without further explanation. This overly simple solution was rightly questioned in 1993 by Adam Łajtar, who, nonetheless, conceded that a verb is expected here.[5] This, however, is only partly correct. As will be seen in more detail below, Nubian stelae of the ἔνθα κατάκειται type, to which category Aroumi's epitaph belongs, normally use one of two verbal forms for describing the event of death, viz. ἐτελεώθη or ἐκοιμήθη, or no verb at all.[6] There is no reason, therefore, to consider ⲁⲣⲁ as an unknown verb for "to die", since no such verb is really demanded in this position.

Another possibility would be to interpret ⲁⲣⲁ either as part of the name or as a surname. The former solution seems to have been adopted by Firth, who transcribed the name of the owner of the stela as ⲁⲣⲟⲩⲙⲡⲁⲣⲁ. However, as Tibiletti Bruno's more reliable re-edition shows, the element ⲁⲣⲁ is singled out as a separate element by a long superlinear stroke, which additionally marks it as an abbreviation or a non-Greek element.[7] Rather than surnames or filiations, stelae of the present type, either from Tafa-Ginari or from adjacent districts, tend to join titles or function-names to the name of the deceased, either ecclesiastical ones like "deacon",[8] or civil ones like *meizoteros* [supervisor].[9] When we compare the textual structure of the numerous epitaphs of the ἔνθα κατάκειται type from Northern Nubia, it appears that the enigmatic ⲁⲣⲁ would exactly fit the position of such a title or function-name, following the name of the deceased and preceding the date of demise.

As a late antique title, the word ⲁⲣⲁ is actually well known. In the area of Thebes, it occurs in a series of sixth-eighth century Coptic documents.[10] Unfortunately, its precise meaning is still obscure.[11] Its use is limited to the domain of local law or finance, where, apparently, it functions as a traditional element, not obviously corresponding to any rank within the familiar hierarchies of the Byzantine Church or state administration. Its etymology, on the other hand, is hardly doubtful: the word is derived from the Ancient Egyptian title *iry-'3*, originally designating a "doorkeeper", later certain purely administrative functions as well.[12] If F.L. Griffith, the founder of Meroitic studies, was right in observing that the Egyptian title may be recognized as *are* in the titulature of late Meroitic functionaries from Karanog and Faras, this would attest its use in Lower Nubia in a period immediately predating the christianization of the country.[13] The stela of Aroumi suggests that the title, in its form ⲁⲣⲁ, survived in both Southern Egypt and Northern Nubia. In fact, this would not be the only example of a traditional Egyptian title remaining in use on both sides of the political frontier. The survival, in the same regions and in the same language domains (Coptic and Greek), of the more familiar religious title ⲟⲩⲏⲏⲃ, ⲟⲩⲉⲉⲡ (Eg. *w'b*), designating a (pagan or Christian) priest, can be cited as a striking parallel.[14] That, among the women of Christian Nubia, Aroumi was not exceptional in playing a role in finance or administration, is otherwise well known.[15]

2. Manna from Kalabsha (*SB* III 6089; *Tibiletti Bruno*, no. 47)

The Leiden Museum of Antiquities houses a small Greek stela from Northern Nubia under inv. no. AM 22.[16] In spite of a rich bibliography, the name of the owner, a woman, is still a source of confusion. It is traditionally read as Manma, a reading accepted by G. Lefebvre in his authorative 1907 *Recueil*, and adopted subsequently by Preisigke, in his *Namenbuch*. Lefebvre's text, including the reading Manma, was reproduced in *SB* V 8737. Already in 1926, however, F. Bilabel had published in *SB* III, under no. 6089, an impeccable edition of the stela

GLEANINGS FROM CHRISTIAN NORTHERN NUBIA

after the reproduction in P.A.A. Boeser's monumental catalogue of the museum's Egyptian antiquities, identifying the owner's name correctly as Manna. It is therefore quite astonishing that, as late as 1964, Mrs. Tibiletti Bruno, who was aware of both *SB* entries, preferred to stick to Lefebvre's less accurate text. Foraboschi's *Onomasticon alterum* adds to the confusion by referring twice to the Leiden stela, once under Manma and again under Manna. The present note aims, first, at rehabilitating Bilabel's 1926 readings and, then, at briefly discussing date and provenance of the stela.

The stela is a plain, well-dressed sandstone tablet,[17] 19.5 × 14 cm, inscribed with ten lines of Greek text. Irregular, slightly sloping uncials, crudely incised; broken-barred symmetrical *A*; the letters were filled with red paint. Simple crosses mark the beginning and the end of the text; the article in 1. 2 has a dot over the н; the ι of the number in 1. 4 and the names in ll. 9–10 bears a diaeresis; καὶ in ll. 9 and 10 is represented by a ϛ-like ampersand; the abbreviations of ll. 3 and 5 are marked by an oblique stroke through the lower right leg of λ and κ.[18] For ease of reference, I give below my own complete text of the epitaph, which does not differ significantly from Bilabel's in *SB* III 6089; the apparatus notes the variant readings of the standard corpora of Lefebvre and Tibiletti Bruno only.

> \+ Ἔνθα κατάκοι-
> τε ἡ μακαρία
> Μάννα · ἐτελ(εώθη)
> 4. μ᾽η᾽(νὶ) Χοίαχ ϊ,
> ἰνδ(ι)κ(τίωνος) ζ · ἀνά-
> παυσον τὴν
> ψυχὴν αὐτοῦ
> 8. [ε]ἰς κώλπης Ἀ-
> [β]ραὰμ (καὶ) Ἰσα-
> [ὰ]κ (καὶ) Ἰακώβ. +

1.-2. κατάκοιτε: read κατάκειται ‖ 3. Μάννα: Lefebvre, Tibiletti Bruno Μανμα; ἐτελ(εώθη): Lefebvre ἐτελ(εύτησεν) ‖ 4. Χοίαχ: Tibiletti Bruno Χοίακ ‖ 7. αὐτοῦ: read αὐτῆς ‖ 8. [ε]ἰς κώλπης: Lefebvre, Tibiletti Bruno ἰς κ.; read εἰς κόλπους (or, ἐν κόλποις).[19]

The text comprises an epitaph of the type ἔνθα κατάκειται ὁ μακάριος N.N., ending with a prayer which asks for rest for the deceased "in the bosom of Abraham, Isaac and Jacob". The name of the owner is undoubtedly to be read Manna, not Manma.[20] The second н of the name does have a small rightward stroke at the upper righthand leg, but similar lapicide's uncertainties can be observed in the preceding н and in those of 1. 6; the letter is clearly distinct from the м as it is written throughout the text. The ghost name Manma should be deleted from the *Namenbuch* and its successors.

Both Lefebvre and Tibiletti Bruno give "Nubia?" as the stela's provenance. It is possible, however, to be more positive. The monument of Manna belongs

281

to a large group of Christian funerary stelae, characterized by a similar opening formula: ἔνθα κατάκειται, "here lies . . .", introducing the name of the deceased. This epigraphic formula can be found, with many variants, all over the Roman and Byzantine world.[21] In Egypt proper, it never became very popular, and examples are few and far between.[22] In contrast, it can frequently be found in Northern Nubia, where it always assumes the precise form which it has in the present stela, i.e. with ensuing ὁ μακάριος/ἡ μακάρια N.N. Indeed, Adam Łajtar's claim that this particular form of the formula was exclusively characteristic of Northern Nubia (Nobadia) may well be justified.[23]

In his classic study of Nubian funerary epigraphy, Junker had observed that stelae of the type ἔνθα κατάκειται ὁ μακάριος N.N. may use, following this opening phrase, one of two verbs describing the event of death, either ἐτελεώθη (as in Manna's epitaph) or ἐκοιμήθη, or no verb at all (as in the stela of Aroumi, discussed above), or, exceptionally, the full formula τέλει τοῦ βίου ἐχρήσατο.[24] Junker also pointed out that the stelae of this type with ἐκοιμήθη (or, with the formula τέλει τοῦ βίου) originated, without exception, from the cemetery of Tafa-Ginari (for which see above, under no. 1). On the other hand, the same cemetery has not produced any stela with the dying-expression ἐτελεώθη among its thirty-seven epitaphs of the ἔνθα κατάκειται type. Tafa is therefore not a likely provenance for the Leiden monument of Manna.

The literature mentions various provenances for stelae of the ἔνθα κατάκειται type which either use the verb ἐτελεώθη or lack a verb for dying altogether, but are also unconnected with Firth's excavations in Tafa-Ginari. Several are assigned to Kalabsha, whereas individual pieces are said to have come from Dakka, Maharraqa and Qasr Ibrim. That is a rather large region, covering most of the ancient Dodekaschoinos plus a site far to the south of it. Upon further consideration, however, most of these provenances are demonstrably spurious: that of the (lost) stone purportedly from Qasr Ibrim has been questioned convincingly by Richter in his recent analysis of the Leipzig collection of monuments from that site;[25] the Maharraqa connection, claimed for a stone acquired in 1838 by Herzog (Duke) Maximillian in Bayern and now in Schloss Banz, is definitely based upon insufficient evidence;[26] finally, the provenance of a third stela, now in Athens, was given as Dakka with a question mark by G. Lefebvre, *Recueil*, under no. 629, but his earlier edition of the stone specifies that it was bought in Luxor.[27] This leaves Kalabsha as the only reliable provenance, represented by four pieces in all: a group of three published in 1821 by Thomas Young (Tibiletti Bruno, *Iscrizioni*, nos. 48, 54, 55) and an isolated one first notified by Arthur Weigall in 1908 (Tibiletti Bruno, *Iscrizioni*, no. 44).[28] The former stelae were acquired in or shortly before 1818 by George Annesley, second Earl of Mountnorris, from Henry Salt, the famous collector, and there is no reason whatever to doubt the trustworthiness of the information provided by Young in his *editio princeps*.[29] The fourth stela was found, by Weigall himself or under his supervision, south of the famous temple of Kalabsha, though apparently not *in situ*.[30] All four use ἐτελεώθη as the verb for "to die".

The name of Henry Salt (1780–1827),[31] the source of the three pieces acquired by the Earl of Mountnorris and published by Young, is associated with some further stelae bearing the formula ἔνθα κατάκειται-plus-ἐτελεώθη. In particular, two specimens presently in the Louvre (Tibiletti Bruno, *Iscrizioni*, nos. 51 and 52) are known to have been acquired from Salt, apparently in 1826.[32] For these Louvre pieces, which closely resemble each other, no provenance is known, but their identical formulary and their common origin with Salt are strongly suggestive of a link with the Mountnorris pieces from Kalabsha. In fact, already in 1925, Junker had claimed for an even greater group of stelae, which all can be traced back to Salt, a common provenance from Kalabsha.[33] Although Junker's claim cannot be accepted unreservedly in all cases, the Salt connection does suggest that the Kalabsha area, rather than any of the neighboring districts, was a major source of Christian epitaphs in the early decades of the nineteenth century.[34]

For the Leiden stela of Manna no precise provenance has ever been given nor am I aware of a connection with Salt. It is nevertheless remarkable that it was acquired in the same period as the other stelae mentioned above. It belongs to the ancient holdings of the Museum, more precisely to the d'Anastasi collection which arrived in Holland in 1828. It is not unlikely, therefore, that the stela of Manna was part of the same stream of antiquities that carried the very similar pieces acquired by Salt from Northern Nubia to Western Europe and one of the sources of which was certainly located in Kalabsha. Extending this argument to the greater group of Nubian stelae of the ἔνθα κατάκειται-plus-ἐτελεώθη type,[35] I would like to hypothesize that, in the absence of contrary evidence, most if not all of these originate from one of the cemeteries of Kalabsha. Given the fact that Talmis was a sizeable town and still important enough in the late sixth century to be the residence of an *exarkhos* [exarch, a military governor],[36] these cannot have been less extensive than those of neighboring Taphis.

Indications for dating the Nubian group of stelae of the ἔνθα κατάκειται type are scarce. Stelae with similar formulae from beyond the Nile Valley are usually not very securely dated either. In relatively recent literature, a closely related ensemble from Asia Minor is ascribed to the fifth-sixth centuries;[37] specimens from Crete are dated to about the same period,[38] which is probably a bit early for Nubia. The very few instances from Egypt, cited above, do not have absolute dates preserved, apart from the apparently unrelated Coptic group from Abydos, which dates to the tenth century. Nubian examples, either from Tafa or from Kalabsha, usually bear only indiction dates.

Providentially, at least one stela of this group, the epitaph of a woman Drosis (Δρώσης), now in Turin (Tibiletti Bruno, *Iscrizioni*, no. 43), does contain an absolute date, viz. A.D. 699. This stela, however, for which no exact provenance is known, exhibits a much more developed textual form than most Nubian monuments of this type, including the Leiden stone of Manna. Apart from an opening formula, which is virtually identical to that of the Manna stela, the prayer section is far richer. God is addressed as "the God of Abraham, Isaac and Jacob" and the abode of the soul is not only situated, as usual, "in the bosom of our holy fathers

Abraham, Isaac and Jacob" (ll. 6–8), but also "in a place of rest (ἀνάπαυσις)" (l. 6). Furthermore, an extensive second prayer ensues, asking for mercy for the deceased's soul, which adds considerably to the length of the text (ll. 9–13). If Martin Krause's theory, based on Henri Munier's material from Aswan, which claims a gradual development from a relatively simple towards an ever-richer formulary, holds here as well, this would assign a typologically late date to the stela of Drosis.[39] Based on that assumption, the bulk of the Nubian ἔνθα κατάκειται group, with its far simpler formulary, may need to be dated before 700, i.e. to the seventh or perhaps even the later sixth century. These unpretentious stelae, like that of Manna, would therefore document the earlier years of Christianity in Nobadia.

3. Paulos from Kalabsha (?), *meizoteros* (*SB Kopt.* I 432)

Whereas no doubt can be possible about the Northern Nubian provenance of the six Greek funerary stelae originating from Henry Salt and published by Thomas Young in 1821, the case is somewhat different for the seventh piece in the lot, an epitaph in Sahidic Coptic. The Greek pieces clearly reflect local traditions in their material (sandstone), layout and formulary. The Coptic stela, on the other hand, is much different. It is a marble slab and bears a text which, though of a type not unfamiliar in Nubia, lacks indubitably local elements. Rather, as will be argued below, its specific combination of textual and formal features is reminiscent of a group of eighth-ninth century stelae that may be assigned to northern Middle Egypt. Possibly, the stone was only secondarily added to the lot from Kalabsha, or Young's information may have been incomplete.[40] Yet it should be emphasized that no single element of the stone is in any way irreconcilable with a Nubian provenance. Below, the first complete edition of the text is given, together with a brief commentary and a photo of the monument, which, hopefully, will stimulate further discussion.

The Coptic stela, together with three of the Greek pieces from the same source, was presented to the British Museum by A. Lyttelton Annesley, heir to the Earl of Mountnorris, in 1854.[41] Young's 1821 publication provides an excellent facsimile as well a partial reading of the text, but his work was virtually ignored by more recent scholarship.[42] My own text is based upon Young's facsimile, which permits an almost complete reconstruction except for the date of demise in the last lines, where the stone is broken off. It was collated with digital photos of the stela.[43]

The stela is a white marble slab, measuring 39 × 33 cm. It is much worn in the center (through re-use as a paving tile or quern) and both lower corners are broken away. There are remains of fourteen lines of text, badly ruled; l. 14 (in smaller letters) is almost certainly the last line of the inscription. The epigraphic field is surrounded by broad plain margins, not preserved in the lower half of the stone. It is written in clear but rather coarse and unelegant uncials, less regular towards

Figure 22.1 A tombstone for Paulos from Kalabsha (?), *meizoteros* (*SB Kopt.* I 432) (photo: © Courtesy of the Trustees of the British Museum)

the end of the text; the script is characterized by a low and wide м and by a big *A* with a left-looking head and a broken bar with a low central curl; both superlinear strokes and dots are used, and a *croix pattée* [a cross with forked ends] opens l. 1 (see plate).

+ ⲡⲛⲟⲩⲧⲉ ⲛ̇ⲛⲉⲡⲛ̇[(ⲉⲩⲙ)ⲁ] ⲁⲩⲱ ⲡⲭⲟⲉⲓⲥ ⲛ̇ⲥⲁⲣⲝ ⲛⲓⲙ ⲁ[ⲣⲓ] ⲟⲩⲛⲁ ⲙⲛ̇ ⲧⲉⲯⲩⲭⲏ ⲙ̇ⲡⲙⲁ[ⲕⲁⲣⲓⲟⲥ ⲡ]ⲁⲩⲗⲟⲥ ⲡⲙⲉⲓ-
4. ⲍⲟⲧⲉⲣⲟⲥ ⲛ̅ⲅⲛⲟ[ⲭ]ϥ̅ ϩⲛ̇ [ⲕⲟⲩⲛ]ϥ ⲛⲁⲃ-
ⲣⲁϩⲁⲙ ⲙⲛ ⲓⲥ[ⲁⲁⲕ] ⲙⲛ ⲓⲁⲕⲱⲃ
ϩⲛ̅ ⲟⲩⲙⲁ ⲛ̇ⲟⲩ[ⲟⲧⲟⲩⲉ]ⲧ ϩⲓⲁⲝⲛ̅
ⲟⲩⲙⲟⲟⲩ ⲛⲙ̇[ⲧⲟⲛ] ⲡⲙⲁ ⲛ̅ⲧⲁ ⲡⲉⲙ-
8. ⲕⲁϩ ⲛϩⲏⲧ ⲡⲱⲧ ⲙⲛ̅ ⲧⲗⲩⲡⲏ ⲙⲛ̅
ⲡⲁϣⲁϩⲟⲙ [ϩⲙ] ⲡⲟⲩⲟⲉⲓⲛ ⲛ̇ⲛⲉⲕ-
ⲡⲉⲧⲟⲩⲁⲁ]ⲃ [ⲛ̅ⲅⲧ]ⲣⲉϥⲙ̇ⲡϣⲁ ⲛ̇ⲥⲱ-
ⲧⲙ ⲉⲧⲉⲓⲥⲙ]ⲏ ⲉⲧϩⲟⲗϭ̅ ϫⲉ ⲁⲙⲟⲩ ⲡϩⲙ̅-

12. ϩⲁⲗ ⲉⲧⲛⲁⲛⲟⲩ]ϥ ⲁⲩⲱ ⲙ̄ⲡⲓⲥⲧⲟⲥ
      [. . . . . ⲙⲧⲟⲛ ⲙ]ⲙⲟϥ . . [. . . .]
      [- - - ⲁⲡⲟ] ⲇⲓⲟⲕⲗ[. . . .]

1. ⲛ̄ⲛⲉⲡⲛ̄[(ⲉⲩⲙ)ⲁ: Hall ⲙ̄ⲛ . . . ‖ 4. beginning: Hall ϩⲟⲧⲉ . . . ‖ 6. ⲟⲩⲙⲁ ⲛ̣ⲟⲩ[ⲟⲧⲟⲩⲉ]ⲧ̄: Young: ⲟⲩⲙⲁ ⲉⲟⲟⲩ . . . ; inconclusive traces in *lacuna* ‖ 8–9: Junker, Hasitzka ϩ︤ⲛ︥ ⲧ]ⲙ̄ⲛ[ⲧⲣⲣⲟ ⲛ̄ⲙⲡⲏⲩⲉ] ‖ 10. [ⲛ̄ⲅ̄ⲧ] ⲣ̣ⲉϥ̣ⲙ̄ⲡⲱ̣ⲁ: Junker, Hasitzka [ⲁⲩⲱ ⲛⲅⲁⲁϥ ⲛ̄]ⲙⲡⲱⲁ; space for ⲁⲩⲱ is lacking; inconclusive traces in *lacuna* ‖ 11. ⲉⲧⲉⲓⲥⲙ]ⲏ̣: Junker, Hasitzka [ⲉⲧⲥⲙⲏ]; -ϩⲟⲗⲟ̄ : Junker, Hasitzka -ϩⲗⲟϭ; final ϩⲛ̄-: written in margin ‖ 12: ⲉⲧⲛⲁⲛⲟⲩ]ϥ: Hall [ⲉⲧⲛⲁⲛⲟⲩϥ] ‖ 13.-14. Junker, Hasitzka ⲛⲅⲃⲱⲕ ⲉϩⲟⲩⲛ ⲉⲡⲟⲩ]ⲛⲟϥ ⲙ̄ⲡ[ⲉ] ⲕ[ⲭ.ⲥ.⁴⁴

+ O, God of the spirits and Lord of all flesh,[45] have mercy upon the soul of the blessed Paulos, the *meizoteros*, and deposit him in the bosom of Abraham and Isaac and Jacob,[46] in a place of verdure, at a water of rest,[47] the place whence sorrow and sadness and sighing have fled away,[48] in the light of Your saints,[49] and make him worthy to hear that sweet voice saying: "Come, o good and faithful servant!"[50] [. . .] went to rest [. . . since Diocl[etian . . .]

This liturgically inspired epitaph in flawless Sahidic consists of a long composite prayer for the deceased (ll. 1–12), followed by a largely lost dating lemma (ll. 13–14). The opening formula offers an instance of the invocation "God of the spirits . . .", which in its Greek guise is perhaps the most characteristic formula of Nubian epigraphy.[51] In Coptic, three other examples of this invocation are known from Nubia. These originate from Sakinya (Toshka West),[52] Faras (a bishop's stela of A.D. 862)[53] and Ghazali,[54] and thus bear witness to its sporadic but geographically wide occurrence, covering both Nobadia and Makuria. However, instead of continuing with the common prayer formula "ⲉⲕⲉⲧ̄ⲙⲧⲟⲛ ⲛ̄-: grant rest to . . .", found in the other Nubian Coptic examples and in many Egyptian ones, for example from Esna, not far from the Nubian border,[55] the present inscription combines the invocation "God of the spirits . . ." with a prayer formula of the type "ⲁⲣⲓ ⲟⲩⲛⲁ ⲙⲛ̄: have mercy upon . . .".

Although funerary prayers with "have mercy upon . . ." are not at all exceptional in Nubian Coptic epigraphy,[56] the typical combination with the present opening invocation is far more reminiscent of a group of funerary stelae from Saqqara in Egypt.[57] To the Saqqara pieces, a number of which bear eighth-ninth century dates, others can be joined for which, in the absence of contrary indications, a similar provenance (northern Middle Egypt) may be postulated.[58] These bear a strong family likeness to the Kalabsha inscription in that they show a certain preference not only for an identical opening formula, but also, in a purely formal sense, for marble as a support and a *croix pattée* at the head of the first line. They are, moreover, undecorated apart from one or more of these crosses and show a similar coarse and unruled script with a preference for big, broken-bar *A*'s. Textually there is great variation, but, as in the Kalabsha piece, extensive and conscious borrowing from liturgical prayer is a conspicuous feature. The "family likeness" with this particular group of Egyptian stelae makes it impossible to

GLEANINGS FROM CHRISTIAN NORTHERN NUBIA

accept an attribution to Kalabsha without reservations. On the other hand, such an impression of similarity alone is insufficient evidence to dismiss out of hand the Kalabsha provenance put forward by Young. No watertight compartments separated the Christian populations of Egypt and Nubia, and this is also apparent in their epigraphic habits. Moreover, towards the end of the first millennium, Christian funerary epigraphy in the Nile Valley at large showed a growing preference for ample liturgical formulae and, simultaneously, a gradual wearing away of regional variety and local pecularities in the text form. A final verdict on the provenance of the stela of Paulos must therefore be postponed.

The remainder of the prayer (ll. 4–12) consists of select phrases from the commemoration of the dead in the Egyptian liturgy.[59] They mostly belong to the standard repertoire of medieval funerary epigraphy both in Nubia and in Egypt.[60] Also the characteristic formula ("make him worthy to hear the sweet voice . . .") which in ll. 10–11 introduces the Matthew quote, is frequently found in later stelae from Nubia and Egypt, only the quote itself is here Mt. 25:21/23, instead of the more usual Mt. 25:34.[61] Unfortunately, the dating lemma of the present epitaph (ll. 13–14) does not survive. Comparison with the stelae from Egypt cited above suggests a date in or about the ninth century.

Should the monument of Paulos really have come from Kalabsha, then it would add one more example to the meager file of Nubian officials bearing the title *meizoteros* (or a related one).[62] As a lower official, the *meizoteros* is familiar from the Greek and Coptic documentary sources of late Byzantine and early Arabic Egypt.[63] Epigraphic attestations from Egypt are far less common.[64] In Nubia, the role and functions of the *meizoteros* prove difficult to define, mainly for lack of helpful contexts. In most cases, but not here, the Nubian titles *meizon, meizoteros* or *proto-meizoteros* receive a further specification, like "of the palace" or "of Nobadia", or are combined with some other rank or title, for example admiral or eparch. Here, however, *meizoteros* appears without any further qualification. Another *meizoteros* without further qualifications is found in a Greek funerary stela, which, though traditionally ascribed to Qasr Ibrim, quite probably originates from the Kalabsha region too.[65] These variations in usage suggest that in Nubia the title *meizoteros* was borne by persons of very different rank and quality whose status was not necessarily in the first place defined by this title, but rather by the context in which they exercised their functions (e.g. whether or not attached to the palace) or by their other functions (e.g. eparch). The owners of the two stelae just mentioned may then represent *meizoteroi* on an unmarked, unspecific level, perhaps indeed, as frequently in Egypt, that of local administration. If the traditional provenance of his stela can be maintained, Paulos may have been an "alderman" of early medieval Talmis.

4. Building the city of Ikhmindi (*SB* VIII 10074)

Already in the autumn of 1843, Richard Lepsius had admired the "Roman" fortifications of the town of Ikhmindi (Mehendi), at a short distance from Maharraqa

(Hiera Sykaminos), just past the southern frontier of the Dodekaschoinos.[66] In 1958, Italian archaeologists discovered a foundation inscription commemorating the erection of these walls under King Tôkiltôeton "of the people of the Nobades".[67] Through indirect evidence, this event can be dated to the late sixth century. In the years following the *editio princeps* by S. Donadoni in 1959, some discussion arose about the text, particularly about the correct rendering of ll. 12–13 which read: ἐπλέρωσεν (for ἐπλήρωσεν) τὴν οἰκοδωμὴν τὴν πολλιν (scil, on 21 Epiphi of a third indiction year). Donadoni had commented upon the latter part of this construction: "naturalmente si puo intendere come una forma di endiadi o un accusativo interno; ma val la pena di ricordare che nelle epigrafi nubiane tale particolare uso di un accusativo è frequente nella epigrafia funeraria . . . (examples follow)".[68] Thus, according to Donadoni, one should correct τὴν πολλιν into τὴν πόλιν (for, more properly, τῆς πόλεως).

In a footnote to a short article published in 1961, Jean Bingen finds this "special use" of an accusative "assez éloigné de la tenue grammaticale de notre inscription". He proposes, with due hesitation, a correction into τὴν πολλήν as "la plus simple et du même ordre que les autres négligences d'orthographe phonétique que présente le texte", in itself a perfectly valid argument.[69] Bingen's cautious footnote met with wide acclaim. The correction τὴν πολλήν found its way into the *Sammelbuch* and the *SEG*, and was still defended with fervor by the most recent commentator of the text, F.W. Deichmann, as late as 1988.[70]

Meanwhile, however, publications by Maria Grazia Tibiletti Bruno and later ones by Tomas Hägg and Adam Łajtar have paved the way for a better understanding of the Greek of Nubia. Even more importantly, the progress made in recent years in the study of linguistic pluralism gradually gave rise to a less Hellenocentric, more sophisticated attitude towards language in medieval Nubia.[71] Christian Nubia, which had never been part of the Byzantine empire, was a multilingual society where Greek was only one of two and later three literary languages and where it was subject, in a varying degree, to the influence of the local vernaculars, one of which, Nobiin, came to be used in written form (Old Nubian). The peculiarities of spelling and syntax manifest in local Greek sources are not in contradiction with the apparent fact that at the same time a generally good standard of written Greek could be maintained, not just in the sixth century but as late as the thirteenth century. The variations that can be observed represent different registers of written communication rather than more or less successful approaches to a classical ideal. In the light of more recent scholarship, it can hardly be denied that Donadoni's comparison with other Nubian inscriptions and their "Nubian" particularisms was entirely pertinent.

The two phenomena which Donadoni had to postulate for his interpretation of ll. 12–13 of the Ikhmindi text, to wit the *accusativus pro genitive* and the doubling of the intervocalic -λ-, are not just "frequente", but part of the normal make-up of Nubian Greek. Of regular occurrence in the naming lemmata of epitaphs, the first of them has almost become a touchstone for the Nubian origin of an inscription.[72] Tibiletti Bruno was able to join even pre-Christian Greek examples from

GLEANINGS FROM CHRISTIAN NORTHERN NUBIA

the Dodekaschoinos to her discussion of the Christian texts.[73] For the doubling of the intervocalic -λ-, it may suffice to quote σελλένη for σελήνη, which in later funerary inscriptions has become more or less a standard spelling, next to e.g. θέλλοντος for θέλοντος (Tibiletti Bruno, *Iscrizioni*, no. 37), or δοῦλλούν for δούλου (Tibiletti Bruno, *Iscrizioni*, no. 22). Judged by the standards of Nubian Greek, Donadoni's proposed correction of τὴν πολλιν into τὴν πόλιν (for τῆς πόλεως) appears not merely justifiable, but quite natural. That it also yields a better sense can hardly be denied.

In addition to Bingen's philological argument, an objection of a more general nature against Donadoni's interpretation was raised in 1988 by F.W. Deichmann. In his opinion, it would be hardly imaginable that Ikhmindi, which he considers to be a mere "Fliehburg", could ever have been called a πόλις, even "im Sinne der Spätantike".[74] Here, again, it would have been preferable to look for information in indigenous, local sources, rather than to apply standards valid elsewhere. A small group of Northern Nubian legal documents in Coptic, datable to the reign of King Chael (around A.D. 800), was apparently drawn up by a priest of the main church in ⲙⲱⲣⲟⲛⲇⲓ or ⲙⲱⲣⲱⲛⲇⲓ for inhabitants of that place.[75] There can be no doubt about the identity of Môhondi with modern Mehendi or Ikhmindi.[76] In the best known of these documents, which show a clear awareness of the distinction between village (ϯⲙⲉ) and city,[77] Môhondi-Ikhmindi is explicitly called a ⲡⲟⲗⲓⲥ (l. 2).[78] Deichmann's argument cannot, therefore, be considered as valid. The Ikhmindi inscription commemorizes how, under the Nobadian king Tôkiltôeton, "the construction of the city was accomplished" on 21 Epiphi of a third indiction year.

Notes

1   For a general picture of the developments see, most recently, D.A. Welsby, *The medieval kingdoms of Nubia: Pagans, Christians and Muslims along the Middle Nile*, London: British Museum Press 2002, 14ff. On the process of christianisation in the area, see S.G. Richter, [*Studien zur Christianisierung Nubiens*, Wiesbaden: Reichert Verlag 2002]; J.H.F. Dijkstra, [*Philae and the end of ancient Egyptian religion: a regional study of religious transformation (298–642 CE)*, Leuven/Paris/Dudley, MA: Peeters 2008].

2   C.M. Firth, *The archaeological survey of Nubia: report for 1908–1909*, Cairo: Government press 1912, 48–50, pl. 2 and 51. For a brief general assessment, U. Monneret de Villard, "Note Nubiane, 2: La chiesa melkita di Nubia", *Aegyptus* 12 (1932), 309–16; cf. H. Junker, "Die christlichen Grabsteine Nubiens", *ZÄS* 60 (1925), 111–48, passim; U. Monneret de Villard, *La Nubia medioevale* I, Cairo: Mission archéologique de Nubie 1935, 25.

3   Listed by A. Łajtar, "Three Greek Christian epitaphs from Lower Nubia in the collection of the Archaeological Museum in Cracow", *Materiały archeologiczne* 27.2 (1994), 60–1.

4   Text: Firth, *Report*, 49 (tomb no. 807); M.G. Tibiletti Bruno, *Iscrizioni Nubiane*, Pavia: Fusi 1964, no. 49: re-edition after a photo and a copy by M. Vandoni (as Elephantine Museum, no. 1038); cf. A. Łajtar, "On the provenance of the four Christian inscriptions: *SB* X 10515–16, M.G. Tibiletti Bruno, *Iscrizioni Nubiane* 49, 56", *ZPE* 95 (1993), 243–44; idem, "Three Greek Christian epitaphs", 61, no. 6.

5   Łajtar, "Provenance", 244.
6   See Junker, "Grabsteine", 126–27, under I, and the discussion below, under no. 2.
7   Cf. Łajtar, "Provenance", 244.
8   E.g. Firth, *Report*, 45, tomb no. 37; cf. A. Łajtar, "A Greek Christian inscription from Ginari, Lower Nubia", *ZPE* 91 (1992), 147–49 (Tafa-Ginari); Tibiletti Bruno, *Iscrizioni*, no. 46 (Kalabsha?).
9   See the discussion, under no. 3, below.
10 For attestations and a brief discussion, see W.E. Crum, *A Coptic dictionary*, Oxford: Oxford University Press 1939, 14.b, s.v.
11 As Crum, loc.cit., despondently remarks: "context never instructive".
12 See, in particular, J. Osing, *Die Nominalbildung des Ägyptischen*, Mainz: Phillip von Zabern 1976, 396–97, who also discusses the semantic development of the title.
13 See F.L. Griffith, "Meroitic funerary inscriptions from Faras, Nubia", in: *Recueil d'études égyptologiques dédiées à Jean-François Champollion*, Paris: É. Champion 1922, 580–81.
14 See Crum, *Coptic dictionary*, 488.a, s.v. (Coptic, Egypt); *FHN* III 313, l. 16 (Greek, Nubia). For similar phenomena, see H. Satzinger, "Sudan-Ägyptisch und Elephantine-Koptisch", *BSEGenève* 4 (1980), 83–7; "Die Personennamen von Blemmyern in koptischen und griechischen Texten: Orthographische und phonetische Analyse", in E. Ebermann, E.R. Sommerauer, and K.É. Thomanek (eds.), *Komparative Afrikanistik (Fs. H.G. Mukarovsky)*, Vienna: Afro-Pub 1992, 313–24.
15 See e.g. A. Łajtar and J. van der Vliet, "Rich ladies of Meinarti and their churches", *JJP* 28 (1998), 35–53 [Study 28].
16 Text: L.J.F. Janssen, *Musei Lugduno-Batavi inscriptiones graecae et latinae*, Leiden: Luchtmans 1842, 63, no. 6 (*editio princeps*); *CIG* IV 9131 (A. Kirchhoff); E. Revillout, "Les prières pour les morts dans l'épigraphie égyptienne", *Revue égyptologique* 4 (1885), 23–4, no. 29; G. Lefebvre, *Recueil des inscriptions grecques-chrétiennes d'Egypte*, Cairo: IFAO 1907, no. 652; *SB* III 6089 (F. Bilabel); *SB* V 8737; Tibiletti Bruno, *Iscrizioni*, no. 47. Photo: P.A.A. Boeser, *Beschrijving van de Egyptische verzameling in het Rijksmuseum van Oudheden te Leiden*, vol. 7: *de monumenten van den Saïtischen, Grieks-Romeinschen, en Koptischen tijd*, The Hague: Nijhoff 1915, no. 32, pl. XVII. Cf. Junker, "Grabsteine", 114, under 3.
17 Not marble, as stated by Boeser, *Beschrijving*, 10.
18 For all details, see the excellent phototype plate in Boeser, *Beschrijving*. I thank Dr. M.J. Raven, curator of the Egyptian Department, for kindly allowing me to inspect the original.
19 Exactly the same orthography in Tibiletti Bruno, *Iscrizioni*, no. 44, a similar woman's stela with a certain provenance from Talmis-Kalabsha (see below).
20 For the not infrequent name Manna, see F. Preisigke, *Namenbuch*, Heidelberg: privately printed 1922, col. 204; G. Heuser, *Die Personennamen der Kopten*, Leipzig: J.C. Hinrichs 1929, 98; D. Foraboschi, *Onomasticon alterum papyrologicum*, Milano: Istituto Editoriale Cisalpino 1967, 186–87, s.v. (examples from sixth-seventh century).
21 Thus C.M. Kaufmann, *Handbuch der altchristlichen Epigraphik*, Freiburg: Herder 1917, 118–19, quotes the Greek epitaph of a Syrian from Trier!
22 Thus already Junker, "Grabsteine", 139; I am unable to add any Greek examples from Egypt to the three mentioned by him (Lefebvre, *Recueil*, nos. 65, 222, 423). Already Revillout, "Prières", 26–7, had referred to Coptic parallels (a small group of Sahidic epitaphs from the region of Abydos, tenth century; *SB Kopt.* I 496, 745, 746); although these open with the same phrase, they are generally much different and probably also much later than the Greek inscriptions.
23 Łajtar, "Three Greek Christian epitaphs", 58.

GLEANINGS FROM CHRISTIAN NORTHERN NUBIA

24 Junker, "Grabsteine", 126–27, under I. For an exception, unknown to Junker and not affecting our further argyment, see Tibiletti Bruno, *Iscrizioni*, no. 59.

25 T.S. Richter, "Die neun Stelen Ägyptisches Museum der Universität Leipzig Inv.-Nr. 680–8 mit der Herkunfstangabe Qasr Ibrim", in: S. Emmel et al. (eds.), *Ägypten und Nubien in spätantiker und christlicher Zeit: Akten des 6. internationalen Koptologen-kongresses, Münster, 20.-26. Juli 1996*, Wiesbaden: Reichert Verlag 1999, II, 296–97, Inv.-No. 680; not in Tibiletti Bruno, *Iscrizioni*, but see Monneret de Villard, *Nubia I*, 115; it is cited again below.

26 See the *editio princeps* by W. Huss, "Eine griechische Grabinschrift aus Nubien in Schloss Banz", *Berichte Historischer Verein Bamberg* 127 (1991), 7–12 (cf. *SEG* XLI 1688); the Duke's summary report of his stay in Maharraqa does not warrant, in my opinion, Huss' conclusions; according to D. Wildung, quoted ibidem, 7, the Duke's acquisitions originated from Karnak, Luxor, Debod, Kalabsha, Dakka and Abu Simbel.

27 G. Lefebvre, "Inscriptions grecques d'Égypte", *Bulletin de correspondance hellénique* 26 (1902), 454–55, no. 17.

28 On the other hand, the provenance of the British Museum stela of ⲁⲕⲕⲉⲛⲇⲁⲣⲡⲉ (no. EA 602; Tibiletti Bruno, *Iscrizioni*, no. 53), traditionally given as Kalabsha, although quite plausible in itself, is not based upon reliable information. See M.L. Bierbrier, "Provenances gained and lost", in: *The unbroken reed (Studies A.F. Shore)*, London 1994, 66. The review of epigraphic material from Kalabsha, by Curto, in S. Curto et al., *Kalabsha*, Rome: Centro per le Antichita e la Storia dell'Arte del Vicino Oriente 1965, 81–91, there especially 85 and 89–91, urgently needs revision and completion.

29 Th. Young, "Observations on a fragment of a very ancient Greek manuscript on papyrus, together with some sepulchral inscriptions from Nubia, lately received by the Earl of Mountnorris", *Archaeologia* 19 (1821), 156–60, with pls. IX–X; the three stelae in question are nos. III, IV and V in Young's publication (Lefebvre, *Recueil*, nos. 624–25; *SB* V 8720–22). Apparently, they were not among the monuments donated to the British Museum by A. Lyttelton Annesley, heir to the Earl of Mountnorris, in 1854 and must have been dispersed with the rest of his collection, two years earlier, see Bierbrier, "Provenances", 66; on the collectors in question, W.R. Dawson et al., *Who was who in Egyptology*, London: Egypt Exploration Society 1995, 17.

30 See A.E.P. Weigall, "Upper Egyptian notes", *ASAE* 9 (1908), 106, no. 2; Weigall was at the time Chief Inspector of Antiquities for Upper Egypt. On the inscription, see further: G. Lefebvre "Egypte chrétienne", *ASAE* 10 (1910), 65, no. 818; *SB* I 1600; Monneret de Villard, *Nubia I*, 41, n. 3; Curto, *Kalabsha*, 85, no. 38.

31 See D. Manley and P. Rée, *Henry salt: artist, traveller, diplomat, egyptologist*, London: Libri Publications Ltd 2001.

32 See now É. Bernand, *Inscriptions grecques d'Égypte et de Nubie au Musée du Louvre*, Paris: Éditions du CNRS 1992, nos. 111 and 113. Perhaps, a third piece, Bernand's no. 128 (Lefebvre, *Recueil*, no. 670; not in Tibiletti Bruno, *Iscrizioni*), also from the Salt collection, can be added; it is, however, much damaged.

33 Junker, "Grabsteine", 114 and 115–16. A. Łajtar, in his review of Bernand, *Inscriptions grecques du Louvre*, in *BiOr* 51 (1994), col. 571, overcautiously extends Junker's attribution to the greater Kalabsha-Tafa region.

34 Note that Monneret de Villard published, in *Nubia I*, 41, a Greek stela recorded at Kalabsha in about the same period (between 1822 and 1834) by James Burton; recently re-edited by A. Łajtar, "Varia Nubica VI-II", *ZPE* 137 (2001), 183–84.

35 Altogether, but excluding the epitaphs from Firth's excavations in Tafa-Ginari, some 14 pieces, plus one with a different verb for "to die", two without such a verb and one much too damaged to permit conclusions.

JACQUES VAN DER VLIET

36   This exarkhate is attested by two inscriptions from Dendur and Ikhmindi (the latter discussed below, under no. 4); cf. T. Hägg, "Titles and honorific epithets in Nubian Greek texts", *Symbolae Osloenses* 65 (1990), 160–61. In spite of Hägg's reservations, both inscriptions quite plainly give the impression that the *exarkhos* of Talmis was the highest representative of the Nobadian king in the area.

37   See C. Mango and I. Ševčenko, "Some recently acquired Byzantine inscriptions at the Istanbul Archaeological Museum", *DOP* 32 (1978), 1–27.

38   A.C. Bandy, *The Greek Christian inscriptions of Crete*, Athens: Éditions du CNRS 1970, *passim*.

39   For details, see M. Krause, "Die Formulare der christlichen Grabsteine Nubiens" in: K. Michałowski (ed.), *Nubia: récentes recherches*, Warsaw: Musée national 1975, 76–82.

40   Note that Young, in his discussion of the Coptic stela, "Observations", 160, does seem a bit vague ("found in the same neighbourhood", *scil.* as the other six stelae).

41   British Museum inv. no. EA 825. For the acquisition, cf. above, n. 29. Not included in recent discussions of the Nubian holdings of the museum by W.V. Davies, " 'Egypt and Africa' in the British Museum", in: idem (ed.), *Egypt and Africa: Nubia from prehistory to Islam*, London: British Museum Press 1991, 314–20 (see 314, under 1854), and Bierbrier, "Provenances", 66.

42   Bibliography: Young, "Observations", 160, with pl. X infra (facsimile; partial transcription of ll. 4–6; partial translation); H.R. Hall, *Coptic and Greek texts of the Christian period from ostraka, stelae, etc. in the British Museum*, London: British Museum 1905, 11, no. 4, pl. 10 (incomplete facsimile and transcription of ll. 8–14; single words from ll. 1–4); Junker, "Grabsteine", 122 (supplements ll. 8–14 after Hall; valuable discussion passim); *SB Kopt.* I 432 (reproduces Junker's text). Briefly mentioned in E.A. Wallis Budge, *British Museum: a guide to the Egyptian galleries (sculpture)*, London: British Museum 1909, 300, no. 1155; Monneret de Villard, *Nubia* I, 41, n. 4; Curto, *Kalabsha*, 85, no. 40 (as "illeggibile").

43   These were kindly provided by Mr. M. Marée, M.A., of the British Museum, Egyptian Department, who also verified my readings of ll. 3–4 on the original. I thank him most warmly for his cooperation.

44   Apart from being irreconcilable with the remaining traces, Junker's ingenious reconstruction is hardly plausible. One would expect: ⲛⲅⲃⲱⲕ ⲉⲅⲟⲩⲛ ⲉⲡⲣⲁϣⲉ ⲙⲡⲉⲕ.ⲭⲟⲉⲓⲥ. The indubitable remains of a date in l. 14 require a reconstruction as proposed here (dying-dating lemma).

45   Cf. *Num.* 16:22; 27:16.

46   Cf. *Lk* 16:22; *Mt.* 8:11; for the verb used here, see Junker, "Grabsteine", 129, under I; 141.

47   Cf. *Ps.* 22:2.

48   Cf. *Isa.* 35:10; 51:11.

49   Cf. *Ps.* 109:3; and Junker, "Grabsteine", 140–41, under 2.

50   Cf. *Mt.* 25:21/23; and Junker, "Grabsteine", 130, under 2; 140, under I.

51   Best known for its occurrence at the beginning of "prayer α" of Junker's classical study, "Grabsteine"; see furthermore, J. Kubińska, *Inscriptions grecques chrétiennes*, Warsaw: PWN–Éditions scientifiques de Pologne 1974, 69–86 (who calls it "prière nubienne"), and A. Łajtar, "Varia Nubica IV. Das älteste nubische Epitaph mit dem Gebet vom sogenannten Typus *Euchologion mega?*", *ZPE* 113 (1996), 101–08, where the designation "prayer of the *euchologion mega* type" is preferred.

52   T. Mina, *Inscriptions coptes et grecques de Nubie*, Cairo: Société d'archéologie copte 1942, no. 78.

53   S. Jakobielski, *A history of the bishopric of Pachoras on the basis of Coptic inscriptions*, Warsaw: PWN–Éditions scientifiques de Pologne 1972, 77, ll. 16ff.

GLEANINGS FROM CHRISTIAN NORTHERN NUBIA

54 J. van der Vliet, *I. Khartoum Copt.* 77. For its distribution in the Nile Valley, see G. Roquet, "Inscriptions bohaïriques du Dayr Abu Maqâr", *BIFAO* 77 (1977), 166–71 (from the list on p. 167, delete no. 3a); cf. C. Wietheger, *Das Jeremias-Kloster zu Saqqara unter besonderer Berücksichtigung der Inschriften*, Altenberge: Oros 1992, 133–34.

55 See S. Sauneron and R.-G. Coquin, "Catalogue provisoire des stèles funéraires coptes d'Esna", in: *Livre du centenaire: 1880–1980*, Cairo 1980, 239–77, nos. 57 and 80.

56 To the single Coptic example quoted by Junker, "Grabsteine", 130 supra (cf. p. 143), many more could now be added, particularly from Sakinya (Toshka West).

57 See Roquet, "Inscriptions", 167; Wietheger, *Jeremias-Kloster*, 134, n. 319.

58 Ibrahim Kamel, *Coptic funerary stelae: catalogue général des antiquités du Musée copte*, nos. 1–253, Cairo 1987, particularly nos. 75 and 76 (inv. nos. 3835 and 3836, both on marble and "bought from T. Flamakrian"; the former is mentioned, with a wrong number, by Roquet, "Inscriptions", 167, no. 16, who refers to earlier editions; the latter also in W. Brunsch, "Koptische und griechische Inschriften in Kairo", *Egitto e Vicino Oriente* 19 (1995), 99, under a wrong number 8636; cf. M.R.M. Hasitzka, A. Łajtar, and T. Markiewicz, "Coptic inscriptions in Egyptian collections: some notes on recent publications", *JJP* 29 (1999), 26–7, no. 51). Though mostly with different opening formulae, the small group presented by Sophia Schaten during the Seventh International Congress of Coptic Studies ("Die Sammlung des Gayer-Anderson Museums in Kairo: Christliche Grabsteine mit Inschriften", in *Abstracts of papers*, Leiden 2000, 91) may be joined to these, mainly on account of peculiarities of script and formulary. The latter date to the ninth century; one of them actually belongs to the Saqqara group with an opening of the "God of the spirits" type (Wietheger, *Jeremias-Kloster*, no. 86; J.E. Quibell, *Excavations at Saqqara (1907–1908)*, Cairo 1909, 31, no. 11, pl. XLIV.I, right; cf. p. 100).

59 See e.g. the references given by Sauneron and Coquin, "Catalogue provisoire", 251–52.

60 For Nubia, see Junker, "Grabsteine", and *I. Khartoum Greek* and *I. Khartoum Copt.*

61 See, however, Junker, "Grabsteine", 140 (the parallel quoted is M. Cramer, *Die Toten-klage bei den Kopten*, Vienna/Leipzig: Hölder-Pichler-Tempsky 1941, no. 8, ll. 20–3); in the present-day Bohairic liturgy, Mt. 25:14–23, is the Gospel reading in the burial service for priests (O.H.E. Burmester, *The Egyptian or Coptic Church*, Cairo: Société d'archéologie copte 1967, 210), and Mt. 25:21/23 is quoted in the commemoration of a deceased bishop (ibidem, 168).

62 Recent discussions: Hägg, "Titles", 162; T.S Richter, "Neun Stelen", 296; A. Łajtar, *I. Khartoum Greek*, under no. 21. Their facts and arguments are not repeated here.

63 See A. Steinwenter, *Studien zur koptischen Rechtsurkunden*, Leipzig 1920: Verlag Haessel, 41–5; G. Rouillard, *L'administration civile de l'Égypte byzantine*, Paris: P. Geuthner 1928, 69–71; E.R. Hardy, *The large estates of Byzantine Egypt*, New York: Columbia University Press 1931, 104–05; A. Grohmann, "Der Beamtenstab der arabischen Finanzverwaltung in Ägypten in früharabischer Zeit", in: *Studien zur Papyrologie und antiken Wirtschaftsgeschichte (Fs. F. Oertel)*, Bonn: R. Habelt 1964, 129–31.

64 I know of no other example than Lefebvre, *Recueil*, no. 62 (stela of *apa* Ôl, A.D. 693, reputedly from Damanhour, but much resembling stelae from the Fayoum); for a useful discussion of epigraphic evidence from outside the Nile Valley, see D. Feissel, "Notes d'épigraphie chrétienne (IV)", *BCH* 104 (1980), 461–62.

65 Stela of Epiphanios, formerly Leipzig Inv.-Nr. 680; see T.S. Richter, "Neun Stelen", 296–97; a stela of the ἔνθα κατάκειται type, the provenance of which was briefly discussed above (under no. 2).

66 R. Lepsius, *Briefe aus Aegypten, Aethiopien und der Halbinsel des Sinai*, Berlin: W. Herz 1852, 113–15.

67 Text: S. Donadoni, "Un'epigrafe greco-nubiana da Ikhmindi", *La parola del passato* 69 (1959), 458–65 (= *Cultura dell'Antico Egitto,* Rome: Università di *Roma* "La Sapienza" 1986, 513–20); *SEG* XVIII 724 (cf. XX 702); J. Bingen, "Un roitelet chrétien des Nobades au VIe siècle", *CdÉ* 36 (1961), 431–33 (= *Pages d'épigraphie grecque: Attique – Egypte [1952–1982]*, Brussels: Epigraphica Bruxellensia 1991, 160–62); *SB* VIII 10074; F.W. Deichmann, "Die Bauinschrift von Ihmîndi", in: F.W. Deichmann and P. Grossmann (eds.), *Nubische Forschungen*, Berlin: Gebr. Mann Verlag 1988, 81–8. Photo: J. Leclant, "Fouilles et travaux en Egypte, 1957–1960", *Orientalia* 30 (1961), fig. 38, pl. XL (cf. p. 193); facsimile: A. Stenico, "Ikhmindi. Una città fortificata medievale delia Bassa Nubia", *Acme* 13 (1960), 62, fig. 4. The stone is kept in the museum of Assouan (Bingen) or Elephantine (*SB*) [Elephantine, no. 2890, where Jacques van der Vliet saw it on December 31, 2004].

68 Donadoni, "Epigrafe", 462.

69 Bingen, "Roitelet", 432, n. I: note that Bingen refers his suggestion to a lengthy footnote and avoids to insert it into his apparatus or translation.

70 Deichmann, "Bauinschrift", 85. Cf., however, the *Bulletin épigraphique*, in *REG* 75 (1962), 217, no. 362, and S. Donadoni, "Les débuts du christianisme en Nubie", in: *Actes du symposium international sur la Nubie, mars 1965*, Cairo: Institut d'Egypte 1969, 29, n. 35.

71 A pioneer study is M. Bechhaus-Gerst, *Sprachwandel durch Sprachkontakt am Beispiel des Nubischen im Niltal*, Cologne: Köppe 1996.

72 Already in 1981, T. Hägg, "Two Christian epitaphs in Greek of the '*Euchologion mega*' type", in: T. Säve-Söderbergh (ed.), *Late Nubian cemeteries*, Solna: Esselte Studium1981, 58, note to l. 23, had quoted the Ikhmindi inscription in this context, but, referring to Bingen's article, remained hesitating to adopt Donadoni's interpretation.

73 M.G. Tibiletti Bruno, "Di alcune caratteristiche epigrafi funerarie cristiane della Nubia", *Rendiconti Istituto Lombarde, Accademia di Scienze e Lettere. Classe di lottere* 97 (1963), 522 and 527.

74 Deichman, "Bauinschrift", 85.

75 Presently in Alexandria, Greco-Roman Museum, no inventory numbers known; partly published by J. Krall, "Ein neuer nubischer König", *WZKM* 14 (1900), 233–42; similar documents in W.E. Crum, *Catalogue of the Coptic manuscripts in the British Museum*, London: British Museum 1905, nos. 447–56.

76 See e.g. Krall, "König", 241; Stenico, "Ikhmindi", 31, n. 1.

77 Cf. Crum, *Catalogue British Museum*, 216, no. 452, doc. 2, 1. 2.

78 Krall, "König", 234; this text reprinted in *SB Kopt.* I 37.

23

FOUR NORTH-NUBIAN FUNERARY STELAE FROM THE BANKES COLLECTION

Jacques van der Vliet and Klaas A. Worp

Introduction

The four funerary stelae edited here for the first time are part of the collections of the Kingston Lacy estate in Dorset (England). This former estate of the Bankes family is now an accredited museum, managed by the British National Trust. The stelae are part of the National Trust collections under inventory nos. 1257703–6 and can be freely accessed on the website of the National Trust at nationaltrust-collections.org.uk. For photographs of the objects, the reader is referred to this website.[1]

The important collection of Egyptian antiquities kept at Kingston Lacy was acquired in the early nineteenth century by William John Bankes (1786–1855), a British collector and amateur orientalist who traveled in Egypt and Nubia between 1815 and 1819.[2] As their textual formularies unambiguously show, the four stelae published below share a provenance in northern Nubia.[3] With a high degree of probability, this can be narrowed down for all four stones to the region of Tafa (Taphis) and Kalabsha (Talmis), the heartland of the former Dodekaschoinos, the stretch of the Nile Valley between Philae and Hiera Sykaminos that had been part of the Roman Empire for four centuries, from Augustus to Diocletian. Bankes traveled to Nubia twice. Once up to Wadi Halfa in the autumn of 1815 and a second time up to Saï Island, south of the Second Cataract, in the winter and spring of 1818/19.[4] Where and when exactly he acquired his four Nubian stelae remains obscure, however.

For the question of the provenance of the Bankes stelae, it is important to bear in mind that they belong typologically to a far larger group of Christian tombstones from northern Nubia that reached Europe in the 1810s and 1820s, many of which can be traced to the person of Henry Salt (1780–1827).[5] Based in Cairo as a diplomat, Salt was active as a dealer and collector on a large scale.[6] Bankes had entered the circle of Salt in Cairo by August 1815.[7] From mid-October 1818, the latter accompanied Bankes on his long second journey through southern Egypt and northern Nubia.[8] Seven stelae with Kalabsha as

295

their stated provenance and very similar to the Bankes stelae were acquired around the same time from Salt by another British aristocrat, George Annesley, Viscount Valentia and second Earl of Mountnorris (1770–1844).[9] They were published in 1821 by Thomas Young, as the first of their kind to receive a scholarly edition.[10] Since Young's "Observations" were presented in an oral form on 11 June 1818,[11] the Annesley stelae were purchased before Bankes' second Nubian expedition, which started only later that year. This opens the possibility that Bankes did not acquire his stelae directly on site. He may simply have been a client of Salt's, on a par with Annesley and other British and European collectors, both private and institutional. If this scenario proves correct, there is no need to connect all four stelae to a single site, even though they undoubtedly share the same general provenance, in the Tafa-Kalabsha area.[12] In any case, reconstructing Henry Salt's much dispersed collection would be an important first step towards recontextualizing the Christian tombstones from Nubia that arrived in Europe in the first half of the nineteenth century.

The following edition of the texts is made after digital photographs provided by the National Trust on its website. We have been unable to collate our readings with the monuments themselves.

1. Funerary stela of a woman Protokia

National Trust inv. no. 1257703.

Round-topped sandstone stela, 33 × 23 cm, inscribed with eleven slightly sloping lines of incised Greek uncials, framed within a single, loosely incised line. The *iota* is almost consistently provided with a trema. The numeral in line 3 is surrounded by a circle of dots; both triple and normal colons are used. Simple crosses precede line 1 and follow lines 10 and 11; a twelfth line consists of three similar crosses only. The letters bear traces of red coloring throughout. The spelling is erratic with many traces of iotacism and various Nubian peculiarities.

Probably Tafa (Taphis),   *ca.* seventh century

   + ἔνθα κατάκηδε
   μηνὴ : Τῖπυ : ιη´ ⫶ ἰνδικ(τίωνος)
   ⫶ ϛ´ ⫶ εἰκιμήθη ἡ μακα-
4. ρία Προτωκία.
   ἀναπαύσῃ, κ(ύρι)ε,
   τὶν ψυχὴν τὶν
   δούλι σου Προτωκία
8. ἐν κόλπις Ἀβραὰμ καὶ
   Ἰσαὰκ ⫶ καὶ Ἰακώβ. ἀμή῀ν´,

FOUR NORTH-NUBIAN FUNERARY STELAE

ἀμήν, ἀ{α}μήν +
ἱ Ἁγία Τριάς +
12. + + +

1. κατάκειται ‖ 2. μηνὶ | Τῖπγ, Τῦβι | ἴΗ | ιΝͅΥϗ, ιΝ- ligatured ‖ 3. ειкͥΝΗΘΗ, ἐκοιμήθη, final η corr. < ? | ἡ corr. < ϊ ‖ 3–4. ΜΑΚΑΡϊΑ ‖ 4. προτωк//Α ‖ 6. τῖΝ, τὴν | τῖΝ, τῆς ‖ 7. ΔογΛϊ, δούλης ‖ 8. коλπϊϲ, κόλποις | καϊ ‖ 9. ϊϲΑΑк | ϊΑκωв ‖ 11. ϊ, ἡ | ΑΓϊΑ τρϊΑϲ

+ Here lies. In the month Tybi 18, indiction 6, fell asleep the blessed Protokia. May You grant rest, O Lord, to the soul of Your servant Protokia in the bosom of Abraham, Isaac, and Jacob. Amen, amen, amen. + Holy Trinity! + + + +

This epitaph for a woman Protokia presents a textual oddity in that the opening formula ("here lies . . .") is not directly followed by the name of the deceased but by the date of her demise, which should logically follow the verb "to fall asleep" (here l. 3).[13] A partial analogy is perhaps provided by the stela of a woman Taria, from Tafa-Ginari, in which the same opening formula is also not followed directly by the name of the deceased but by what looks like a spelling of the verb ἐκοιμήθη, "she fell asleep".[14]

Nubian epitaphs with the present opening formula are generally believed to come from the territory of the former Dodekaschoinos. Apart from a few spurious exceptions, all appear to derive more specifically from either Kalabsha or Tafa.[15] A further precision has been suggested by Hermann Junker, who observed that stelae of this type with "to fall asleep" as the verb for "to die" are without exception from the huge cemetery of Ginari, our primary source for Tafa epitaphs.[16] This observation, together with the other indications noted here, assign the stela of Protokia with a high degree of probability to Tafa. Similar stelae have been plausibly dated to around about the seventh century.[17]

2. The spelling Τῖπυ for Τῦβι seems unattested elsewhere in Nubia; cf. G. Ochała, *Chronological systems of Christian Nubia*, Warsaw: University of Warsaw/Raphael Taubenschlag Foundation 2011, 229. The π/β interchange is found occasionally in Nubia, for instance in *I. Qasr Ibrim* 20, l. 4 (πίου for βίου).

4. The name Protokia (here and l. 7) seems, in its present spelling, unattested in Egypt and no Greek noun thus written is known. Undoubtedly the same name, spelled Protokeï (προτοκεϊ) and likewise a woman's name, is found in a Coptic funerary stela from Sakinya (Toshka West): T. Mina, *Inscriptions coptes et grecques de Nubie*, Cairo: Société d'archéologie copte 1942, no. 220 (= U. Monneret de Villard, *Le iscrizioni del cimitero di Sakinya (Nubia)*, Cairo: IFAO 1933, no. 120), ll. 5–6 (DBMNT 300 = TM 141421). Protokia/Protokeï could derive from either (A) Prototokia (Πρωτοτοκία, "Primogeniture"; cf. the Greek adjective πρωτοτόκος and a number of personal names starting with an element πρωτο-, "first", see www.trismegistos.org/nam/list_all.php?selection=P&p=138) or (B) Prosdokia (Προσδοκία, "Expectation"; cf. TM_Nam ID 24095). Since Προσδοκία is well attested as a proper name in Egypt, our option (B) seems preferable. Omission of the *sigma* before a

JACQUES VAN DER VLIET AND KLAAS A. WORP

dental stop is not uncommon in Egypt (Gignac, *Grammar* I, 130, quotes various examples, ranging from the second to the late sixth centuries); the interchange of δ/τ before the ending in -οκια is very frequent. The name fits in with the predilection for fanciful Greek names that is apparent in other Tafa epitaphs as well (cf. U. Monneret de Villard, "Note nubiane: I. Articula [Plin., *N.H.*, VI, 184]. 2. La chiesa melkita di Nubia", *Aegyptus* 12 [1932], 305–16 at 314). It is oddly spelled here with a double oblique stroke instead of an *iota*, as in l. 7.

5. Ἀναπαύσῃ, instead of the otherwise far more frequent ἀνάπαυσον (thus in our stela no. 2, l. 6), occurs in nearly half of the epitaphs from Firth's excavation of the cemetery of Tafa-Ginari, which is likewise indicative of a Tafa provenance.

5–6. The continuation of the prayer with an accusative instead of a genitive after τὴν ψυχὴν is a typical feature of Nubian Greek; see *I. Khartoum Greek*, p. xx. For the omission of the final consonant in δούλ, compare *I. Khartoum Greek* 9, l. 19: τὸν δοῦλο σου (DBMNT 8 = TM 99098).

10. For the double *alpha* in the third Amen, see *I. Qasr Ibrim* 29, ll. 6–7 (DBMNT 657 = TM 141742); for other instances of word-initial vowel doubling from Nubia, *I. Khartoum Copt.*, p. 23, n. 113.

11. The same acclamation of the Holy Trinity occurs frequently at the end of seventh-eighth century Coptic letters from Thebes; see Anneliese Biedenkopf-Ziehner, *Untersuchungen zum koptischen Briefformular unter Berücksichtigung ägyptischer und griechischer Parallelen*, Würzburg: G. Zauzich 1983, 106, 252. We are unaware of other examples from Nubian funerary epigraphy (but cf. *I. Khartoum Copt.* 12 [DBMNT 41 = TM 141208]: an epitaph with invocation of the Holy Trinity in the prayer section).

2. Funerary stela of a woman Edra

National Trust inv. no. 1257704.

Rectangular sandstone stela, 18.5 × 14 cm, inscribed with eleven lines of incised Greek uncials. In the dating lemma (ll. 3–5), high points are used. Crosses precede line 1 and follow line 11. The letters bear traces of red coloring throughout.

Probably Kalabsha (Talmis),   *ca.* seventh century

     + ἔνθα κατάκοι-
     τε ἡ μακαρία
     Ἔδρα · ἐτελεώ-
4.   θη · μ`η´(νὶ) · Φαρμοῦ-
     θης · κα´ · ἰνδ(ικτίωνος) · ζ´ ·
     ἀνάπαυσον τὴ(ν)
     ψυχὴν αὐτοῦ
8.   ἐν κώλποις
     Ἀβραὰμ καὶ
     Ἰσὰκ κ(αὶ) Ἰακώβ.
     Ἀμήν, (ἀμήν), (ἀμήν) +

1–2. κατάκειται ‖ perhaps read Ἔδρα ‖ 4–5. Φαρμοῦθι ‖ 6. τῆ ‖ 7. αὐτῆς ‖ 8. κόλποις ‖ 10. ϊсακ, Ἰσαὰκ | ϊακωв ‖ 11. twice ϥθ

298

+ Here lies the blessed Edra. She came to perfection in the month Pharmouti 21, indiction 7. Grant rest to her soul in the bosom of Abraham, Isaac, and Jacob. Amen, amen, amen. +

This epitaph shows roughly the same formulary as our no. 1, only in the present case the use of the verb ἐτελεώθη, for "to die" (here ll. 3–4), would seem to favor a Kalabsha provenance.[18] Instead of Edra, one may be tempted to read the name of the owner as Hedra, with a *spiritus asper*, on the assumption that it is related to the Greek noun ἕδρα, "seat, foundation", as was suggested to us by Adam Łajtar. However, the name seems otherwise unknown in either form and may not be Greek at all.

3. Funerary stela of a priest Severus

National Trust inv. no. 1257706.

Round-topped sandstone stela, 27.5 × 20 cm, inscribed with seventeen lines of incised Greek uncials, framed within a single incised line. Part of the righthand margin and the lower right corner are broken, causing some loss of text. The top of the stela is decorated with a squarish cross, flanked on either side by the abbreviated names of Jesus Christ. A series of simple crosses brings the text to a close (in l. 17). The letters bear traces of red coloring throughout.

Cf. James, "Egyptian antiquities", 30: tiny photo with caption dating the stela to the seventh-eighth century and quoting the name of the owner as Snerou.

Region of Tafa (Taphis) – Kalabsha (Talmis),   *ca.* seventh-ninth century?

```
     Ἰ(ησοῦ)ς + Χ(ριστό)ς
     ὁ θ(εὸ)ς ὁ τῶν πν(ευμ)άτων καὶ
     πάσης σαρκὸς, τῶν ὁρω-
4.   μένων καὶ ἀοράτων, <ὁ> κ[α-]
     τὰ τὴν ἀπέρρητων β[ου-]
     λὴν ἑνώσας ψυχὴν σώ-
     ματι καὶ πάλιν κατὰ [τὸ]
8.   θέλημα τῆς σῆς ἀγαθό[τη-]
     τος διαλίον τὸ πλάσμ[α]
     ὃ ποίσας αὐτός, ἀνάπαυ[σον]
     τὴν ψυχὴν τοῦ θεοφυλ[εστάτου]
12.  Σευήρου πρ`ε´(σβυτέρου) ἐν κόλποις [τῶν]
     ἁγίων πατέρων ἡμῶ[ν Ἀ-]
     βραὰμ (καὶ) Ἰσὰκ (καὶ) Ἰακ[ῶβ. τέ-]
     λει δὲ τοῦ βίου ἐχρ[ήσατο μη-]
16.  νὶ Παεῖνη ιθ΄, ἰνδ(ικτίωνος) [. .].
     ἀμήν, (ἀμήν), (ἀμήν) + + [+ +]
```

5. ἀπόρρητων ‖ 9. διαλύων ‖ 10. ἐποίησας ‖ 11. θεοφιλεστάτου ‖ 14. ϊϲακ, Ἰσαὰκ ‖ 16. Παῦνι | ιθ ‖ 17. twice q̅θ̅

Jesus + Christ. God of the spirits and all flesh, of the visible and the invisible, who according to Your mysterious counsel united soul and body and according to the will of Your goodness again dissolved the creature that You made Yourself, grant rest to the soul of the most pious Severus, priest, in the bosom of our Holy Fathers, Abraham, Isaac and Jacob. He ended his life in the month Payni 19, indiction [. .]. Amen, amen, amen. + + [+ +]

The very characteristic and relatively rare formulary of this epitaph is that of Hermann Junker's prayer β.[19] Its liturgical model had already been identified by Wilhelm Weissbrodt in 1909, in a manuscript from Grottaferrata.[20] Its occurrence on tombstones seems to be limited to northernmost Nubia. The prayer's apparent absence from Egyptian epitaphs, however, cannot be used in support of Junker's claim of a postulated "Byzantine", as opposed to Egyptian or "Coptic", inspiration of Nubian funerary epigraphy.[21] As Junker himself was the first to point out, a mutilated form of the same prayer survives till today in the Bohairic Coptic funerary rite for nuns.[22] In addition to the present stela, the formulary is attested in the following seven epitaphs:[23]

(1) Tibiletti Bruno, *Iscrizioni*, no. 24 (= Lefebvre, *Recueil*, no. 659 [DBMNT 497 = TM 102865]): limestone stela of a woman Pista, an early acquisition of the Egyptian Museum in Turin (no. 23), of unknown provenance;[24]

(2) Tibiletti Bruno, *Iscrizioni*, no. 25 (DBMNT 536 = TM 102425): stela of a priest Pistoteknos, Elephantine Museum no. 1075, of unstated provenance;

(3) Tibiletti Bruno, *Iscrizioni*, no. 26 (DBMNT 2639 = TM 102426): stela of Stephen, Elephantine Museum no. 1039, of unstated provenance;

(4) Ginari (Tafa), Grave 79 (DBMNT 405 = TM 141513): stela of John, present whereabouts unknown to us;[25]

(5) Ginari (Tafa), Grave 839 (= *SB* X 10516 = *SEG* XLIII 1177 [DBMNT 438 = TM 99246]): stela of Mary, Amsterdam, University Library, Schriftmuseum J.A. Dortmond, inv. no. 1055, with a much abbreviated text;[26]

(6) Bab-Kalabsha no. 1 (DBMNT 2465 = TM 370968[27]): a fragmentary stela, name of the owner lost, present whereabouts unknown to us;[28]

(7) Bab-Kalabsha no. 2 (DBMNT 2466 = TM 370969): fragmentary stela of a cleric, name lost, present whereabouts unknown to us.[29]

As the list shows, the epigraphic use of the formulary appears to be a very local phenomenon. For four out of the eight occurrences (nos. 4–7) provenances are known, which assigns this group of Nubian epitaphs again to the region of Tafa and Kalabsha. The two pieces in the Elephantine Museum, which holds many pieces from northern Nubia, among them several stelae from Cecil Firth's excavations at Tafa-Ginari,[30] may come from the same general region. All stelae of this

FOUR NORTH-NUBIAN FUNERARY STELAE

group that have a date preserved include an indiction year as was common practice in the Tafa-Kalabsha region.[31] None of them bears an absolute date, however, which makes dating the present stela a bit hazardous. In any case, paleography excludes a date after the ninth century.

With the exception of the truncated version of the Amsterdam stela, our no. 5, the stelae from this group offer a remarkably identical text. Differences in spelling and scribal errors are numerous, but real textual variants are rare, which again argues in favor of the group's coherence in terms of time and space. After the word τὸ πλάσμα (l. 9 in the Bankes stela), most stelae of the group as well as the liturgical source quoted by Weissbrodt add σου ("*Your* creature"). The lacuna at the end of line 9 of the Bankes stela seems too short for this addition, which appears to be lacking also in no. 7 of the list above. Furthermore, nos. 3 (in l. 10) and 4 (in ll. 10–11) of the list have a redundant insertion ὁ πλάσας, "Your creature *that You created*", not found in Weissbrodt's liturgical source.[32] Instead of διαλύων (Bankes stela, l. 9: διαλίον), no. 1, l. 8, has διαλήγον. Although perhaps not impossible as such,[33] it is considered a spelling of διαλέγων by Weissbrodt.[34] Following Tibiletti Bruno, *Iscrizioni*, no. 24, *ad loc.*, we interpret it as a mere orthographic variant of διαλύων, reflecting the realization of both *eta* and *upsilon* as /i/ and the shift of intervocalic *gamma* to a fricative in later Greek.[35] The Bankes stela of Severus omits ὁ before κατὰ in line 4, where no. 6 of the list has καὶ. On the whole, the epitaph of Severus appears to be textually closest to the complete nos. 1 and 2 (unknown provenances) and the fragmentary nos. 6 and 7 (from Bab-Kalabsha), as far as the latter are preserved. No. 2, moreover, another priest's stela, would appear to have a similar decorative element (the abbreviated names of Jesus Christ on either side of a cross) in top of the stela.

9.　　　After πλάσμα, space for σου (see above) seems to be lacking.

11.　　　Stela Bab-Kalabsha no. 2 (no. 7 in the list above), l. 5 (same spelling with *upsilon*), shows that the title, which must have been abbreviated here, has to be θεοφιλέστατος, not θεοφύλακτος, which seems unattested in Nubia. The same clerical title, θεοφιλέστατος, is given to another priest in a Greek votive inscription from Maharraqa (Hiera Sykaminos), see A. Łajtar, "Varia nubica VIII-IX", *JJP* 34 (2004), 87–94 at 89–94 (DBMNT 1535 = TM 104307); for the title in general, see O. Hornickel, *Ehren- und Rangprädikate in den Papyrusurkunden: Ein Beitrag zum römischen und byzantinischen Titelwesen*, Borna-Leipzig: Noske 1930, 16–17.

14–16.　This formula for introducing the date of decease is very common in Christian Greek epitaphs from Nubia, in particular northern Nubia; see Junker, "Grabsteine", 127; A. Łajtar, "The epitaph of Iesousinkouda: eparch of Nobadia (?), *domestikos* of Faras and *nauarchos* of the Nobadae, died AD 1102 (from the collections of the Sudan National Museum, Khartoum)", *Gdańsk Archaeological Museum African Reports* 1 (1998), 73–80 at 77 [= *I. Khartoum Greek* 5]. The formula must have been abbreviated in the lacuna at the end of l. 15, since the full text would not fit the space.

301

JACQUES VAN DER VLIET AND KLAAS A. WORP

4. Funerary stela of a deacon Kolothos

National Trust inv. no. 1257705.

Rectangular sandstone stela, 18 × 14 × 8 cm, inscribed with twenty lines of incised Greek (l. 1) and Sahidic Coptic (ll. 2–20) text. Minor marginal damage at the bottom of the slab affects the reading of the last line. A simple cross precedes line 1; three similar crosses are interspersed in the text of the last line (l. 20).

The writing is a sloping and irregular uncial, carelessly incised and sometimes difficult to decipher. The mason (or his model) added a redundant м in the end of line 11 and omitted letters in lines 10 and 16; the second т in line 19 is written as an oblique stroke as if an abbreviation were intended.

Kalabsha (Talmis) or its vicinity,   *ca.* ninth century.

```
     + ἐγράφη μενὸς Χοίχ β'
     ϩιτν τεπρονοιа ᴅε м-
     πνογτε аϥмτον ммоϥ
4.   νϭι πᴅιаκ(ονος) κολο`θ´(ος) πνογτε
     παναϩοορ ммντϣανε-
     ϩτης νιм εϥετι ναϥ νογ-
     αναπαγсιс мн νεϥπετογ-
8.   ааβ τηρογ αγω ον νεϥ-
     τι мτον ναϥ ϩν κογνϥ
     ναβραϩам (мн) <ι>саκ (мн) ιак-
     ωβ νϥсаνογϣϥ ϩν π{м}-
12.  ма νογοτογετ ϩιχν ο-
     γмоог нмтон пма ете
     мн емкаϩ нϩнтϥ ог-
     ᴅе λγπη огᴅе аϣаϩ-
16.  ам νκᴅιτϥ εϩογν ετη<π>с
     ннекπετογααβ αγω νκ-
     ааϥ нмпϣа нтесмн е-
     τεсмамаат ϩамнн
20.  εϥ+ε +ϣопе + ϩам[нн]
```

1. μηνὸς; Χοίακ ǁ 4. пᴅιаκ, ǁ 5–6. read ммнтϣανεϩτηϥ ǁ 10. (мн): twice ς | read ιсаак ǁ 15–16. read аϣаϩом ǁ 18–19. ετεсмамаа,

+ Written in the month Choiak 2.

Through the providence of God went to rest the deacon Kolothos.

May God, the treasury of all mercy, grant him rest together with all His saints and may He also grant him rest in the bosom of Abraham and Isaac and Jacob, and feed him in the place of verdure, at a water of rest, the place where sorrow, sadness and sighing are not, and may You include him in the number of Your saints and make him worthy of hearing the blessed voice. Amen. + So be + it. + Amen.

Apart from the Greek dating formula (l. 1), this stela is in Sahidic Coptic. Following the typically Nubian opening formula of the *pronoia* type,[36] which introduces the name of the deceased (ll. 2–4), the remainder of the text is taken up by a long composite prayer of a type current in both Egypt and Nubia. It comprises seven distinct formulaic units, to wit prayers (1) to grant rest to the deceased "with all (God's) saints", (2) similarly in the bosom of the Patriarchs Abraham, Isaac and Jacob, (3) to feed him in a place of verdure, (4) at a water of rest and (5) in a land that does not know sorrow (etc.), (6) to include him in the number of God's saints and, finally, (7) to make him worthy of being welcomed by the voice of the divine judge. From line 16, the series of prayers switches from the third to the second person. The partial redundancy of the text (the formulae of units 1 and 6 almost overlap) and the switch in grammatical person in line 16 demonstrate the composite character of the prayer, which is put together from stock liturgical phrases.[37] Of the seven units distinguished here, nos. 2–5 have a near-literal correspondence in the *memento* of the dead in the Egyptian *Liturgy of Saint Basil*.[38] Several textual peculiarities have been noted below.

In the prayer section, summarized above, the text is very close to the epitaph of the *meizoteros* Paul (British Museum EA 825), which was acquired by the British collector George Annesley, Earl of Mountnorris, from Henry Salt before June 1818, with Kalabsha as its stated provenance.[39] The stela of Kolothos may therefore come from the same region as our nos. 1–3, as seems likely in any case. The developed prayer section suggests a date in or around the ninth century.

1.   This Greek dating formula is a frequent component of Coptic legal documents. Its occurrence in epitaphs is slightly incongruous, but likewise frequent, in particular in the numerous Coptic stelae from Aswan (*SB Kopt.* I 498–675; see also R. Dekker, "The memorial stone of Bishop Joseph III of Aswan", [in: *Nubian voices* II, 5–25]; in Nubian epitaphs it is less commonly found, for instance at Sakinya (Toshka West): Mina, *Inscriptions* (see above, under no. 1, commentary to l. 4), nos. 131 (DBMNT 213 = TM 102307), 132 (DBMNT 616 = TM 141659), in Greek, and 236 (DBMNT 316 = TM 141437), 261 (DBMNT 341 = TM 141462) in Coptic, but also at Tafa: Firth, *Archaeological survey*, 46, no. 249 (DBMNT 413 = TM 141521), 48, nos. 483 (DBMNT 428 = TM 141536) and 486 (2; cf. pl. 51a, no. 4 [DBMNT 450 = TM 141555]), 50, left row, third stela (DBMNT 444 = TM 141549), all Greek. Its funerary use is witness to the influence of notarial practice in the domain of mortuary epigraphy, where – of course – the date of death not that of the drafting of the text is important. The spelling of the month name is attested only once for Nubia in a stela from Sakinya dated to A.D. 954, Mina, *Inscriptions* (see above, under no. 1, commentary to l. 4), no. 318, l. 7 (DBMNT 396 = TM 141509; see Ochala, *Chronological systems*, 228).

4.   Since the name of the deceased is abbreviated, it could be reconstructed alternatively as Kolouthos or Kolothe; cf. *NB Kopt.*, s.v. Kollouthos.

5–6. The epithet that qualifies God as "the treasury of all mercy" is, to the best of our knowledge, unique in Nubian funerary epigraphy. It occurs occasionally in Coptic homiletic texts, for instance in the opening lines of Ps.-Celestinus of Rome, *Encomium on Victor the General* (Sahidic, ed. E.A. Wallis Budge, *Coptic martyrdoms etc. in the dialect of Upper Egypt*, London: British Museum. 1914, 47, ll. 13–14): ⲡⲁⲛⲁϩⲱⲣ ⲉⲧⲙⲉϩ ⲙⲙⲛⲧϣⲁⲛϩⲧⲏϥ ⲛⲓⲙ, "the treasury full of all mercy"). The expression must have a liturgical background. The spelling ⲙⲛⲧϣⲁⲛⲉϩⲧⲏⲥ (for ⲙⲛⲧϣⲁⲛⲉϩⲧⲏϥ) is most likely influenced by the feminine gender of the word.

11–12. The standard form of the phrase has "*a* place (ⲟⲩⲙⲁ) of verdure"; thus in the stela of Paul from Kalabsha, *SB Kopt.* III 1645.6 (which lacks the verb, however).

12–13. Similarly in the stela of Paul from Kalabsha, *SB Kopt.* III 1645.6–7.

13–16. Also this formula is usually phrased differently: "the place whence sorrow (etc.) *have fled*", with the verb ⲡⲱⲧ ϩⲛ, Greek ἀπέδρα; thus for instance in the stela of Paul from Kalabsha, *SB Kopt.* III 1645.7–9; also *SB Kopt.* I 719.20–1 (stela of Bishop Thomas, Faras, A.D. 862 [DBMNT 89 = TM 101786]).

16–17. Here the stela of Paul from Kalabsha, *SB Kopt.* III 1645.9–10, has "in *the light* of Your saints", which is the standard formula expected in this place (cf. A. Budde, *Die ägyptische Basilios-Anaphora: Text – Kommentar – Geschichte*, Münster: Aschendorff 2004, 190, Greek and Sahidic); the present phrase (ⲛⲕϫⲓⲧϥ ⲉϩⲟⲩⲛ ⲉⲧⲏ<ⲡ>ⲥ ⲛⲛⲉⲕⲡⲉⲧⲟⲩⲁⲁⲃ) corresponds to the Greek prayer συναριθμήσον αὐτὸν (etc.), found in Lefebvre, *Recueil*, no. 607, ll. 9–10 (DBMNT 473 = TM 102836; from Ghazali; cf. Junker, "Grabsteine", 120–21).

17–19. The formula has been truncated for lack of space. It ends at the same point in the stela of Eparch Marianou from Qasr Ibrim, of A.D. 1092, ed. T.S. Richter, "Koptische und griechische Grabstelen aus Ägypten und Nubien", in: Suzana Hodak, T.S. Richter, and F. Steinmann (eds.), *Coptica: Koptische Ostraka und Papyri, koptische und griechische Grabstelen aus Ägypten und Nubien, spätantike Bauplastik, Textilien und Keramik*, Berlin: Manetho Verlag 2013, no. 66, ll. 14–16 (DBMNT 568 = TM 110999). Its continuation can be filled in differently, with either Matthew 25:21–3 ("Come, good and faithful servant . . ."), thus in the stela of Paul from Kalabsha, *SB Kopt.* III 1645.10–12, or Matthew 25:34 ("Come, O blessed ones of my Father . . ."), as in a group of stelae from Debeira and Meinarti (*I. Khartoum Copt.* 17 [DBMNT 44 = TM 141211], 18 [DBMNT 45 = TM 141212], 19 [DBMNT 46 = TM 141213], 20 [DBMNT 47 = TM 141214], all eleventh century). The present formula seems to occur in neither form in Greek epitaphs from Nubia. For Egyptian parallels, both Greek and Coptic, see Junker, "Grabsteine", 140, and Bianca Tudor, *Christian funerary stelae of the Byzantine and Arab periods from Egypt*, Marburg: Tectum Verlag 2011, 185–86; cf. Van der Vliet, "Gleanings", 189 [Study 22].

20. The crosses appear to have been incised before the text, which they interrupt.

The four epitaphs edited here are quite modest witnesses to Christianity in northernmost Nubia. Yet they represent a twofold interest. First, the textual peculiarities of all four, but in particular the rare prayer of our no. 3, attest to the persistence of local traditions in the former Dodekaschoinos, which distinguish this erstwhile Roman territory from both the Aswan region to its north and central Nobadia, around Qasr Ibrim and Faras, to the south. Secondly, the four stelae belong to a greater group of similar monuments that arrived in European collections in the early decades of the nineteenth century, presumably all hailing from one or two sites in the same region. Thereby they document an early chapter in the rediscovery of Christian Nubia, a chapter that deserves more scholarly attention.

[See also Jacques van der Vliet and Klaas A. Worp, "A fifth Nubian funerary stela from the Bankes collection: An addendum to *CIEN* 3, 26–29", *JJP* (forthcoming).]

Notes

1 We thank Ian Barnes, MCIfA, FSA, Head of Archaeology of the National Trust, for his permission to publish the stelae here.

2 For Bankes, see M.L. Bierbrier, *Who was who in egyptology*, London: Egypt Exploration Society 2012, 4, 38–9, and in particular Patricia Usick, *Adventures in Egypt and Nubia: the travels of William John Bankes (1786–1855)*, London: The British Museum Press 2002. Bankes himself published the autobiography of the Italian adventurer Giovanni Finati (*Narrative of the life and adventures of Giovanni Finati*, 2 vols., London: John Murray 1830; cf. Usick, *Adventures*, 31, n. 3; Bierbrier, *Who was who*, 190) as well as a series of prints showing the obelisk from Philae, which he had transported to England and now adorns the park of Kingston Lacy, published in London in 1821 (cf. Usick, *Adventures*, 32, n. 7, with pl. 6, and p. 154).

3 Thus already T.G.H. James, "Egyptian antiquities at Kingston Lacy, Dorset: the collection of William John Bankes", *KMT: a modern journal of ancient Egypt* 4/4 (Winter 1993–1994), 20–32, who mentions, at 27, "a small group of simple Greek and Coptic grave-stelae collected by Bankes somewhere in Nubia" (with a photo on p. 30), a statement copied in Usick, *Adventures*, 195. James announced their imminent publication by the Egyptologist A.F. Shore (1924–1994), a publication that did not materialize, as far as we are aware.

4 See Usick, *Adventures*, 31–52 (first journey) and 76–148 (second journey).

5 See J. van der Vliet, "Gleanings from Christian Northern Nubia", *JJP* 32 (2002), 175–94, especially 178–91 [Study 22].

6 For Salt and his connections, see Deborah Manley and Peta Rée, *Henry Salt: artist, traveller, diplomat, egyptologist*, London: Libri Publications Ltd 2001; for his financial motives, at 154–57; cf. Usick, *Adventures*, 76–7.

7 See Usick, *Adventures*, 28.

8 See Manley and Rée, *Henry Salt*, 158–68; Usick, *Adventures*, 97–8.

9 For Annesley, see Bierbrier, *Who was who*, 22; for his long-standing relation with Salt, see Manley and Rée, *Henry Salt*.

10 T. Young, "Observations on a fragment of a very ancient Greek manuscript on papyrus, together with some sepulchral inscriptions from Nubia, lately received by the Earl of Mountnorris", *Archaeologia* 19 (1821), 156–60, pls. IX–X.

11 Young, "Observations", 156.

12 For Bankes' visits to the area, see Usick, *Adventures*, 49, 104–08, who nowhere mentions the acquisition of inscriptions; at some stage, though, Bankes seems to have entertained the idea of publishing a volume of – presumably Greek – inscriptions from Egypt and Nubia; see ibidem, 206, n. 25.

13 For the word order in similar epitaphs, see Kalle Korhonen, "Between meaningful sentences and formulaic expressions: fronted verbs in Christian epitaphs", *Glotta* 87 (2011), 95–125.

14 C.M. Firth, *The archaeological survey of Nubia: report for 1908–1909*, Cairo: Government press 1912, 50, third text in the righthand row (DBMNT 448 = TM 141553).

15 See the discussion in Van der Vliet, "Gleanings", 180–83.

16 H. Junker, "Die christlichen Grabsteine Nubiens", *ZÄS* 60 (1925), 111–48 at 126–27.
17 See Van der Vliet, "Gleanings", 184.
18 For a discussion, following Junker, see Van der Vliet, "Gleanings", 180–82.
19 See Junker, "Grabsteine", 125 and 137.
20 W. Weissbrodt, "Ein aegyptischer christlicher Grabstein mit Inschrift aus der griechischen Liturgie im Königlichen Lyceum Hosianum zu Braunsberg und ähnliche Denkmäler in auswärtigen Museen. Zweiter Teil", in: *Verzeichis der Vorlesungen am Königlichen Lyceum Hosianum zu Braunsberg*, Sommer-Semester 1909, 3–32 at 20–3. The manuscript is Grottaferrata *Γ.β*. X, f. 81; its prayers for the deceased have been edited in full by T. Christodoulou, Ἡ νεκρώσιμη ἀκολουθία κατά τους χειρογράφους κώδικες 10ου-12ου αἰῶνός, Thera 2005, II, 36–69 (*non vidimus*; we owe this information to the kindness of Prof. Stefano Parenti, Rome); for a general discussion, see Elena Velkovska, "Funeral rites according to the Byzantine liturgical sources", *DOP* 55 (2001), 21–45 at 31–7.
21 See the methodological remarks in H. Brakmann, "*Defunctus adhuc loquitur*: Gottesdienst und Gebetsliteratur der untergegangenen Kirche in Nubien", *Archiv für Liturgiewissenschaft* 48 (2006), 283–333 at 302–03.
22 Junker, "Grabsteine", 137, n. 2. For a modern printed text of the Bohairic rite, see H. Becker, and H. Ühlein (eds.), *Liturgie im Angesicht des Todes: Judentum und Ostkirchen* I, Sankt Otilien: EOS Verlag 1997, 99–100; English translation: R.M. Woolley, *Coptic offices*, London: Society for Promoting Christian Knowledge 1930, 139.
23 Cf. A. Łajtar, "On the provenance of the four Christian inscriptions: *SB* X 10515–16, M.G. Tibiletti Bruno, Iscrizioni nubiane 49, 56", *ZPE* 95 (1993), 245–47 at 243.
24 Cf. Weissbrodt, "Grabstein", 20–3; Junker, "Grabsteine", 114, *sub* 2; the unusual material, limestone, does not preclude a Nubian provenance, already surmised by Lefebvre, *Recueil*; cf. Junker, "Grabsteine", 123.
25 Only edition: Firth, *Archaeological survey*, 45, pl. 51a, no. 2; partly corrected by Junker, "Grabsteine", 125.
26 See G.J.M.J. te Riele, in: P.J. Sijpesteijn and K.A. Worp (eds.), "Greek texts in the possession of the Amsterdam University Library", *Talanta* 8–9 (1977), 100–18 at 116, and Łajtar, "On the provenance", 243. Cf. M. Kuhn, *Coptic texts and artifacts hidden in Amsterdam*, Leiden: Faculty of Arts/Amsterdam: Allard Pierson Museum 2000, 39, no. 29.
27 But note that Bab-Kalabsha is *not* Contra-Taphis. The site is south of Tafa on the west bank.
28 L.V. Žabkar, "Three Christian grave stelas", in: H. Ricke (ed.), *Ausgrabungen von Khor-Dehmit bis Bet el-Wali*, Chicago: University of Chicago Press 1967, 16–20 at 19, with fig. 28, pl. 30a.
29 Ibidem, 19, with fig. 29 (right), pl. 30b.
30 See Łajtar, "On the provenance", 243–44.
31 Compare the table in Ochała, *Chronological systems*, 106–07.
32 See Junker, "Grabsteine", 125, for a discussion of this addition, plausibly interpreted as the echo of another liturgical prayer.
33 G.W.H. Lampe, *A patristic Greek lexicon*, Oxford: Clarendon Press 1961, 356, notes a rare verb διαλήγω, "to cease". The reading διαλήγον (for διαλήγων) is accepted without discussion in Lefebvre, *Recueil*, no. 659, *ad loc*.
34 Weissbrodt, "Grabstein", 20; he even postulates διαλέγων as the original (at p. 21), which seems very unlikely, however.
35 For the latter phenomenon, see Gignac, *Grammar* I, 71–5.
36 See J. van der Vliet, "'What is man?' The Nubian tradition of Coptic funerary inscriptions", in: *Nubian voices* I, Warsaw: University of Warsaw/Raphael Taubenschlag Foundation 171–224 at 215–20 [Study 31].

FOUR NORTH-NUBIAN FUNERARY STELAE

37 For the underlying principle of "centonization", see Van der Vliet, "What is man?" 197–201; generally on the liturgical background of Nubian funerary epigraphy: Brakmann, "*Defunctus adhuc loquitur*", 300–10.

38 A. Budde, *Die ägyptische Basilios-Anaphora: Text – Kommentar – Geschichte*, Münster: Aschendorff 2004, 190 (Sahidic).

39 See above. The stela of Paulos (*SB Kopt.* III 1645 [DBMNT 626 = TM 101499]) was re-edited in Van der Vliet, "Gleanings", 185–91 [Study 22]; the present stela shows the editor's reservations about its Kalabsha provenance to be unfounded.

24

CHURCHES IN LOWER NUBIA, OLD AND "NEW"

Jacques van der Vliet

Introduction

"Ah! tous les cierges sont éteints".

Ad. Retté, *En déshérence*

The following notes discuss philological evidence for churches in three former Nubian districts: Ibrim, Toshka West and Faras. In spite of their modest scope, they may help us to fill in the picture of a Christian landscape which has disappeared forever.

I. Ibrim Cathedral: Church of St. Mary, or Church of the apostles?

In a recent synthesis of the history and archaeology of late medieval Qasr Ibrim, W.Y. Adams discusses, among other things, the famous cathedral within the hilltop citadel.[1] In his opinion, this cathedral is to be identified "unquestionably" with the Church of St. Mary, described in a late twelfth–early thirteenth-century handbook of Coptic *topographia sacra*, attributed formerly to one Abu Salih, and lately to a Cairo priest, Abu 'l-Makarim.[2] This source states that, in the town of Ibrim, "there is a splendid, big church, beautifully designed, named after Our Lady, the Pure Virgin Saint Mary, on top of which there is a lofty dome carrying a large cross".[3] However, two objections can be made against this identification. The first objection is of an architectural nature, and actually already raised by Adams himself; the second one concerns a well-known pair of documents, to wit the so-called "scrolls of Bishop Timotheos".

As Adams points out, the remains of the cathedral building indicate that it cannot possibly have supported a dome, not even of wood.[4] In this respect it differs from the majority of later Nubian churches, which actually did carry a dome.[5] At Qasr Ibrim a late medieval church of this type survived, including its dome, until quite recently.[6] An earlier example is provided by the so-called Church on the Point which, situated high above the citadel, was visible from afar.[7] The unusual

309

iconography of some bishops' stelae, too, suggests that similar domed churches were a conspicuous feature of the Ibrim townscape.[8] There can be no doubt, furthermore, that Ibrim did have a sanctuary dedicated to the Virgin Mary. Its existence is well attested by texts found within the citadel.[9] Thus far, Abu 'l-Makarim's description has nothing unlikely. It is, moreover, only part of a longer passage which deals with the capture of Ibrim by Shams al-Dawla, the brother of Salah al-Din, in A.D. 1173. The author relates, among other things, how the Muslim conqueror had the cross removed from the dome and ordered the muezzin to chant from its top.[10] The fall of Ibrim made a profound impression on its contemporaries.[11] Abu 'l-Makarim's report cannot be much posterior to the actual event, which increases its credibility.[12] However, if Abu 'l-Makarim's information on the Church of St. Mary is trustworthy, as seems probable, this church cannot, according to Adam's architectural analysis, have been the well-known cathedral.

This negative conclusion is confirmed by the "scrolls of Bishop Timotheos" of A.D. 1372. These two parallel documents, one in Bohairic Coptic and one in Arabic, proclaim the consecration of Timotheos as bishop of Faras, the ancient capital of Lower Nubia.[13] They were issued by the patriarchal chancery in Cairo and the texts of both retain, as of old, the titles and institutions of the important diocese of Faras.[14] However, owing to historical vicissitudes which, most probably, led to a fusion of episcopal sees, Timotheos had been made bishop of a "double" diocese, of Faras *and* Ibrim. The Coptic scroll does not acknowledge this new situation. It clearly reflects the time-honored usage of Faras, whereas in all probability Ibrim was to become the new bishop's actual residence.[15] The Arabic version of the same document, on the other hand, reflects the new situation. Thus, it speaks about "sees" in the plural.[16] Moreover, the wording of the Arabic scroll was adapted to the needs of the moment by "last minute" insertions into a text which, obviously, had already been written. Thus, every time Faras is mentioned in the Arabic text, the addition "and Ibrim" is supplemented in a cramped hand, above the line.[17] The reason for this is obvious; as Arabic had become the current language of communication, at least the Arabic document should correctly reflect the actual situation.[18]

The addition "and Ibrim", however, is not the only sign of adaptation to the new situation. Another insertion into the Arabic text, which hitherto went unnoticed,[19] concerns the patronage of Timotheos' "double" episcopal see. As is well known, the Faras diocese and its cathedral church were dedicated to the Virgin Mary. Among others, inscriptions discovered in the Cathedral of Faras itself bear witness to this dedication.[20] Also the Coptic scroll situates the entire career of Timotheos within "the Church of the Holy Mother of God, St. Mary of Pkhoras (i.e. Faras) and Nubia" (ll. 63–8).[21] Both scrolls are addressed, in the words of the Coptic text, to "the entire orthodox people in the diocese of the Christ-loving town of Pkhoras and Nubia and all its provinces" (ll. 25–8).[22] At this point, the Arabic scroll is phrased differently; it addresses "the entire orthodox people of Abakhuras (Faras) *and Ibrim* belonging to the provinces of Nubia, who received the seal (of baptism) in the Church of Our Lady *and the Apostles* and what pertains to it" (ll. 8–10).[23] As the photograph in Plumley's edition shows, not only the group "and

310

Ibrim" has been inserted secondarily, but the phrase "and the Apostles" as well. What was initially the last word of l. 9 (ما) has been clumsily erased and transformed into the article of ²⁴الرسل, which is written in the lefthand margin. In the righthand margin of l. 10, وما ("and what") has been added again.

Thus, updating the Arabic text did not only involve the secondary insertion of the name of the new see, Ibrim, but also of that of its titular saints, the Apostles. Hence, there can be little doubt that the Cathedral of Ibrim was dedicated to the Holy Apostles.[25] In point of fact, a Church of the Apostles is frequently mentioned in Old Nubian documents from the citadel.[26] Which of the town's sanctuaries was Abu 'l-Makarim's "big church . . . named after our Lady, the Pure Virgin Saint Mary", with its conspicuous dome, remains to be discovered.

II. Toshka West (Sakinya): the Church of St. Gabriel

The Coptic and Greek inscriptions unearthed in 1933 in a big cemetery near the hamlet of Sakinya (Nagʻ Sakanya), in the southern part of the Toshka West district,[27] still represent the greatest single corpus of Christian epitaphs from Nubia, consisting of over 300 pieces.[28] This richness contrasts oddly with the paucity of other sources of information concerning the medieval history of this particular part of Nubia.[29] Very few Christian sites in Toshka West have been excavated and the results of these excavations have been unsatisfactorily published.[30] In particular, no town sites have ever been recorded.[31] Also secondary sources are scarce and in fact a great part of the area was flooded without being documented. In 1970, B.G. Trigger, in an assessment of the remains from Toshka and Arminna, had to conclude that "the rise in water level . . . destroyed much of the evidence on which a detailed study of the local environment during the Christian period might have been based".[32]

Our scanty information suggests that the district of Toshka West was a relatively populous and prosperous part of Christian Nubia. This impression is confirmed by the numerous inscribed tombstones which, moreover, contain important material for the language and prosopography of its medieval inhabitants,[33] for the chronology[34] and iconography[35] of Nubian stelae, etc. They have, however, remarkably little to offer in the field of local topography.[36] That Toshka West must have possessed several churches, one of which was a parish church, a ⲕⲁⲑⲟⲗⲓⲕⲏ, is about all we know, thanks to E. Drioton's improved translation of Togo Mina's stela no. 35, ll. 7–8.[37]

To this single record of a Toshka West church, one more can, in my opinion, be added. It is provided by an interesting, but incomplete terracotta stela which Mina edited under no. 321.[38] The main body of its text, which although fairly corrupt can easily be understood, is in Coptic. In l. 7, however, where the funerary formula requires the name of the deceased, following the epithet ⲙⲁⲕⲁⲣⲓⲟⲥ, the text switches abruptly from Coptic to Greek:

ⲩⲡⲉⲣ ⲕⲩⲙⲓⲥⲉⲟⲥ ⲕⲁⲓ̈ ⲁⲛⲁⲡⲁⲩⲥⲟ ⲓ̈ⲱⲁⲛⲛⲟⲩ ⲡ˅ⲣ´(ⲉⲥⲃⲩⲧⲉⲣⲟⲩ) ⲅⲁⲃⲣⲓ̈(ⲏⲗ) ⲩ(ⲓ)ⲟⲩ ⲡⲉⲧⲏⲕⲉⲥⲁⲣⲏ (ll. 7–9), i.e.:

ὑπὲρ κοιμήσεως καὶ ἀναπαύσεως Ἰωάννου πρ(εσβυτέρου) Γαβρι(ὴλ) υ(ἱ)οῦ Πετηκεσαρη.[39]

The text continues again in Coptic.

Mina relegates the first four words of the Greek passage, as an intercalation, to his apparatus and translates the remainder as: "le bienheureux prêtre Ioannou, fils de Gabri(el), l'homme de Césarée (?)". This interpretation obscures the fact that the naming lemma of this predominantly Coptic epitaph takes, as a whole, the form of the Greek commemorative formula ὑπὲρ κοιμήσεως (etc.), a formula which occurs on other Sakinya stelae as well.[40] Accordingly, the entire phrase requires the following translation: "for the falling asleep and the rest of Ioannes, priest of (St.) Gabriel, son of Petekesare". The father's name, although reminiscent of a familiar type of Egyptian theophorous names,[41] has perhaps to be split up and read: Πετη κεσαρη (probably = Καισάριος).[42] His son, Joannes, is designated, in the much abbreviated way which is habitual in Nubian epigraphy,[43] as a priest serving a church (or, perhaps, a monastery) dedicated to the Archangel Gabriel.

Gabriel must have been a popular patron in the region. Somewhat more to the north, in Ibrim, at least one church bore his name.[44] As the epitaph of Joannes shows, also Toshka West had its Church of St. Gabriel. Owing to our meager knowledge of the site, it cannot, regrettably, be further identified or located.

III. Faras: the Church of the Four Living Creatures

The Coptic funerary stela of a priest Marianou (Marianos), dated A.M. 671 (A.D. 955), was found during the second campaign of the Polish excavations in Faras, in 1961–62.[45] It has since been edited several times.[46] In spite of considerable surface damage and some cracks, the text can be read relatively easily, apart from a short passage in ll. 5–6. Since it follows the deceased's name and a common abbreviation for πρεσβύτερος, this passage must have contained the name of the church where Marianou held his office, as was rightly surmised by the first editor.[47] In fact, the title in question can still be deciphered with considerable certainty, as follows: ⲡⲙⲁⲕⲁⲣⲓⲟⲥ ⲙⲁⲣⲓ[ⲁ]ⲛⲟⲩ ⲡ`ⲣ´(ⲉⲥⲃⲩⲧⲉⲣⲟⲥ) ⲙⲡϥⲧⲟⲟⲩ ⲛⲍⲟⲱⲛ (ll. 4–6):[48] ". . . the blessed Marianou, priest of (the Church of) the Four Living Creatures".

The cult of the Four Living Creatures from Rev. 4 was as popular in Nubia as it was in Coptic Egypt. Both literature and the visual arts bear witness to this devotion.[49] In fact, several Nubian sanctuaries are known to have been dedicated to these angelic beings: one, a monastery, may have been situated in or near Sonqi Tino,[50] and another one, on an island called Teme, which I am unable to locate.[51] Faras, too, as it appears from the epitaph of the priest Marianou, could boast a church or monastery dedicated to the Four Living Creatures. Unfortunately, the stela was not found in its original context, but on the east side of Faras Cathedral, "in a pile of drifted sand", where no traces of a tomb could be discovered.[52] As

CHURCHES IN LOWER NUBIA, OLD AND "NEW"

no Christian buildings seem to have been excavated east of the cathedral and as the alleged monastic complex to its northeast appears to have been dedicated to St. George,[53] there will probably be no way of ever ascertaining the precise location of the Faras Church of the Four Living Creatures.

Notes

1 W.Y. Adams, *Qaṣr Ibrîm: the late mediaeval period*, London: Egypt Exploration Society 1996, 73–8.

2 Adams, *Qaṣr Ibrîm*, 66. For recent assessments of this work see U. Zanetti, "Abû l-Makârim et Abû Ṣâliḥ", *BSAC* 34 (1995), 85–138, and J. den Heijer, "Coptic historiography in the Fâṭimid, Ayyûbid and early Mamlûk periods", *Medieval encounters* 2 (1996), 67–98 at 77–81. Although the question of authorship is far from settled, I refer to it here by the name of Abu 'l-Makarim.

3 See B.T.A. Evetts, (ed.), *The churches and monasteries of Egypt and some neighbouring countries attributed to Abû Ṣâliḥ, the Armenian*, Oxford: Clarendon Press 1895, 121 (Arabic); cf. 266 (English), and G. Vantini, *Oriental sources concerning Nubia*, Heidelberg: Heidelberger Akademie der Wissenschaften/Warsaw: Polish Academy of Sciences 1975, 327.

4 Adams, *Qaṣr Ibrîm*, 77. Earlier in the century, Somers Clarke and U. Monneret de Villard had made similar observations.

5 Cf. P. Grossmann, *Elephantine* II: *Kirche und spätantike Hausanlagen im Chnumtempelhof*, Mainz: Phillip von Zabern 1980, 86–111; P.M. Gartkiewicz, "An introduction to the history of Nubian church architecture", *Nubia christiana* 1 (1982), 43–133 at 91–2; Adams, *Qaṣr Ibrîm*, 67–8.

6 Cf. Adams, *Qaṣr Ibrîm*, 79–81, and especially Grossmann, in: F.W. Deichmann and P. Grossmann, *Nubische Forschungen*, Berlin: Gebr. Mann Verlag 1988, 22–5. Of course, this "miniscule structure" (Adams) cannot be well identified with Abu 'l-Makarim's "splendid, big church".

7 Cf. B. Kjølbye-Biddle, "The small early church in Nubia with reference to the church on the Point at Qasr Ibrim", in: K. Painter (ed.), *'Churches built in ancient times': recent studies in early Christian archaeology*, London: Society of Antiquaries of London 1994, 17–47 (I owe this reference to Dr. P. Grossmann). Its remarkable position, but not its size, could be taken to favor identification with St. Mary's Church.

8 Thus J.M. Plumley, "Some examples of Christian Nubian art from the excavations at Qasr Ibrim", in: E. Dinkler (ed.), *Kunst und Geschichte Nubiens in christlicher Zeit*, Recklinghausen: Verlag Aurel Bongers 1970, 129–40 at 131–32 and figs. 103, 106, 108 (cf. p. 132: domes "might represent a church building or buildings and be symbolic of the see of Ibrim").

9 See G.M. Browne, *Old Nubian texts from Qaṣr Ibrîm* III, London: Egypt Exploration Society 1991, nos. 37, l. 15; 43, l. 1; 62, I, l. 2; cf. Adams, *Qaṣr Ibrîm*, 66 and 252, table 21.

10 Ed. Evetts, 121–22 (Arabic); cf. 266–67 (English); Vantini, *Sources*, 327–28.

11 Cf. the Coptic inscription, written only two weeks afterwards, in F.L. Griffith and G.M. Crowfoot, "On the early use of cotton in the Nile Valley", *JEA* 20 (1934), 5–12 at 7–8. The rich echoes of the event in Oriental historiography have been assembled by Vantini, *Sources*, 357–58, 368–69, 422, 436, 673; cf. J. Cuoq, *Islamisation de la Nubie chrétienne: VIIe-XVIe siècles*, Paris: P. Geuthner 1986, 65–7. Also modern historians tend to consider it as a turning point in Nubian history, cf. W.Y. Adams, *Nubia: corridor to Africa*, London: Penguin 1977, 510.

12 For dating this source, see Zanetti, "Abû l-Makârim", 122–25, cf. 133–34, and Den Heijer, "Coptic historiography", 78.

JACQUES VAN DER VLIET

13 See J.M. Plumley (ed.), *The scrolls of Bishop Timotheos: two documents from medieval Nubia*, London: Egypt Exploration Society 1975 (where they are given the Egyptian Museum nos. 90 223 and 90 224); cf. R.G. Coquin, "A propos des rouleaux coptes-arabes de l'évêque Timothée", *BiOr* 34 (1977), 142–47.

14 For a recent discussion of the later history of the Faras diocese, see W. Godlewski, "The bishopric of Pachoras in the 13th and 14th centuries", in: C. Fluck, L. Langener, S.G. Richter (eds.), *Divitiae Aegypti: Koptologische und verwandte Studien zu Ehren von Martin Krause*, Wiesbaden: Reichert Verlag 1995, 113–18 (115–16 study the evidence of the scrolls).

15 The scrolls were found together with the bishop's body, buried in a crypt under the Cathedral of Qaṣr Ibrim; cf. Plumley, *Scrolls*, 3, pls. I-III; J.M. Plumley, "Qaṣr Ibrîm 1963–1964", *JEA* 50 (1964), 3–5 at 3–4, pl. I/1–2.

16 Plumley, *Scrolls*, 30, ll. 24–5, and 31, l. 56.

17 Cf. Plumley, *Scrolls*, 22, note ad l. 6; however, Plumley's supposition that Ibrim had taken over not only Faras' role as a see, but also its name, is contradicted precisely by this procedure.

18 Thus Godlewski, "Bishopric", 116; cf. M. Krause, "Zur Kirchengeschichte Nubiens", in T. Hägg (ed.), *Nubian culture: past and present*, Stockholm: Almquist & Wiksell 1987, 293–308 at 299–300 and 301–02.

19 But cf. Coquin, "Rouleaux", 146.

20 Cf. S. Jakobielski, *A history of the bishopric of Pachoras on the basis of Coptic inscriptions*, Warsaw: PWN–Éditions scientifiques de Pologne 1972, 176–86.

21 Plumley, *Scrolls*, 10, pl. VIII.

22 Ibidem, 8, pl. VII.

23 Ibidem, 29, pl. XVII; cf. Coquin, "Rouleaux", 146. I accept Coquin's reading of المرسوم or المرشوم, but prefer a slightly different translation.

24 The traces of the *mîm* led the first editors (*apud* Plumley, *Scrolls*, 29) to transcribe بالرسل ; corrected by Coquin, "Rouleaux", 146.

25 It may be noted, incidentally, that the earliest cathedral of Faras, too, may have been dedicated to the Apostles; cf. [Study 26:] "The Church of the Twelve Apostles: the earliest Cathedral of Faras?" [*Orientalia* 68 (1999), 135–42].

26 Cf. Browne, *Old Nubian texts* III, nos. 30, l. 24; 36.I, l. 19; 40, l. 32; 60, l. 17 (ad l. 9); 62.I, l. 1; cf. Adams, *Qaṣr Ibrîm*, 66 and 252, table 21.

27 For its situation, see U. Monneret de Villard, *Le iscrizioni del cimitero di Sakinya (Nubia)*, Cairo: IFAO 1933, pp. I and III, fig. 1; W.B. Emery, and L.P. Kirwan, *The excavations and survey between Wadi Es-Sebua and Adindan, 1929–1931*, Cairo: Government press 1935, maps in vol. II, pls. 66–7 (as Nag' Sakânya); B.F. Trigger, "The cultural ecology of Christian Nubia", in: E. Dinkler, *Kunst und Geschichte Nubiens*, 347–68 at 362 and 360, fig. 40. Cf. W.K. Simpson, art. "Toschqa", in: W. Helck, and W. Westendorf (eds.), *Lexicon der Ägyptologie* 6, Wiesbaden: Harrassowitz 1986, col. 637–39.

28 Edited by T. Mina, *Inscriptions coptes et grecques de Nubie*, Cairo: Société d'archéologie copte 1942; cf. W. Brunsch, "Konkordanz zu T. Mina, Inscriptions coptes et grecques . . .", *GM* 100 (1987), 85–90, replacing an incomplete and less reliable edition by U. Monneret de Villard, *Iscrizioni*.

29 Cf. Trigger, "Cultural ecology", at 359–62.

30 Sakinya cemetery: cf. U. Monneret de Villard, *La Nubia medioevale* I, Cairo: Mission archéologique de Nubie 1935, 123–28, figs. 103–10; vol. IV, pls. CXXVI.b and CXXX–CXXXII; cf. vol. III, 63–78 ("Le tombe"). Somewhat more to the north, an older cemetery ("A" or "TWA") containing Christian tombs was excavated in the sixties, cf. W.K. Simpson, "Nubia: 1962 excavations at Toshka and Arminna", *Expedition* 4/4 (Summer 1962), 36–46 at 39, with photo at 41; W.K. Simpson, *Heka-nefer*

CHURCHES IN LOWER NUBIA, OLD AND "NEW"

and the dynastic material from Toshka and Arminna, New Haven: The Peabody Museum of Natural History of Yale /Philadelphia: University Museum of the University of Pennsylvania 1963, 48, n. 1; idem, "Toshka-Armina 1962: the Pennsylvania – Yale archaeological expedition to Nubia", in: *Fouilles en Nubie (1961–1963)*, Cairo: Organisme général des impr. gouvernementales 1967, 169–83 at 173–74; Trigger, "Cultural ecology", 362, fig. 335. Still more to the north, cemetery 259, near Nag' Ambukab al-Qibli (on the map in Emery and Kirwan, *Excavations and survey*, pl. 66: Nag' Amberkâb al-Qibli), perhaps of C-group date, produced "surface sherds of Byzantine . . . date", see H.S. Smith, *Preliminary reports of the Egypt exploration society's Nubian survey*, Cairo: General Organisation for Govt. Print Offices 1962, 49.

31   Cf. Trigger, "Cultural ecology", 362; Simpson, "Toschqa", col. 638.

32   Trigger, "Cultural ecology", 360–61; meanwhile some new information became available through the edition of documents from Qaṣr Ibrim; thus, in Browne, *Old Nubian Texts* III, nos. 37, l. 34 and 43, l. 7, a ⲧⲟⲧ of Toshka acts as a witness. For his title, see ibidem, 80, note to no. 30, l. 17; G.M. Browne, *Old Nubian dictionary*, Louvain: Peeters 1996, 180–81, s.v.

33   Cf. Mina, *Inscriptions*, VI–VII, with the reviews by E. Drioton, *BSAC* 8 (1942), 227–29 at 228, and W.C. Till, *Orientalia* 17 (1948), 357–58; W. Vycichl, "Coptic dialect geography based on inscriptions", *Enchoria* 8 (1978), Sonderband, 63* (109)–65* (111) at 64* (110).

34   See especially M. Krause, "Die Formulare der christlichen Grabsteine Nubiens", in: K. Michałowski (ed.), *Nubia: récentes recherches*, Warsaw: Musée national 1975, 76–82.

35   Cf. especially Monneret de Villard, *Iscrizioni*, pls. I–II.

36   Thus already Mina, *Inscriptions*, VI.

37   In his review of Mina, *Inscriptions*, in *BSAC* 8, 228–29, quoted in J. Vergote's review, *BiOr* 3 (1946), 109–10 at 110, and in E. Wipszycka, "Καθολική et les autres épithètes qualifiant le nom ἐκκλησία: contribution à l'étude de l'ordre hiérarchique des églises dans l'Égypte byzantine", *JJP* 24 (1994), 191–212 at 203, n. 17.

38   *Inscriptions*, 145–46; no. 222 in Monneret de Villard, *Iscrizioni*, 23, who offers a facsimile at p. VIII, fig. 4; presently in the Coptic Museum, Old-Cairo, inv. no. 6833.

39   The original writes the abbreviation ⲡ̅ⲣ̅ as a monogram; ⲅⲁⲃⲣⲓ is a common abbreviation for ⲅⲁⲃⲣⲓⲏⲗ, cf. G. Heuser, *Die Personennamen der Kopten*, Leipzig: J.C. Hinrichs 1929, 110.

40   In fact, the present stela inserts Krause's Greek formula no. 3 within his Coptic formula no. 6, and can accordingly be dated to the ninth or, more probably, tenth century; cf. Krause, "Formulare", 78–81. For a comparable, though not identical, example of code-switching in the naming lemma of the formula, see the terracotta stela of Michaelkouda from Ukma East, ed. R. Kasser, in: Ch. Maystre, *Akasha* II, Geneva: Université de Genève 1996, 24–7, pl. XII, ll. 6–8 (probably tenth century). For an inverse example (Greek-Coptic), cf. *SB Kopt.* I 734 (A.D. 913).

41   For Coptic names formed on the ancient Πετ(ε)-*cum*-divinity model, see Heuser, *Personennamen*, 49; however, an interpretation along these lines would encounter several difficulties (which Dr. B. Muhs, Papyrological Institute, Leiden University, was kind enough to point out to me).

42   This on the assumption that ⲡⲉⲧⲏ is an orthographic variant of ⲡⲉⲧⲓ, a common name in Lower Nubia; compare e.g. ⲡⲉⲧⲓ̈ (Mina, *Inscriptions*, 141, no. 316, l. 5), ⲡⲉⲧⲓ̈ ⲁⲗⲗⲁⲙ (ibidem, 140, no. 315, l. 7; both from Sakinya), ⲡⲉⲧⲓ ⲡⲉⲥⲛⲏⲑ (J. Krall, "Ein neuer nubischer König", *WZKM* 14 (1900), 233–42 at 235, l. 16; from Ikhmindi). Mina, *Inscriptions*, in his apparatus, proposed to read ⲡⲉⲧⲛⲕⲉⲥⲁⲣⲏ for ⲡⲉⲧⲛⲕⲉⲥⲁⲣⲏ and took -ⲕⲉⲥⲁⲣⲏ for the toponym Caesarea (cf. ibidem, p. VI).

43   See note 47.

JACQUES VAN DER VLIET

44 Browne, *Old Nubian texts* III, no. 60, l. 10; Adams, *Qaṣr Ibrîm*, 252, table 21, lists two Gabriel-Churches for the Ibrim district.

45 Presently in the National Museum, Warsaw, inv. no. MN 149397 (excavation no. F 22/61–2).

46 By S. Jakobielski, in: K. Michałowski, *Faras: fouilles polonaises 1961–1962*, Warsaw: PWN–Éditions scientifiques de Pologne 1965, 171–72, no. 6, fig. 90 (*ed. princeps*); again in: Jakobielski, *Bishopric of Pachoras*, 125–21, fig. 35; photo and French translation only in J. Kubińska, *Inscriptions grecques chrétiennes*, Warsaw: PWN–Éditions scientifiques de Pologne 1974, 54, fig. 15.

47 Jakobielski, *Bishopric of Pachoras*, 127, with a reference to p. 169, where he discusses the usual Nubian way of briefly indicating a clergyman's institutional affiliation (name + office + name of church/monastery) and its value for local topography. Exactly the same device was used in the Sakinya stela of Joannes, discussed above, under no. II.

48 Thus my reading after the reproduction in Kubińska, *Inscriptions grecques* (as I was kindly informed by Prof. W. Godlewski, Warsaw, the original has much deteriorated since its discovery). The abbreviation ⲡˋⲣˊ is written as a monogram. Earlier readings of the last two words: *ed. princeps*: ⲙⲡⲛϥ. ⲥⲟ. . |. . . ⲟϥⲛ; Jakobielski, *Bishopric of Pachoras*, 125: ⲡϥ. . . . |. ϥⲍⲟⲱⲛ̣. The ⲱ/ⲟ interchange, here in ⲍⲟⲱⲛ, is ubiquitous, cf. W.A. Girgis, "Greek loan words in Coptic", *BSAC* 18 (1965–66), 71–96 at 81 and 89; F.T. Gignac, *A grammar of the Greek papyri of the Roman and Byzantine periods* I, Milan: Cisalpino- La Goliardica 1976, 325–26.

49 Literature: see the references in G.M. Browne, "An old Nubian version of Ps.-Chrysostom, In quattuor Animalia", *Altorientalische Forschungen* 15 (1988), 215–19; add: G.M. Browne, *Old Nubian texts from Qaṣr Ibrîm* II, London: Egypt Exploration Society 1989, 22–5, no. 16; the Pierpont Morgan Coptic version of the Ps.-Chrysostom homily has meanwhile been edited by C.S. Wansink, in: L. Depuydt, *Homiletica from the Pierpont Morgan Library: seven Coptic homilies*, Louvain: Peeters 1991, 27–46. Art: see P.P.V. van Moorsel, art. "Christ, Triumph of", in: *CoptEnc* 2, 525–26, where further literature is mentioned.

50 Cf. S. Donadoni, "Les graffiti de l'église de Sonqi Tino", in: Michałowski (ed.), *Nubia: récentes recherches*, 31–9 at 35 (several mentions).

51 Cf. the Qaṣr Ibrîm stela of Bishop Marianos (ed. Kubińska, *Inscriptions grecques*, 38–40, no. 9, fig. 9), ll. 8–9[= *I. Qasr Ibrim* 22, ll. 11–12]: τέσσαρα ζῷα νῆσος Τημε. J.M. Plumley's interpretation of the name Τημε ("The stele of Marianos bishop of Faras", *Bulletin du Musée national de Varsovie* 12 (1971), 77–84 at 83–4) is certainly erroneous.

52 Jakobielski, *Bishopric of Pachoras*, 125, n. 106. The *ed. princeps* and the archaeological report for the season 1961–62 do not record its provenance.

53 According to Jakobielski, *Bishopric of Pachoras*, 189, no. 14, on the authority of the Coptic stela of Philotheos, which was found there (cf. 186–88).

25

TWO COPTIC EPITAPHS FROM QASR IBRIM*

Jacques van der Vliet

Introduction

A first edition and brief discussion of two Coptic funerary stelae from the medieval cemeteries of Qasr Ibrim in Northern Nubia (ancient Nobadia).

1. Funerary stela of a woman called Martha

This modest stela is kept in the Rijksmuseum van Oudheden (National Museum of Antiquities), Leiden, under the inventory number F 1985/4.2 (Figure 25). It was discovered on 17 March 1966, during the EES excavations at Qasr Ibrim, directed by J.M. Plumley, and donated by the EES to the Leiden Museum in 1985.[1] The register number, QI 66 A/107, is still visible on the object. It was found in Tomb 2, the second of the rock cut "bishop's tombs" south of the Cathedral.[2]

The stela is a small block of reddish sandstone, rudely hewn in an irregular trapezoid form, narrower towards the bottom. Its largest measurements are 22 × 22 × 6.5 cm. The backside is of irregular shape, whereas the sides are roughly dressed. The object is essentially complete, in spite of some marginal chips that cause minor loss of text in l. 1; some surface wear owing to secondary use as a tile or a quern has affected the text in the middle of ll. 3–7.

The stone is inscribed with seven lines of text in Sahidic Coptic, written in irregular incised uncials and not well aligned. The script is characterized by a low three-stroke м, a left-looking ⲁ and regularly placed superlinear strokes. The letters have preserved remains of plaster coating and dark paint. Decoration consists of simple crosses (l. 7). The lower part of the stone is uninscribed.

Unpublished. The stone will be included in the publication of the Greek and Coptic inscriptions from the EES-excavations in Qasr Ibrim, prepared by Adam Łajtar and the present author [*I. Qasr Ibrim* 29].

Qasr Ibrim, *ca.* eighth century?

[+] ⲡ̄ⲣⲡⲙⲉⲩⲉ ⲛ̄ⲧⲙⲁⲕⲁ-
ⲣⲓⲁ ⲙⲁⲣⲑⲁ \<ⲡⲉϩⲟⲟⲩ\> ⲛ̄ⲧⲁⲥ̄ⲛ-

317

ⲧⲟⲛ ⲙ̄ⲙ[ⲟⲥ] ⲛ̄ϩⲏⲧϥ̄
4. ⲥⲟⲩ ⲙⲏⲧ [ⲛⲉ]ϣⲑ ϩⲛ
ⲟⲩⲉⲓⲣⲓⲛⲏ ⲛ̄ⲧⲉ
ⲡⲛⲟⲩⲧ[ⲉ] ⲁⲁ-
ⲙⲏⲛ +++

1. Lacuna: no room for ϩⲁ | - ⲙⲉⲩⲉ: l. - ⲙⲉⲉⲩⲉ ‖ 2. ⲡⲉϩⲟⲟⲩ: demanded by following relative clause ‖ 6–7. ⲁⲁⲙⲏⲛ: l. ϩⲁⲙⲏⲛ or ⲁⲙⲏⲛ.

+ The commemoration of the blessed Martha, (the day) on which she went to rest: the tenth of Thoth. In the peace of God. Amen. + + +

The simple opening formula of this epitaph, characterized by the phrase ⲡⲣⲡⲙⲉⲉⲩⲉ ⲛ-/ⲙ-: "the commemoration of (the blessed N.N.)", serves to introduce the name of the deceased and the date of her death, information indispensable for her periodical commemoration. No prayer follows but a mere *pax*-formula. With minor variations and also in Greek forms, this formulary is frequently found in Qasr Ibrim itself as well as elsewhere in Northern Nubia and southernmost Egypt (Aswan).[3] A nearly identical Coptic example from Qasr Ibrim is the stela of a man

Figure 25.1 Coptic stela of Martha, Qasr Ibrim
(photo: © National Museum of Antiquities, Leiden)

named Stephanos, from the South Cemetery.[4] The language of the present stela shows hardly any irregularities; -ⲙⲉⲩⲉ for -ⲙⲉⲉⲩⲉ (l. 1) is occasionally found elsewhere in Lower Nubia and Upper Egypt,[5] whereas the rare form ⲁⲁⲙⲏⲛ (ll. 6–7), if not a scribal error, may plausibly be explained as an instance of word-initial vowel-doubling, for which see no. 2, below, l. 6 (ⲁⲁⲁⲁⲙ for ⲁⲁⲁⲙ).

Epitaphs of this type usually do not bear an absolute date. In M. Krause's classification, based upon the material from Sakinya (Toshka West), the present text would correspond to formulary no. 1, in its second redaction.[6] Accordingly, the Leiden stela may be dated tentatively to the eighth century.[7] The monument was found in one of the "bishop's tombs" in the Cathedral area, apparently in association with eleventh-twelfth century episcopal epitaphs. However, its modest character and early date as well as traces of secondary use make it less likely that this was its original context.[8] Presumably, like the bulk of similar Nubian stelae, it originally adorned a free-standing tomb that may have been situated on one of the extensive Christian cemeteries on the outskirts of the town. At some later time it may have been re-used for building works on the citadel.

2. Funerary stela of Patarmoute, priest of the Catholic Church

The present location of the second stela is unknown. Also no information is available about material, measurements, script or find circumstances (South Cemetery? – see below). From comparison with other Nubian stelae it may be inferred that it is an oblong sandstone slab. It is inscribed with twenty lines of text in Sahidic Coptic and Greek, following what is presumably a decorative headline. The beginning and end of the text are marked by crosses.

The text is reproduced here after a handwritten transcription by Reginald Engelbach, now in the Griffith Institute, Oxford, with some pencil annotations by W.E. Crum.[9]

Unpublished.

Qasr Ibrim, *ca.* ninth-tenth century?

[top line]
+ ⲓ(ⲏⲥⲟⲩ)ⲥ ⲭ(ⲣⲓⲥⲧⲟ)ⲥ ⲛⲁ ⲛⲁⲛ
ⲅⲓⲧⲛ <ⲧⲉ>ⲡⲣⲟⲛⲓⲁ ⲙⲡⲛⲟⲩⲧⲉ
ⲡⲁⲛⲙⲏⲟⲩⲣⲅⲟⲥ
ⲉⲡⲧⲏⲣϥ̄ ⲕⲁⲧⲁ
4. ⲡϣⲁⲭⲉ ⲛⲧⲁ[ϥ-]
ⲭⲟⲟⲥ ⲙⲡⲉⲛⲡⲣⲟⲡ-
ⲁⲧⲱⲣ ⲁⲁⲁⲁⲙ ⲭⲉ
ⲭⲡⲟ ⲅⲁⲣ ⲛⲧⲥⲓⲙⲉ
8. ⲛϣⲁⲣⲁⲅⲉ ⲛⲉ ⲗⲟⲓ-
ⲡⲟⲛ ⲉϥⲕⲏ ⲉⲅⲣⲁⲓ

319

JACQUES VAN DER VLIET

ⲙⲡⲉⲓⲙⲁ ⲛϭⲓ ⲡⲗⲓ-
ⲯⲁⲛⲟⲛ ⲙⲡⲙⲁⲕⲁⲣ-
12. ⲣⲓⲟⲥ ⲡⲁⲧⲁⲣⲙⲟⲩⲧⲉ
ⲡⲉⲡⲣⲉⲥⲃⲩⲧⲉ-
ⲣⲟⲥ <ⲛ>ⲧⲕⲁ̄ⲑ̄ ⲉⲕⲕⲗⲏⲥⲓ[ⲁ]
ⲡⲉϩⲟⲟⲩ ⲛⲧⲁϥⲙⲧⲁⲛ
16. ⲙⲙⲟϥ ⲛ̄ϩⲏⲧϥ ⲙⲡⲉ-
ⲃⲟⲧ ⲡⲁⲱⲛⲉ ⲕ̄ⲏ̄
ὁ θ(εό)ς, ἀνάπαυσεν τ-
ὴν] ψυχὴ(ν). ϩⲁⲙⲏⲛ
20. ϥ̄ⲑ̄ +

Top line. ⲛⲁ ⲛⲁⲛ: sic? – 1. ⲛⲓⲕⲁ? || 1. ϩⲓⲧⲛ <ⲧⲉ>ⲡⲣⲟⲛⲓⲁ: pseudo-haplography || 4. ⲉⲡⲧⲏⲣϥ̄: 1. ⲙⲡⲧⲏⲣϥ̄ || 5. ⲙⲡⲉⲛ-: Engelbach reads ϫⲉⲛ and notes: "3 blundered letters", here restored to the expected four-letter group; perhaps the lapicide mechanically continued ϫⲟⲟⲥ with ϫⲉ and tried to correct his error afterwards || 6. ⲁⲁⲁⲙ: 1. ⲁⲁⲁⲙ || 6–8. ϫⲉ ϫⲡⲟ ⲅⲁⲣ ⲛⲧⲥⲓⲙⲉ ⲛ̄ϣⲁⲣⲁϩⲉ ⲛⲉ: 1. ⲛⲉϫⲡⲟ ⲅⲁⲣ ⲛⲧⲥⲓⲙⲉ <ϩⲣⲉ>ⲛ̄ϣⲁⲣⲁϩⲉ ⲛⲉ (? – see below); ⲛⲧⲥⲓⲙⲉ: properly ⲛⲧⲉⲥϩⲓⲙⲉ || 8–9. ⲗⲟⲓⲡⲟⲛ: Engelbach ⲛ̄ⲣ̄|ⲡⲟⲛ, corr. in margin ⲗⲟⲛ-? ⲗⲟⲓ-? || 10–11. ⲡⲗⲓⲯⲁⲛⲟⲛ: Crum (in margin); Engelbach ⲡⲏ|ⲕⲁⲛⲟⲛ || 15. -ⲙⲧⲁⲛ: properly -ⲙⲧⲟⲛ || 18–19. ὁ θ(εό)ς, ἀνάπαυσεν τ[ὴν] ψυχὴ(ν): Engelbach ⲟ ⲟ̄ⲥ̄ ⲁⲛⲁⲡⲁⲩⲥⲉ ⲛⲧ⳿[ϥ]ⲯ̄ⲩⲭⲏ; ἀνάπαυσεν: sic? – 1. ἀνάπαυσον || 20. ϥ̄ⲑ̄: isopsephic spelling ("99") of Amen.

+ Jesus Christ, have mercy upon us (?).[10]
Through the providence of God, the creator of the universe, in accordance with the word which He had spoken to our forefather Adam, for those born of a woman are short-lived, now then here lie the mortal remains of the blessed Patarmoute, the priest of the Catholic church.
The day on which he went to rest: in the month Paone, 28.
[Greek] O God, rest the soul (of your servant).
Amen, amen. +

The text as presented above is the slightly retouched version of a handcopy made by the British Egyptologist Reginald Engelbach (1888–1946). Engelbach was Chief Inspector for Upper Egypt of the Egyptian Antiquities Service from 1920 until 1924 and served the Egyptian Museum in Cairo in various capacities from 1924 until 1946.[11] During these years he published many recent discoveries from all periods of Egyptian history, mainly in the *Annales* of the Service. He may have transcribed the text of this stela with a view to a similar publication, which never appeared. Engelbach submitted his transcription to W.E. Crum and this copy is presently kept among the Crum papers in the Griffith Institute. It bears the laconic heading "2. Stela from Qaṣr Ibrîm", without further details, and was joined to the transcription of another Coptic inscription, the epitaph of a priest Mena, stated by Engelbach to be found at Sakinya and now in the Coptic Museum in Old Cairo.[12] This circumstance suggests that the present stela, like the well-known lot of Sakinya inscriptions, had come to light during the second archaeological campaign in Nubia, in the early 1930s. Accordingly, one would expect the stela

320

TWO COPTIC EPITAPHS FROM QASR IBRIM

to have been deposited in the Egyptian Museum in Cairo and subsequently transferred to the Coptic Museum, as was the case with the stela of the priest Mena.[13] Apparently, however, it is not among the collections of the latter museum, nor was it possible to trace it elsewhere.

The opening lines of the epitaph of Patarmoute as transcribed by Engelbach show a strong similarity to a stela discovered in the early 1930s by W.B. Emery at the South Cemetery of Qasr Ibrim, and partially copied by U. Monneret de Villard in his compendious *La Nubia medioevale* of 1935. Actually, the published text of the Emery stela, about which otherwise nothing is known, is practically identical with that of the Engelbach stela and one might suspect them to be one and the same stela, if it were not for some minor but distinctive differences of orthography. Monneret de Villard was able to read only the first few lines of this stela, which he qualified as "quasi illeggibile", and reproduced them as follows:[14]

[top line]   + ι(нсоγ)с х(ρісто)с ніка
ᒿітє пронна мп[ноүтє]
плнмноүргос
єптнрϥ [. . . . .]

Although the similarities in wording and layout are striking, it cannot be concluded on the basis of this evidence that the two inscriptions are identical, unless their editors are to be blamed for gross errors of transcription. Even so the likelihood remains that they share a common date and provenance. The Engelbach stela, too, might well originate from the South Cemetery, Emery's Cemetery 193, surveyed in the early 1930s by Emery[15] and Monneret de Villard[16] and again in 1961 by Emery.[17]

The Coptic of this, actually bilingual, stela is a fairly regular Sahidic with only a few peculiarities that are common in informal Sahidic throughout the Nile Valley: reduction of prepositional ন-/м- (resulting in the pseudo-haplography ᒿітн <тє>пронıа in l. 1[18] and the spelling є- for м- in l. 3),[19] сıмє for сᒿıмє (in l. 7)[20] and /a/ for /o/ in a closed syllable (-мтаν in l. 15).[21] The doubling of the word-initial vowel in l. 6 (ааλам) occurs not infrequently in both Nubian and Southern Egyptian Sahidic,[22] and can perhaps be found in no. 1, above, as well (ll. 6–7: аамнν). Also the insertion of a brief prayer in Greek, in ll. 18–19, is not rare in epitaphs from Nubia. It is usually found, as it is here, towards the end of the text, following the date of decease.[23] Engelbach took this prayer for Coptic, but it is certainly Greek. The drop of the final -ν in ψυχῆ(ν) is a common feature of Nubian Greek.[24]

For the acclamation in top of the stela, Engelbach's spurious readings (на наν) have been retained in the text above, but it is in fact far more probable that ι(нсоγ)с х(ρісто)с ніка should be read, as in the Emery stela partly transcribed by Monneret de Villard.[25]

Following the commonplace opening formula, which recalls the providence and omnipotence of God and the prototypical fate of Adam (ll. 1–6),[26] the text

321

itself exhibits several unusual features. The most remarkable of these is without any doubt that, instead of the standard quotation from Gen. 3:19 ("for dust thou art . . ."), which normally should have followed the reference to Adam in l. 6, a version of Job 14:1 ("for those born of a woman are short-lived") is quoted (ll. 7–8). I know of no parallels for this either in Nubian or in Egyptian epigraphy. The text as it stands (with ⲭⲉ, in l. 6, which normally serves to introduce the Genesis quote) could create the impression that the scribe is falsifying Holy Scripture. Probably, however, he is merely combining stock phrases of liturgical origin.[27] This is confirmed by a late ninth-century Ethiopian funerary stela which quotes the same Job verse in an obviously liturgically inspired context.[28] As for the textual form of the quote (which seems slightly garbled, see the apparatus above), it may be noted that the Sahidic Bible (following the standard LXX text of Job) has, instead of the plural which is apparently intended here, a generic singular: ⲡⲣⲱⲙⲉ ⲅⲁⲣ ⲛ̄ⲭⲡⲟ ⲛ̄ⲥ̄ϩⲓⲙⲉ ⲟⲩϣⲁⲣⲁϩⲉ ⲡⲉ (Ciasca), "for man, born of a woman, is short-lived".

Furthermore, the continuation of the text, introducing the deceased, is not a standard one in Nubia. The transition from the biblical prototype to the individual decease is made not by a common expression like ⲛⲧⲉⲓϩⲉ ⲇⲉ . . ., "thus then . . .", but by ⲗⲟⲓⲡⲟⲛ, literally "further" (ll. 8–9). This is rather an epistolographic feature, introducing a (new or first) subject in letters.[29]

The following statement of death: "here lie the mortal remains (λείψανον)[30] of the blessed N.N." (ll. 9–12), is not entirely standard either. In Lower Nubia, an analogous Coptic formula occurs in a stela from Sakinya (Toshka West): ⲉⲥⲟⲩⲏϩ (for ⲉϥⲟⲩⲏϩ) ϩⲙ ⲡⲓⲙⲁ ⲛⲟ̄ⲓ ⲡⲥⲱⲙⲁ ⲛⲧⲙⲁⲕⲁⲣⲓⲁ N.N., "here dwells the body (σῶμα) of the blessed N.N.".[31] Even more closely similar phrases are used in the mid-eighth-century stela of an abbess, of unknown provenance (possibly Middle Egypt): ⲉⲣⲉ ⲡⲗⲓ†ⲁⲛⲟⲛ ⲙ̄ⲡⲥⲱⲙⲁ ⲛ̄ⲧⲉⲛⲙⲁⲕⲁⲣⲓⲁ ⲙ̄ⲙⲁⲁⲩ N.N. ⲕⲏ ⲉϩⲣⲁⲓ ⲙ̄ⲡⲓⲙⲁ, "here lie the mortal remains (λείψανον) of the body (σῶμα) of our blessed mother N.N.",[32] and in a small group of tenth-century funerary stelae from the region of Abydos: ⲉϥⲕⲏ ⲉϩⲣⲁⲓ ⲙ̄ⲡⲉⲓⲙⲁ ⲛⲟ̄ⲓ ⲡⲉⲥⲕⲏⲛⲱⲙⲁ ⲙ̄[. . .] ⲡⲙⲁⲕⲁⲣⲓⲟⲥ N.N., "here lies the mortal dwelling (σκήνωμα) of . . . the blessed N.N.".[33] All look like variants of the familiar Greek formula ἔνθα κατάκειται ὁ μακάριος N.N., "here lies the blessed N.N.", which in a slightly earlier period was popular in northernmost Nubia, the former Dodekaschoinos.[34] In fact, quite recently the exact Greek equivalent of the present formula, incorporating precisely the term λείψανον, has come to light in Old Dongola.[35]

The owner of the stela bears the Egyptian name Patarmoute (properly, Patermoute), which was extremely popular in the Theban region, where also several saints of this name were venerated.[36] This is, to the best of my knowledge, its first attestation in Nubia. He was not a commoner, but a priest of the Catholic Church (ll. 13–14), that is the episcopal church, the Cathedral of Qasr Ibrim.[37] His status as a minister to the principal church of the town may explain the exceptional form of his epitaph, with its quote from Job 14. It apparently did not entitle him to be

322

buried in the Cathedral area itself, as were the bishops of Qasr Ibrim whose tombs came to light during the EES excavations of the citadel in the 1960s. Judging from the text of his epitaph alone, in particular its developed liturgical formulary, the priest Patarmoute may have served the Cathedral in or about the ninth or tenth century.

Concluding remarks

The two funerary stelae published here are part of the rich but sadly scattered epigraphic record of Qasr Ibrim, one of the principal political, military and ecclesiastical centers of Christian Nubia. In spite of uncertainty over their find circumstances, it can be assumed that both pieces were meant to decorate considerable tombs, situated in cemeteries outside the town proper. The first stela is that of a woman called Martha, whose profession is not mentioned, the other one of a cleric named Patarmoute, who was a priest of the Cathedral. Both must have belonged to the literate upper strata of eighth- to tenth-century Qasr Ibrim society, where for formal and monumental writing purposes Coptic was used in addition to Greek. Peculiarities of language and onomastics (the name Patarmoute) give the impression, conveyed by other sources as well, of a shared Southern Egyptian–Northern Nubian literary culture. The older epitaph of Martha is short and very simple and follows patterns widespread in the entire region. The more considerable length of Patarmoute's epitaph, its bilingualism, and its unusual, liturgically inspired text were undoubtedly meant to reflect the status and erudition of its owner. Carefully read, each in its own way illustrates the culture of the early medieval Nobadian elites.

Notes

*   I wish to thank Dr. M.J. Raven, Keeper of the Egyptian Department of the National Museum of Antiquities in Leiden, and the staff of the Griffith Institute, Oxford, for their assistance and for authorizing me to publish the texts here. I am also much obliged to Mrs. Sofia Schaten, Münster; Dr. A. Lajtar, Warsaw; Dr. T.S. Richter, Leipzig; and Dr. Pamela J. Rose, Cambridge, for providing various information. Dr. Richter kindly put at my disposal his useful list of published Coptic and Greek inscriptions from Qasr Ibrim, due to appear in the forthcoming catalogue of the Leipzig Coptic collections [Suzana Hodak, T.S. Richter, and F. Steinmann, *Coptica: Koptische Ostraka und Papyri, koptische und griechische Grabstelen aus Ägypten und Nubien, spätantike Bauplastik, Textilien und Keramik*, Berlin: Manetho Verlag 2013].

1   See H.D. Schneider, "Rijksmuseum van Oudheden", *Nederlandse Rijksmusea* 107 (1985), 253–67 at 261.

2   See J.M. Plumley, "Qasr Ibrîm 1966", *JEA* 52 (1966), 9–12 at 11 with pl. V, no. 2 (the tomb in question is the second from the left); the present stela is not mentioned. For these tombs, cf. W.Y. Adams, *Qaṣr Ibrîm: The late mediaeval period*, London: Egypt Exploration Society 1996, 82–3.

3   See H. Junker, "Die christlichen Grabsteine Nubiens", *ZÄS* 60 (1925), 111–48 at 131; M. Krause, "Die Formulare der christlichen Grabsteine Nubiens", in K. Michałowski

(ed.), *Nubia: récentes recherches*, Warsaw: Musée national 1975, 76–82, esp. 78–9 (Sakinya).

4   See U. Monneret de Villard, *La Nubia medioevale* I, Cairo: Mission archéologique de Nubie 1935, 112–15; cf. idem, "Rapporto preliminare dei lavori della missione per lo studio dei monumenti cristiani della Nubia, 1930–1931", *ASAE* 31 (1931), 7–18 at 10.

5   Nubia: T. Mina, *Inscriptions coptes et grecques de Nubie*, Cairo: Société d'archéologie copte 1942, nos. 70–2 and 182 (Sakinya); Upper Egypt: P.E. Kahle, *Bala'izah: Coptic texts from Deir el-Bala'izah in Upper Egypt*, London: Oxford University Press 1954, I, 67, sub f.

6   See Krause, "Formulare", 78–9.

7   Ibidem, 79–80.

8   Little is known about the precise find circumstances of the stela within Tomb 2. In any case, the entire context seemed much disturbed; cf. J.M. Plumley, "Qasr Ibrim and the Islam", *ÉtudTrav* 12 (1983), 157–70 at 163.

9   Crum MSS VII 6 (f). Mentioned in J. Malek, and D.N.E. Magee, "Nubian and Meroitic material in the archives of the Griffith Institute, Oxford", *BSF* 3 (1988), 49–55 at 53.

10   Or perhaps: "+ Jesus Christ prevails (νικᾷ)!" See the discussion below.

11   See W.R. Dawson et al., *Who was who in egyptology*, London: Egypt Exploration Society 1995, 141–42.

12   "1 Stela from Sakiniya (Nubia)". This is Mina, *Inscriptions*, no. 89 (Coptic Museum, no. 6744), which, actually, was found not in Sakinya, but in al-Ramal; cf. Monneret de Villard, *Nubia medioevale* I, 135 (no. II). A note by Crum ("omitted by Monneret") could imply that he received Engelbach's copy between the publication of Monneret de Villard's unsatisfactory first edition of the Sakinya stelae in *Le iscrizioni del cimitero di Sakinya (Nubia)* (1933) and the stela's publication in *La Nubia medioevale* (1935).

13   Note, however, that the Sakinya lot was divided between several museums, including the Musée gréco-romain in Alexandria and the Museo delle Terme in Rome; see Mina, *Inscriptions*, II and 147, and S. Pernigotti, "Stele cristiane da Sakinya nel Museo di Torino", *Oriens antiquus* 14 (1975), 21–55 at 22.

14   Monneret de Villard, *Nubia medioevale* I, 115, lowermost (lines numbered differently).

15   See W.B. Emery, and L.P. Kirwan, *The excavation and survey between Wadi es-Sebua and Adindan, 1929–1931*, Cairo: Government press 1935, 268–77 (where only pre-Christian material is dealt with).

16   Monneret de Villard, *Nubia* I, 112–15; idem, "Rapporto preliminare", 9–10. Cf. W.B. Emery, *Egypt in Nubia*, London 1965, 55–6.

17   A.J. Mills, *The cemeteries of Qaṣr Ibrîm: a report of the excavations conducted by W.B. Emery in 1961*, London: Egypt Exploration Society 1982, 47–67.

18   Almost similarly in the stela Monneret de Villard, *Nubia* I, 115, lowermost (quoted above), where in l. 1: ⲣⲓ<ⲧⲛ> ⲧⲉⲡⲣⲟⲛⲛⲁ should be read; -ⲧⲛ and ⲧⲉ- must have sounded nearly identical. In Nubia, ⲡⲣⲟⲛⲓⲁ, is a normal spelling in both Greek and Coptic, see *I. Khartoum Copt.* 4, l. 1; for ⲡⲣⲟⲛⲛⲁ (Monneret de Villard), see Mina, *Inscriptions*, no. 89, ll. 1–2.

19   Correct in the stela Monneret de Villard, quoted above, l. 3.

20   A frequent spelling; see Kahle, *Bala'izah*, 128, sub 108.

21   See Kahle, *Bala'izah*, 80–1; -ⲙⲧⲁⲛ for -ⲙⲧⲟⲛ in another Nubian stela (Sakinya): Pernigotti, "Stele", no. 19, l. 9 (cf. Tav. XV/2 = Mina, *Inscriptions*, no. 311).

22   See *I. Khartoum Copt.*, 23, n. 113, ad no. 3, l. 13.

23   Representative examples (also for the prayer's laconic form): Mina, *Inscriptions*, nos. 63, 88 and 92b (Sakinya); *I. Khartoum Copt.* 18 and 19 (Meinarti).

24   *I. Khartoum Greek*, 257.

TWO COPTIC EPITAPHS FROM QASR IBRIM

25   For another Nubian example, see *I. Khartoum Copt.* 31, with further references.

26   Cf. *I. Khartoum Copt.*, 26–7. For the epithet προπάτωρ, see ibidem, no. 27, l. 5, with note 392.

27   The ⲭⲉ in l. 6, if not added entirely mechanically, could be an error for the article ⲛⲉ, which correct grammar demands before ⲭⲡⲟ (see the apparatus ad ll. 6–8).

28   See M. Kropp, " 'Glücklich, wer vom Weib geboren, dessen Tage doch kurzbemessen, . . .!' Die altäthiopische Grabinschrift von Ham, datiert auf den 23. Dezember 873 n. Chr.", *Oriens christianus* 83 (1999), 162–76 at 170–73. Other verses from Job 14 are echoed e.g. in the Coptic "Totenklage"-epitaphs (Middle Egypt, eighth-ninth century); see M. Cramer, *Die Totenklage bei den Kopten*, Vienna/Leipzig: Hölder-Pichler-Tempsky 1941, 101.

29   See H. Förster, *Wörterbuch der griechischen Wörter in den koptischen dokumentarischen Texten*, Berlin/New York: de Gruyter 2002, 482, s.v.; cf. A. Biedenkopf-Ziehner, *Untersuchungen zum koptischen Briefformular unter Berücksichtigung ägyptischer und griechischer Parallelen*, Würzburg: G. Zauzich 1983, 32.

30   Cf. Förster, *Wörterbuch*, 467–68, s.v.

31   Mina, *Inscriptions*, no. 285, ll. 1–5.

32   Alexandria, Musée gréco-romain, A 11751, ll. 1–9; ed. W. Brunsch, "Koptische und griechische Inschriften aus Alexandria", *WZKM* 84 (1994), 9–33 at 18 (with ill.). The piece is dated to A.D. 742.

33   *SB Kopt.* I 486, 745, and 746; dates ranging from A.D. 932 to 946.

34   Cf. Junker, "Grabsteine", 126–27; J. van der Vliet, "Gleanings from Christian Northern Nubia", *JJP* 32 (2002), 175–94 at 180–83 [Study 22]. The provenance of the single stela with this formula attributed to Qasr Ibrim (Leipzig, Inv.-Nr. 680) is now disputed; see T.S. Richter, "Die neun Stelen Ägyptisches Museum der Universität Leipzig Inv.-Nr. 680–8 mit der Herkunftsangabe Qasr Ibrîm", S. Emmel et al. (eds.), *Ägypten und Nubien in spätantiker und christlicher Zeit: Akten des 6. Internationalen Koptologenkongresses, Münster, 20.-26. Juli 1996*, Wiesbaden: Reichert Verlag 1999, II, 295–304 at 296–97; Van der Vliet, "Gleanings", 181.

35   Bilingual (Greek-Coptic) inscription commemorating Bishop Joseph of Aswan, who died in Dongola in A.D. 668; see, provisionally, D. Gazda, "Monastery church on kom H in old Dongola: third and fourth season of excavations (2004, 2004/5)", *PAM* 16 (2005), 285–95 at 292–93 [see now Study 29].

36   See A. Papaconstantinou, *Le culte des saints en Égypte des Byzantins aux Abbasides: l'apport des inscriptions et des papyrus grecs et coptes*, Paris: Éditions du CNRS 2001, 168–70.

37   For this interpretation, see E. Wipszycka, "Καθολική et les autres épithètes qualifiant le nom ἐκκλησία: contribution à l'étude de l'ordre hiérarchique des églises dans l'Égypte byzantine", *JJP* 24 (1994), 191–212 at 202–12; reprinted in: eadem, *Études sur le christianisme dans l'Égypte de l'antiquité tardive*, Rome: Institutum Patristicum Augustinianum 1996, 157–75. For the Cathedral of Qasr Ibrîm, see Adams, *Qasr Ibrîm*, 73–8; cf. for its denomination, J. van der Vliet, "Churches in Lower Nubia, old and 'new'", *BSAC* 38 (1999), 135–42 at 135–38 [Study 24].

26

THE CHURCH OF THE TWELVE APOSTLES
The earliest Cathedral of Faras

Jacques van der Vliet

Introduction

Twenty-five years after the publication of Stefan Jakobielski's synthetic monograph on the "history of the bishopric of Pachoras",[1] the preparation of a catalogue of the Coptic inscriptions in the National Museum of Khartoum enabled me to inspect in person some of the most important epigraphic sources for the history of Christian Faras.[2] Foremost among them is the long Coptic building inscription of Bishop Paulos. My re-examination of this text led to the revision of some current interpretations not only of the text, but also of part of the history of the main ecclesiastical complex of Faras, viz. the Cathedral and its immediate vicinity. The following observations are a modest contribution to the *topographia sacra* of a drowned country. Naturally, they represent the epigraphist's point of view, not that of the archaeologist. The reader will, moreover, appreciate the debt which I owe, in spite of occasional disagreement, to Jakobielski's pioneering study.

The monumental building inscription of Bishop Paulos, dated 423 of the Diocletian era (A.D. 707), is certainly the longest and most ambitious of all surviving Coptic inscriptions from Nubia.[3] This text and its more concise Greek counterpart of the same year[4] are still widely believed to be documents commemorating the expansion and renovation of the Cathedral of Faras, famous for its splendid wall paintings.[5] However, against this point of view, fervently advocated by the text's excavator, K. Michałowski, and by its first editor, Jakobielski, several objections are possible. These concern both the internal (textual) and the external (contextual) evidence. I will first discuss the external, then the internal evidence.

Under the sign of the cross

Both the Coptic and the Greek building inscriptions of Bishop Paulos were found *in situ*. They did not adorn the body of the Cathedral or any of its entrances, but the outer northwestern corner of a largely destroyed edifice to the south of it.[6] Since the two stones appeared to be perfectly adapted, both technically and esthetically, to the walls in which they were inserted, this was most probably their original position.[7] The nature and function of this latter building has proved difficult to define from its archaeological remains. It has been described at first as a monastery,[8] then as "a residence for bishops",[9] or "Bishops' Palace".[10] Its identification with a monastery of the Virgin Mary is a mere guess.[11]

Whatever its exact purpose may have been, this building clearly represents a unit distinct from the Cathedral. It is physically separated from it by a small triangular "square" which is partly filled in by an episcopal mortuary complex (from the eighth-ninth century, so postdating Bishop Paulos)[12] and an unidentified structure at the back of it.[13] Moreover, its orientation (SSE-NNW) is markedly different from that of the Cathedral (SE-NW). Its excavator, K. Michałowski, acutely observed that by its deviating axis it is orientated towards the monumental cross-post situated at its northwestern corner.[14] Thus, it appears to be associated not with the Cathedral, but with the pedestal of the wooden cross which marked the conversion of Faras to Christianity.[15] Precisely in the northwestern corner of the building, next to the cross-post, Paulos' two inscriptions were situated.

It is, of course, conceivable that this southern building formed, from a certain period, an integrated part of the variegated buildings which made up the Cathedral precinct or parvis.[16] It is, furthermore, equally well conceivable that both buildings, the Cathedral and the southern building adjacent to the cross-post, were constructed or – rather better – reconstructed in more or less the same period, as has been argued from an architectural point of view.[17] Still, even if these theories could be verified, further evidence would be needed in order to designate Paulos' commemorative texts, found as part of the walls of a building outside the actual Cathedral church, as the foundation inscriptions of this Cathedral church itself. Not only is such evidence lacking, but there exists, in my opinion, evidence to the contrary.

Before we turn to the textual evidence, another remarkable archaeological fact deserves to be noticed. The Coptic text of A.D. 707 and its Greek counterpart of the same year are not the only surviving inscriptions to commemorate the building activities of Bishop Paulos. Other epigraphic evidence of his architectural interests includes: (1) an inscribed lintel, presently in Warsaw;[18] (2) an (according to report) identical lintel, in a British private collection;[19] (3) a keystone with Paulos' monogram.[20]

The first of these pieces, the Warsaw lintel, bears a Greek building inscription which commemorates how this "good work" ($\dot{\alpha}\gamma\alpha\theta\dot{o}\nu$ $\ddot{\epsilon}\rho\gamma o\nu$), which usually designates an architectonic achievement of some kind, was accomplished ($\dot{\epsilon}\gamma\acute{\epsilon}\nu\epsilon\tau o$), at a certain date (which is lost), under Bishop Paulos.[21] It was discovered, partly damaged, as a re-used element, at the eastern side of the Cathedral.[22] Even in its original, complete state, however, this lintel must have been, according to Jakobielski's estimate, too short to fit above any of the doorways of the Cathedral. Instead, the same author plausibly suggests that it may originally have come from his "Bishop's Palace", the building to the south of the Cathedral.[23]

The second piece, a fragmentary lintel resembling the former, bears Bishop Paulos' name and the same date as his big foundation inscriptions, viz. 423 of the

THE CHURCH OF THE TWELVE APOSTLES

Diocletian era (A.D. 707).[24] Its dimensions, too, are reported to be similar to those of the Warsaw lintel. Since it was apparently acquired long before the discovery of the Cathedral, it cannot originate from there. Again, Jakobielski argued convincingly that it might have been found on the much disturbed south slope of the Faras Kom, the area south of the Cathedral.

The origin of our third piece, a keystone with Paulos' monogram, is well established.[25] It was found during the Polish excavations of the 1962–63 season, on the south slope of the Kom.

To these, altogether five, monuments from the episcopacy of Paulos, a sixth can be added on purely formal grounds. Another lintel from Faras carrying a building inscription presents a striking resemblance to the first of our Paulos lintels.[26] Both are sandstone lintels, decorated with a central raised panel representing a sculptured cross, surrounded by vine leaves; above the panel, a cornice runs over the entire length of the lintel. The cornice and the undecorated surface at both sides of the central panel carry a building inscription. The general layout of the lintels and especially their almost identical sculptured decoration suggest a common workshop and a common patron. The Khartoum lintel, however, does not mention Bishop Paulos: it bears the Coptic inscription of an Eparch Iesou who flourished some two centuries later, since the text is dated to 646 of the Diocletian era (A.D. 930). This leaves the possibility, as Jakobielski was the first to notice, that Iesou re-used a far older lintel and even that his inscription replaces an older one.[27] A closer comparison of both stones permits us to be more precise.

Formally, two striking differences can be observed: firstly, on the Paulos lintel the cornice protrudes about 10 mm over the panel with the cross,[28] whereas on the Iesou lintel it is flat and marked off from the panel by an incised horizontal line only, not by any difference in surface relief; secondly, a deep groove, running immediately under the cornice, is a conspicuous feature of the Iesou lintel, but entirely absent from the Paulos lintel. Apart from a monogram to the left of the central cross, the Paulos lintel bears two lines of text only: one on the cornice and a second one immediately under it. The particularities which distinguish the Iesou lintel from that of Paulos, viz. the flattened cornice and the curious groove running immediately under it, would be exactly accounted for by the intentional and careful removal of an earlier text consisting of two identically situated lines.[29] If the distinctive features of the Iesou lintel are indeed due to such a treatment, as seems probable, then both lintels must have been, in their original state, near twins.

An additional proof of their intimate relationship is perhaps furnished by a textual similarity. Both building inscriptions, the Coptic one of Iesou and the Greek one of Paulos, use the same formula: "this good work (scil. the present construction) was accomplished (under/by)", followed by the name of the builder (Paulos [*I. Varsovie* 101], l. 1: τοῦτο τὸ ἀγαθὸν ἔργον ἐγένετο ἐπ[ὶ; Iesou [*I. Khartoum Copt.* 2], ll. 3–3*: ⲛⲧⲁ ⲡⲉⲓϩⲱⲃ ⲛⲁⲅⲁⲑⲟⲛ ϣⲱⲡⲉ ⲉⲃⲟⲗ ϩ[ⲓ]ⲧⲟⲟⲧϥ̅ ⲙ̅-). To commemorate his efforts, the Eparch Iesou may not only have borrowed a lintel, decorated and inscribed two centuries earlier by Bishop Paulos, but perhaps its textual formula as well.[30] Thus, all indications (layout and workmanship of the lintels' decoration; placement and nature of their inscriptions) concur to suggest that both lintels date back to the episcopacy of Bishop Paulos, who, to judge from their strong similarity, may very well have intended them for one and the same monumental building.[31]

Finally, there can be little doubt about the original emplacement of the lintel re-inscribed by Iesou: it was excavated among the rubble of the so-called "Church on the south slope of the Kom", "opposite its north entrance, over which it had undoubtedly once been".[32] This church was situated, again, to the south of the Cathedral, where it was built right next to and, partly, into Jakobielski's "Bishop's Palace".[33]

The lintel re-used by the Eparch Iesou raises the number of inscribed monuments that can be attributed to Bishop Paulos to six. Of these six, three bear the same date (423 Diocletian era = A.D. 707), while four of them can be associated with certainty, and two others tentatively, with buildings in the area south of the Cathedral square. This clustering of evidence around one year and in one particular area of the town strongly suggests that around A.D. 707 Paulos' building activities were concentrated at the southern end of the Kom. Whatever his contribution to the expansion, restoration or decoration of the Cathedral may have been is not borne out by the surviving monuments of his episcopacy.

Instead of pointing to the Cathedral, the epigraphic heritage of Bishop Paulos invites us to turn southward, to the buildings on the south slope of the Kom. Unfortunately, this quarter of the town suffered much from a landslide which ruined all but the northernmost part of the buildings in question. Moreover, the excavations in this area have never been properly published. Still, three monumental units can be distinguished which line the southern façade of the Cathedral square. These are, from (local) west to east: the cross-post, the "Bishop's Palace" and the so-called "Church on the south slope of the Kom". Their common deviating axis (SSE-NNW) marks them off from the Cathedral complex at the opposite side of the "square".[34] Besides, the published plans show that the church and the "Palace" have been built more or less "into each other".[35] The distinctive common situation of these three units on the south slope of the Kom, combined with their apparent architectural interlocking, points at a certain conceptual coherence. Actually, a mere look at the ground plan of the area strongly suggests that, at least from a certain period, these three structures, whatever their original relationship might have been, had come to be considered as one complex.

Originally, of course, they may have been separate units dating from different periods. The cross-post, in particular, is thought to be older than the rest. If this is correct, it may – as we saw – explain the deviating axis of the other, posterior parts of the complex which appear to be orientated towards it.[36] The "Palace" walls seem to be a bit out of line with those of the church, perhaps because they had to link up two already existing elements, the church and the cross-post.[37] The church, moreover, was apparently rebuilt or restored by the Eparch Iesou in the tenth century, to an extent which remains unknown to us.[38] That it was an entirely

THE CHURCH OF THE TWELVE APOSTLES

new foundation of Iesou is rather improbable on account of the mere wealth of Paulos' material from the area, material which Iesou partly re-inscribed. Considering Bishop Paulos' emphatic presence in this area, I would find it much more attractive to suppose that it was the latter who welded various given elements into one complex, rebuilding the church and adding or much expanding the central "Palace" structure which connected the church with the existing cross-post.[39] As we will presently see, the existence of such a coherent complex, distinct from the cathedral, is unequivocally confirmed by the textual evidence.

The Twelve Apostles

Among the textual witnesses, Paulos' two great building inscriptions from the "Bishop's Palace" take, of course, pride of place. According to Jakobielski, the evidence for associating them with the Cathedral, instead of with the building of which they were part, is to be found in the text of the inscriptions themselves.[40] In order to appreciate this so-called evidence, the texts themselves, and particularly their statements about the edifice for which they were destined, have to be examined afresh. Two questions seem especially relevant here: do they contain information about the nomenclature of the edifice, and do they contain information concerning its status or function?

The first question is certainly the easier to answer. In fact, the Coptic building inscription of Paulos explicitly mentions the titular saints of the edifice. Already in 1972, in his extensive discussion of the text, Jakobielski had for a moment considered the possibility that Paulos' foundation (in his opinion, the Cathedral) was dedicated to the Twelve Apostles.[41] He went on to reject this hypothesis since there is indubitable evidence, from within the building itself and from elsewhere, that the Cathedral of Faras had been dedicated to the Virgin Mary.[42] Still, in support of his alternative suggestion, favoring the Twelve Apostles, Jakobielski was able to adduce two passages from the Coptic text [*I. Khartoum Copt.* 1], to wit lines 7 and 17.

The first of these passages, which will be more fully discussed below, is far from conclusive. The term ⲁⲡⲟⲥⲧⲟⲗⲓⲕⲏ, "Apostolic", in the phrase ⲧⲕⲁⲑⲟⲗⲓⲕⲏ ⲁⲩⲱ ⲛⲁⲡⲟⲥⲧⲟⲗⲓⲕⲏ ⲉⲕⲕⲗⲏⲥⲓⲁ (l. 7), is certainly too vague and generalized to serve as testimony for a dedication to the Apostles.[43]

Jakobielski's second reference, however, is fully pertinent and to this passage we will presently turn our attention. Before entering upon the final doxology, the long central part of the Coptic text winds up with the following phrase: ϩⲓⲧⲛ ⲛⲉⲡⲣⲉⲥⲃⲉⲓⲁ ⲙⲛ ⲛⲥⲟⲡⲥ ⲙⲡⲙⲛ̅ⲧⲥⲛⲟⲟⲩⲥ ⲛⲁⲡⲟⲥⲧⲟⲗⲟⲥ ⲙⲛ ⲛϭⲟⲙ ⲧⲏⲣⲟⲩ ⲛ̅ⲛⲟⲉⲣⲟⲛ ⲉⲧⲥⲱⲟⲩϩ ⲉϩⲟⲩⲛ ⲉⲡⲧⲟⲡⲟⲥ ⲉⲧⲟⲩⲁⲁⲃ ⲛϥ̅ⲟⲩⲡⲣⲟⲥⲕⲩⲛⲉⲓ ⲛⲁϥ ⲉⲧⲙ̅ⲙⲁⲩ, ". . . through the intercession (plur.) and the prayers of the Twelve Apostles and all the spiritual powers which assemble in this holy place, worthy of veneration" (ll. 17–19; no parallel in the Greek text).

This rather unexpected appearance of the Twelve Apostles could perhaps be interpreted as a means of emphasizing the apostolic character of Paulos'

331

founding act. Such an interpretation could, moreover, be thought to fit the tendency of the preceding body of the text where, not unexpectedly, Saint Peter figures prominently (ll. 12–14; actually, Mt. 16:18 is quoted). However, by line 17, a marked shift in the text's perspective occurs. From line 9 onwards, the text can be described as a long prayer for Bishop Paulos. Drawing its inspiration from liturgical sources, it seeks to obtain several favors for the bishop by emphasizing the theological, or, better, ecclesiological significance of his founding act.[44] Lines 17–19 bring this prayer to a liturgically correct end by asking for the intercession of a particular group of saints ("may God grant . . . through the intercession and the prayer of . . ."). The choice of these saints must definitely have been a logical one, fitting the place and circumstances. In fact, the text explicitly connects them with the worship within "this holy place", viz. the church presently built (or, rebuilt) by Paulos himself. The "spiritual powers who assemble in this holy place" are no pious rigmarole, but can be quite precisely identified as the angels who, according to common Coptic belief, attend the holy liturgy which was to be celebrated there.[45] The Twelve Apostles, too, who are invoked on a par with these, must have a quite specific relation with this individual church as a place of worship. Doubtless, they are the actual titular saints of the sanctuary.

This conclusion is confirmed by a second important Coptic text, whose connection with the south slope of the Kom is, as we saw above, equally well established. The lintel (re-)inscribed by the Eparch Iesou in A.D. 930, but – in all probability – originally commissioned by Bishop Paulos, was part of the northern entrance of the "Church on the south slope of the Kom" [I. Khartoum Copt. 2]. The text commemorates Iesou's work in this church by a formula which has been quoted above. Following this formula and the name and title of the builder, the dating lemma is introduced by the following curious phrase: ⲁⲡⲛⲟⲩⲧⲉ ⲛ̄ⲛⲁⲡⲟⲥⲧⲟⲗⲟⲥ ⲧⲱⲃⲥ̄ ⲙ̄ⲙⲟϥ ⲁϥⲟⲩⲱϩ ⲧⲉⲛ̄ⲧⲉ ⲙ̄ⲡⲉⲓⲧⲟ`ⲡ´(ⲟⲥ) ⲉⲧⲟⲩⲁⲁⲃ ϩⲛ̄ ⲥⲟⲩ . . ., "the God of the Apostles incited him (i.e. Iesou) and he laid the foundation of this holy place (τόπος) on day . . . (etc.)" (ll. 4–5).

Here, the words ⲡⲛⲟⲩⲧⲉ ⲛ̄ⲛⲁⲡⲟⲥⲧⲟⲗⲟⲥ, inconspicuous as they may seem, are in fact a significant feature. As an invocation, the phrase "God of (St.) so-and-so . . ." is commonly used in e.g. Coptic inscriptions (funerary and non-funerary)[46] and oracular questions.[47] In a formulaic way, it invokes God in order to obtain some favor (e.g. heavenly reward or divine inspiration) and, at one and the same time, asks for the intercession of the titular saint(s) of the local sanctuary or, in any case, of a saint much venerated on the spot. A characteristic example of this formula, in the epitaph of a Deacon Theophilus, has actually been found in Faras itself and strongly suggests that the so-called Church on the Mastabafield had been dedicated to Saint Isidore.[48] Another Nubian epigraphic occurrence, a commemorative inscription, attests to the cult of Saint Andrew in the church which occupies the former temple of Bayt al-Wali.[49] Also in the formula's not infrequent appearances in Coptic literary texts and prayers for donors of books, the name of the saint is always topically relevant.[50]

THE CHURCH OF THE TWELVE APOSTLES

Although in Iesou's text the same formula is not part of an invocation, but rather of a statement about a past event, the topicality of the Apostles in the pictured event cannot be doubted. Indeed, on account of the idea of inspiration suggested by the verb ⲧⲱⲃⲥ, "to incite", the oracular use of the formula could well be particularly pertinent for the interpretation of Iesou's inscription.[51] It is not Iesou himself who deserves praise for the project of erecting this "holy place", but God's inspiration, granted through the intercession of the building's patrons, the Apostles. Thus, the inscription of the Eparch Iesou establishes without a shadow of doubt that the "Church on the south slope of the Kom" was dedicated to the Apostles whose "intercession and prayers" had been requested by Bishop Paulos on the neighboring wall of the so-called "Bishop's Palace" two centuries earlier.[52] Apparently, both buildings shared the same titular saints, the Twelve Apostles. This cultic link confirms the conclusions reached earlier on the basis of the archaeological evidence: they both belonged to one architectural complex and this complex is to be distinguished from the Cathedral to the north of it, which was dedicated to the Virgin Mary.

Another cathedral?

How, it may presently be asked, is this complex designated in the inscriptions of Bishop Paulos? Only twice does the Coptic text [*I. Khartoum Copt.* 1] refer explicitly to the building itself. In the passage from ll. 17–19, which we discussed above, it is called "this holy place, worthy of veneration" (ⲡⲧⲟⲡⲟⲥ ⲉⲧⲟⲩⲁⲁⲃ ⲛ̄ϣⲟⲩⲡ̄ⲣⲟⲥⲕⲩⲛⲉⲓ ⲛⲁϥ ⲉⲧⲙ̄ⲙⲁⲩ; ll. 18–19),[53] and, moreover, a place where "spiritual powers" assemble (l. 18). Earlier on in the text, the following designation appeared: ⲡⲉⲓ̈ⲧⲟⲡⲟⲥ ⲉⲧⲟⲩⲁⲁⲃ ⲛ̄ϣⲟⲩⲱ̄ⲙ̄ϣⲉ ⲛ̄ϩⲏⲧϥ ⲛ̄ⲧⲉ ⲧⲕⲁⲑⲟⲗⲓⲕⲏ ⲁⲩⲱ ⲛ̄ⲁⲡⲟⲥⲧⲟⲗⲓⲕⲏ ⲉⲕⲕⲗⲏⲥⲓⲁ, "this holy place – worthy to worship in it[54] – of the Catholic and Apostolic Church" (l. 7). Both passages leave no doubt about the primary function of the building in question. It is unequivocally characterized as a church, a place of worship destined for the celebration of the holy liturgy. Accordingly, the appellation "Bishop's Palace", for the complex adjacent to the cross-post, is less appropriate. Indeed, there is no epigraphic or archaeological evidence whatever to suggest that it was used as such at any time.

A closer look at the terminology permits us, perhaps, to be more precise. The expression ⲡⲉⲓ̈ⲧⲟⲡⲟⲥ ⲉⲧⲟⲩⲁⲁⲃ (l. 7, with a variant in l. 18) is rather colorless. It is the habitual "self-designation" of a church or a monastery in Coptic inscriptions.[55] On the other hand, the second half of the phrase in l. 7: ". . . this holy place . . . of the Catholic and Apostolic Church" (ⲛ̄ⲧⲉ ⲧⲕⲁⲑⲟⲗⲓⲕⲏ ⲁⲩⲱ ⲛ̄ⲁⲡⲟⲥⲧⲟⲗⲓⲕⲏ ⲉⲕⲕⲗⲏⲥⲓⲁ), is open to divergent interpretations. At first sight, the words "of the Catholic and Apostolic church" seem to be a mere echo of the time-honored Constantinopolitan creed: the sanctuary ("this holy place") is said to pertain to the Church universal, of which the epithets ⲕⲁⲑⲟⲗⲓⲕⲏ and ⲁⲡⲟⲥⲧⲟⲗⲓⲕⲏ stress the orthodoxy.[56] In fact, this is a quite natural and plausible explanation. Alternatively, however, the phrase could be taken as a means of specifying the nature of "this holy place" as a ⲕⲁⲑⲟⲗⲓⲕⲏ, i.e. as the "main church" and, hence, the episcopal church of the town.[57] The second epithet, ⲁⲡⲟⲥⲧⲟⲗⲓⲕⲏ, would then, with little or no consideration for its original meaning, have been added merely to enhance the status of the sanctuary.[58]

Here, the phrasing of the Greek counterpart of this Coptic passage [*I. Varsovie* 101] is of some help. It definitely favors the second, more specific interpretation. The Greek text runs: . . . τὸν σεπτὸν τόπο(ν) τῆς καθολ(ικῆς) κ(αὶ) ἀποστολικ(ῆς) τοῦ Θ(εο)ῦ ἐκκλη(σίας)», "the venerable place of the Catholic and Apostolic Church of God" (ll. 8–9). The use of the qualification "of God" (τοῦ Θεοῦ) appears to be especially significant. It does not fit well into a formulaic echo of the creed. Rather, in the usage of the Greek papyri from Egypt, this seemingly inconspicuous phrase is reserved for churches which come directly under a bishop.[59] Its use here strongly suggests that, indeed, both the Greek (ll. 8–9) and the Coptic (l. 7) building inscriptions of Paulos specifically commemorate the renovation of a bishop's church, a ⲕⲁⲑⲟⲗⲓⲕⲏ.

If this interpretation is correct, it would mean that the church lying next to the cross-post and dedicated to the Twelve Apostles was considered, at least in A.D. 707, to be the episcopal church of Faras. Indeed, its location in the shadow of the great wooden cross, the symbol of the establishment of Christianity in Faras, would seem a quite suitable and logical place for the town's earliest cathedral. What then was its relation to the building currently known as "the cathedral"? Here, various answers are possible, all of them necessarily hypothetical. A plausible reconstruction of the facts would be that, somewhere between A.D. 707 and 930, the (existing) church of the Virgin Mary took over the function of episcopal church from the earlier cathedral, dedicated to the Twelve Apostles, which may have become too small, antiquated or simply ruinous.[60] Perhaps significantly, it was renovated in 707 by the bishop himself, whereas it was restored or rebuilt in 930 by a civil functionary, the Eparch Iesou. Before the latter year, the new cathedral must have been inaugurated. A plausible date for such a transfer would be the episcopacy of Bishop Kyros, in the second half of the ninth century.[61] Kyros was the first bishop to have his portrait painted on the walls of the "second" cathedral.[62] Shortly after his death, the first part of the famous list of Faras bishops, up to and including Kyros, must have been copied onto the walls of the same church from an existing document, perhaps from a similar list in the "first" cathedral.[63]

Alternatively, the possibility can be envisaged that both cathedrals functioned side by side. Indeed, double or even triple churches and cathedrals are well attested in later antiquity, both in the East and in the West. As it appears from the papyrological documentation, the Egyptian town of Hermopolis must, from a certain moment, have had several episcopal churches at the same time, all called ⲕⲁⲑⲟⲗⲓⲕⲏ and each of them bearing distinctive names.[64] Archaeologically, this phenomenon is well attested for the entire Christian world.[65] Special mention may be made here of the three episcopal churches of late antique Geneva.[66] Faras could very well represent a Nubian example of the same phenomenon.[67]

Conclusions

The foregoing observations lead to several important conclusions. First of all, the epigraphic sources confirm the impression conveyed by the available archaeological evidence for the area south of the Saint Mary Cathedral. Not only their ground plan, but also their common dedication to the Twelve Apostles, bear witness to the fact that both the so-called "Bishop's Palace" and the "Church on the south slope of the Kom" were part of one and the same architectural complex. Secondly, the dates and other information gained from Paulos' inscriptions bear upon this particular ecclesiastical complex and cannot be applied, without further qualification, to the Saint Mary Cathedral. Hence, the latter building may be less firmly dated than some scholars believe. Thirdly, the modern appellation "Bishop's Palace" for the western part of this complex is demonstrably inaccurate and misleading. The Coptic inscription of Bishop Paulos characterizes the sanctuary of the Twelve Apostles as a place of worship, destined for the celebration of the holy liturgy. Fourthly, to judge from, especially, the Greek inscription of Bishop Paulos, this church of the Twelve Apostles may have been, at least in A.D. 707, an episcopal church. Its location near the cross-post would favor the hypothesis that it was the first cathedral of Faras. As such it was replaced or joined, perhaps as late as the second half of the ninth century, by the nearby sanctuary of the Virgin Mary, which is currently known as "Faras Cathedral". Finally, the obvious historical importance of the area south of the Saint Mary Cathedral makes the full publication and renewed study of its archaeological record an urgent task.

Notes

1 S. Jakobielski, *A history of the bishopric of Pachoras on the basis of Coptic inscriptions*, Warsaw: PWN–Éditions scientifiques de Pologne 1972.

2 [*I. Khartoum Copt.*] The project was supervised by Prof. W. Godlewski (Warsaw) and supported by the Sudan Antiquities Service, UNESCO and the Netherlands Foundation for the Advancement of Tropical Research (WOTRO).

3 Presently in Khartoum, National Museum, excavation no. FA 134/63–4; [= *I. Khartoum Copt.* 1] in my forthcoming catalogue (which includes a new discussion of the text). The richest source of information on this inscription is still Jakobielski, *Bishopric of Pachoras*, 37–48 (edition with discussion and bibliography). My personal readings of the Coptic monuments in Khartoum, as quoted below, may occasionally differ from those of Jakobielski; for all details the reader is referred to my catalogue.

4 Presently in Warsaw, National Museum, inv. no. MN 234848; for a rather indifferent edition: J. Kubińska, *Inscriptions grecques chrétiennes*, Warsaw: PWN–Éditions scientifiques de Pologne 1974, 14–19, no. l, fig. 1 [= *I. Varsovie* 101].

5 Of course, occasional doubts have been voiced, e.g. as early as 1969 by F.W. Deichmann, in a review of two books by K. Michałowski, *Byzantinische Zeitschrift* 62, 108–11 at 110 (for "Petrus", read: Paulos).

6 Building N in Jakobielski, *Bishopric of Pachoras*, plan II; some *in situ* photographs: K. Michałowski, *Faras: Die Kathedrale aus dem Wüstensand*, Einsiedeln: Benziger Verlag 1967, pl. 14; T. Dzierżykray-Rogalski, *The bishops of Faras: an anthropological-medical study*, Warsaw: PWN-Editions scientifiques de Pologne 1985, 17, no. 3; P.M. Gartkiewicz, "Cathedral in Faras in the light of an architectural re-analysis",

JACQUES VAN DER VLIET

in: M. Krause (ed.), *Nubische Studien*, Mainz: Phillip von Zabern 1986, 245–68 at fig. 31.

7 Only rarely have they been considered as *spolia* (thus Gartkiewicz in: Krause (ed.), *Nubische Studien*, 251–52); as the published photographic record shows, there is little to recommend this theory. For technical details, cf. Jakobielski, *Bishopric of Pachoras*, 39.

8 K. Michałowski, "Polish excavations at Faras, 1962–63", *Kush* 12 (1964), 195–207 at 195 and 197–99.

9 S. Jakobielski, "Two Coptic foundation stones from Faras", in: M.L. Bernhard (ed.), *Mélanges offerts à Kazimierz Michałowski*, Warsaw: PWN–Éditions scientifiques de Pologne 1966, 103–09 at 104.

10 Jakobielski, *Bishopric of Pachoras*, 37, with discussion in n. 17; cf. Michałowski, *Kathedrale*, 73 and 95.

11 Jakobielski, *Bishopric of Pachoras*, 189, no. 9: "The so-called Bishops' Palace, probably a monastery from 8th to 12th century . . . went under the name of St. Mary of Pachoras". Such a monastery, probably closely linked with the Cathedral, did exist, but there is not a shred of evidence to connect it with the so-called "Bishop's Palace".

12 For this complex (C in Jakobielski, *Bishopric of Pachoras*, plan II), see Dzierżykray-Rogalski, *The Bishops of Faras*, 16–25.

13 Building O in Jakobielski, *Bishopric of Pachoras*, plan II; Michałowski, *Kathedrale*, 95, mentions "Wandelhallen" from the eleventh century: is this the corridor between C and O?

14 Michałowski, *Kush* 12, 197–98; idem, *Kathedrale*, 73.

15 For the cross-post (D in Jakobielski, *Bishopric of Pachoras*, plan II), which in its present form seems to pre-date the Paulos building (it is usually dated to the second half of the seventh century), see further: Michałowski, *Kathedrale*, 54 and 64, pl. 14; Jakobielski, *Bishopric of Pachoras*, 29–32; F.W. Deichmann and P. Grossmann, *Nubische Forschungen*, Berlin: Gebr. Mann Verlag 1988, 106. For "setting up the cross" (ογωϩ πⲥⲧⲁγρος) on newly Christianized sites in Nubia, cf. G. Roquet, "Le morphème (E) TAH- et les graffites coptes de Kalabcha", *BIFAO* 78 (1978), 533–38, with further references.

16 Jakobielski, *Bishopric of Pachoras*, 39, proposes several ingenious explanations to this effect. This supposition might have been substantiated by the full excavation of the Kom which, however, could not be completed.

17 Michałowski, *Kush*, 12, 198–99; Jakobielski, *Bishopric of Pachoras*, 37; for extensive discussions of masonry techniques, which will not be taken up here, see Gartkiewicz in: Krause (ed.), *Nubische Studien*; Deichmann and Grossmann, *Nubische Forschungen*, 95ff.

18 MN 234680 (quoted as 149236 by Jakobielski); excavation number F 56/61–2 [= *I. Varsovie* 102].

19 Provenance and present location unknown; a publication by A.F. Shore was announced, but – to the best of my knowledge – has never appeared.

20 Excavation number FA 86/62–3; now probably in Khartoum. [It is referred to in *I.Khartoum Greek*, 121, but on the basis of U. Monneret de Villard, *La Nubia medio-evale* I, Cairo: Service des antiquités de l'Égypte 1935, 49–50.]

21 See Jakobielski, in: K. Michałowski, *Faras: fouilles polonaises 1961–1962*, Warsaw: PWN–Éditions scientifiques de Pologne 1965, 164, no. 2, fig. 86 (*ed. princeps*); Kubińska, *Inscriptions grecques chrétiennes*, 20, no. 2, fig. 2 (revised edition); Jakobielski, *Bishopric of Pachoras*, 48–9 (discussion) [= *I. Varsovie* 102].

22 Reports about its provenance are imprecise and contradictory: "tief im Sand bei der Nordostecke der Kathedrale"; (Michałowski, *Kathedrale*, 65); "by the south-east comer of the Cathedral in a kind of path around the church in which it had probably

THE CHURCH OF THE TWELVE APOSTLES

been re-used as a threshold" (Jakobielski, *Bishopric of Pachoras*, 48); "dégagé dans l'angle Est de la Cathédrale" (B. Idzikowska, "La décoration en pierre des portes dans les églises et édifices profanes de Faras", *ÉtudTrav* 12 (1983), 195–237 at 221).

23 Jakobielski, *Bishopric of Pachoras*, 48; cf. Idzikowska, *ÉtudTrav* 12, 221. Contrary to Jakobielski's opinion, the text's ἀγαθὸν ἔργον, although basically a neutral term, can very well apply of the erection or decoration of a church; cf. G.W.H. Lampe, *A patristic Greek lexicon*, Oxford: Clarendon Press 1961, 547b, s.v. ἔργον, B.5; Deichmann and Grossmann, *Nubische Forschungen*, 82–3, and further discussion below.

24 See Michałowski, *Kathedrale*, 65, but especially the discussion in Jakobielski, *Bishopric of Pachoras*, 48.

25 See the edition and discussion by Jakobielski, *Bishopric of Pachoras*, 49–50, fig. 4; cf. Idzikowska, *ÉtudTrav* 12, 236; Gartkiewicz in: Krause (ed.), *Nubische Studien*, 251 (with fig. 27).

26 National Museum, Khartoum, excavation no. FA 100/62–63 [= *I. Khartoum Copt.* 2]. See cf. Jakobielski, *Mélanges Michalowski*, 107–09, fig. 2 (*ed. princeps*); idem, *Bishopric of Pachoras*, 110–14, fig. 28 (revised text with commentary and bibliography). For the following comparison the reader is referred to the photos in the publications mentioned here and in note 21 above.

27 Cf. Jakobielski, *Mélanges Michalowski*, 108, n. 14; *Bishopric of Pachoras*, 111, n. 53 [*I. Khartoum Copt.* 2, p. 16].

28 Kindly checked on the original by Prof. W. Godlewski.

29 It should be added that my personal inspection of the Iesou monument, in February 1997, revealed no visible traces of an earlier text.

30 The correspondence between the Coptic and the Greek formulae was first noted by H. Quecke, in his review of Jakobielski, *Bishopric of Pachoras*, in *Orientalia* 43 (1974), 135–41 at 136. Admittedly, this formula is not a unique one. In several variants, Coptic and Greek, it occurs elsewhere, both in Nubia and in Egypt, to commemorate civil and ecclesiastical foundations; for further references, see my forthcoming re-edition of the Iesou lintel [= *I. Khartoum Copt.* 2].

31 For similar opinions: Idzikowska, *ÉtudTrav* 12, 221; W. Godlewski, "Remarks on the art of Nobadia (V–VIII century)", in: Krause (ed.), *Nubische Studien*, Mainz: Philipp von Zabern, 269–79 at 273.

32 Thus Jakobielski, *Bishopric of Pachoras*, 111.

33 The Church is building P in Jakobielski, *Bishopric of Pachoras*, plan II. Its excavation records are unpublished, but cf. Michałowski, *Kush* 12, 203; idem, *Kathedrale*, 82. For a partly preserved wall painting from this church, see K. Michałowski, *Faras: Die Wandbilder in den Sammlungen des National Museums zu Warschaw*, Warsaw: Wydawnictwo Artystyczno-Graficzne/Dresden: VEB Verlag der Kunst 1974, 174–76, no. 31; I doubt whether the painting represents Saint Joseph, particularly since the inscription (ibidem, 301 n. 15) is certainly a graffito, not a legend. Against the north wall of the church, a late tenth-century bishop's tomb was built; see Dzierżykray-Rogalski, *The bishops of Faras*, 32–5; cf. the photos in Z. Jeżewska, *Na krańcach czasu*, Warsaw: Czytelnik 1966, 48 and 96 (over against page, below).

34 Also Michałowski, *Kush*, 12, 203, underlines the common axis of the church on the south slope and his "monastery" (i.e. the "Bishop's Palace").

35 Michałowski, *Kathedrale*, 82: "sie (i.e. the Church) greift auf den Ostteil dieses Gebäudes (i.e. the Bishop's Palace) über" (some more precise information would have been welcome).

36 See notes 15 and 16 above.

37 Alternatively, this phenomenon could be ascribed to secondary revisions of the ground plan (perhaps to be connected with the activity of the Eparch Iesou) which affected the church (see below).

JACQUES VAN DER VLIET

38 Michałowski, *Kathedrale*, 82, suggests that the "Bishop's Palace" lay in ruins by the time of Iesou.

39 This point of view is possibly confirmed by Michałowski's statement that the "Palace" and the church were "constructed from the same materials", in *Kush* 12, 195–96. It should be borne in mind that, according to his Greek inscription, Paulos *"renovated and built"* the "venerable place" in question (1. 8), which therefore cannot have been the earliest sacred building on the spot.

40 Jakobielski, *Bishopric of Pachoras*, 37–9, with (in n. 18) a reference to p. 46, where, however, no such evidence is adduced.

41 Jakobielski, *Bishopric of Pachoras*, 44.

42 See Jakobielski, *Bishopric of Pachoras*, 176–86. Here, Jakobielski's line of reasoning is not only wholly convincing, but actually confirmed by the fourteenth-century "scrolls of Bishop Timotheos", where the diocesan Church of Faras is called "the Church of the Holy Mother of God, Saint Mary of Pkhôras (Faras) and Nubia" (Coptic scroll, ll. 64–5; ed. J.M. Plumley, *The scroll of Bishop Timotheos: two documents from medieval Nubia*, London: Egypt Exploration Society 1975, 10, pl. VIII). On the other hand, evidence from the complex on the south slope of the Kom which could suggest a special connection with the Virgin Mary is lacking entirely.

43 If the text had meant to imply a dedication to the Apostles, it would have been phrased quite differently; see, for an example, E. Wipszycka, "Καθολική et les autres épithètes qualifiant le nom ἐκκλησία: contribution à l'étude de l'ordre hiérarchique des églises dans l'Égypte byzantine", *JJP* 24 (1994), 191–212 at 205.

44 See, for further commentary, *I. Khartoum Copt.* 1.

45 See e.g. C.D.G. Müller, *Die Engellehre der koptischen Kirche*, Wiesbaden: Harrassowitz 1959, 68, n. 529 and 69, n. 535; J. Michl, art. "Engel IV (christlich)", *RAC* 5 (1962), 109–200 at 161–63. For "spiritual powers" = "angels", cf. G.W.H. Lampe, *A Patristic Greek Lexicon*, Oxford: Clarendon Press 1961, 916a, s.v. νοερός, II.C. Actually, the concept of the dual nature of the church as a terrestrial and as a spiritual, celestial entity is a (secondary) theme of the text (see *I. Khartoum Copt.* 1).

46 See e.g. M. Cramer, *Koptische Inschriften im Kaiser-Friedrich-Museum zu Berlin*, Cairo: Société d'archéologie copte 1949, 32–3; H. Munier. "Stèles chrétiennes d'Antioé", *Aegyptus* 29 (1949), 126–36 at 129–30 and 135–36 (all funerary). For examples of ⲡⲛⲟⲩⲧⲉ ⲛⲛⲁⲡⲟⲥⲧⲟⲗⲟⲥ from Coptic epigraphy, see G. Roquet, "Les graffites coptes de Bagawât (Oasis de Kharga): Remarques préliminaires", *Bulletin de la Société française d'égyptologie* 76 (1976), 25–49 at 33, with n. 40.

47 Cf. L. Papini, "Biglietti oracolari in copto dalla Necropoli Nord di Antinoe", in T. Orlandi, and F. Wisse (eds.), *Acts of the Second International Congress of Coptic Studies*, Rome: C.I.M. 1985, 245–55; A. Papaconstantinou, "Oracles chrétiens dans l'Égypte byzantine: Le témoignage des papyrus", *ZPE* 104 (1994), 281–86, esp. 282–84.

48 As was convincingly argued by F.L. Griffith, "Oxford excavations in Nubia", *LAAA* 14 (1927), 57–116 at 62 and 104–05, with pl. XLII, no. 1, and Jakobielski, *Bishopric of Pachoras*, 170–71.

49 Thus already Monneret de Villard, *Nubia* I, Cairo 1935, 32; text: C.R. Lepsius, *Denkmaeler aus Aegypten und Aethiopien* VI, Berlin: Nicolai s.a.; reprinted Geneva: Éditions de Belles-Lettres 1973, pl. 103, no. 38, ll. 1–2; an incompetent copy, read: ⲡⲛⲟⲩⲧⲉ ⲙ̄ⲡⲣⲁⲅⲓⲟⲥ ⲁⲛⲇⲣⲉⲁⲥ ⲡⲁⲡⲟⲥⲧⲟⲗⲟⲥ ⲙ̄ⲡⲉⲭ(ⲣⲓⲥⲧⲟ)ⲥ (sic prob.); cf. G. Roeder, *Der Felsentempel von Bet el-Wali*, Cairo: IFAO 1938, 82–3 and 173–74 (re-edition by C. Schmidt after Lepsius).

50 Literary texts, a random typical example: the *Miracles of saint Menas*, ed. J. Drescher, *Apa Mena*, 24.a. Prayers for donors, see A. van Lantschoot, *Recueil des colophons des manuscrits chrétiens d'Égypte* I: *les colophons coptes des manuscrits sahidiques*, Louvain: J.B. Istas 1929, no. XLV, ll. 30–9 (and *passim*).

THE CHURCH OF THE TWELVE APOSTLES

51 Was the church renowned for incubation or oracular practices?

52 In fact, T. Orlandi, in his review of Jakobielski, *Bishopric of Pachoras*, in *BiOr* 31 (1974), 266–69, on 267, had already pointed out that both inscriptions (the Coptic stone of Paulos and the Iesou lintel) mention the Apostles. However, accepting a priori Jakobielski's thesis, he unwarrantedly speaks of "formule rituali senza altri significati".

53 For another Nubian example of the same epithet "worthy of veneration" (with typically wrong gender selection), see Van Lantschoot, *Recueil*, no. CXXII, l. 11, with notes 7 and 9 (fasc. 2, 87–8). There it applies to "the church: ⲉⲕⲕⲗⲏⲥⲓⲁ, of Our Saviour Jesus Christ" in Illarte. The expression may render Greek προσκυνητός, as was plausibly suggested by an anonymous reviewer of *Orientalia* (cf. the following note).

54 This expression may correspond to σεπτός in the Greek text, l. 8; for compounds with ϣⲟⲩ- rendering Greek words on -τός, cf. W.E. Crum, *A Coptic dictionary*, Oxford: Oxford University Press 1939, 600b. Again, I follow a suggestion of the *Orientalia* reviewer.

55 Thus, too, Jakobielski, *Bishopric of Pachoras*, 42, n. 30; it translates Greek ἅγιος τόπος, for which see A. Łajtar, and E. Wipszycka, "Deux καθολικαὶ ἐκκλησίαι dans le Mons Porphyrites", *JJP* 24 (1994), 71–85 at 80–1. The Greek text calls the building once a "venerable place: σεπτὸς τόπος" (l. 8), once "this edifice: τὸ κτίσμα τοῦτο" (l. 17).

56 Cf. H. Denzinger, and A. Schönmetzer, *Enchiridion symbolorum*, thirty-second edition, Freiburg im Breisgau: Herder 1963, 66–7, no. 150. For a recent discussion of the different meanings of the epithet καθολική applied to ἐκκλησία, see Wipszycka, *JJP*, 24, 202–04; for its combination with ἀποστολική, "d'habitude dans des contextes très solennels", like the present inscription, at 202–04.

57 This possibility is actually considered, with great hesitation ("perhaps"), by Jakobielski, *Bishopric of Pachoras*, 42, n. 31, and again on p. 44.

58 Wipszycka, *JJP* 24, 204, quotes an appropriate Coptic example of this usage.

59 Thus according to Wipszycka's discussion of τοῦ Θεοῦ ἐκκλησία, *JJP*, 24, 195–96.

60 Cf. Michałowski, *Kathedrale*, 82.

61 See Jakobielski, *Bishopric of Pachoras*, 84–86. However, the last line (l. x+21) of the funerary stela of Bishop Thomas (died A.D. 862; Jakobielski, *Bishopric of Pachoras*, 75–80; *SB Kopt.* I 719.22) could be taken to indicate that already around the middle of the century the "Theotokos Maria" was the titular saint of the Faras diocese.

62 See Michałowski, *Kathedrale*, 117, pl. 37; Jakobielski, *Bishopric of Pachoras*, 86–8.

63 Cf. Jakobielski, *Bishopric of Pachoras*, 191.

64 Cf. the discussion in Wipszycka, *JJP*, 24, 207–08.

65 *Antiquité tardive* 41 (1996) was entirely devoted to the theme of "les églises doubles et les familles d'églises"; apart from case studies, current interpretations of the phenomenon are extensively discussed. I thank Prof. P.V.V. van Moorsel for this valuable reference.

66 Cf. Ch. Bonnet, "Les installations liturgiques du baptistère et des trois églises épiscopales de Genève durant l'Antiquité tardive", *Antiquité tardive* 41, 101–03.

67 It should be added that the double reference to the "Church of Our Lady and of the Apostles" in the Arabic "scroll of Bishop Timotheos" from A.D. 1371 (ed. Plumley, *Scrolls*, 29, pl. XVII, l. 9; cf. R.G. Coquin, "A propos des rouleaux coptes-arabes de l'évêque Timothée", *BiOr* 34 (1977), 142–47 at 146) *cannot*, in all probability, be cited in support of this theory. Here, "and of the Apostles" is a secondary insertion into the Arabic text alone, and therefore relates in my opinion to Ibrim and not to Faras; cf. my miscellany "Churches in Lower Nubia, old and 'new'", *BSAC* 38 (1999), 135–42 at 138 [Study 24].

27

EXIT TAMER, BISHOP OF FARAS (*SB* V 8728)*

Jacques van der Vliet

Introduction

In the last few decades considerable progress has been made in our understanding of the language, the formal conventions and the onomasticon of the non-literary and epigraphic texts from medieval Nubia. As a result, many earlier text editions now require revision. The present squib takes another look at the original owner of one of the first monuments from Christian Nubia that reached Europe, a Greek funerary stela acquired in 1820 by the Italian nobleman Carlo Vidua at Faras, which is now kept in the Egyptian Museum in Turin, under inv. no. 7147.[1] In the course of the last two centuries this extensive epitaph has been published, re-published and discussed very often.[2] Thus, the text was included in G. Lefebvre's *Recueil des inscriptions grecques-chrétiennes d'Égypte*, Cairo: IFAO 1907, as no. 636, in the *Sammelbuch* as no. 8728 (by F. Bilabel), and in the collection of Maria G. Tibiletti Bruno, *Iscrizioni Nubiane*, Pavia: Fusi 1964, as no. 8.[3] For the problematic date in l. 27, an interpretation as A.M. 900 (i.e. A.D. 1184) has been proposed.[4]

The text follows the standard formulary of many later Greek epitaphs from Nubia, opening with the invocation "God of the spirits and all flesh". This formulary normally allows the introduction of the name of the deceased at two places: in the beginning of the prayer for eternal repose in Paradise, here ll. 5–6, and in the verses that lead up to the concluding doxology, here ll. 18–20. For the first instance, where the stone is slightly damaged, no entirely satisfactory reading has ever been proposed. In the second passage, a long-standing scholarly tradition recognizes, in l. 19, the name of a certain "Tamer (ⲧⲁⲙⲏⲣ), bishop of Faras", unknown from other sources. The reading of the name Tamer has never been questioned, and apparent difficulties, like the feminine articles in ll. 5 and 18, were attributed to the scribe's imperfect knowledge of Greek. Accordingly, Bishop Tamer entered the history of Faras as one of the last incumbents of that see.[5]

The revised readings that are proposed here have been made after a photo of the monument published in 1990.[6] Below, only lines 5–6 and 18–20 are reproduced in a diplomatic transcription that retains the case endings and the orthography of the original.

341

JACQUES VAN DER VLIET

5. ⲁⲛⲁⲡⲁⲩⲥⲟⲛ ⲧⲏⲛ ⲯⲩⲭⲏⲛ ⲧⲏⲛ ⲩⲁ′(ⲟⲩⲗⲏⲛ) [ⲡ]ⲁⲡ-
6. ⲥⲓⲛⲉ ⲑⲩⲅ′(ⲁⲧⲏⲣ) ⲉⲡⲓⲥ[ⲕ(ⲟⲡⲟⲩ) ⲉⲛ] ⲕⲟⲗⲡⲟⲥⲓ ⲁⲃ[ⲣⲁ]ⲙ   (etc.)

18. ⲥⲩ ⲅⲁⲣ ⲟⲓ ⲁⲛⲁⲡⲁⲩⲥⲏⲥ ⲧⲏⲛ ⲩⲁ′(ⲟⲩⲗⲏⲛ) ⲇⲟⲩⲗ̄ⲥ-
19. ⲧⲁ ⲙⲏ(ⲧⲏ)ⲣ ⲉ̇ⲡⲓⲥⲩⲕ′(ⲟⲡⲟⲩ) ⲡⲁⲭⲱⲣⲁⲥ ⲑⲩⲅ′(ⲁⲧⲏⲣ) ⲧ̄ⲧ̄-
20. ⲧⲁ ⲥ̀ ⲉ̇ⲭⲱⲛ ⲭ̄ⲡ̄ⲑ̄ ⲡⲁⲭⲱⲣⲁⲥ   (etc.)

5–6. (Lord . . .) grant rest to the soul of (your) servant Papsine, bishop's daughter, in the bosom (l. ἐν κόλποις) of Abraham . . . (etc.)

18–20. For you are the rest (l. σὺ γὰρ εἶ ἡ ἀνάπαυσις) of (your) servant Doulista, mother of the bishop of Faras, daughter of Titta, and owner of (the church of St.) Michael (in) Faras, . . . (etc.).

The lines transcribed above contain three Old Nubian names and several conventional abbreviations that have not or incorrectly been identified by the earlier editors of the text. The name of the deceased is preceded, both in l. 5 and in l. 18, by the commonplace abbreviated expression that characterizes her as God's "servant", with a raised ⲁ above the article ⲧⲏⲛ. As usual it is given in the accusative case, instead of the correct genitive (a normalized resolution of the abbreviation would be τῆς δουλῆς σου).[7] In both cases, the abbreviation was correctly identified by most editors, but not the name of the deceased, which must follow immediately. In fact, in ll. 5–6, Papsine can be read; in ll. 18–19, Doulista. On account of other occurrences of the same name (see below), the reading of the former name can be considered certain, in spite of the small lacuna in l. 5.[8] The first part of the latter name, ⲇⲟⲩⲗ̄ⲥ-, in the end of l. 18, was hitherto taken for an unexplained (and, in fact, further unattested) repetition of δουλῆς, in spite of the indubitable presence of the ⲁ above the preceding article ⲧⲏⲛ.[9]

Both names are followed by stereotypical filiation formulae. In l. 6, this is the abbreviation ⲑⲩⲅ′, for θυγάτηρ (or, in the correct genitive case: θυγατρός), "daughter".[10] It is followed by a slightly doubtful group, interpreted here as the same abbreviated form of the word ἐπίσκοπος that is found in l. 19.[11] In l. 19, the name Doulista is followed by the standard abbreviations ⲙ̄ⲏ̄ⲣ, for μήτηρ, "mother",[12] and, again, ⲑⲩⲅ′, "daughter". Only this last reading was established by F.L. Griffith already in 1925.[13] The following Old Nubian name Titta, however, was not recognized by him or any other editor of the text.[14] Finally, the last epithet of the deceased, the phrase that may be resolved as: (καὶ) ἔχων (Μιχαὴλ) Παχωρας, was successfully deciphered by Adam Łajtar already in 1994.[15] It is paralleled by similar statements in other Greek, Coptic and Old Nubian inscriptions from all over Nubia that record church ownership.[16] It may be noted that ἔχων is not inflected, and also the kinship terms that follow the names of the deceased, like μήτηρ, were apparently used in the – strictly spoken – incorrect nominative case.[17]

Obviously, the owner of the stone is not a bishop of Faras, but a woman who, according to the text of l. 19, was the mother of a Faras bishop. There is no more

EXIT TAMER, BISHOP OF FARAS

need, therefore, to attribute the feminine articles in l. 5 and 18 to the Nubians' weak sense of grammatical gender,[18] or to claim the influence of Nubian word order.[19] The name Tamer has been made up by joining the final syllable of the Old Nubian name Doulista, in l. 19, to the following unrecognized abbreviation of μήτηρ. Tamer is definitely a phantom and should be deleted from the list of Faras bishops.

Even though the readings proposed above appear to do justice to both the text as it can be read from the published photo and the conventions of medieval Nubian funerary epigraphy, it cannot be denied that they raise new problems as well. The logic of the formulary demands that the names given in ll. 5–6 and ll. 18–20 refer to the same person, the deceased woman commemorated by the inscription. Yet, both passages seem to provide conflicting information. According to our practically certain reading of ll. 5–6, she is called Papsine (a variant of more common Papasine). She is furthermore said to be a bishop's daughter. In the extensive filiation formula of ll. 18–20, on the other hand, she is called Doulista, and she is said to be the mother of an unnamed bishop of Faras as well as the daughter of an untitled Titta. The latter name can in no way be recognized in the filiation formula of l. 6, where, following ⲟⲩ`ⲣ´(ⲁⲧⲏⲣ), "daughter", a word or name beginning with an unquestionable ⲉ- can be seen, read as ⲉⲡⲓⲥ̣[ⲕ(ⲟⲡⲟⲥ)], "bishop", by me.

Since it is hardly plausible that serious scribal errors in the name of the owner of such an expensive funerary stela were allowed to pass uncorrected, the apparent lack of agreement between the first and the second mention of the deceased woman must be explained in another way. Thus it can be assumed that she bore a double name, Papsine Doulista. Double names of this kind were actually quite common in medieval Nubia,[20] but in the present case it is somewhat disturbing not to see them combined even once.[21] Likewise, the filiations that are given in both passages are not necessarily incompatible. In all likelihood Titta (ll. 19–20) is a feminine name (see below), and it is therefore most probably the name of the deceased's mother, her father, a bishop according to l. 6, remaining anonymous. Furthermore, in Nubia, where bishops often were married and had children, it is theoretically possible that someone could be at the same time the daughter and the mother of a bishop. In fact, little "dynasties" of bishops are known from tenth-twelfth-century Faras documents.[22] The person commemorated by the Turin stela could be a member of such an episcopal family.

Thus, if the information from both passages is accepted at face value and taken as complementary instead of contradictory, the following pedigree would be obtained:

Bishop of ?, N.N. x Titta

|

husband N.N. x Papsine a.k.a. Doulista

|

Bishop of Faras, N.N.

Regrettably, as the bishops remain anonymous and the women named are unknown from other sources, there is no way to test this hypothetical reconstruction of episcopal family life. Moreover, even if the genealogy could be proven correct, this would not explain the striking differences between two passages that were supposed to refer to one and the same person, for not only the names, but also the filiations given are different in both instances.

If the family tree sketched above would apply nonetheless, it is noteworthy that only the female members of the family are named, not their husbands, sons or fathers, even if they were bishops.[23] Did the mother of a bishop hold an influential position similar to that of the mothers of kings that are known from Nubian history? Another bishop's mother is in fact mentioned in a Coptic letter from Qasr Ibrim, although she is merely cited as greeting the addressee.[24] In any case, the owner of the Turin stela clearly counted among the elite of her time, providing another instance of a woman who owned a church.[25]

The women's names themselves are exclusively Nubian. Papsine (or Papasine) and Titta are frequently found elsewhere, in particular in the Old Nubian documents from Qasr Ibrim that date from the same general period as the Turin stela. Apparently, the name Pap(a)sine could be borne by both women and men.[26] Titta, on the other hand, would correspond to Greek Χάρις,[27] which seems more suitable for a woman than for a man, and, in fact, in all cases where the gender of its bearers can be deduced, these are women.[28] Doulista is not attested in its present form by other sources, but could well be a variant form of Doulousta, which is also known from contemporary Qasr Ibrim.[29] The name could possibly mean "root" (a similar Old Nubian noun translates Coptic ⲛⲟⲩⲛⲉ).[30]

If the readings proposed above are correct, we lose a Bishop Tamer, but possibly gain some new insight into the status of women in medieval Nubian society. Papsine Doulista, if she is really one and the same person, was certainly a wealthy woman, the patron of a local church, and possibly the central figure of a family that had provided the Church of Nubia with several bishops.

Notes

* The following pages owe much to the incisive criticisms of Adam Łajtar and Joost Hagen, whom I thank for their vigilance.

1 For Vidua, see M. Dewachter, "Le voyage nubien du comte Carlo Vidua (fin Février – fin Avril 1820)", *BIFAO* 69 (1971), 171–89; S. Donadoni, S. Curto, and A.M. Donadoni Roveri, *Egypt from myth to Egyptology*, Milan/Turin: Fabbri 1990, 233–34; W.R. Dawson et al., *Who was who in Egyptology*, London: Egypt Exploration Society 1995, 426.

2 For a fairly complete bibliography, see A. Łajtar, "Varia Nubica (I-II)", *ZPE* 104 (1994), 201–04 at 201.

3 The most recent edition is J. Kubińska, *Inscriptions grecques chrétiennes*, Warsaw: PWN–Éditions scientifiques de Pologne 1974, 42–5, no. 10; a re-edition based on a study of the original stone is a desideratum.

4 By R.S. Bagnall, and K.A. Worp, "Dating by the moon in Nubian inscriptions", *CdÉ* 61 (1986), 347–57 at 353–54, followed by Łajtar, "Varia Nubica (I-II)", 201.

5 See S. Jakobielski, *A history of the bishopric of Pachoras on the basis of Coptic inscriptions*, Warsaw: PWN–Éditions scientifiques de Pologne 1972, 166–67; W. Godlewski,

"The bishopric of Pachoras in the 13th and 14th centuries", in: C. Fluck, L. Langener, and S.G. Richter (eds.), *Divitiae Aegypti: Koptologische und verwandte Studien zu Ehren von Martin Krause*, Wiesbaden: Reichert Verlag 1995, 113–18 at 114.

6   Donadoni, Curto, and Donadoni Roveri, *Egypt from myth to Egyptology*, 234.

7   See A. Łajtar, *I. Khartoum Greek* 3, *ad* l. 6.

8   It is usually read, together with the following group, as: . . . π|σίν, thus e.g. Tibiletti Bruno, *Iscrizioni*.

9   Thus Lefebvre, *Recueil*, no. 636, l. 18, transcribes την δ δουλ(ην) σ̄(ου), and desperately notes: "δ = δέ?".

10 An example of an exactly similar lemma: *I. Khartoum Greek* 20, l. 8.

11 Earlier readings seem to be prompted by the desire to read the preposition ἐν here; thus Tibiletti Bruno, *Iscrizioni*, read: ἐν τ.

12 The same abbreviation is found e.g. in *I. Khartoum Copt.* 2, l. 1 (a king's mother, Faras, A.D. 930), and *passim* in legends of representations of the Virgin Mary, cf. S. Jakobielski *apud* K. Michałowski, *Faras: Die Wandbilder in den Sammlungen des Nationalmuseums zu Warschau*, Warsaw: Wydawnictwo Artystyczno-Graficzne/ Dresden: VEB Verlag der Kunst 1974, 302–03.

13 "Pakhoras – Bakharâs – Faras in geography and history", *JEA* 11 (1925), 259–68 at 262. Only, Griffith took the phrase of l. 19 to mean "daughter of Tamer the bishop of Pakhoras", for which he had to claim the influence of Nubian word order.

14 Łajtar, "Varia Nubica (I-II)", 201–03, following a suggestion of G.M. Browne, took the group read here as ⲟⲩⲅⲣ′(ⲁⲧⲏⲣ) ⲧ̄ⲧ̄ⲧⲁ for an Old Nubian rendering of the Greek ἐλέει θεοῦ.

15 Łajtar, "Varia Nubica (I-II)", 201–03.

16 For further examples, see A. Łajtar, and J. van der Vliet, "Rich ladies of Meinarti and their churches", *JJP* 28 (1998), 35–53 [Study 28].

17 Further examples: Łajtar, *I. Khartoum Greek*, 258.

18 Cf. Jakobielski, *History*, 166: "illogical use of both masculine and feminine genders".

19 Thus Griffith, "Pakhoras", 262, n. 6.

20 For examples, see A. Łajtar, *I. Khartoum Greek*, 63.

21 On the other hand, the Greek stela of Marianta a.k.a. Asta from Qasr Ibrim (edition forthcoming by Adam Łajtar [= *I. Qasr Ibrim* 57]) shows that both names need not always be used together.

22 See M. Krause, "Bischof Johannes III von Faras und seine beiden Nachfolger: Noch einmal zum Problem eines Konfessionswechsels in Faras", in: *Études nubiennes: colloque de Chantilly, 2–6 juillet 1975*, Cairo: IFAO 1978, 153–64 at 154–58; cf. A. Łajtar, "Varia Nubica VI–VII", *ZPE* 137 (2001), 183–86 at 186, n. 5.

23 Of course, the identity of the son of Papsine Doulista, as the local bishop, would have been generally known.

24 Letter from a bishop writing from the south (Faras? Saï?) to his colleague at Qasr Ibrim, using the same abbreviations as the Turin stela; publication forthcoming by J.L. Hagen.

25 For more examples, see Łajtar and van der Vliet, "Rich ladies".

26 See R.S. Bagnall, and A. Łajtar, "*Collection Froehner* 81 once again", *JJP* 24 (1994), 11–12; the form Papsine is found in G.M. Browne, *Old Nubian texts from Qasr Ibrîm* III, London: Egypt Exploration Society 1991, no. 44, l. 7.

27 See G.M Browne, *Old Nubian dictionary*, Louvain: Peeters 1996, 175.

28 See Browne, *Old Nubian texts*, index, 121, s.v.

29 Ibidem, no. 44, l. 10 (gender unspecified).

30 Browne, *Old Nubian dictionary*, 54, s.v.

28

RICH LADIES OF MEINARTI AND THEIR CHURCHES
With an appended list of sources from Christian Nubia containing the expression "having the Church of so-and-so"

Adam Łajtar and Jacques van der Vliet

Introduction

The position of women in medieval Nubia has never, as far as we are aware, been the object of systematic investigation. Nevertheless, indications of their independent status are numerous and appear in different social contexts. Thus, in the sphere of royalty, the queen mother enjoyed a privileged position, reflected in paintings and inscriptions.[1] Usually, this phenomenon is related to the supposed principle of matrilineal succession.[2] As for the less exalted strata of society, business documents from Lower Nubia show a remarkable number of women acting independently in financial transactions.[3] Equally conspicuous is the percentage of female donors of wall paintings in major centers likes Faras and Dongola.[4] Finally, also among those who could afford inscribed tombstones with a Greek or Coptic text, obviously only a small portion of Nubian society, we find quite a number of women.

The present squib discusses a small group of funerary stelae of women, written in Coptic and discovered at the Nubian site of Meinarti. It will, more particularly, focus on the expressions used in two of them to describe these women as "owners" of an ecclesiastical establishment, either a church or, perhaps, a monastery. For a better understanding of these expressions, an inventory of their occurrences in medieval Nubian texts has been appended. As a first attempt to contextualize and interpret the Nubian phenomenon of church ownership, the following pages do not, however, pretend to be definitive, neither in their analysis of the sources nor in their conclusions.

Three Coptic stelae from eleventh-century Meinarti

Meinarti was a village-size settlement on an island which occupied a strategic positon at the northern end of the Second Cataract area.[5] Before being flooded, in the sixties of this century, the site has been the object of large-scale excavations. Unfortunately, these excavations have never been fully published and their results have been only summarily evaluated.[6] For the medieval period, the archaeological evidence appears to point at a period of "urban renewal" perhaps towards the end of the tenth century A.D., when the village of Meinarti received its definitive shape. This phase of urban renewal, characterized by spacious houses, was followed by a "long period of stability".[7] Apparently, these years of renewal and prosperity produced the main group of surviving text material associated with Meinarti: a small series of funerary stelae, in Greek, Coptic and Arabic, part of them bearing eleventh- and twelfth-century dates.[8]

These written sources, too, received very little scholarly attention and have remained partly unpublished. One of the texts, a Greek stela of A.D. 1161, commemorates Ġoassi, son of Sentiko(l), eparch of Nobadia and Choiakeikšil.[9] This piece of evidence suggests that, if Meinarti was not at the time itself a center of political power, it was at least in contact with such centers.[10] This holds certainly true if Meinarti has to be identified with the "Island of Michael", repeatedly mentioned by Arab historians from the end of the thirteenth century onwards. Closely linked with the highest political authority in Lower Nubia, the eparch, this town played an important role in the military struggle of the Nubians with Mameluke Egypt, during the thirteenth and fourteenth centuries.[11] It should be noted, however, that the identification of Meinarti with the "Island of Michael" rests upon an etymology which, although plausible, is by no means certain.[12]

To medieval Meinarti belongs a small but interesting group of three Coptic funerary stelae, now in the National Museum of Khartoum. They are remarkable for their size and material, a rather good-quality sandstone, for the length and fullness of their almost identical inscriptions and for the comparatively rich decoration of at least one of them. Moreover, they were all made for women. Two of our three stelae still carry a date which situates them in the eleventh century. The third one should almost certainly be dated to the same period.

The first of these three stelae has been known since 1935, when its text was published by U. Monneret de Villard.[13] It belongs to Michaeliko (ⲙⲓⲭⲁⲏⲗⲓⲕⲟ),[14] the daughter of a priest Iohannou, and is dated to 11 December 1037 A.D. (Khoiak 15, 754 A.M.). The stone lacks real sculptured decoration, although the text-field is surrounded by an embossed border of about 2.5–3 cm in width.

A second stela is as yet unpublished and bears the name of Iesousyko (ⲓ̈ⲏⲥⲟⲩⲥⲩⲕⲟ), who died 9 January 1046 A.D. (Tybi 14, 762 Diocletian era).[15] Among stelae from Christian Nubia, where unembellished plaques highlighting a plainly incised text are more or less the standard,[16] it stands out by its nicely sculptured and painted decoration, which clearly continues late antique patterns. Its text is framed by an *aedicula* consisting of two columns which support an arched canopy containing a conch.[17] Within Nubia, it has close iconographical parallels at i.a. Ibrim (twelfth-century bishops' stelae),[18] Toshka West (Sakinya),[19] Faras (stelae

348

of Bishops Aaron, A.D. 972, and Georgios, A.D. 1097),[20] Serra West[21] and Debeira (Komangana; stela of high official Staurophoros, A.D. 1069; our no. 11, below), sites which all are situated in the area north of Meinarti.

A third Meinarti stela survives, unfortunately, only as a fragment.[22] Remains of an embossed border suggest that it may have been decorated too. However, just enough is left to obtain a general idea of its funerary formulae and to read the name of its owner, a woman πετρον[ια], Petronia (1. 6).[23] No further information about her person or lifetime has been preserved.

All three stelae bear an elaborate epitaph which, next to Coptic, incorporates elements in Greek (Michaeliko, Iesousyko) and Old Nubian (Iesousyko). The structure of these three texts is remarkably similar. They bear, moreover, a striking ressemblance to the epithaph of a Deacon Petrou, the son of a bishop of Kourte, which originates from the same period (A.D. 1029) and region (Debeira-West).[24] Thus, the stela of Michaeliko contains, following an extensive opening formula of the προνοια type, which quotes God's judgment on Adam (Gen. 3:19; ll. 1–4), and lemmas containing the name of the deceased and the date of her demise (ll. 4–8), a long composite prayer (ll. 8–21). This is followed by a lemma stating the age of the deceased (ll. 21–2) and by another prayer, now in Greek, for resting her soul (ἀνάπαυσον . . . ; ll. 23–5). Especially, the prayer of ll. 13–14 (absent from Iesousyko): "may He (i.e. God) cause the Archangel Michael to watch over her bones" is a noteworthy feature, which, to the best of our knowledge, occurs outside Meinarti only in the stela from Debeira West, mentioned above. [25]

The general impression which these three ladies' stelae convey is of conforming closely to a model which was current among the clerical and lay élite of the region during the tenth-eleventh centuries. As we will see presently, in the cases of Michaeliko and Iesousyko, something more can be said about their socio-economic status.

"Owning" a church in medieval Nubia

The stela of Iesousyko, who died at the age of thirty-four, mentions no parents. She is, however, qualified as: ἐχων φιλοθεϲ ἐγο῁π῾μογλογε (ll. 6–7).[26] Although part of this phrase, as it switches to Old Nubian, is difficult to interpret, it clearly begins with the Greek loanword ἔχων. The participle ἔχων, in combination with church names and toponyms, occurs several times more in Nubian epigraphy, both on stelae and in graffiti or *dipinti*, as well as in a documentary text (see the provisonal list of sources given below).[27] It qualifies the name of a person who, often next to other qualifications, is said to "have" a certain ecclesiastical establishment in a certain place. A rather well-known example is provided by the Greek funerary stela of Abba Marianos (our no. 2, below), who is styled "orthodox bishop of Faras and archimandrite of Pouko and envoy to Babylon (= Cairo) and ἔχων τέσσαρα ζῷα νῆσος Τημε: having (the Church of) the Four Living Creatures (on) the Island (of) Teme" (ll. 8–9). We may, therefore, expect the same ἔχων in the stela of Iesousyko to introduce the name of an ecclesiastical establishment. In fact, there can be little doubt that the ϕιλοθε<ο>c of the Iesousyko stela stands for St. Philotheos of Antioch, the well-known martyr who was so widely venerated in Egypt that it is not astonishing to find traces of his cult in Nubia too.[28] What follows appears to contain a form of the Old Nubian verb ⲙⲟⲩⲗⲟⲩ-, "to lead, to guide", but remains obscure otherwise. It almost certainly serves to identify the St. Philotheos Church in question[29] and may therefore either specify its location or give its (or, the saint's) local surname.[30]

Until quite recently, the expressions with ἔχων (etc.) seemed limited to Greek or Old Nubian, resp. Greco-Nubian, texts and the person thus styled always a man.[31] Iesousyko's stela is conspicuous for the first attested occurrence of this Greco-Nubian expression in a predominantly Coptic context.[32] Moreover, Iesousyko was a woman.

Here, the stela of Michaeliko offers an interesting parallel. Michaeliko, who died at the age of seventy-five, is qualified as: ⲉⲩ͑ⲧ͑ ⲛⲓⲱⲩ ⲡⲉⲡⲣⲉⲥ ⲡⲉⲧⲟⲩⲁⲛⲧⲥ · ⲙⲓⲭⲁⲏⲗ · ⲛⲁⲣⲅⲓⲛⲏ, "the daughter (θυγάτηρ) of Iohannou the priest (πρεσβύτερος), she (or, he) who has (the Church of) Michael at Argine" (ll. 6–7). There can be no doubt that ⲡⲉⲧⲟⲩⲁⲛⲧⲥ (which in standard Sahidic should be either ⲧⲉⲧⲉⲟⲩⲛ̄ⲧⲁⲥ or ⲡⲉⲧⲉⲟⲩⲛ̄ⲧⲁϥ)[33] is the Coptic equivalent of the Greek loanword ⲉⲭⲱⲛ used on the stela of Iesousyko. Owing to the Nubian uncertainty with respect to grammatical gender, which can also be noticed in the present stela, one might hesitate whether this epithet, "having", goes with Michaeliko or with her father Iohannou: the head of the epithet is masculine, the possessor suffix being feminine. However, supposing the feminine to represent the marked choice, we would prefer to recognize in Michaeliko herself the owner of St. Michael's Church.[34] That this is perfectly possible for a woman is shown by the stela of Iesousyko.

Unlike the Church of St. Philotheos owned by Iesousyko, Michaeliko's church is identified by a known toponym. It is situated not in Meinarti itself, but in a place called Argine, which occurs as Argini in our no. 12 below. This has been quite plausibly identified with modern Argin, a district on the west bank north of the Second Cataract, a few kilometers from Meinarti.[35] There, the ruins of a considerable church, perhaps that of St. Michael, could be seen even quite recently.[36]

What, we may presently ask, is the exact meaning of the expression ἔχων (etc.) and its equivalents? Especially since we are dealing with women in the Meinarti stelae, it seems less plausible that these could be "titulars" or "prebendaries" of any church or monastery, or that they could have a church "in consegna, in cura", as S. Donadoni proposed.[37] They might, of course, be the founders of the churches in question, but this is not what the texts say. Both the Greek ἔχων and its Coptic and Old Nubian equivalents are clearly denoting a relation of "possession": the persons in question are said to "have" a sanctuary. For a parallel it seems appropriate to look where we can best expect it, in Byzantine Egypt.

Actually, in Byzantine Egypt private persons could, in spite of Justinianian legislation, be the full juridical owners of an ecclesiastical establishment.[38] As

RICH LADIES OF MEINARTI AND THEIR CHURCHES

its owners they might or might not be the founders of their sacred property.[39] The rights of ownership over a church or a monastery could be transferred by testament. They could, moreover, be exercised by women as well.[40] Although the precise juridical and economic implications of their ownership remain obscure,[41] we may suppose our ladies from Meinarti to have been the wealthy patrons of ecclesiastical establishments. They may have either founded these establishments themselves or acquired them by inheritance. As the case of Michaeliko shows, their property could be situated at some distance from the place where they lived or, at least, died.[42]

The record of the other Nubian examples of the expression ἔχων (etc.) and its Coptic and Old Nubian equivalents (see the list below) may serve to supplement this picture. First of all, Iesousyko and Michaeliko are not the only ladies said to "own" a church: our nos. 12 and, perhaps, 17 are additional examples. Secondly, in as far as our examples are dated, it may be seen that they all belong to the eleventh and twelfth centuries, dates running from A.D. 1036 to 1184. Thirdly, as our no. 17, from the Old Dongola region, shows, examples are geographically not limited to Lower Nubia alone. Finally, where professions or titles are mentioned, we count among the "owners" of churches a deacon, priests (as well as the daughter of a priest), bishops, high civil authorities and even a king. From all this we may conclude that "owning" a church was considered *de bon ton* among the upper classes of later medieval Nubia. In styling themselves patrons of a church, rich Nubians of both sexes proclaimed their status as well as their piety.

Appendix

List of sources from Christian Nubia containing the expression "having the Church of so-and-so"

The present list bears a provisional character and is merely intended as a preliminary for more detailed studies. It is arranged geographically, from north to south, after the find-spots of the texts. Every entry contains the following elements:

- character of the source;
- find-spot and present whereabouts of the artifact in question;
- date;
- basic bibliographical data;
- language of the source;
- quotation of the fragment with the expression "having the Church of so-and-so"; if this expression is part of the more elaborate titulature of a person, we quote the whole titulature for prosopographical reasons; while quoting texts we occasionally propose corrections for the readings, sometimes important ones; we do not correct orthographic and syntactic mistakes abounding in Nubian Greek; Greek texts are normally printed in minuscles; when a text is a mixture of Greek and Old Nubian we give it in majuscles;

351

ADAM ŁAJTAR AND JACQUES VAN DER VLIET

- translation;
- commentary.

1. Graffito in the temple of Ramses II in al-Sebu'a, which was turned into a church in the Christian period; date unknown. Published by F.L. Griffith, *The Nubian texts of the Christian period*, Berlin: Verlag der Königl. Akademie der Wissenschaften 1913, 60–2, with pl. 3. Republished by G.M. Browne, "Griffith's Old Nubian Graffito 4", *ÉtudTrav* 17 (1995), 17–21. The graffito contains a prayer to St. Peter the Apostle in Old Nubian (ll. 1–7), set up by one Peter the priest as is stated in a subscript written in a mixture of Greek and Old Nubian (ll. 7–8). Here follows the reading of the subscript, as established by G.M. Browne:

ⲉⲅⲱ ⲡⲉⲧⲣⲟ ⲓ̅ⲥ̅ ⲭ̅ⲥ̅ ⲡⲁⲭⲱⲣⲁⲥ ⲉ ⲓⲟⲛ ⲕ`ⲡⲁ´ ⲥⲧⲁⲩⲣⲟⲥ ⲡⲁⲭⲱⲣⲁⲥ
ⲉⲛⲫⲓⲁ ⲉ̄ⲗⲓ ⲉⲧⲁⳅⲁ

I, Peter, being priest of Christ of Pachoras and . . . of the cross of Pachoras, today have set (this) up.

We suggest to read ⲉⲭⲱⲛ after ⲡⲁⲭⲱⲣⲁⲥ. In Old Nubian majuscles ⲱ and ⲓⲟ look very much alike and could have been easily mistaken by Griffith, whose copy serves as a basis for any attempt at reading. The sense of the subscript would be: "I, Peter, the priest, having the Christ-Church of Pachoras and also being (something) of the Cross-Church of Pachoras, today have set (this) up".

The Jesus Christ-Church of Pachoras is also mentioned in some graffiti in the cathedral and in the Rivergate Church at Faras. Stefan Jakobielski (*Bishopric of Pachoras*, 170–75 and 189) proposed to identify it with the Rivergate Church. The Cross-Church of Pachoras is attested also in a graffito near Ashkeit, below, no. 13.

2. Funerary stela of Marianos, Bishop of Faras (died 11 November 1036 A.D.), found in Qasr Ibrim, now in the Coptic Museum at Cairo. The stela bears a Greek epitaph with a prayer of the Euchologion Mega type. It was edited by J.M. Plumley, "The Stele of Marianos Bishop of Faras", *Bulletin du Musée national de Varsovie* 12 (1971), 77–84, and reproduced by J. Kubińska, *Inscriptions grecques chrétiennes*, Warsaw: PWN–Éditions scientifiques de Pologne 1974, 38–40, no. 8 [= *I. Qasr Ibrim* 22, where the bishop is called Marianou].

In ll. 6–9 the epitaph gives Marianos the following titles (here and in other epitaphs of the Euchologion Mega type, the genitives depend on either ἀνάπαυσον τὴν ψυχὴν τοῦ δούλου σου or σὺ γὰρ εἶ ἀνάπαυσις τῆς ψυχῆς τοῦ δούλου σου):

ἄββα Μαριανοῦ ὀρθ(οδόξου) ἐπισκ(όπου) Παχωρας
κ(αὶ) ἀρχιμανδρίτης Πουκω κ(αὶ) ἀπόστολος τοῦ

RICH LADIES OF MEINARTI AND THEIR CHURCHES

8 Βαβυλὼν κ(αὶ) ἔχων τέσσαρα ζῷα νῆσος
Τημε

(Rest the soul of Your servant) Abba Marianos, orthodox bishop of Faras and archimandrite of Pouko and envoy to Babylon (= Cairo) and having (the Church of) the Four Living Creatures (on) the Island (of) Teme.

The Island of Teme is otherwise unknown. A Church of the Four Living Creatures is attested three times elsewhere: (1) in a *dipinto* in the Faras cathedral, also in connection with ἔχων (see *infra*, no. 6); (2) in a Coptic epitaph of a priest Marianou from Faras, dated A.D. 955 (cf. J. van der Vliet, "Churches in Lower Nubia, old and 'new'", *BSAC* [38 (1999), 141–42 = Study 24]); 3) in graffiti in the church of Sonqi Tino, again in connection with ἔχων (see *infra*, no. 16). Nos. 1 and 2 may well refer to the same church, which was probably situated in Faras or its near vicinity; for the sanctuary of the Island of Teme and for the one mentioned in Sonqi Tino, which may have been a monastery, this must remain uncertain.

3. Old Nubian document from Qasr Ibrim from the second half of twelfth century: a contract of sale and cession of land by a certain Poṅitta, his daughter Persi as well as Ṅonnen, daughter of Mena to Maššuda, Choiakeikšil: G.M. Browne, *Old Nubian texts from Qaṣr Ibrîm* III, London: Egypt Exploration Society 1991, no. 34ii. Among witnesses to the contract, the document mentions in line 18: "Eiñitta, who has the Jesus-Church of the Mountain". The church in question is otherwise unattested and its location is unknown. It is most probably to be situated somewhere in the vicinity of Qasr Ibrim.

4. Graffito on a wall of the Church of the Archangel Raphael in Tamit; Greek with Old Nubian intercalations; date unknown: S. Donadoni, *Tamit (1964): Missione archeologica in Egitto dell'Universita di Roma*, Rome: Istituto di Studi del Vicino Oriente Universita 1967, 67, no. 13:

[εγ]ω φιλο(θεoc) ⲇⲓⲁⲕⲟⲛⲟⲩ ⲩ̅ⲥ̅
[.]ⲁⲣⲓⲁⲛⲧⲁ ⲉⲭⲱⲛ ⲡⲟⲩⲣⲅⲟⲩⲛⲇⲓ
[ⲉ]ⲡⲁⲅⲱⲙⲉⲛⲟⲥ ⲉⲟⲣⲑⲏⲛ ⲣⲁⲫⲁⲏⲗ
4   [ⲉ]ⲅⲣⲁ⳿ⲥⲁ

I, Philotheos, deacon, son of [.]arianta, having Purgundi, have written (this) during the feast of the Archangel Raphael (of the) epagomenal day.

The name of Philotheos' father should most probably be read [ⲙ]ⲁⲣⲓⲁⲛⲧⲁ. Donadoni (p. 67) identifies Purgundi in this inscription with modern Furgundi, a locality on the East bank of the Nile, ca. 5 km north of Abu Simbel, opposite of Tamit.

In the majority of cases cited in this list, the participle ἔχων or its Old Nubian and Coptic equivalents are connected with the name of a church, sometimes additionally identified by a place-name or another topographical specification. However, here and probably also in a *dipinto* from Faras cathedral (below, no. 6) and in the epitaph of Angelosko from Sheikh Arab Hag (below, no. 17), merely a place-name appears. In our opinion, also these three toponyms which appear in connection with ἔχων, viz. Pourgoundi, []touri and Komati, should be understood as names of churches. We imagine the situation thus: Pourgoundi, []touri and Komati were small villages boasting one church each. The names of the villages were for some reason so distinctive that their churches were called, at least in everyday speech, not by their official names ("The Church of St. N.N."), but by the name of the village. The transfer of toponyms to cult places (temples and chapels) is well attested for Greco-Roman Egypt. We can also cite a parallel from present-day Poland. The village of Drelów in East Poland has a parish church of Our Lady that occupies the center of the village.[43] Three kilometers away, still within the administrative borders of Drelów, in a locality called Horodek (a Ukrainian loanword meaning "a small fortified town"), on a small hill among fields, meadows and trees, lies the Church of St. Onnophrios. The church is never or very rarely referred to as the Church of St. Onnophrios but only as Horodek. Even the parson, announcing the mass at St. Onnophrios in the parish church, says: "the mass will take place in Horodek". Pourgoundi, Komati and []touri could have been, *mutatis mutandis*, something similar to Horodek.

5. Graffito in the Church of the Angels in Tamit; date unknown. U. Monneret de Villard, *Nubia* I, 157, no. 26; S. Donadoni in: *Tamit (1964)*, Rome 1967, 71, no. 29 b. The language of the graffito is Old Nubian with an admixture of Greek.

+ ⲙⲉⲡⲟⲗⲉ ⲟⲩⲙⲁⲉ · ⲉⲝⲱⲛ
. ⲧⲏⲙⲧ′ ⲟⲧⲟⲩ ⲧⲉⲗⲓⲣⲉ .. ⲛⲟⲃ′ ⲥⲡ .
ⲭⲁⲣⲧ′ · ⲧⲟⲩⲧ′ ⲓⲥ ⲭⲉ ⲟⲛⲟⲙ′ ⲉⲓⲧⲱⲛ

The graffito does not permit a continuous translation; it apparently begins with names. ⲟⲩⲙⲁⲉ might be an incorrect rendering of ⲟⲩⲙⲁⲥ. This name appears, in a slightly different orthography, in another graffito from Tamit; Donadoni, *Tamit (1964)*, 65–6, no. 7. ⲉⲝⲱⲛ seems to be connected with. ⲧⲏⲙⲧ′.

6. Faras cathedral, east pilaster to the right of the *haykal. Dipinto* E203 a3 written in black ink to the left of a wall painting depicting Bishop Georgios (died 1097). The painting and the inscription are at present in the Sudan National Museum, Khartoum. The inscription remains unpublished. It can be seen on the photo of the painting reproduced in: Jakobielski, *Bishopric of Pachoras*, 159 and fig. 48. Contrary to what Jakobielski says, the inscription is in Greek and not in Old Nubian. It begins with a prayer to the Virgin Mary and then continues with the personal data of a man whose name has not been preserved. Adam Łajtar reads from the photo in the archive of the Research Centre for Mediterranean Archaeology of the Polish Academy of Sciences:

RICH LADIES OF MEINARTI AND THEIR CHURCHES

6   [   ] . ⲚⲔⲟ ⲓⲥ ⲋ ⲡⲁ`ⲗ´ ⲧⲉⲥⲥⲁ`ⲣⲁ´ ⲍⲱⲁ̀ ⲉ̀ⲭⲱⲛ

7   [   ]ⲧⲟⲩⲣⲓ ⲉⲭⲱⲛ etc.

For the Church (or, Churches) of the Four Living Creatures see above, no. 2.
[ ]touri looks very much like an Old Nubian place-name. For toponyms used as
names of churches see above, no. 4.

7.   Faras cathedral. Legend to a painting showing a man in a richly decorated
garment and wearing a royal crown of Nubia (King Moses George). The language of the legend is a mixture of Greek and Old Nubian. According to
stylistic criteria, the painting may be dated to the end of the twelfth century.
The Moses George named in the legend was King of Dotawo in the second
half of the same century.[44]

Facsimile of the inscription in: K. Michałowski, *Faras: Die Kathedrale aus dem
Wüstensand*, Einsiedeln: Benziger Verlag 1967, 38. The first, still incomplete
attempt at reading in: S. Jakobielski, "Some new data to the history of Christian
Nubia as found in Faras' Inscriptions", *Klio* 51 (1969), 503, fig. 6. The reading of
the whole text in: F. Altheim and R. Stiehl, *Christentum am Roten Meer* I, Berlin/
New York: de Gruyter 1971, 487–508. Critical remarks to the latter with a new
proposal for the reading of the text in: S. Jakobielski, "Inscriptions from Faras
and the problems of the chronology of murals", in: *Études nubiennes: colloque de
Chantilly, 2–6 juillet 1975*, Cairo: IFAO 1978, 147–51, pl. XL and XLIIIB.
    Below we reproduce Jakobielski's transcription from *Études nubiennes*, 148,
with some changes resulting from comparison with the facsimile. Also Jakobielski's translation has been modified.

    + ⲟⲩⲧⲟⲥ ⲉⲥⲧⲓⲛ ⲁ̀ⲇⲁⲩⲉ̄ⲗ ⲃⲁ̄ⲥ

    ⲙⲱⲩⲥⲉⲥ ⲅⲉⲱⲣⲅⲓⲟⲩ ⲃⲁ̄ⲥⲓⲗⲉⲩ ⲛⲟⲩⲃ`ⲁ´ⲏ̄ⲥ

    ⲋ ⲁ̀ⲣⲟⲩⲁ`ⲇ´ⲋ ⲙⲁⲕⲣⲟ̀ : . . ⲩ ⲙⲁⲣⲁⳁⲁ ⲅⲉⲱⲣ`ⲅⲓ´ ⲁ̀ⲙⲛ̄

4   ⲟⲓⲕⲟⲇⲟⲙⲓⲥ ⲉⲅⲅⲟⲛⲟⲥ ⲋ ⲡⲁⲗ`ⲗ´ ⲍⲁⲭⲁⲣⲓⲥⲟⲗ`ⲙⲉⲛ´ . ⲧⲟ`ⲛ´

    ⲋ ⲡⲁⲗ`ⲗ´ ⲇⲁ̄ⲇ ⲃⲁⲥⲓⲗⲉⲩ ⲉⲅⲅⲟⲛ[ⲟⲥ ⲁ]ⲙⲛ̄ ⲣⲛ̄ⲃ ⲉ̀ⲭⲱⲛ : —

This is the Great King Moses George, King of (the) Nobadians and Alwa
and Makuria. Being descendant of Maraña George, the Builder and also
(something) of Zacharias and also being descendant of King David, having (the Church of) the Virgin Mary.

Although this inscription remains obscure on many points, the reading of the very
end of l. 5 seems obvious to us. ρνβ is a very well known numerical cryptogram
for Μαρία (40 + 1 + 100 + 10 + 1 = 152) and ρνβ ἔχων must mean "having (the
Church of) the Virgin Mary"; a numerical cryptogram in connection with ἔχων
is also used in the epitaph of Bishop Tamer, below, no. 9: ἔχων χπθ Παχωρας.
It is not quite clear to whom the title ρνβ ἔχων in this inscription refers. Two

355

possibilities exist: (1) King Moses George, (2) King David who was an ancestor of Moses George. The former possiblity appears to be more probable.

The Church of the Virgin Mary mentioned here was in all probability situated in Faras or its immediate vicinity. One should note the absence of topographical qualifications as if it was obvious to everybody which particular church of the Virgin Mary was concerned; note, furthermore, that the painting and its accompanying inscription adorned a wall of Faras cathedral. Under this circumstances one is tempted to identify the Church of the Virgin Mary from this inscription with the Cathedral of Faras itself which, as is well known, was dedicated to the Virgin Mary; cf. Jakobielski, *Bishopric of Pachoras*, 176–86. If this identification is correct we obtain a very nice and significant picture: a member of the Nubian royal family, actually the king himself, "owned" the main church of the capital of Nobadia.

8. Greek graffito in the Rivergate Church at Faras; date unknown. F.L. Griffith, *LAAA* 13 (1926), 74, no. 14, pl. LXIV 2.

    ϵⲅⲱ ⲏⲗⲓ ⲡⲁⲭ[ⲱ-]
    ⲣⲁⲥ ⲧⲱⲙⲁ ⲉⲭⲱ[ⲛ]

    I, Eli, having Toma of Faras.

The word ⲧⲱⲙⲁ appears also in an unpublished *dipinto* E251 h4 from a wall of Faras cathedral:

    ⲭⲁⲓⲣⲉ ⲙⲁⲣⲓⲁ ⲕⲉⲭⲁⲣⲓⲧⲟⲙⲉⲛⲛ[ⲏ - - -]
    ⲕⲁ`ⲙ´ ⲱⲩⲅⲣⲁⲙⲱⲛⲏ ⲓ̅ⲥ̅ ⲙⲉ`ⲅ´ ⲧⲱⲙⲁ [- - -]
    ⲡⲁⲭⲱⲣⲁⲥ ⲋ ⲁⲩ`ⲑ´ . . ⲅⲣⲁⲫⲏ

ⲧⲱⲙⲁ may be identical with ⲧⲟⲛⲙⲁ, which is found, in combination with ⲉⲭⲱⲛ, in a subscript to an unpublished *dipinto* G61 h3, again from Faras cathedral:

                    ⲱⲩⲅⲣⲁⲙⲱⲛⲏ
    ⲕ/ ⲙⲉ`ⲅ´ ⲧⲟⲛⲙⲁ ⲉⲭⲱ [*blank*] ⲛ [*blank*] ⲅⲣⲁϯⲟⲛ

We are unable to identify this ⲧⲱⲙⲁ (ⲧⲟⲛⲙⲁ).

9. Funerary stela of Tamer, Bishop of Faras who died on 30 March 1184 A.D. The stela was found around 1820 by Carlo Vidua in Kolasuča (Faras); since, it is kept in the Museo Egizio at Turin. The stela is inscribed with a Greek epitaph of the Euchologion Mega type. It has been edited several times; for the complete bibliographical lemma, see A. Łajtar, *ZPE* 104 (1994), 201, n. 2; cf. also idem, *ZPE* 113 (1996), 105, no. 6. The most recent edition of the full text is: Kubińska, *Inscriptions grecques chrétiennes*, 42–5, no. 10. The correct

RICH LADIES OF MEINARTI AND THEIR CHURCHES

reading of ll. 20–1, which are crucial for our present interest, was established by A. Łajtar, *ZPE* 104 (1994), 201–03. The deceased is styled thus:

20 Ταμηρ ἐπισκ(όπου) Παχωρας θ(εο)ῦ τ͞τ-
21 τα (καὶ) ἔχων χπθ Παχωρας

(You are the rest of) Tamer, by the grace of God Bishop of Faras, and having (the Church of the Archangel) Michael in Faras.

[N.B. J. van der Vliet corrected the reading of the stela in *JJP* 37 (2007), 185–91 ll. 18–19: ⲁⲟⲩⲗ͞ⲥ ⲧⲁ ⲙⲏ(ⲧⲏ)ⲣ ⲉ̇ⲡⲓⲥˋⲕ′(ⲟⲡⲟⲩ) ⲡⲁⲭⲱⲣⲁⲥ ⲑⲩˋⲅ′(ⲁⲧⲏⲣ) ⲧ͞ⲧⲧⲁ, "Doulista, mother of the bishop of Faras, daughter of Titta"; the abbreviations are solved here by the editors.]

Until now, the Church of the Archangel Michael in Faras has not been attested elsewhere besides this inscription. Jakobielski (*Bishopric of Pachoras*, 188) tentatively identifies it with the church ruins in present day Nabindiffi, not far from Kolasuča where the stela was discovered.

10. Epitaph of Iesousinkouda (died October 27th, 1102); provenance unknown, probably Wadi Halfa-region; Sudan National Museum Khartoum, without number; A. Łajtar, "The epitaph of Iesousinkouda, eparch of Nobadia (?), domestikos of Faras and nauarchos of the Nobadae, died A.D. 1102", *Gdańsk Archaeological Museum African Reports* 1 (1998), 73–80 [= *I. Khartoum Greek* 5]. This epitaph is in Greek with Old Nubian intercalations. It is not of the Euchologion Mega type, which is surprising for the beginning of the twelfth century. Iesousinkouda, who died at the age of thirty-six, is thus designated:

ιηϲⲟⲩϲⲛ͞ⲕⲟⲩⲁ
α υ(ἰὸ)ς ἔχων Γ(α)β(ριὴ)λ ἐξ-
8 ουσία{ι} δὲ (καὶ) δω-
μίστικος Παχ-
ωρας ναυάρχ(ος)
Νο(βάδων)

Iesousinkouda, son of the one who has (the Church of the Archangel) Gabriel, power[45] and *domesticus* of Faras, *nauarches* of the Nobadae.

L. 20, which is a later addition to the epitaph but done by the same lapicide who is responsible for the rest of the inscription, specifies that the church "owned" by the father of Iesousinkouda was that of Phrim (Ibrim):

20 ἔχων Γαβριὴλ ⲫⲣⲙ.

357

The Church of the Archangel Gabriel of Phrim is attested in an Old Nubian document from Qaṣr Ibrim from the second half of the twelfth century; G.M. Browne, *Old Nubian texts from Qaṣr Ibrîm* III, London: Egypt Exploration Society 1991, no. 60, l. 10. There it is stated that the land of the Gabriel-Church in Ibrim bordered to the south on the possessions of the Jesus-Church of Touggili.

11. Funerary stela of Staurophoros (died 4 December 1069) found by the Scandinavian Joint Expedition in Komangana (Debeira), now kept in Uppsala. The stela is inscribed with an epitaph of the Euchologion Mega type, edited by T. Hägg in: T. Säve-Söderbergh (ed.), *Late Nubian cemeteries*, Solna: Esselte Studium 1981, 56–9. The deceased is presented twice in this epitaph: ll. 7–8 list his secular titles, in ll. 24–5 his ecclesiastical title is given:

6   Σταυροφόρου
7   μεί(ζονος), ναυάρχ(ου) Νοβ(άδων), ναυ(κράτου) Ἑπτὰ χῶραι

24   Σταυροφόρου ἔχων Σευῆρος
25   ⲁ̄ⲡⲡⲉⲁⲱ

(Rest the soul of Your servant) Staurophoros, *meizon, nauarches* of the Nobadae, *naukrates* of the Seven Lands.

(You are the rest of the soul of Your servant) Staurophoros, having the Church of Severus in the town.

In l. 24 the stone has ⲉⲭⲱⲏ́. In his edition, Hägg left this in majuscules, but suggested the reading ἔχων, which, in our opinion, is indubitable. Severus, whose church is concerned here, must have been the famous anti-chalcedonian patriarch of Antioch from the first half of the sixth century. He had close ties with the anti-chalcedonian church of Alexandria and after his death was much venerated in the Nile valley; cf. H. Brakmann, "Severos unter den Alexandrinern. Zum liturgischen Diptychon in Boston", *JbAC* 26 (1983), 54–8.

The words ⲁ̄ⲡⲡⲉⲁⲱ in l. 25 were recognised by Hägg as Old Nubian but not read. We suggest the reading "in the town", from ⲁ̄ⲡⲡ- "town, city" and ⲁⲱ (ⲁⲟ) – "upon, at, to, for, over, etc.". A similarly precise designation in an Old Nubian document from Qasr Ibrim, cited above as our no 3: "having the Jesus-Church of the Mountain". The precise wording of the expression may suggest that there were at least two churches dedicated to Severus: one inside the town, another one outside it. Which town is meant here remains obscure.

12. Epitaph of a woman Eikkir (died 6 August 1084 A.D. or later; for a discussion of the date, see *I. Khartoum Greek* 7). Sudan National Museum Khartoum, inv. 3727. Found at Saqia 43, Ashkeit, near Wadi Halfa. Presented by

Mohamed Mohamed Werdie to the Halfa Museum in 1940, whence it was transferred to Khartoum in 1941. The epitaph is of the Euchologion Mega type. It is in Greek, but the personal data of the deceased given in the end are rendered in Old Nubian. The epitaph was edited by J.W.B. Barns, *Kush* 2 (1954), 28–9, who misunderstood the Old Nubian part. The text was revised by Adam Łajtar during his work on the catalogue of Greek Christian epitaphs in the Sudan National Museum at Khartoum. Below we present the text as established by Łajtar [after *I. Khartoum Greek* 7]:

σύ γὰρ <εἶ> ἀνά-
παυσις · τὴν δ(οὐλην) · εκκιρ ᴀ . .
. τι · ᴀгρινι κονᴀгᴀ ἀπὸ
24 μαρτ(ύρων) ω´ [. .]. · τᴀννᴀ ᴀψεν ου-
κριгоуλλον · οв `λο´ мεсо-
ρεᴀ ïг λο ᴀᴀппонᴀ

You are the rest of Your servant Eikkir, having . . . of Argini. (She died in the year) since the Martyrs 800. The years of her life were seventy-two. She departed on Mesore 13th.

κονᴀгᴀ is an imperfect participle of the verb κον (a variant of κο) = "to have, to owe, to possess"; cf. E. Zyhlarz, *Grundzüge der nubischen Grammatik im christlichen Frühmittelalter (Altnubisch)*, Leipzig: Kraus 1928, 55. It is then an exact equivalent of the Greek ἔχων; compare also cᴀλ ᴀψοιν κοᴀгᴀ (= λόγον ζωῆς ἐπέχοντες) in *Griffith's Old Nubian lectionary* 101, ll. 7–8 (ed. by G.M. Browne, Rome/Barcelona: Papyrologica Castroctaviana 1982).

For Argini see above, note 35. ᴀргινι from this inscription and ᴀргιнн from the epitaph of Michaeliko are merely two forms of the same toponym, "η" and "ι" both being pronounced as "ī" at the time.

The name of the church owned by Eikkir was contained in ll. 22–3, between the name of the deceased woman and the place-name ᴀргινι. We are unable to reconstruct any reasonable word from the preserved rests of letters. In any case, мιхᴀнλ, as in the epitaph of Michaeliko, cannot be read here.

13. Greek graffito on a rock near Ashkeit; date unknown; see J.H. Dunbar, "Some Nubian rock pictures", *SNR* 17 (1934), pl. II (infra; cf. p. 164); U. Monneret de Villard, *Nubia* I, 211.

εгω мερκн ᴃε оθс
πᴀ`х´ θε`г´ стᴀуρος εхω
. гρᴀψον`ε´

ᴃε in l. 1 is the ligature for πρε(σβύτερος).

359

ADAM ŁAJTAR AND JACQUES VAN DER VLIET

The graffito does not allow a continuous translation. It should probably be understood: "I, Merke, the priest . . . having the Cross-Church of Faras, have written (this)". For the Cross-Church in Faras, see above, no. 1.

14. Coptic epitaph of a woman Michaeliko (died 11 December 1037 A.D.) found in Meinarti, now in the Sudan National Museum, Khartoum [=*I. Khartoum Copt.* 18].

ⲧⲘⲁⲕⲁⲣⲓⲁ
6   ⲘⲓⲬⲁⲉⲗⲓⲕⲟⲗ ⲉⲩˋⲅˊ Ⲛⲓⲱⲩ ⲡⲉⲡⲣⲉⲥ ⲡⲉⲧⲟⲩⲁⲛⲧⲥ · ⲘⲓⲬⲁⲏⲗ · ⲚⲁⲣⲅⲓⲚⲏ

The blessed (μακάρια) Michaeliko(l), the daughter (θυγάτηρ) of Iohannou the priest (πρεσβύτερος), she who has (the Church of St.) Michael at Argine.

For a detailed discussion see above.

15. Coptic epitaph of a woman Iesousyko (died 9 January 1046 A.D.) found in Meinarti, now in the Sudan National Museum, Khartoum [*I. Khartoum Copt.* 19]. For a detailed discussion see above.
16. The expression "having the Church of so-and-so" is attested in graffiti on the walls of the church in Sonqi Tino, too. Unfortunately, the epigraphic material from this site, apparently rich in data of different kinds, remains unpublished. It would appear from a somewhat unclear presentation by S. Donadoni, in: K. Michałowski (ed.), *Nubia: récentes recherches*, Warsaw: Musée national 1975, 35, that this expression occurs at least two times there, namely: ⲉⲭⲱⲚ ⲧⲉⲥⲥⲉⲣⲁ ⲍⲱⲁ and ⲉⲭⲱⲚ [. . .] ⲫⲩˋⲗˊ (or, in a reversed order: ⲧⲉⲥⲥⲉⲣⲁ ⲍⲱⲁ ⲉⲭⲱⲚ and [. . .] ⲫⲩˋⲗˊ ⲉⲭⲱⲚ). According to Donadoni, "the Four Living Creatures" mentioned here was a monastery. See, too, above, no. 2. We are unable to suggest anything for [. . .] ⲫⲩˋⲗˊ.
17. Greek epitaph of the Euchologion Mega type, commemorating one Angelosko. The epitaph was found in Sheikh Arab Hag near Old Dongola; now in the Sudan National Museum at Khartoum. Its date is unknown. See J. W. Crowfoot, *JEA* 13 (1927), 228–29, no. 2 (*SB* IV 7429; *SEG* VIII 872; Tibiletti Bruno, *Iscrizioni nubiane*, no. 2).

In ll. 17–18, the epitaph reads: σὺ γὰρ εἶ ἀνάπαυσις τῶν σῶν δοῦλόν σου Ἀγγελοσκω ΚΟΜΑΤΙΕΧΩΝ. The letters ΚΟΜΑΤΙΕΧΩΝ have been variously interpreted by previous editors. Crowfoot transcribed them κοματι εχων but translated "(Thy servant Angelosko) Komatiekhon". In his commentary he put forward a hypothesis, however with a question mark, "that the name was Komati and that the last four letters are the present participle of the verb ἔχων, governing δοῦλον". Tibiletti Bruno printed Ἀγγελόσκω Κοματιέχων and suggested in her critical apparatus the reading τοῦ σοῦ δούλου (σου) . . . Κοματιέχου. According to this suggestion, Komatiechon would have been either the second name of Angelosko or the name of his (her) father. We are convinced that the correct reading of the passage in question is: Ἀγγελοσκω Κοματι ἔχων. Komati must have been a toponym just like Purgundi in the graffito from Tamit (no. 4 above). It is probably to be situated somewhere in the vicinity of Old Dongola, but we are unable to identify it with any modern toponym.

It would be interesting to know whether the deceased person was a man or a woman. The name ⲁⲅⲅⲉⲗⲟⲥⲕⲱ, an Old Nubian formation (Greek substantive ἄγγελος + Old Nubian predicative -ⲕⲟ), says nothing in this respect. Old Nubian did not distinguish grammatical gender and, consequently, names which are morphologically Old Nubian could have been borne, in principle, by both sexes. On the other hand, the data supplied by the Greek text of the epitaph are contradictory. In line 6, after ἀνάπαυσον τὴν ψυχὴ(ν) and before Ἀγγελοσκω, the feminine article τήν (for τὴν δούλην) appears, while in ll. 17–18, already quoted above, we have the masculine substantive τὼν σὼν δοῦλον (for τὸν σὸν δοῦλον). Facing this situation we have to pronounce a *non liquet*.

Notes

1 See S. Donadoni, "ΜΗΤΗΡ ΒΑΣΙΛΕΩΣ", *SCO* 18 (1969), 123–25; reprinted in: idem, *Cultura dell'antico Egitto*, Rome: Università degli Studi di Roma "La Sapienza" 1986, 567–69; cf. A.Osman, "The post-medieval kingdom of Kokka: a means for a better understanding of the administration of the medieval Kingdom of Dongola", in: J.M. Plumley (ed.), *Nubian studies: proceedings of the symposium for Nubian studies, Cambridge 1978*, Warminster: Aris & Phillips 1982, esp. 188–89, 193 and 196.

2 S. Jakobielski, *A history of the bishopric of Pachoras on the basis of Coptic inscriptions*, Warsaw: PWN–Éditions scientifiques de Pologne 1972, 113, referring to A. and W. Kronenberg, "Parallel cousin marriage in mediaeval and modern Nubia, I", *Kush* 13 (1965), 241–60; cf. W.Y. Adams, *Nubia: corridor to Africa*, London: Penguin 1977, 260 (pre-Christian) and 463 (medieval). This model of analysis, however, has now been challenged by J. Spaulding, "Medieval Nubian dynastic succession" (paper read at the Ninth International Conference of Nubian Studies, Boston 1998).

3 Cf. W.E. Crum, "La Nubie dans les textes coptes", *RdTr* 21 (1899), 224.

4 Faras: Jakobielski, *Bishopric of Pachoras*, 180; Old Dongola, monastic complex of Kom H: unpublished inscriptions accompanying wall paintings in the southwestern part of the monastery.

5 Cf. S. Clarke, "Ancient Egyptian frontier fortresses", *JEA* 3 (1916), 155–79. For the spelling Meinarti, which is retained here, see W.Y. Adams, "Sudan Antiquities Service excavations in Nubia: fourth season, 1962–63", *Kush* 12 (1964), 223, n. 18.

6 See: Adams, *Kush* 12, 216–48; idem, "Sudan Antiquities service excavations at Meinarti, 1963–64", *Kush* 13 (1965), 148–76; idem, "Settlement pattern in microcosm: the changing aspect of a Nubian village during twelve centuries", in: K.C. Chang (ed.), *Settlement archaeology*, Palo Alto, CA: National Press 1968, 174–207; idem, *Nubia: corridor to Africa*, London: Penguin 1977, esp. 488–94. For an earlier account of the site: U. Monneret de Villard, *La Nubia medioevale* I, Cairo: Mission archéologique de Nubie 1935, 217–21; ibidem II, pl. XCIV.

7 Thus Adams, *Nubia*, 489–92.

8 A provisonal list of funerary stelae ascribed to Meinarti, is given by Adams, *Kush* 12 (1964), 248; for later finds, see Adams, *Kush* 13 (1965), 172–73. Both factually and

ADAM ŁAJTAR AND JACQUES VAN DER VLIET

bibliographically, Adams' list is very deficient. Besides, of those pieces which Monneret de Villard (*Nubia* I, 220–21) ascribed to Meinarti on formal grounds alone, the connection with this particular site seems doubtful.

9   Khartoum no. 3726 (formerly Wadi Halfa, H. 184); see Monneret de Villard, *Nubia* I, 218–19, and, better, J.W.B. Barns, "Christian monuments from Nubia", *Kush* 2 (1954), 26–7, pl. Va. This stela is to be republished by Adam Łajtar in his forthcoming catalogue of the Greek stelae of the Sudan National Museum in Khartoum [*I. Khartoum Greek* 8]. It is interesting to note that from the textual point of view the epitaph of Goassi has two close parallels: (1) the epitaph of Papasa, a priest and Choiakeikšil (died A.D. 1181) now in the Louvre (cf. E. Bernand, *Inscriptions grecques d'Egypte et de la Nubie au Musée du Louvre*, Paris 1992, no. 115); (2) the epitaph of a woman Eikkir (died A.D. 1084 or later) found in Ashkeit, now in the Sudan National Museum Khartoum (see below, our no 12 [*I. Khartoum Greek* 7]). The epitaph of Papasa is of unknown provenance. On account of its similarity with the epitaph of the eparch Goassi it might be ascribed to Meinarti or its vicinity.

10   Close by, on the west bank of the Nile, an eparch of Nubia appears as the patron of the late church of Abd al-Qadir; cf. B. Rostkowska, "Patronage of the arts in Nobadia on the basis of archaeological and written sources", in J.M. Plumley (ed.), *Nubian studies: proceedings of the symposium for Nubian studies, Cambridge 1978*, Warminster: Aris & Phillips 1982, 211. Monneret de Villard, *Nubia* I, 220, beliefs that the Greek stela of yet another eparch of Nobadia, Ioannes (M.G. Tibiletti Bruno, *Iscrizioni nubiane*, Pavia: Fusi 1964, no. 6; A.D. 1006), could also originate from Meinarti; this is highly uncertain.

11   Cf. U. Monneret de Villard, *Storia della Nubia cristiana*, Rome: Pontificium Institutum Orientalium Studiorum 1938 (reprint 1962), 135 and 140; Adams, *Nubia*, 467. For the Arabic sources in question, see G. Vantini, *Oriental sources concerning Nubia*, Heidelberg: Heidelberger Akademie der Wissenschaften /Warsaw: Polish Academy of Sciences 1975, 813, index, s.v. "Mika'il, island(s)".

12   For the etymology in question, see esp. F.L. Griffith, "Oxford excavations in Nubia", *LAAA* 14 (1927), 103; but cf. Adam's cautious remarks in *Kush* 12 (1964), 223, n. 18. It may be added that there is no positive reason whatever to suppose that either the Island of Michael or Meinarti was the site of a Monastery of Saints Michael and Cosmas (*pace* G.S. Mileham, *Churches in Lower Nubia*, Philadelphia: University of Pennsylvania Museum 1910, 5; Adams, *Kush* 12, 223; "Settlement pattern", 191, n. 18). Indeed, it seems doubtful whether Meinarti ever possessed a monastery (cf. P. Jeute, "Monasteries in Nubia: an open issue", *Nubica* 3/1 (1989–1993), 66–8 and 83). Significantly, monastic titles, well attested in Faras, Dongola and elsewhere, are lacking in the published epigraphic material from Meinarti.

13   In *Nubia* I, 219–20 (cf. p. 212). The stela is Khartoum, National Museum, old inv. no. 14; it was collected on the site in the beginning of this century by J.W. Crowfoot. For a reedition, see J. van der Vliet's forthcoming catalogue of the Coptic inscriptions of the Sudan National Museum in Khartoum [= *I. Khartoum Copt.* 18].

14   Written thus in l. 24, ⲙⲓⲭⲁⲏⲗⲓⲕⲟⲗ in l. 6. Another occurrence of the same name (as ⲙⲓⲭⲁⲗⲓⲕⲟ) in Old Nubian graffito no. 15 in Tamit (Monneret de Villard, *Nubia* I, 152; S. Donadoni, in: *Tamit (1964): Missione archeologica in Egitto dell'Università di Roma*, Rome: Istituto di Studi del Vicino Oriente Universita 1967, 67–8). Names ending in -iko are Nubian names formed with an adjectival suffix -ⲕⲟ (cf. G.M. Browne, *Introduction to Old Nubian*, Berlin: Akademie-Verlag 1989, 11, sub 3.4.2); the final -ⲗ in l. 6 is a case-ending (cf. idem, 13).

15   Khartoum, inv. no. 18098, to be published in the catalogue referred to in note 13, above [= *I. Khartoum Copt.* 19]. It was found *in situ* in "one of the two or three oldest surviving graves" of Meinarti's cemetery during the Sudan Antiquities Service-UNESCO excavations, 1963–64 campaign (Adams, in *Kush* 13, 172; for the cemetery, see ibidem, 169–71, pl. XXXVII). The name Iesousyko (-iko, -ikol) is formed

RICH LADIES OF MEINARTI AND THEIR CHURCHES

on the same pattern as Michaeliko; it is found in Ibrim as well, always for women: the daughter of a bishop (Coptic stela Leipzig no. 687, A.D. 1035; see T.S. Richter, "Die neun Stelen Ägyptisches Museum der Universität Leipzig Inv.-Nr. 680–88 mit der Herkunftsangabe Qasr Ibrim", [in: S. Emmel et al. (eds.), *Ägypten und Nubien in spätantiker und christlicher Zeit: Akten des 6. Internationalen Koptologenkongresses, Münster, 20.-26. Juli 1996*, Wiesbaden: Reichert Verlag 1999, I, 295–304]) and the queen mother (see G.M. Browne, *Old Nubian texts from Qaṣr Ibrim* III, London 1991, nos. 35, l. 5; 38, l. 3; 40, l. 3; A.D. 1188–1200).

16   H. Junker, "Die christlichen Grabsteine Nubiens", *ZÄS* 60 (1925), 122–24, but cf. M. Krause, "Die Formulare der christlichen Grabsteine Nubiens", in: K. Michałowski (ed.), *Nubia: récentes recherches*, Warsaw: Musée national 1975, 77, n. 16.

17   The late antique Egyptian models of this particular type of stelae have been variously interpreted, see *e.g.* A. Badawy, "La stèle funéraire copte à motif architectural", *BSAC* 11 (1945), 5–6 (type 2a); E.R. Goodenough, *Jewish symbols in the Greco-Roman period* IV, New York 1954, 136–39; M. Cramer, *Archäologische und epigraphische Klassifikation koptischer Denkmäler*, Wiesbaden 1957, 1–11.

18   Cf. J.M. Plumley, "Some examples of Christian Nubian art from the excavations at Qasr Ibrim", in: E. Dinkler (ed.), *Kunst und Geschichte Nubiens in christlicher Zeit*, Recklinghausen: Verlag Aurel Bongers 1970, pl. 104, 105 and 107.

19   See Monneret de Villard, *Nubia* I, 123, fig. 103 (his type 6); cf. ibidem, IV, Cairo 1957, pl. CXXX–CXXXII.

20   See Jakobielski, *Bishopric of Pachoras*, 120–23 and 154–57.

21   A particularly well sculptured but fragmentary stela (Khartoum, National Museum, inv. no. 17041; unpublished [=*I. Khartoum Copt.* 16]); unfortunately, the greater part of its text is lost.

22   Khartoum, inv. no. 18105 [=*I.Khartoum Copt.* 20]; found "on the floor of a house" during the Sudan Antiquities Service-UNESCO excavations of Meinarti, 1963–64 campaign; cf. Adams, in *Kush* 13 (1965), 172.

23   A not uncommon name in Egypt, cf. F. Preisigke, *Namenbuch*, Heidelberg: privately printed 1922, col. 321 and 328; D. Foraboschi, *Onomasticon alterum papyrologicum*, Milano/Varese: Istituto Editoriale Cisalpino 1967, 256; in Nubia attested at Ginari, Ibrim and Toshka West (Sakinya).

24   Khartoum, National Museum, inv. no. 3990 [=*I. Khartoum Copt.* 17]; ed. W.E. Crum, in: G.S. Mileham, *Churches in Lower Nubia*, Philadelphia 1910, 21 (with p. 19 and pl. 7).

25   Cf. Junker, *ZÄS* 60, 130. For more commentary, see the catalogue announced in n. 13, above.

26   The group ⲣⲟ (beginning of l. 7) looks rather like ⲡ; the present reading was suggested by Adam Łajtar. Between ⲟ and ⲙ some space is left open, above which a small and thin ⲡ is added, close to the ⲙ. [See the adapted reading in *I. Khartoum Copt.* 19, ll. 6–7: ⲫⲓⲗⲟⲑⲉⲥⲉ̄ⲅⲟ, with a suffixal ⲉ̄ⲅⲟ-.]

27   Up to now this epithet has only rarely been discussed, viz. by S. Donadoni, in: *Tamit (1964)*, Rome 1967, 67, no. 13, l. 2 (cf. the same, "Les graffiti de l'église de Sonqi Tino", in: K. Michałowski (ed.), *Nubia: récentes recherches*, Warsaw: Musée national 1975, 35), and A. Łajtar, "Varia Nubica (I-II)", *ZPE* 104 (1994), 202–03.

28   Cf. M. van Esbroeck, art. "Philotheus of Antioch, Saint", in: *CoptEnc* 6, 1960–61. The proper name Philotheos occurs in Nubia at Tamit (cf. our no. 4, below), Toshka West (Sakînya) and Faras.

29   Next to the stela of Bishop Marianos, just quoted, cf. below our nos. 3 ("the Jesus-Church of the mountain") and 11 ("the Church of Severus in the town").

30   A perhaps comparable epithet is to be found in a Greek–Old Nubian graffito at Assuan, see F.L. Griffith, "Christian documents from Nubia", *PBA* 14 (1928), 145, l. 4, where it follows the (enciphered) name of St. Michael: ⲭ̅ⲡ̅ⲑ̅ (i.e. Michael) ⲁⲓ̈ⲕⲁⲙⲟⲩⲇⲟⲩⲉ.

31 This latter aspect was earlier emphasized by Łajtar, *ZPE* 104 (1994), 202–03, but see now our list of sources below.

32 And for a rare example of Coptic–Old Nubian code-switching!

33 The double shift between schwa and -ⲁ- which marks off ⲡⲉⲧⲟⲩⲁⲛⲧⲥ̄ from the standard Sahidic is certainly no mere slip, but a dialectical variant; cf. Crum in: H.E. Winlock and W.E. Crum, *The monastery of Epiphanius at Thebes* I, New York: Metropolitan Museum of Art 1926, 236–37, and P.E. Kahle, *Bala'izah: Coptic texts from Deir el-Bala'izah in Upper Egypt*, Oxford/London: Oxford University Press 1954, I, 52 and 57: examples esp. from the Theban area.

34 The inappropriate masculine forms used in both the Michaeliko and Iesousyko stelae (ⲉⲭⲱⲛ, ⲡⲉⲧ-) may, quite apart from the Nubians' weak sense for Coptic and Greek grammatical gender, reflect the predominant influence of a masculine standard form.

35 By Monneret de Villard, *Nubia* I, 212 (following F.L. Griffith, *The Nubian texts of the Christian period*, Berlin: Verlag der Königl. Akademie der Wissenschaften 1913, no. 8, p. 130), and P.L. Shinnie, in a footnote in Barns, *Kush* 2 (1954), 28, *ad* our no. 12, below.

36 See Monneret de Villard, *Nubia* I, 212; H-Å. Nordström, "Excavations and survey in Faras, Argin and Gezira Dabarosa", *Kush* 10 (1962), 44 (the habitation sites mentioned seem not to have been excavated). It is needless to insist on the popularity of St. Michael in Christian Nubia, profusely attested by archaeological, prosopographical and literary material.

37 In: *Tamit (1964)*, 67; cf. the same, in: K. Michałowski (ed.), *Nubia: récentes recherches*, 35: " 'titulaire' (ou quelque chose de semblable. . .) ". Already Łajtar, *ZPE* 104 (1994), 202–03, pointed out that also non-clerics could be styled ἔχων (etc.).

38 See A. Steinwenter's fundamental study, "Die Rechtsstellung der Kirchen und Klöster nach den Papyri", *Zeitschrift der Savigny-Stiftung* 50 (1930), 1–50; E. Wipszycka, *Les ressources et les activités économiques des églises en Égypte du IVe au VIIIe siècle*, Bruxelles: Fondation égyptologique Reine Elisabeth 1972, 26–7, 80–3, 90–1; both draw also on Coptic material. Private churches (εὐκτήριοι οἶκοι) were a common phenomenon in the Byzantine Empire. It was extensively studied by E. Herman: "Die kirchlichen Einkünfte des byzantinischen Niederklerus", *OCP* 8 (1942), 378–442, esp. 419–25; " 'Chiese private' e diritto di fondazione negli ultimi secoli dell'impero bizantino", *OCP* 12 (1946), 302–21; "The secular Church", in: *Cambridge medieval history* IV/2, Cambridge: Cambridge University Press 1967, 104–33, esp. 116–25. Also monasteries could belong to private persons (χαριστικάριοι); cf. E. Herman, "Ricerche sulle istituzioni monastiche bizantine: typika ktetorika, caristicarie, monasteri 'liberi'", *OCP* 6 (1940), 293–375; P. Lemerle, "Un aspect du rôle des monastères à Byzance: les monastères donnés à des laïcs, les charisticaires", *CRAI* 1967, 9–28.

39 On the rights of the κτίστης and its heirs, see particularly Steinwenter, "Rechtsstellung", 5–23.

40 See Steinwenter, "Rechtsstellung", 7–8, 16–19.

41 Thus, the question may be raised how it fits into a picture of Nubian economy as sketched e.g. by L. Török, "Money, economy and administration in Christian Nubia", in: *Études nubiennes: colloque de Chantilly, 2–6 juillet 1975*, Cairo: IFAO 1978, 287–311.

42 Compare, for even greater geographical distances, our nos. 1 and 10 below.

43 It is this very church where Adam Łajtar was baptized and made his the First Communion.

44 For Moses George, see J.M. Plumley, "New light on the kingdom of Dotawo", in: *Études nubiennes: colloque de Chantilly, 2–6 juillet 1975*, Cairo: IFAO 1978, 236–39 (with a facsimile of the Faras inscription on p. 237).

45 The term ἐξουσία, "power" most probably describes Iesousinkouda as an eparch of Nobadia.

29

FROM ASWAN TO DONGOLA
The epitaph of Bishop Joseph (died A.D. 668)*

Stefan Jakobielski and Jacques van der Vliet

Introduction

This first edition of the bilingual (Greek-Coptic) stela of Bishop Joseph, discovered in Old Dongola in 2004, is the result of the combined efforts of both authors. A certain division of labor was observed, however. Stefan Jakobielski is responsible for the description of the find and its archaeological context at the beginning of this paper as well as for the comments on its historical and religious context that occupy the final section. The presentation of the text and the philological commentary are written by Jacques van der Vliet.

The find and its context

The funerary stela of Joseph, a former bishop of Aswan (Syene), was discovered in January 2004, during the thirty-eighth season of the Polish excavations in Old Dongola, directed by Stefan Jakobielski. It was found inside the presbytery of the church of the monastic complex on Kom H, the so-called Monastery of the Holy Trinity, near the southern end of the altar screen (Figure 29.1).[1]

Although the stela was not found *in situ*, but lying on a layer of sand and rubble above floor level, it can be assumed to have belonged to one of the tombs that were situated in its immediate vicinity (Figure 29.2). Taking into consideration the find-spot and the width of the slab, it may originally have been inserted in the southern pilaster of the apse wall, facing a column dividing the nave and the south aisle. It is far less probable that it could originate from the east wall of the south aisle, located to the south of the doorway to the *diakonikon*. The stela is likely to be connected with tomb G.3, situated in the eastern end of the south aisle. According to archaeological indications, this must have been dug after the earliest paving was done in the area, but before the second layer of the floor was laid, which belongs to a rebuilding of the church undertaken at the end of the seventh or the beginning of the eighth century, which can be deduced on the basis of other archaeological evidence, including pottery. As it is obvious from the text (see below) that the church had already been erected before Joseph died, in A.D. 668, the period of construction of the tomb would perfectly match the archaeological

Figure 29.1 The stela of Bishop Joseph at its find-spot
(photo: Wojciech Chmiel)

data. Another possibility would be tomb G.2 situated nearby, in the presbytery on the axis of the church. The latter burial can undoubtedly be classified as an Early Christian tomb but, unfortunately, the evidence for the date of its construction is less clear, and it appears difficult to establish whether it was dug before or after the rebuilding of the church.

The text

The stela of Bishop Joseph belongs to the collection of the Sudan National Museum in Khartoum (inv. no. 31481). It bears the field register no. D I/04.

The monument consists of a rectangular slab of originally grayish-yellow, now grey-brownish sandstone, measuring 81 × 62 × 9 cm. When found, it was broken into many fragments, with cracks and other surface damage causing minor losses to text. The present edition was based on the restored stone, which has been exhibited in the National Museum in Warsaw since August 2006 (on a loan basis), and with the help of the photo and facsimile reproduced here, which show the state of the stone after its preliminary conservation (Figures 29.3–4).

The epigraphic field is contained within a raised border that follows the outline of an expanded, almost square cross, arranged in such a way that it occupies nearly the entire surface of the slab. It is inscribed by a single hand with a text in

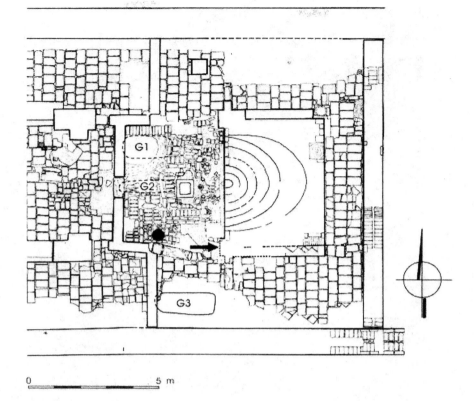

Figure 29.2 Plan of the eastern part of the Church showing the place of the discovery of the stela and its suggested original position
(drawing by Marek Puszkarski; updated by Marta Momot)

Greek and Sahidic Coptic, consisting of thirty-one lines of incised late antique uncials, well-shaped and fairly regular. Both the left-looking ⲁ and broken-bar symmetrical *A* are used without apparent distinction. Superlinear strokes are frequently and logically placed; the *iota* occasionally bears a diaeresis. A small number of standard abbreviations are marked by strokes over the letters or by following dashes. The letters were apparently painted in red originally, traces of which can be seen throughout.

The text proper is preceded by the monograms for Jesus and Christ inscribed in the top of the cross in a somewhat larger script. The upper left and lower right corners bear a small incised cross (possibly lost in the other outer corners). A similar small cross marks the beginning and the end of the Coptic text in ll. 4 and 28 respectively, and precedes the dating formula in l. 29 as well as the Diocletian year-date inscribed at the bottom of the cross (l. 31). In addition to small crosses,

Figure 29.3 The stela of Bishop Joseph after its preliminary conservation on site (photo: Wojciech Chmiel)

a long stroke in l. 28 separates the dating formula in ll. 29–31 from the body of the text; a short filler stroke ends l. 5. Clause division is occasionally marked by blanks and high points (the latter only in ll. 23 and 24), both logically placed.

Bibliography: see above, n. 1.

28 April A.D. 668 (or 670?)

[top]   Ἰ(ησοῦ)ς Χ(ριστό)ς
    + Ἐνθάδε κεῖται τὸ λείψανον τρ[ῦ ἐν ἁ-]

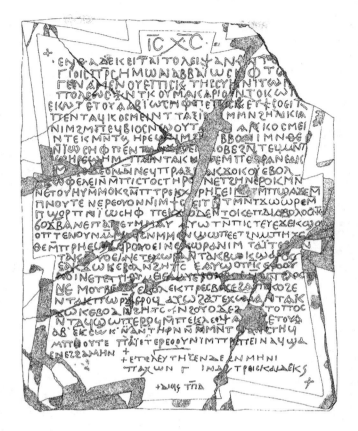

Figure 29.4 The stela of Bishop Joseph
(tracing by Stefan Jakobielski)

γίοις π(ατ)ρ(ὸ)ς ἡμῶν ἄββα Ἰωσῆφ το[ῦ]
γεναμένου ἐπισκ(όπου) τῆς Συηνιτῶν
4. πόλεως. + ⲛⲧⲕ ⲟⲩⲙⲁⲕⲁⲣⲓⲟⲥ ⲛⲧⲟⲕ ⲱ ⲡ[ⲉⲛ-]
ⲉⲓⲱⲧ ⲉⲧⲟⲩⲁⲁⲃ ϊⲱⲥⲏⲫ ⲡⲉⲡⲓⲥⲕⲟ(ⲡⲟⲥ) ⲉⲧ†ⲥⲟⲉⲓⲧ
ⲡⲉⲛⲧⲁϥⲕⲟⲥⲙⲉⲓ ⲛ{ⲧ}ⲧⲁⲝⲓⲥ [ⲛ]ⲓⲙ ⲙⲛ ϩⲏⲗⲓⲕⲓⲁ
ⲛⲓⲙ ϩⲙ ⲡⲉϥⲃⲓⲟⲥ ⲛϣⲟⲩⲧⲁ[ⲓⲟ]ϥ *vac* ⲁⲕⲕⲟⲥⲙⲉⲓ
8. ⲛⲧⲉⲕⲙⲛⲧϣⲏⲣⲉ ϣⲏⲙ ϩⲛ [ⲧ]ⲃⲃⲟ ⲛⲓⲙ ⲛⲑⲉ
ⲛϊⲱⲥⲏⲫ ⲡⲉⲛⲧⲁϥⲭⲣⲟ ⲉⲡⲛⲟⲃⲉ ϩⲛ ⲧⲉϥⲙⲛⲧ-
ϣⲏⲣⲉ ϣⲏⲙ *vac* ⲡⲁⲓ ⲛⲧⲁⲕⲙ[ⲁ]ⲧⲉ ⲙⲡⲉϥⲣⲁⲛ ⲉⲁⲕ-
ⲙ[ⲁ]ⲧⲉ ⲇⲉ ⲟⲛ ⲛⲛⲉϥⲡⲣⲁⲝⲓ[ⲥ] ⲁⲕⲭⲟⲕⲟⲩ ⲉⲃⲟⲗ
12. ⲁⲕⲱⲫⲉⲗⲉⲓ ⲛⲙⲡⲓⲥⲧⲟⲥ ⲧⲏⲣⲟ[ⲩ] ⲛⲉⲧϩⲏⲛ ⲉⲣⲟⲕ ⲙⲛ
ⲛⲉⲧⲟⲩⲏⲩ ⲙⲙⲟⲕ ϩⲙ ⲡⲧⲣⲉⲕⲭ[ⲟ]ⲣⲏⲅⲉⲓ ⲛⲁ[ⲩ] ⲙⲡϣⲁϫⲉ ⲙ̄-

STEFAN JAKOBIELSKI AND JACQUES VAN DER VLIET

ⲡⲛⲟⲩⲧⲉ ⲛⲉⲣⲉ ⲟⲩⲟⲛ ⲛⲓⲙ †ⲥ[ⲟ]ⲉⲓⲧ ⲛ̄ⲧⲙ̄ⲛ̄ⲧϫⲱⲱⲣⲉ ⲙ̄-
ⲡϣⲟⲣⲡ ⲛ̄ⲓ̈ⲱⲥⲏⲫ *vac* ⲡⲉⲕϫ[ⲣ]ⲟ ⲇⲉ ⲛ̄ⲧⲟⲕ ⲉⲡⲇⲓⲁⲃⲟⲗⲟⲥ ⲛⲏ̄ϥ-

16. ϭⲟⲗϫⲃ ⲁⲛ ⲉⲡⲁ ⲡⲉⲧⲙ̄ⲙⲁⲩ *vac* ⲁⲩⲱ ⲧⲛ̄ⲡⲓⲥⲧⲉⲩⲉ ϫⲉ ⲕϣⲟ-
ⲟⲡ ⲧⲉⲛⲟⲩ ⲛⲙ̄ⲙⲁϥ ϩⲛ̄ ⲙⲙⲁⲛ̄ϣⲱⲡⲉ ⲉⲧϩⲛ ⲙⲡⲏⲩⲉ ⲛ̄-
ⲑⲉ ⲙ̄ⲡⲣⲏ ⲉϣⲁϥⲣⲟⲩⲟⲉⲓⲛ ⲉⲭⲱⲣⲁ ⲛⲓⲙ *vac* ⲧⲁⲓ̈ ⲧⲉ ⲑ[ⲉ] ⲛ̄-
ⲧⲁⲕⲣ̄[ⲟ]ⲩⲟⲉⲓⲛ ⲉⲧⲉⲭⲱⲣⲁ ⲛ̄ⲧⲁⲕⲃⲱⲕ ϣⲁⲣⲟⲥ

20. ⲉⲁⲕϫⲱⲕ ⲉⲃⲟⲗ ⲛ̄ϩⲏⲧⲥ̄ *vac* ⲉⲁⲩϣⲟⲡⲕ̄ ⲉⲣⲟⲟⲩ
ⲛ̄ϭⲓ ⲛⲉⲧⲉⲡⲓⲑⲩⲙⲉⲓ ⲉⲛⲁⲩ ⲉⲣⲟⲕ ⲉⲧⲉ ⲛ̄ⲁⲅⲅⲉⲗⲟⲥ
ⲛⲉ *vac* ⲙⲟⲩⲛ ⲛ̣[ⲁⲕ] ⲉⲃⲟⲗ ⲉⲕⲡⲣⲉⲥⲃⲉⲩⲉ ϩⲁ ⲡⲉⲕⲟϩⲉ
ⲛ̄ⲧⲁⲕⲡⲱⲣⲝⲭ̄ ⲉⲣⲟϥ · ⲁⲩⲱ ϩⲁ ⲧⲉⲭⲱⲣⲁ ⲛ̄ⲧⲁⲕ-

24. ϫⲱⲕ ⲉⲃⲟⲗ ⲛ̄ϩⲏⲧⲥ̄ · ⲛ̄ϩⲟⲩⲟ ⲇⲉ ϩ[ⲁ] ⲡⲧⲟⲡⲟⲥ
ⲛ̄ⲧⲁϥϣⲱⲡ ⲉⲣⲟϥ ⲙ̄ⲡⲉⲕⲗⲉⲓⲯⲁ[ⲛⲟⲛ] ⲉⲧⲟⲩⲁⲁⲃ *vac* ⲉⲕⲥⲱⲕ ⲛⲁⲛ ⲧⲏⲣⲛ ⲛ̄ⲙ̄ⲙⲛ̄ⲧϣⲁⲛϩⲧⲏϥ
ⲙ̄ⲡⲛⲟⲩⲧⲉ *vac* ⲡⲁⲓ̈ ⲉⲧⲉⲣⲉⲟⲟⲩ ⲛⲓⲙ ⲡⲣⲉⲡⲉⲓ ⲛⲁϥ ϣⲁ

28. ⲉⲛⲉϩ ϩⲁⲙⲏⲛ +
+ Ἐτελεύτησεν δὲ ἐν μηνὶ
Παχὼν γ΄ ἰνδ(ικτίονος) τρεισκαιδεκ(άτης)
+ Διοκ(λητιανοῦ) τπδ΄. +

Top. ⲓ̄ⲥ̄ ⲭ̄ⲥ̄|| 1. cross in upper left corner || 2. ⲡⲣ̄ⲥ | ⲓ̈ⲱⲥⲏⲫ || 3. ⲉⲡⲓⲥⲕ || 5. ⲉⲡⲓⲥⲕˋⲟˊ || 15. ⲛⲏ̄ϥ-: ⲛ cramped
(*ex corr.*?) || 22. ⲛ̣[ⲁⲕ]: traces inconclusive, but not ⲇⲉ || 27. ⲉⲧⲉⲣⲉⲟⲟⲩ with contraction for ⲉⲧⲉⲣⲉ ⲉⲟⲟⲩ
|| 30. ⲓⲛⲇˌ ⲧⲣⲉⲓⲥⲕⲁⲓⲇⲉⲕⲥ̄ || 31. ⲇⲓⲟⲕ ⲧⲡⲇ.

[Greek]

Jesus-Christ.

+ Here lie the mortal remains of our saintly father Abba Joseph, the former bishop of the city of the Syenites.

[Coptic]

+ Blessed you are, thou [our] holy (5) father Joseph, the illustrious bishop, who adorned all stations and all ages of his praiseworthy life!

You adorned your youth with every kind of purity, just like Joseph, who had overcome sin in his (10) youth, him whose name you acquired, while you have also acquired his achievements and brought them to perfection.

You have bestowed benefit on all the faithful, those near to you and those far away from you, while you ministered to them the word of God.

Everybody celebrated the prowess of the (15) first Joseph, but your own triumph over the devil was not inferior to that of him, and we believe that you are now with him in the dwellings that are in heavens.

Like the sun that illuminates every country, thus you illuminated the country where you went (20) and where you reached perfection (i.e. died).

When they who long to see you, namely the angels, have received you, remain interceding for your flock from which you became separated and for the country where you reached perfection and in particular for the place (25) which received

your holy remains, attracting for all of us the mercies of God, He to whom all glory is due till eternity. Amen. +
[Greek]
+ He died in the month (30) Pachon (day) 3, of the thirteenth (year of the) indiction. + (Year) of Diocletian 384. +

Commentary

The funerary stela of Bishop Joseph is a bilingual monument. The formulaic parts, containing standard elements that can be found in most late antique epitaphs (the introduction of the deceased, ll. 1–4, and the dating lemma, ll. 29–31), are in Greek. The non-standard, personal parts, which address the deceased himself and proclaim his merits, are in Coptic. The Greek parts frame the Coptic text, which makes up both the core and the bulk of the composition (ll. 4–28). In both languages, the standing of the text is excellent and the execution almost flawless. The redundant article in l. 6 and the cramped end of l. 15 are minor blemishes only. Otherwise, the entire stela with its intricate layout reflects very high standards of execution. The habit of projecting a form of cross on a rectangular slab is also found in later funerary stelae from Dongola.[2]

Apart from the designation of the deceased as "the former bishop of the city of the Syenites" in ll. 2–4, the Greek parts of the text show little that is remarkable. The introduction of the deceased's name and dignity with its characteristic opening "here lie the mortal remains of N.N." (l. 1) is known in Greek and Coptic variants from various periods and regions of Egypt and Nubia.[3] The epithet ὁ ἐν ἁγίοις (ll. 1–2), which has recently been discussed by Adam Łajtar,[4] corresponds to a simple ετογααв, "holy", in the Coptic text of l. 5.[5] The spelling γεναμένου for classical γενομένου (l. 3) reflects the replacement of second aorist by first aorist endings, which was extremely common in later Greek.[6] In ll. 3–4, the designation of Aswan as ἡ Συηνιτῶν πόλις, "the city of the Syenites", is a late witness to its status as a nome capital in late antiquity.[7] The date in ll. 29–31, introduced by the ubiquitous ἐτελεύτησεν, contains an error: the Diocletian year 384 (A.D. 667/668) is an eleventh indiction year, not a thirteenth.[8] If the indiction date proves correct and the era date wrong, the bishop died on 28 April A.D. 670, instead of A.D. 668. In either case, the text is the earliest precisely dated document found in Christian Dongola so far.

The Coptic part of the text is of an entirely different nature. It is a personalized prose poem in praise of the deceased, which clearly betrays the author's intention to produce a sophisticated and original literary composition. In this respect it can be compared to the Coptic building inscription of Bishop Paulos of Faras, which dates from some forty years later (*I. Khartoum Copt.* 1). In Egypt, such personal and ambitious epitaphs are rarely found, and hardly ever in Coptic. The slightly later group of the so-called *Totenklage* stelae, which date from the late seventh to the middle of the ninth century, is a notable exception, but these texts are very different in character.[9] They lament the ephemeral nature of human life and the hardship of death, whereas the present text is panegyrical in style throughout. No comparable episcopal epitaphs are known from Egypt, where mortuary monuments for bishops are rare altogether. Nor does the epitaph of Joseph resemble other bishop's stelae from Nubia, whether in Greek or in Coptic. These are as a rule dominated by stock liturgical formulae, and they usually provide information about the age of the deceased and the length of his episcopate, which is lacking here.[10] It is true that the Greek tombstone of Archbishop Georgios of Dongola (died A.D. 1113) devotes considerable space to a eulogy of the deceased, but its formulaic and list-like style is far distant from the stately periods of Joseph's epitaph.[11] Therefore, the monument can be called unique for the Nile Valley without any exaggeration.

The actual panegyric, which addresses the deceased bishop in the second-person singular throughout, consists of six lengthy phrases. The first of these is an elaborate acclamation, couched in general terms (ll. 4–7). The second (ll. 7–11) and the fourth (ll. 14–17) both compare Joseph to his namesake, the biblical patriarch. The third (ll. 11–14) and the fifth (ll. 17–20) also make up a pair, since both emphasize his activity as a teacher. A long final prayer asks for his continued intercession following his welcome among the angels (ll. 20–6). It is unclear whether any metrical organization underlies this composition, which by its rhetorical style recalls the encomiastic prooemia of contemporary saints' lives, for example the Sahidic *Life of Bishop Pesynthios of Koptos* (ed. E.A.W. Budge, *Coptic apocrypha in the dialect of Upper Egypt*, London: British Museum 1913, 89–91, 100–05), and was undoubtedly meant to be performed orally.[12]

Biblical influence is particularly apparent in the opening *makarismos* (ll. 4–5), the references to the story of Joseph (see below), and the image of the sun (ll. 17–20; cf. Sir. 42:16, LXX). The influence of commemorative liturgy, strongly felt in later Nubian funerary epigraphy, is remarkably absent here, apart from the motif of the reception of the righteous deceased among the angels, "who long to see" him (ll. 20–2).[13] In rather general terms, the text sketches a fairly standard ideal picture of a late antique bishop.[14] According to his epitaph, Joseph was an outstanding personality in all the stages of his career; he was pure from his youth, immune against sin, a benefactor of the faithful, a minister of God's word, a hero in his combat against the devil, an illuminator of the country, and a powerful interceder with God. In several respects, however, the text is more than a collection of hackneyed phrases.

Even though practically no hard facts are given about the deceased, the panegyric is clearly a personal text. It alludes to events from the life of Bishop Joseph and draws upon his name in order to compare his merits to those of his Old Testament namesake, the Patriarch Joseph. The main biographical element that receives attention is his status as a foreigner who was separated from his own people and died in a foreign country. This is a recurrent theme in the panegyrical part of the epitaph (found in ll. 12–13, 18–20 and 22–4), announced by his earlier qualification as "the former bishop of the city of the Syenites" (ll. 2–4). At first sight, the term γενάμενος, "former", used in the Greek opening lines, might be taken to refer to the bishop's death rather than the loss of his diocese, but since the Coptic text clearly supposes a prolonged sojourn in his second homeland, the latter interpretation is undoubtedly correct. The notion "deceased, late", moreover, is already expressed in these same lines by the epithet ὁ ἐν ἁγίοις (ll. 1–2).[15]

Being a foreigner may have given an extra dimension to the analogy with his namesake, the patriarch, for both the biblical Joseph and the bishop were expatriates. It is tempting to push the analogy a bit further and to suppose that, as in the case of the patriarch, the bishop's stay abroad was not entirely voluntary. It is indeed remarkable that, however he may have been appreciated as someone who "illuminated" his new homeland, "the country where (he) went" (l. 19), he bears no titles that connect him with local ecclesiastical life. The only formal dignity that his epitaph assigns to Joseph is that of "former bishop" of Aswan (ll. 2–4), and also the text of ll. 22–3 clearly states that his flock was not to be found in Dongola, but left behind in Aswan ("your flock from which you became separated"). Again, the context makes it unlikely that this phrase should be interpreted as "separated by death".

The text's references to the biblical patriarch should not be judged as biographical information, however, but primarily as literary tropes. Although the Patriarch Joseph is practically absent from the epigraphical record,[16] he was one of the heroes of ancient parenetical literature, both Jewish and Christian.[17] Thus, an unknown Christian homilist praises the paradigmatic value of his life in terms that vividly recall ll. 14–16 of the Dongola epitaph: "For great, o my brothers and my beloved, is the story of Joseph, the son of Jacob, and not to be abridged. Rather, we shall commemorate him and what befell him and his prowesses (ⲙⲛ̄ⲧϫⲱⲱⲣⲉ) and his combats (ἀγών). Rightly did David say: "Numerous are the trials of the righteous ones; the Lord will save them from all of these" (Ps. 33:20, LXX)".[18] Among his many roles, the Patriarch Joseph is frequently quoted as a model of virginity and virtue, but also as a paradigm for the triumph over evil influences, in particular envy.[19] In his Dongola epitaph, Bishop Joseph appears to be compared to his biblical namesake in both these qualities. Once, he is said to have "adorned (his) youth with every kind of purity, just like Joseph, who had overcome sin in his youth" (ll. 6–10), which is clearly an allusion to the events of Gen. 39 as well as an echo of 4 Macc. 2:1–4.[20] A second time, it is stated that his "triumph over the devil was not inferior" to that of the patriarch (ll. 15–16), which most likely refers to the story of Joseph and his brothers, in Gen. 37, whose envy was generally believed to be inspired by the devil in later literature.[21] Both passages present Bishop Joseph as a paragon of virtue and charismatic power, but their evocation of the Patriarch Joseph reflects stock themes in the literature of late antiquity. They cannot, therefore, be used to reconstruct Bishop Joseph's biography, even if they do bear witness to the rhetorical skills of the author and the high regard in which Bishop Joseph was held by his entourage.

The most striking element in Bishop Joseph's epitaph remains the fact that he is represented as a "former bishop" of Aswan who had become separated from his flock and died in exile. Although several other epitaphs of bishops buried outside of their diocese are known,[22] the stela of Joseph is unparalleled for the explicit way in which it presents the bishop's exile as a turning point and, so it would seem, the most important single event in his entire career. Regrettably, however, the text does not offer the precise kind information that could serve to interpret this event. Nor do other Nubian or Egyptian sources shed light on the circumstances of Joseph's life. The surviving records of the episcopal see of Aswan are very meager indeed and practically silent about the seventh century.[23] An unpublished Greek ostracon from Elephantine does mention a bishop Appa Josephios of Aswan, but he appears to have lived much earlier than his Dongola namesake.[24] Although a few more Egyptian bishops with the same name are known, none of them can be assigned to the Aswan diocese with any certainty.[25] Thus the inscription of a bishop Appa Josephios, who dedicated a brass foundry to a "Holy Church" in an unspecified town, provides no clue as to his see (*SEG* XVIII 720). A "holy Bishop Abba Joseph", invoked on lamps from Upper Egypt and Lower Nubia (Faras),[26] is more plausibly to be identified with the saintly bishop of Sbeht/Isfaht (Apollinopolis Parva), who was a contemporary of bishops Constantine of Asyut and Pesynthios of Koptos (late sixth/early seventh century), than with our Joseph.[27]

Likewise very little is known about the political and military situation in Aswan following the Arab invasion of Egypt in the end of A.D. 639. Apparently, the town was touched by the ongoing skirmishes between Arabs and Nubians, for it had to be reconquered by the Arabs in A.D. 651/652.[28] For this liminal region, a long period of turmoil came to an end only when in the summer of A.D. 652 the famous *baqt*, a bilateral peace treaty between Muslim-ruled Egypt and the Makurian kingdom, was concluded. Situated on a contested frontier, Aswan soon gained considerable importance for the new Muslim rulers of Egypt, and became second only to Fustat as a military and administrative center.[29] Although it is tempting to link Bishop Joseph's fate to the dramatic political events of the period, his highflown epitaph leaves the question open.

Historical and religious context

Turning to the political and religious situation of the sixth and seventh centuries in Nubia itself, one more question must be raised in the margin of Joseph's stela. It appears from various sources, both literary[30] and archaeological,[31] that the Kingdom of Makuria was converted around A.D. 568 to the Melkite faith.[32] How long this religious persuasion dominated the kingdom, before all Nubia became Monophysite, is a matter for debate, as is the exact date of the unification of Makuria and its northern neighbor Nobadia. For both facts, the acceptance of Monophysitism in all parts of Nubia and the unification of both kingdoms under the king of Dongola, the only trustworthy documents belong to the reign of King Merkourios (from A.D. 696 till after A.D. 710),[33] while the last mention of Nobadia as an independent state dates from A.D. 580.[34] On the other hand, if the *baqt* was

FROM ASWAN TO DONGOLA

signed at Dongola in A.D. 652 with only one king of the Nubians, Qalidurut, there is a good deal to suggest that the incorporation of Nobadia into Makuria took place before that date, possibly in direct response to the Arab threat from Egypt.[35] Whether the shift from the Melkite to the Monophysite creed in Dongola happened at the same time remains obscure. In Coptic sources, however, much stress was laid on the merits of King Merkourios for the Monophysite Church.[36] Could it be that this was on account of the role he had played in the religious unification of an already politically united Nubia?

If Dongola was still Melkite at the time Bishop Joseph lived there, he must have found himself in a somewhat difficult position. This perhaps made the author of the inscription stress that Joseph was separated from his flock (living in a country of rival persuasion) while at the same time he "triumphed over the devil" by continuing his mission, presumably in the company of a few compatriots.[37] If we follow the suggestion of Stuart Munro-Hay that the change of religious persuasion was not the merit of King Merkourios but of his predecessor,[38] Joseph may even have lived to see the end of Melkite Dongola. At this point, the testimony of al-Makin may prove significant, who quotes an early eighth-century source mentioning that from the caliphate of Umar ibn al-Khattab (A.D. 634–44) onwards there was no Melkite patriarch in Alexandria for ninety-seven years, which resulted in Monophysite bishops being appointed in the place of Melkite ones throughout the whole of Egypt, including Nubia.[39] The religious change in Makuria is perhaps mirrored in archaeological facts as well. Most if not all excavated early church buildings at Dongola were rebuilt in the second half or towards the end of the seventh century.[40] In the "Church of the Stone Pavement", also called the first Dongola Cathedral (EC), the columns were replaced by brick piers supporting a dome over the central bay.[41] Following a somewhat similar principle the "Monastery Church" (HC), in which Bishop Joseph was buried, was rebuilt. The "Church of the Granite Columns", the Second Cathedral, was erected to replace the "Old Church", which was leveled purposely, and the so-called "Mosaic Church" (EEC) was also rebuilt and enlarged at that time.[42]

Facing the religious and political turmoil of the second half of the seventh century in Nubia, we have to admit that the text itself of Joseph's stela does not provide any answers to the questions raised above. As it is predominantly of a panegyrical nature, it forbids speculations about the nature and background of Joseph's Dongolese exile.

*

In conclusion, the intricately designed and carefully executed funerary monument of Bishop Joseph attests to the high status that this former bishop of Aswan enjoyed in the Makurian capital in the 660s. He was buried in a prominent place, near the altar of the church of an important monastery, while his epitaph explicitly praises him as an illuminator of his second homeland, and implores his continuing intercession not only for his former diocese, but also for the country where he had died and the monastery where he was buried. His

375

elaborate epitaph is an ambitious bilingual composition in Greek and Sahidic Coptic that bears witness to a high level of literary culture. Whereas Greek is used for the formulaic parts, the Coptic body of the text contains a prose poem that is unique in the epigraphy of the Nile Valley on account of its rhetorical structure and panegyrical style. The inscription is of special interest, moreover, as it is the earliest precisely dated document found in Christian Dongola so far. As such it is due to play an important role in future discussions of the relations between Egypt and Nubia in the middle of the seventh century and the early history of Christian Makuria.

Notes

* Tasha Vorderstrasse and Brian Muhs kindly revised our English text. We also thank Heinzgerd Brakmann, Jelle Bruning, Adam Łajtar, Sofia Schaten, Klaas Worp, Khaled Younes and Ewa Zakrzewska for their assistance and helpful comments.

1 The work in the Monastery Church was supervised by D. Gazda; see his report, "Monastery church on kom H in Old Dongola: third and fourth season of excavations (2004, 2004/5)", *PAM* 16 (2005), 285–95 at 292–93 (photo and brief description of the find); idem, "Excavations at the Monastery Church on Kom H at Old Dongola: 2002/3–2006", *Gdańsk Archaeological Museum and Heritage Protection Fund African reports* 6 (2010), 46, with fig. 10 (photo of the stela); see also the exhibition catalogue S. Jakobielski (ed.), *Polish excavations at Old Dongola: 45 years of the archaeological co-operation with the Sudan*, Warsaw: National Museum in Warsaw 2006, 19 (photos) and 55–6, no. 14 (brief description).

2 E.g. in the stela of Kel, A.D. 758 (*I. Khartoum Greek* 20), and the stela of Archbishop Georgios (see A. Łajtar, "Georgios, archbishop of Dongola (+ 1113) and his epitaph", in: T. Derda, J. Urbanik, and M. Węcowski (eds.), *Εὐεργεσίας χάριν: studies presented to B. Bravo and E. Wipszycka*, Warsaw: University of Warsaw/Raphael Taubenschlag Foundation 2002, 159–92, especially 163 with n. 9).

3 Examples are listed in J. van der Vliet, "Two Coptic epitaphs from Qasr Ibrim", *JEA* 92 (2006), 217–23 at 222 [Study 25]; for background and examples from elsewhere in the Roman and Byzantine world, see *I. Qasr Ibrim* 28 (commentary).

4 *I. Khartoum Greek*, 90–1.

5 A literal Coptic rendering (ⲡⲉⲛⲉⲓⲱⲧ ⲉⲧⲡⲛ ⲛⲉⲧⲟⲩⲁⲁⲃ) can be found e.g. in W.E. Crum, *P. Lond. Copt.* I 468, vo, l. 3 (letter from a bishop John to a priest Merkourios).

6 See Gignac, *Grammar* II, 336 and 344.

7 For which see now J.H.F. Dijkstra and K.A. Worp, "The administrative position of Omboi and Syene in Late Antiquity", *ZPE* 155 (2006), 183–87.

8 Curiously, the same error is found in a Greek stela from the northern fringes of Egypt, the region of al-Arish (Rinokoloura), which was dated only three days later (Pachon 6); ed. H. Verreth, "Epigraphic notes on the Sabkhat Bardawil and el-Arisch region in the northern Sinai", *Ancient society* 28 (1997), 107–19 at 117; cf. *SEG* XLVII 2134.

9 See M. Cramer, *Die Totenklage bei den Kopten*, Vienna/Leipzig: Hölder-Pichler-Tempsky 1941; H.D. Schneider, "The lamentation of Eulogia: a Coptic dirge in the Leiden Museum of Antiquities", *Oudheidkundige mededelingen* 50 (1969), 1–7.

10 Some relatively early examples in Coptic: stela of Bishop Kerikos of Sai (probably ca. A.D. 800), ed. J. van der Vliet in G.T. Martin (ed.), *Stelae from Egypt and Nubia in the Fitzwilliam Museum, Cambridge, c. 3000 BC–AD 1150*, Cambridge: Cambridge University Press 2005, no. 115); stela of Bishop Thomas of Faras (A.D. 862; ed. S. Jakobielski, *Bishopric of Pachoras*, 75–80, cf. *SB Kopt.* I 719).

FROM ASWAN TO DONGOLA

11 See the extensive commentary by Łajtar, "Georgios, Archbishop of Dongola", especially 166–74.

12 See A. Papalexandrou, "Echoes of orality in the monumental inscriptions of Byzantium", in: L. James (ed.), *Art and text in Byzantine culture*, Cambridge: Cambridge University Press 2007, 161–87.

13 This motif, which has New Testament roots, is already fully developed in the early Alexandrian Fathers, see A. Recheis, *Engel, Tod und Seelenreise: Das Wirken der Geister beim Heimgang des Menschen in der Lehre der alexandrinischen und kappadokischen Väter*, Roma: Edizioni di storia e letteratura 1958, 60–1, 102–06 and *passim*. It is also reflected in the funerary epigraphy of late antique Egypt: Lefebvre, *Recueil*, no. 48 (*SB* I 1540; Alexandria, A.D. 409), *SB* I 3913 (Antinoopolis; *editio princeps*: G. Lefebvre, "Égypte chrétienne", *ASAE* 10 (1910), 260–84 at 280–82; cf. G. Roquet, "Inscriptions bohaïriques de Dayr Abû Maqâr", *BIFAO* 77 (1977), 163–79 at 173–74).

14 As found in other episcopal epitaphs of the period; see D. Feissel, "L'évêque, titres et fonctions d'après les inscriptions grecques jusqu'au viie siècle", in: *Actes du XIe Congrès international d'archéologie chrétienne* I, Rome: École Française de *Rome* 1989, 801–28 at 802. For general background, see C. Rapp, *Holy bishops in late antiquity: the nature of Christian leadership in an age of transition*, Berkeley: University of California Press 2005.

15 The word γενάμενος is used with regard to a deceased bishop in H. Grégoire, *Recueil des inscriptions grecques-chrétiennes d'Asie mineure*, Paris: E. Leroux 1922 (and later reprints), no. 94 (N.N. "our former/erstwhile bishop"), but this is not a funerary inscription and does not appear to reflect the normal practice of episcopal epitaphs. Otherwise, γενόμενος/γενάμενος in the sense of "former, erstwhile" is commonplace in Greek epigraphy.

16 His popularity in Egypt did not leave traces in epigraphy, cf. A. Papaconstantinou, *Le culte des saints en Égypte des Byzantins aux Abbasides: l'apport des inscriptions et des papyrus grecs et coptes*, Paris: Éditions du CNRS 2001, 120–21; the occurrence of his name in the legends of the Madaba mosaic map (see P.-L. Gatier, *Inscriptions grecques et latines de la Syrie XXI: inscriptions de Jordanie* 2, Paris: P. Geuthner 1986, nos. 153–35, 37 and 43) is of an entirely different order.

17 See the useful reviews by H.W. Hollander, "The portrayal of Joseph in Hellenistic Jewish and early Christian literature", in: M.E. Stone and Th.A. Bergren (eds.), *Biblical figures outside the Bible*, Harrisburg, Pa.: Trinity Press International 1998, 237–63; P. Pilhofer and U. Koenen, "Joseph I (Patriarch)", *RAC* 18 (1998), col. 715–48; J. Dochhorn and A. Klostergaard Petersen, "*Narratio Joseph*: a Coptic Joseph-apocryphon", *Journal for the study of Judaism in the Persian, Hellenistic and Roman Period* 30 (1999), 431–63 at 436–44.

18 Coptic fragment, *CAVT*, no. 112 ("Historia Joseph et fratrum eius, sahidice"), quoted after *Stud. Pal.* XVIII, 269c (with minor emendations); cf. Dochhorn and Klostergaard Petersen, "*Narratio Joseph*", 443–44.

19 In the wake of *Acts* 7: 9–10 and 1 *Clem.* 4: 9, this latter quality determined his popularity as an apotropaic icon; see H. Maguire, "Magic and the Christian image", in: idem (ed.), *Byzantine magic*, Washington, DC: Dumbarton Oaks 1995, 51–71, especially 53–4; J. van der Vliet, "Satan's fall in Coptic magic", in: M. Meyer and P. Mirecki (eds.), *Ancient magic and ritual power*, Leiden: Brill 1995, 401–18 at 412.

20 Quoted in Hollander, "Portrayal", 241. Among Coptic sources, this "encratite" theme is much elaborated in *CAVT*, no. 110 ("Legenda Joseph, fragmentum copticum"); cf. Dochhorn and Klostergaard Petersen, "*Narratio Joseph*", 443.

21 See in particular J. Zandee, "*Iosephus contra Apionem*: an apocryphal story of Joseph in Coptic", *Vigiliae christianae* 15 (1961), 193–213 at 200–03 (on *CAVT*, no. 111

377

["Historia apocrypha de Joseph, sahidice"]; for a different view of this text, see Dochhorn and Klostergaard Petersen, "*Narratio Joseph*") and 212–13 (on *CAVT*, no. 112 ["Historia Joseph et fratrum eius"]).

22 See Feissel, "L'évêque, titres et fonctions", 812–13; for bishops of Faras and Kourte buried at Qasr Ibrim in a far later period, see *I. Qasr Ibrim* 18–26 (commentary). In none of these cases, the deceased is said to be the "former" bishop of his diocese.

23 For a recent list of late antique bishops of Aswan, see J.H.F. Dijkstra, *Philae and the end of ancient Egyptian religion: a regional study of religious transformation (298–642 CE)*, Leuven/Paris/Dudley, MA: Peeters 2008, 359, "Appendix 4".

24 The ostracon is dated to the fifth century paleographically; see R. Duttenhöfer, "Greek ostraca", in: D. Raue et al., "Report on the 34th season of excavation and restoration on the Island of Elephantine" (at [www.dainst.org/documents/10180/384618/Elephan tine+-+Report+on+the+34th+Season+(ENGLISH)/9b1f2a31-367a-48fd-8f83-c7bfb 0a5fea1;jsessionid=2BFB5D43DBB6F5287FF80A0D96F8DAD1?version=1.1]), 19–20; a connection was first suggested by H. Brakmann, "*Defunctus adhuc loquitur*: Gottesdienst und Gebetsliteratur der untergegangenen Kirche in Nubien", *Archiv für Liturgiewissenschaft* 48 (2006), 283–333 at 297, n. 63.

25 Cf. K.A. Worp, "A checklist of bishops in Byzantine Egypt (AD 325-c. 750)", *ZPE* 100 (1994), 283–318 at 291 and 312.

26 See Papaconstantinou, *Culte des saints*, 119, cf. G. Nachtergael, "Remarques sur des lampes et d'autres objets inscrits d'Égypte", *ZPE* 119 (1997), 185–88 at 187. For the Faras lamp, see F.L. Griffith, "Oxford excavations in Nubia", *LAAA* 14 (1927), 57–116, pl. LXI, no. 7 (the identification proposed by W.E. Crum on p. 111 is unconvincing).

27 See G. Garitte, "Constantin, évêque d'Assiout", in: *Coptic studies in honor of Walter Ewing Crum*, Boston: Byzantine Institute 1950, 287–304 at 300–02.

28 Ibn Hawqal, quoted in G. Vantini, *Oriental sources concerning Nubia*, Heidelberg: Heidelberger Akademie der Wissenschaften/Warsaw: Polish Academy of Sciences 1975, 153; see further J. Cuoq, *Islamisation de la Nubie chrétienne: VIIe-XVIe siècle*, Paris: P. Geuthner 1986, 9–15; D. Ayalon, "The Nubian dam", *Jerusalem studies in Arabic and Islam* 12 (1989), 372–90; reprinted in idem, *Islam and the abode of war: military slaves and Islamic adversaries*, Aldershot: Variorum 1994, no. XII.

29 J.-Cl. Garcin, in J.-C. Garcin and M. Tuchscherer, "Uswân", *Encyclopaedia of Islam* 10, new edition, Leiden: Brill 2000, 938.

30 John of Biclar, in *Chronica minora. Saec. IV – VII*, 2, ed. Th. Mommsen, Berlin: Weidmann 1894, 212 (English translation in Vantini, *Oriental sources*, 27–8); cf. S.G. Richter, *Studien zur Chistianisierung Nubiens*, Wiesbaden: Reichert Verlag 2002, 113, 183.

31 See W. Godlewski, "A new approach to the Christianization of Makuria: an archaeological note", in: C. Berger, G. Clerc, and N. Grimal (eds.), *Hommages à Jean Leclant* 2, Cairo: IFAO 1994, 169–76. For further discussion, cf. D.A. Welsby, *The medieval kingdoms of Nubia: Pagans, Christians and Muslims along the Middle Nile*, London: British Museum Press 2002, 34–5, 100–01; W. Godlewski, "The rise of Makuria (late 5th–8th century AD)", in: T. Kendall (ed.), *Nubian studies 1998: proceedings of the Ninth Conference of the International society of Nubian studies, August 21–6, 1998*, Boston, MA: Museum of Fine Arts and Northeastern University 2004, 52–73. See also note 42 below.

32 But cf. M. Krause, "Zur Kirchengeschichte Nubiens", in: T. Hägg (ed.), *Nubian culture past and present: main papers presented at the Sixth International Conference for Nubian studies in Uppsala, 11–16 August, 1986*, Stockholm: Almquist & Wiksell 1987, 293–308 at 296–97; L. Török, *Late antique Nubia: history and archaeology of the southern neighbour of Egypt in the 4th–6th c. AD*, Budapest: Archaeological Institute of The Hungarian Academy of Sciences 1988, 70–1; L.P. Kirwan, "Some

FROM ASWAN TO DONGOLA

thoughts on the conversion of Nubia to Christianity", in: J.M. Plumley (ed.), *Nubian studies: proceedings of the symposium for Nubian studies, Cambridge 1978*, Warminster: Aris & Phillips 1982, 143–44.

33 These are the foundation inscriptions from Faras (Jakobielski, *Bishopric of Pachoras*, 37–48; *I. Khartoum Copt.* 1) and Tafa (J. Maspero, "Le roi Mercure à Tâfah", *ASAE* 10 (1910), 17–20; *SB* I 1594).

34 The date refers to the missionary activity of Bishop Longinos described by John of Ephesus (*Church history*, ch. 53), see Vantini, *Oriental sources*, 19–23; extensively commented by Richter, *Christianisierung Nubiens*, 93–8; cf. now Dijkstra, *Philae*, 271–304.

35 See S. Lane-Poole, *A history of Egypt in the Middle ages*, London: Methuen 1901, 22; L.P. Kirwan, "Notes on the topography of the Christian Nubian kingdoms", *JEA* 21 (1935), 57–62 at 61; Welsby, *Medieval kingdoms*, 83–4.

36 The main source is the *Life of Patriarch Michael I* by John the Deacon quoted in the *History of the Patriarchs of Alexandria*, ed. B.T.A. Evetts, Paris 1909, 140; cf. Vantini, *Oriental sources*, 40. See also U. Monneret de Villard, *Storia della Nubia cristiana*, Rome: Pontificium Institutum Orientalium Studiorum 1938, 80.

37 To put the case in a somewhat blunt way, there must have been a highly educated Coptic monk in his retinue who composed the stela after Joseph's death, as the Coptic language was not generally used at Dongola at the time. This perhaps explains the use of two languages in the text of the stela, the better-known Greek being used in the parts that contain the most important information about the deceased.

38 S. Munro-Hay, "Kings and kingdoms of ancient Nubia", *Rassegna di studi etiopici* 29 (1982–1983), 87–137 at 98; Welsby, *Medieval kingdoms*, 100, n. 164.

39 Vantini, *Oriental sources*, 375; quoted also by al-Qalqashandi: ibidem, 575–76. Cf. U. Monneret de Villard, "Note nubiane", *Aegyptus* 12 (1932), 305–16 at 309–16; idem, *Storia*, 62–3. Cf. the Coptic (Bohairic) version of the *Life of Isaac*, Monophysite patriarch of Alexandria in 686–9 a.d., by Mena of Nikiou, ed. E. Porcher, *Vie d'Isaac, patriarche d'Alexandrie de 686 à 689, écrite par Mina, évêque de Pchati*, Paris: Firmin-Didot 1915, [79]–[80].

40 Doubts concern only the "Early Church on Kom D" (EDC), cf. J. Dobrowolski, "The first church at Site 'D' in Old Dongola (Sudan)", *Archéologie du Nil moyen* 5 (1991), 29–40; on its site another church (DC) was built, but apparently at a later date.

41 In this case, however, the rebuilding is a quite natural consequence of the destruction of the church during the Arab raid on Dongola in 651/652, cf. W. Godlewski, "The Cruciform Church site in Old Dongola: sequence of buildings from 6th to 18th century", *Nubica* I/2 (1990), 511–34 at 524–27; idem, "The rise of Makuria", 72.

42 For general information on Dongolese churches, see especially S. Jakobielski, "35 years of Polish excavations at Old Dongola: a factfile", in: S. Jakobielski and P.O. Scholz (eds.), *Dongola-Studien: 35 Jahren polnischer Forschungen im Zentrum des makuritischen Reiches*, Warsaw: ZAŚ PAN 2001, 1–28, and W. Godlewski, "The Churches of Dongola: their origin and importance in the general line of development of church architecture in Makuria", in: I. Caneva, and A. Roccati (eds.), *Acta Nubica: proceedings of the X International Conference of Nubian studies, Rome, 9–14 September 2002*, Rome: Istituto Poligrafico e Zecca dello Stato 2006, 263–86.

30

ROME – MEROE – BERLIN
The southernmost Latin inscription
rediscovered (*CIL* III 83)*

Adam Łajtar and Jacques van der Vliet

Introduction

Habent sua fata et saxa. In January 2003, one of the present authors was allowed access to the storerooms of the Skulpturensammlung und Museum für Byzantinische Kunst of the Staatliche Museen in Berlin (hereafter SMBK) for a collation of the surviving pieces of a lot of Coptic and Greek funerary stelae from the medieval monastery of Ghazali (Wadi Ghazal, northern Sudan).[1] Richard Lepsius had collected these inscriptions on the spot in 1844, and had them shipped to the Prussian capital.[2] In 1849, he published drawings of most of them in his *Denkmaeler aus Aegypten und Aethiopien*.[3] In Berlin, this lot of nearly thirty pieces, mostly quite fragmentary, was deposited in the Egyptian Department of the Königliches Museum, which later became the Ägyptisches Museum.[4] Almost a century later, in the 1930s, the collection was split up into two more or less equal parts, one of which remained in the Ägyptisches Museum, whereas the other was transferred to the Kaiser-Friedrich-Museum, the predecessor of the present-day SMBK.[5] Shortly afterwards, the Coptic inscriptions of the Kaiser-Friedrich-Museum, including those from Ghazali, were inventoried and described by Maria Cramer. Her catalogue, however, although written in 1937–1938, appeared only in 1949.[6] Meanwhile, the Second World War had intervened. Several of the Ghazali pieces from both Berlin museums disappeared during the war and have not re-emerged since.

In her catalogue, Cramer mentioned three Coptic items from the Ghazali lot that, according to her knowledge, had not been included in Lepsius' *Denkmaeler* one century earlier, and had remained unpublished.[7] Two of these are tiny fragments bearing a few letters only (SMBK inv. nos. 9682 and 9684); a third one, SMBK inv. no. 9675 (formerly Ägyptisches Museum, inv. no. 1504), is described by her as being of reddish sandstone "mit leicht eingeritzter Inschrift". She added: "ein Zusammenhang ist nicht mehr herzustellen". Neither a text nor a photo is given. This latter piece has survived the war, and its Ghazali provenance seems to be confirmed by the entry in the SMBK inventory, which calls the stone a "Grabstein aus Wadi Gazal".[8] As it appeared upon inspection in the

381

museum storerooms, however, it does not fit very well into the collection. Its material and manufacture are quite different from those of the characteristic Ghazali pieces. There is nothing to mark it as a funerary inscription or even as a Christian monument. Moreover, the text itself, although difficult to decipher, is clearly conceived neither in Coptic, Greek or Old Nubian. Instead, this unsightly block of stone turned out to bear one of the most curious texts from the ancient Sudan, the Latin graffito left by a visitor at the Meroitic temple site of Musawwarat al-Sufra, over thousand kilometers away from the nearest Roman frontier.[9] Best known as the southernmost Latin inscription, it was considered lost for more than a century.

Although both the general content of the inscription and its bibliographical lemma are reasonably well known,[10] we briefly recapitulate here the main facts from its history before presenting the text as it can be read now, together with a first-ever photographic reproduction of the original piece (Figure 29).

The text was discovered in 1822 by Louis Linant de Bellefonds, the first European visitor of the site since antiquity. He copied it in his notebooks, which remained unpublished during his lifetime.[11] In the same year, it was noticed by Frédéric Cailliaud, who published his own reading soon afterwards.[12] According to both travelers, the stone was part of a wall in one of the ceremonial stairways of the central temple complex of Musawwarat al-Sufra.[13] These early reports virtually exclude the possibility of the inscription being a fake.[14] In 1844 it was found

Figure 30.1 Latin visitor's inscription, Musawwarat al-Sufra (Nubia), SMBK inv. no. 9675 (photo: © Skulpturensammlung und Museum für Byzantinische Kunst)

by the expedition of Richard Lepsius[15] and taken to Berlin, together with many other monuments, among which were the stelae from Ghazali. Lepsius published the inscription twice – once, immediately upon its discovery, in one of his reports that appeared in the Prussian national gazette,[16] and a second time, in 1849, in his *Denkmaeler*, Abt. VI, Blatt 101, no. 56. For his work on the latter publication, Lepsius had a squeeze made, which still exists in the archives of the Berlin Ägyptisches Wörterbuch in Berlin.[17] The text was subsequently republished by Th. Mommsen in *CIL* III (Berlin 1873), presumably after autoptic examination of the stone.[18] A fine study of the inscription, based on the text in *CIL* III, was published in 1911 by the Flemish scholar Josué De Decker.[19] He intended to publish a photo of the piece, but was informed from Berlin that the stone had been lost ("anscheinend gestohlen") for already over twenty years.[20] In the 1960s, Fritz Hintze again tried to retrieve the inscription, but with no better result. According to his information, the stone had been destroyed during the Second World War. He did, however, publish a photo of the Lepsius squeeze, together with a facsimile, made after the squeeze by K.-H. Priese.[21] Subsequent discussions are all dependent on Hintze's article. Although the squeeze published by him remains a valuable witness for the text's earlier state, future studies may henceforth be based upon the original.

Far from being stolen or destroyed, the inscription had been hiding under the disguise of a medieval epitaph at least since about 1890. How exactly the monument's erroneous association with the finds from Ghazali has come about is not entirely clear. It may be due either to a superficial physical likeness or, perhaps more likely, to the number it was assigned. In the inventory of the Ägyptisches Museum it bore the no. 1504, a number that would range it with the series of Christian stelae from Ghazali, which, almost without exception, had received numbers between 1480 and 1510.[22] In any case, the resulting confusion accounts for its later transfer to the Kaiser-Friedrich-Museum as well as for its classification as a Coptic inscription by Maria Cramer.

The inscription is presently part of the collection of the Skulpturensammlung und Museum für Byzantinische Kunst, Staatliche Museen zu Berlin – Stiftung Preußischer Kulturbesitz, under inventory no. 9675. It is inscribed on a small block of very hard and bright white sandstone, covered by a reddish-brown patina. This is, in fact, the standard local building material in Musawwarat al-Sufra.[23] The greatest measurements of the stone are $28 \times 32 \times 10$ cm. The block is not entirely square: the carefully dressed righthand side of the stone is made to slant inwards. This feature may be connected with its original position near the entrance of the ramp where it was seen by the early nineteenth-century travelers. In the back, which is only roughly dressed, a modern hanging device is let in. The smoothened front of the stone bears the remains of six lines of text. The text occupies only part of the block, leaving a margin of about 3.5 cm to its left and an uninscribed surface of about 8 cm high, below it; at the right, the text continues up to the rim of the stone. The stone is damaged at several spots and the lower lefthand corner was restored in modern times. Damage affects the legibility of l. 1 (several chips, some of which are recent, postdating the squeeze) and, in particular, of l. 5 (an apparently old slash obliterated part of the writing).

The inscription is properly a graffito rather than the work of a professional stonecutter.[24] The letters are squarish and inelegant, about 2–2.5 cm high, scratched with visible difficulty into the hard surface; no traces of ruling are visible and, from l. 3 onwards, the lines tend to slope upward. Towards the ends of the lines, the writing is often cramped. Word division and the end of the text are marked by high points. On account of its informal character, it is practically impossible to date the script paleographically; for the proposed dating, see the commentary below. The elements of a bibliography have been given above.

SMBK inv. no. 9675, Musawwarat al-Sufra (Sudan),   third/fourth century A.D.

BONA · FORTVN[A] · DOMINAE ·
REGINAE · IN · MVLTOS · AN-
3.   NOS · FELICITER · VENIT
B VRBE · MENSE · APR(ILI)
DIE · XV [·. .]IDIT · ACV-
6.   TVS ·

1. Text much damaged, but readings quite certain. FORTVNA: A certain from squeeze, the chip is very recent. Final point merely possible ‖ 2. IN: the damage in N seems to be recent (cf. the squeeze) ‖ 4. B for AB, with loss of the vowel in anlaut; all earlier editors: E for EX; uncertain whether a high point follows. APR: cramped ‖ 5. Following XV only scratches can been seen (according to Hintze, "Latin inscription", 297, these would belong to a stonemason's mark), but the number seems complete as it stands. Mommsen: VIDI, noting: "litterae VID valde incertae". Actually, two or three letters are required before]IDIT; perhaps ET preceded. [· ET V]IDIT would fill the space and yields a good sense; compare e.g., at another monumental site in the Nile Valley, *I. Syringes* 1620: *vidi et miravi*, and 1504: *vidi et miravi locum* | End of line, – T · ACV-: thus certain, with Hintze. The group CV is cramped: V looks like I but is slightly more inclined to the right; its left-hand jamb is inscribed in the C. This definitely excludes Mommsen's reading TACITVS ‖ 6. End of text.

Good fortune to the Lady Queen for many years in happiness!
Acutus came from the City in the month of April, day 15, (and) he. . . . [25]

The rediscovery of the stone shows that the text as edited in *CIL* III 83, and the squeeze published by Hintze are generally trustworthy. The crucial problem in reading and interpreting the inscription remains the verb in l. 5, which for the most part is lost. Nevertheless, the re-emergence of the original also permits some positive conclusions about the general character of the text.

The inscription, brief as it is, consists of two parts. The first is a wish for good luck and long life, addressing a female sovereign (ll. 1–3). It is composed of a series of three conventional acclamatory formulae that have here been rendered in one phrase: *bona fortuna, in multos annos, feliciter*.[26] Such acclamations, celebrating authorities and patrons, enjoyed an increasing popularity in the epigraphy of the later Roman Empire, where they could acquire a real political significance.[27] Here, they address an unnamed "Lady Queen" (ll. 1–2), who must have been the local ruler. Neither the nature of the acclamations nor any other feature of the text suggest that she might have been a goddess, as has been claimed. The adherents of this latter thesis generally favor an identification of the "Lady Queen" with Isis[28] on the basis of an analogous title in the Greek inscription *SB* V 7944 (*FHN* II 168), from Dakka in Lower Nubia.[29] Indeed, the κυρία βασίλισσα in l. 3 of that inscription offers a striking parallel to the *domina regina* of the present text.[30] Contrary to a widespread opinion, however, the Dakka text does not designate Isis, but in all probability the ruling "Kandake" of Meroe.[31] The same interpretation is likely to be correct here.[32] Recent scholarship emphasizes the meaningful relationship between inscriptions, in particular acclamations, and the space where they are situated.[33] One way to account for the original position of the present text would be to assume that official apparitions of the Kandake took place on the ceremonial stairway where it was found. Perhaps the impressive cult center of Musawwarat al-Sufra, rather than a palace in Meroe itself, was the place where she received foreign visitors and their homages.

The second half of the inscription (ll. 3–6) generally conforms to the type of the Latin traveler's graffito that can be found also elsewhere in the Roman East.[34] Its author states to have arrived from somewhere at the present spot, and to have done something or simply to have been there at a certain date. In a similar way, earlier travelers coming back from India had marked their passage in the desert near Berenike.[35] In the present case, the traveler came from Rome, as is clear from the indubitable text of l. 4. An alternative reading, such as VRBE MEROE, proposed by L. Török,[36] can be definitely ruled out. The author's provenance is also apparent from his name, Acutus, which was common in Italy and Rome, but quite rare in Egypt.[37] The same applies to his use of Latin and of the Roman calendar (ll. 4–5), both relatively uncommon in the Nile Valley.[38] Apparently, Acutus was a Roman from Rome on some mission to the Meroitic court. The purpose of his mission may have been diplomatic, the kind of mission documented by the earlier Greek inscription from Dakka, or commercial, like that of the Red Sea navigators returning from India.[39] In any case, the multiple acclamations addressing the local ruler (ll. 1–3) are rather not the expression of the fervor of a mere tourist, but bear out the interested, if not official character of his visit.

The date of Acutus' mission is a final point of interest. On the one hand, the absence of Christian elements in the text and the honors paid to the Kandake of Meroe do not favor a date posterior to the fourth century A.D. On the other hand, several features of the text do suggest a relatively late date. Already the use of a single personal name, Acutus, instead of the traditional threefold Roman name, can be seen as significant. The most telling feature, though, is undoubtedly the use of the "modern" month date "April, day 15", instead of the traditional Roman system that reckons with Kalendae, Nonae and Idus. This way of dating became a general practice only after the middle of the sixth century A.D., but earlier examples can be found in inscriptions from the third-fourth centuries onwards.[40] The Musawwarat al-Sufra inscription is therefore considered a witness of direct diplomatic or commercial contacts between Rome and Meroe in the latter days of the Kushite monarchy, towards the end of the third and the first half of the fourth century A.D.

Notes

* The find that we publish here could not have been made without the generous assistance of Prof. Dr. A. Effenberger, director of the Skulpturensammlung und Museum für Byzantinische Kunst, Staatliche Museen Berlin; Dr. Gabriele Mietke, keeper; and Dr. Cäcilia Fluck, consultant at the same museum. The latter also provided us with a set of useful working photos. Our discussion of the text much profited from the remarks of Drs. T. Plóciennik and K. Rzepkowski, University of Warsaw. Dr. Tasha Vorderstrasse, Leiden, kindly advised us in matters of English language.

1 The biggest monastic site of Makuria outside of the Old Dongola region and a rich source of inscriptions, most of them now in the National Museum of Khartoum (*I. Khartoum Copt.* 43–116, *I. Khartoum Greek* 31–55). We intend to publish an updated list and critical revision of the Ghazali inscriptions that are not kept in Khartoum.

2 See C.R. Lepsius, *Briefe aus Aegypten, Aethiopien und der Halbinsel des Sinai*, Berlin: W. Herz 1852, 234–36, where he describes his visit of the monastery ruins.

3 C.R. Lepsius, *Denkmaeler aus Aegypten und Aethiopien*, Sechste Abteilung, Berlin: Nicolai 1849; reprinted Geneva 1973: Éditions de Belles-Lettres, Blatt 99 (the Greek pieces) and Blatt 103 (the Coptic ones).

4 For a long time, some of them have been on show in the Nubian rooms of the museum, see e.g. *Ausführliches Verzeichnis der aegyptischen Altertümer und Gipsabgüsse*, second edition, Berlin: W. Spemann 1899, 412–13.

5 According to the museum inventory the transfer took place in 1934 or 1935 (both dates are mentioned).

6 M. Cramer, *Koptische Inschriften im Kaiser-Friedrich-Museum zu Berlin*, Cairo: Société d'archéologie copte 1949.

7 Ibidem, 10.

8 An old cardboard inventory card of the Ägyptisches Museum (after 1871) describes the piece also as "Inschrift aus Wadi Gazal", but calls it Greek: "Fragment einer griechischen Inschrift von sechs wagerechten vertieften Zeilen. Unlesbar" (all information kindly provided by Dr. G. Mietke).

9 Roman occupation never extended beyond Qasr Ibrim, in Lower Nubia; for a recent discussion, see R.B. Jackson, *At empire's edge: exploring Rome's Egyptian frontier*, New Haven/London: Yale University Press 2002, 143–46; cf. J. Locher, "Die Anfänge der römischen Herrschaft in Nubien und der Konflikt zwischen Rom und Meroe", *Ancient society* 32 (2002), 73–133, and T. Stickler, *"Gallus amore peribat"? Cornelius Gallus und die Anfänge der augusteischen Herrschaft in Ägypten*, Rahden/Westfalen: Leidorf 2002.

10 See most recently *FHN* III 297, and S. Bersina, "L'inscription latine de Moussawwarat es-Sofra", *CRIPEL* 20 (1999), 97–104 (cf. *AE* 1999, no. 1718; Locher, "Anfänge", 124, n. 139); both the *FHN* and Bersina provide extensive bibliographies that are not repeated here.

11 See (L.M.A.) Linant de Bellefonds, *Journal d'un voyage à Méroé dans les années 1821 et 1822*, ed. M. Shinnie, Khartoum: Sudan Antiquities Service 1958, pl. XXIV, cf. 119; P.L. Shinnie, "A late Latin inscription", *Kush* 9 (1961), 284–86.

ROME — MEROE — BERLIN

12 In his *Voyage à Méroé, au Fleuve Blanc* (etc.), III, Paris: l'Imprimerie royale 1826, 375 (with notes by A.J. Letronne).

13 Linant de Bellefonds, *Journal*, pl. XXIV: "en dedans du grand escalier III"; cf. 119: "à l'entrée du grand escalier III en dedans"; Cailliaud, loc. cit.: "sur le mur d'une rampe dans la partie postérieure du monument central".

14 Thus already Shinnie, "Late Latin inscription", 285.

15 It is not mentioned in Lepsius' description of his visit of the site, *Briefe aus Aegypten*, 153–54, but cf. *Denkmaeler*, Text, V, Leipzig: J.C. Hinrichs 1913; reprinted Geneva: Éditions de Belles-Lettres 1975, 345: "an der Rückseite der L.D. VI, 11, No. 57 genannten Treppen steht die lateinische Inschrift". On the visit of the Royal Prussian Expedition to Musawwarat al-Sufra, see recently St. Wenig, "Die Erforschungsgesschichte des Apedemak-Tempels von Mussarawat es-Sufra", *Der antike Sudan* 14 (2003), 25–42 at 33–5. He reproduces as fig. 15 the photo of the inscription left by the Expedition on the back wall of the main temple. The two lines text reads as follows: KOENIGL. PREUSS. EXPEDITION | März 1844.

16 Best known in a French translation, with notes by A.J. Letronne, "Extraits des dernières lettres du docteur Lepsius", *Revue archéologique* 1 (1844), 573–80 at 576–77.

17 Published by F. Hintze, "The Latin inscription from Musawwarat es Sufra", *Kush* 12 (1964), 296–98. On the Lepsius squeezes, see S. Köpstein, "Das Abklatscharchiv beim Wörterbuch der Ägyptischen Sprache", *Mitteilungen aus der Arbeit am Wörterbuch der ägyptischen Sprache* 3 (1994), 13; the piece in question, however, is not listed at 55–60 ("Berlin, Ägyptisches Museum"), perhaps because no museum inventory number was known (see 53).

18 No inventory number is mentioned, though. Of course, Mommsen may have used the Lepsius squeeze, as did A. Kirchhoff for the Greek Ghazali pieces published in *CIG* IV (Berlin 1877), but he does not mention it. Hintze, "Latin inscription", 296, n. 4, suggests that Mommsen prepared his edition after the plate in Lepsius' *Denkmaeler*; on account of the (otherwise unimportant) divergences between both, this is less probable.

19 "Le culte d'Isis à Méroé en Éthiopie", *Revue de l'instruction publique en Belgique* 54 (1911), 293–310 at 300–9; De Decker is now mainly remembered for his political activities, see S. De Groote, "Josué De Decker (1878–1953): Activist en hoogleraar", *Het Land van Nevele* 23/1 (March 1992), 3–56.

20 Letter by H. Dessau, quoted in De Decker, "Culte", 301, n. 1. Dessau held his information from H. Schäfer, then director of the Ägyptisches Museum; he also briefly comments on the inscription, using Lepsius' squeeze and diaries.

21 Hintze, "Latin inscription", the facsimile at 297. Hintze was also able to consult unpublished notes by members of the Lepsius expedition; cf. 296, n. 4.

22 Actually ranging from 1482 to 1508, according to our present knowledge. The old inventory card mentioned above shows that no. 1504 bore no. 125 in a catalogue of 1871; it also refers to another (earlier?) inventory by a number S.41.14x, and mentions its placement in "Saal III".

23 For a brief characteristic, see K.-H. Priese, in F. Hintze et al., *Musawwarat es Sufra*, Bd. I/1: *Der Löwentempel*, Berlin: Akademie-Verlag 1993, 20–1.

24 There is no reason whatever, though, to doubt the competence of the scribe, *pace* L. Török in *FHN* III, p. 1093, who does not take the quality of the stone into account.

25 Perhaps, "saw (the place)", as suggested in the apparatus above.

26 For acclamations in general, see the classic study by E. Peterson, *ΕΙΣ ΘΕΟΣ: Epigraphische, formgeschichtliche und religionsgeschichtliche Untersuchungen*, Göttingen: Vandenhoek & Ruprecht 1926; Th. Klauser, art. "Akklamation", *RAC* 1 (1950), 216–33, and the article cited in the next note. For *bona fortuna*, compare the Greek ἀγαθῇ τύχῇ, frequently found at the head of dedicatory inscriptions, in the Nile Valley e.g. in

ADAM ŁAJTAR AND JACQUES VAN DER VLIET

I.Hermoupolis 12 (in honor of Emperors Marcus Aurelius and Commodus) and *passim* in *I.Akôris* (second-third century), cf. Peterson, *ΕΙΣ ΘΕΟΣ*, 21 and 26; for *in multos annos* (Greek πολλὰ τὰ ἔτη), see Peterson, *ΕΙΣ ΘΕΟΣ*, 167–8; Klauser, "Akklamation", 231, no. 25; for *feliciter*, Peterson, *ΕΙΣ ΘΕΟΣ*, 223–24; Klauser, "Akklamation", 229, no. 15; in Nubia: *I.Dakka* (Ruppel) Gr. 61 (Latin graffito), with note.

27   See the seminal essay by C. Roueché, "Acclamations in the later Roman empire: new evidence from Aphrodisias", *JRS* 74 (1984), 181–99.

28   Only Bersina, "Inscription latine", (unconvincingly) proposes to recognize a *Dea Caelestis* here.

29   Thus, first, De Decker, "Culte", and still recently e.g. Török in *FHN* III, 1094.

30   As was first observed by U. Wilcken, "Kandake", *Hermes* 28 (1893), 154–56.

31   See, most recently, Locher, "Anfänge", 123–24.

32   Thus already Wilcken, "Kandake", 156; Hintze, "Latin inscription", 298.

33   See, in particular, C. Roueché, "Looking for late antique ceremonial: Ephesos and Aphrodisias", in: H. Friesinger and F. Krinzinger (eds.), *100 Jahre österreichische Forschungen in Ephesos: Akten des symposions Wien 1995*, Vienna: Verlag der Österreichischen Akademie der Wissenschaften 1999, 161–68; D. Feissel, "Les inscriptions des premiers siècles byzantins (330–641): documents d'histoire sociale et religieuse", in: [N.N.] (ed.), *XI Congresso Internazionale di Epigrafia Greca e Latina: Atti* II, Rome: Edizioni Quasar 1999, 577–89 at 583–84.

34   E.g. in Egypt, *I.Thèbes-Syène* 321 and 323 (Philae); H. Cuvigny and A. Bülow-Jacobsen, "Inscriptions rupestres vues et revues dans le désert de Bérénice", *BIFAO* 99 (1999), 133–93 at 137–41 (*paneion* of Wadi Minayh).

35   Compare e.g. Cuvigny and Bülow-Jacobsen, "Inscriptions rupestres", 140, no. 5: (N.N.) *exs India | redie<n>s hic fuit* (+ year + month).

36   But see now his retractation in *FHN* III, 1094.

37   A single example is mentioned in *NB*, 16 (P. Berlin IV 1173.3 [reign of August]: Lucius Cornelius Acutus, clearly a Roman), another one in *Onomasticon*, 24, both s.v. Ἀκοῦτος. See furthermore, I. Kajanto, *The Latin cognomina*, Helsinki: Keskuskirjapaino 1965, 69 (cf. 93 and 249).

38   Note, however, that at the time of Diocletian the use of Roman month names became somewhat more frequent in Egypt; see P.J. Sijpesteijn, "Some remarks on Roman dates in Greek papyri", *ZPE* 33 (1979), 229–40.

39   See above, notes 32 and 35–6.

40   For a discussion, see A. Ferrua, "Il giorno del mese", *RivArchCrist* 61 (1985), 61–75.

"WHAT IS MAN?"
The Nubian tradition of Coptic funerary inscriptions*

Jacques van der Vliet

Introduction

Inscriptions are traditionally classed as documentary texts, texts that are supposed to provide primarily matter-of-fact quantifiable information. In *Sammelbücher* and dictionaries they are lumped together with accounts, contracts and letters, as "Urkunden". Coptic funerary inscriptions have also most often been studied as documents, for example as a source of onomastic material for the reconstruction of a particular community or episcopal see,[1] or the diffusion of particular cults,[2] if not simply for the sake of classification along topographical or chronological lines.[3] Usually their contents are taken to coincide with the information they contain about a deceased person and the date of his or her demise. The funerary inscription is treated as a kind of death certificate ("Sterbeurkunde") that happens to be chiseled in stone.

The present contribution takes a radically different view of funerary inscriptions. Its main contention is that funerary inscriptions are products of literary culture that do not coincide with the factual information they may happen to contain and deserve to be taken seriously as texts. Furthermore, in addition to being a textual artifact, inscribed tombstones are also monuments. For that reason alone, their role in society must have been markedly different from other textual artifacts, such as letters and contracts. They belong as much to the domain of art history and archaeology as to that of philology and textual criticism.[4] In so far as this paper is focused on funerary inscriptions as texts, it presents an outline of the development of the Coptic-language epitaph in early-medieval Nubia, not as a mechanical succession of formulae through time but as the reflection of shifting intellectual attitudes. The discussion will be based on material from a single significant site, Qasr Ibrim in Lower Nubia.

The Arabicized name Qasr Ibrim ("Castle of Ibrim") preserves the ancient toponym Phrim, Latin Primis.[5] It designates a strategically situated natural citadel, over 200 km south of Aswan, as well as its surrounding habitation quarters and cemeteries that are now submerged by the waters of Lake Nasser. Qasr Ibrim was the political center of an independent kingdom, Nobadia, in late antiquity.

Following the Christianization of the region, it was made the see of a bishop, for whom a magnificent stone-built cathedral was erected, still visible on site.[6] Presumably shortly after the year 600, Nobadia became politically integrated into the southern kingdom of Makuria, which had its capital in Old Dongola.[7] Qasr Ibrim retained its political significance and came to be – at some time after 652 – the residence of a high official, the eparch of Nobadia, who played a pivotal role in the contacts between Christian Nubia and Islamic Egypt. The Coptic funerary inscriptions from this religious and political center make up a rather modest but representative corpus that has become available for study only very recently.[8] It can be usefully compared to similar corpora from nearby sites in Nobadia as well as material from elsewhere in Nubia and Egypt. The contemporary Greek-language tombstones from Nubia are also used for comparison but – for practical reasons – not discussed for their own sake here, even though they are part of the same cultural and religious setting.

The following pages will first present a brief sketch of the role of Coptic and in particular Coptic epigraphy in Christian Nubia, in an attempt to situate the material in time and place. This will be followed by some methodological and generalizing considerations in order to better understand the nature of the material. Then the four textual types that can be distinguished in the Qasr Ibrim material will each be presented and discussed. The conclusions will not only summarize the results of the preceding discussion but also indicate some directions and questions for further research.

Coptic in Nubia[9]

The Christianization of Nubia came about in the prolonged period of political turmoil that followed the withdrawal of the Roman frontier to the First Cataract area under Diocletian in 298 and the fall of the Meroitic empire in the early fourth century.[10] When Christianity started to spread, presumably in the fifth century, this coincided with the formation of new political entities that were culturally centered on neighboring Egypt and, in a broader sense, the Roman East. The new kingdoms of Nubia, Nobadia in the north and Makuria and Alodia further south, became officially Christian in the sixth century, when missionary efforts came to sanction an already ongoing process of acculturation. In the course of this process the two literary and ecclesiastic languages of late antique Egypt, Greek and Coptic had come to replace Meroitic as written languages. The place of Coptic in the Christianization of Nubian culture is illustrated by various sources, among which the so-called Tantani letters, a small group of fifth-century letters addressed in Coptic from Egypt's southern frontier to a Nobadian chieftain Tantani, based at Qasr Ibrim,[11] and the sixth-century Dendur inscription, for which see below, are undoubtedly the most telling. In fact, Greek and Coptic, not the indigenous Nubian, were to remain the literary languages of Christian Nubia for many centuries. The Coptic of Nubian monuments is always Sahidic, often with peculiarities that connect it with southernmost Egypt, which is hardly surprising. Yet it should be emphasized that Coptic literacy in Nubia was borne by an indigenous elite and was not a mere in-group language of Egyptian immigrants.

From the literary and documentary sources that are now gradually being made known, it appears that Qasr Ibrim was a center of Coptic literary culture of equal importance to the cities of Upper Egypt. Yet, in spite of significant text finds from elsewhere as well, such as the Qasr al-Wizz Codex, found in the neighborhood of Faras,[12] literacy in Nubia in general seems to have been much more limited than in Egypt. The Nubian Church, dependent on the anti-Chalcedonian Church of Alexandria, and the royal or eparchial chancelleries must have been the main markets for professional literacy. Moreover, one gets the impression that the further south one goes, the thinner the spread of literacy becomes (excepting, of course, the Makurian capital at Dongola). The impression is perhaps deceptive since the climatic conditions are less favorable for the preservation of manuscripts in the south than in the north. Yet, also the majority of Nubian Coptic stone inscriptions stem from Nobadia, south of the former Dodekaschoinos (Qasr Ibrim, Toshka and Arminna, the Faras-Meinarti region and Saï), even though Makurian sites like Mesho, Ghazali and al-Koro also produced substantial quantities.[13] The numerically most important lot was brought to light during excavations in the 1930s in the cemetery of Sakinya in the district of Toshka West, to the south of Qasr Ibrim.[14] Those from Qasr Ibrim itself are far less numerous (about fifty items, including many fragmentary pieces), but these are sufficiently interesting to warrant a separate discussion.

The Coptic epigraphic tradition in Nubia was a very vital one and spans a period of about six centuries. The Dendur inscription, which commemorates the conversion of the temple of Dendur into a church during the reign of the Nobadian King Eirpanome, is actually the earliest datable dedicatory inscription in Coptic.[15] The mention of the well-known Bishop Theodore of Philae (ca. 525–after 575) assigns the text to roughly the middle of the sixth century.[16] At the other end of the chronological range, extensive epitaphs in Coptic were still being produced for civil and ecclesiastical authorities in such Nubian centers as Faras and Saï till the very end of the eleventh century, at a time when Sahidic Coptic in Egypt was rapidly giving way to Arabic and, for ecclesiastical purposes, Bohairic Coptic. The latest precisely dated Nubian epitaph in Coptic is that of Bishop Georgiou of Faras (*I. Khartoum Copt.* 5), who died in 1097. The bulk of the more modest Coptic epitaphs from Nubia, although undated in most cases, may therefore originate from the period between about 550 and 1100.

The chronological lines can be drawn somewhat more precisely, however. The funerary monument of Bishop Joseph of 668, discovered in Old Dongola in 2004, contains a long personalized encomium of the deceased in Sahidic Coptic, whereas the funerary formulae that frame the main text are in Greek.[17] The Greek formulae serve to introduce the basic information about the deceased, his name and dignities and the date of his demise. This procedure suggests that whereas Coptic could be used for epigraphic purposes, as it in fact had been already a hundred years earlier in the Dendur foundation inscription, in the sixties of the seventh century the composition of a funerary stela still demanded opening and closing formulae in Greek. If the inference would prove to be correct, the development and diffusion in Nubia of a Coptic funerary formulary, in the sense of an established textual format (see below), cannot be dated before the middle of the seventh century. This would agree with the situation in Egypt, where the incipience of a Coptic-language epigraphic tradition appears to have been likewise slow and late.[18]

Monuments for the living

The normal way of presenting an ancient epitaph, printed on paper in a normalized font, accompanied by an apparatus and a commentary that are primarily focused on the editor's understanding of the text's form and contents, is utterly misleading. Before being a text, a funerary inscription is part of a monumental setting, devised for public visibility. Thanks to archaeology, we are reasonably well informed about the material aspects of this setting.

The most typical form of the Nubian monumental tomb consisted of a low bench-like superstructure with a rounded top.[19] One of the short sides could incorporate a stela let into the superstructure. Below the stela, a niche was often introduced to accommodate a lamp. Both lamps and stelae have been found *in situ*, for example in the cemeteries of the suburbs of Faras.[20] The tombs were whitewashed and so were the stelae, which were additionally provided with a lively polychromy, meant to bring out the text and whatever usually rather sober decoration they bore. A Greek epitaph discovered in Old Dongola in February 1997 was incised into the wet plaster surface of the superstructure and subsequently filled in with black paint.[21] Usually, however, the texts were inscribed on slabs immured in the superstructure. The material could be marble, terracotta or, exceptionally, limestone, but sandstone stelae are by far the most common. Only quite rarely did such stelae have sculptured decoration. This almost invariably took the form of an *aedicula*, the small-scale representation of an architectural device, which sets the text between two pillars, within a niche crowned by a conch.[22] This type of decoration, with its overt Hellenistic associations, perpetuated late antique models till well into the twelfth century. The *aedicula* is the symbolic rendering of a shrine or a sanctuary and a traditional means of demarcating sacred space. Framing a text in this way automatically turns it into an agent in a ritual configuration.[23]

In the case of "privileged burials", such as those of ecclesiastical dignitaries, the ritual configuration is primarily provided by their situation. In Nubia, as elsewhere in the Christian world, they were singled out by their proximity to either the liturgical space of a church or chapel or the relics of a saint.[24] Thus, the bishops of Faras were buried within the precincts of the cathedral compound. In the eleventh-twelfth centuries, several had a mortuary chapel erected above their tombs with their epitaphs let into its rear wall.[25] In Dongola, stelae of bishops were likewise found immured in visible places within the liturgical space above their tombs.[26] On the citadel of Qasr Ibrim no epitaphs have been found *in situ*, but higher clergy appear to have been buried in both rock-cut burial chambers and above-ground brick tombs conspicuously situated on a terrace immediately adjacent to the cathedral.[27] The tombs of commoners, on the other hand, were situated in extensive cemeteries outside of the city.[28]

In each of these cases, a sacral setting provided the first interpretational framework for the text. The way in which funerary stelae are nowadays exhibited in museums, as freestanding monuments that usually have lost any trace of color and have become isolated from their original setting, is as deceptive as the printed version of their texts. They were originally part of a sepulchral or even liturgical environment that was publicly accessible and the texts were highlighted by vivid colors that made them stand out against a predominantly whitewashed background.

Within this monumental setting, funerary inscriptions were not merely meant to be seen, but also to be admired. Quite apart from their setting, decoration or other material aspects, textual artifacts in an only marginally literate society such as medieval Nubia represented by themselves a form of symbolic capital. As conspicuously visible texts, funerary inscriptions in particular were much less "data carriers" than vehicles of power, displaying the wealth and literacy of their owners. Irrespective of their legibility as texts, their visibility as textual artifacts that indexed literate culture and social status must have impressed even the illiterate.[29] In the words of Armando Petrucci, providing funerary deposits with writing "is a substantially and profoundly 'political' practice aimed at celebrating and recording the power and social presence of the group, corporate or familial, to which the deceased belonged and is directed at consolidating its wealth, prestige, endurance over time, vitality, and capacity for reproduction and expansion".[30] In the particular case of medieval northeastern Africa, the use of Greek script, common also to Coptic and Old Nubian, is in itself a way of referring to the shared values of Mediterranean Christian culture, which implied cultural Hellenism and religious orthodoxy, however defined.[31]

In spite of these symbolic connotations, however, there can be no doubt that the full meaning of the texts was only accessible to the literate. Paradoxically, precisely the decorated tombstones of the rich – with their sculptured frames that enclose and highlight the text as a *sacrum* – show most clearly that it is the text, not any other formal feature, which is of focal importance. And, just like any other genre of text, funerary inscriptions are expressions of literary culture, composed with a particular aim in mind, for a particular occasion and for a particular audience. Simply defining these parameters brings us closer to the texts themselves.

The occasion seems obvious at first sight: someone's death. Yet tombstones are produced for the living by the living. Even though it is undoubtedly the death event that occasions the manufacture of a funerary stela, the stone itself serves the commemoration of this event by posterity. This commemoration is, moreover, not primarily a private act of remembrance, but a social process. Elena Velkovska rightly emphasizes the social character of post-mortem commemorations in the Byzantine world, in particular when the deceased had been an important person.[32]

In the small-scale, "face-to-face" society of late antique and medieval Nubia their social and public character can indeed be taken for granted. Inscriptions are public text and commemoration is a liturgical and thereby communal rite. The actual agents in this rite are not the deceased, although he or she is central to it, but a varied group of people that will consist typically of the deceased's kin, his or her class group, and – in a Christian society – the local clergy. Within this social setting, the epitaph is bound to become something totally different than the functional equivalent of a label that identifies the owner of a tomb or a death certificate in stone.

Texts in search of a body

Martin Krause, in his contribution to the international conference of Nubian studies held in Warsaw in 1972, was the first to base the study of the Coptic and Greek epitaphs from Nubia in a systematic way on their "formularies" ("Formulare").[33] It is true that Hermann Junker in his still fundamental 1925 study of Nubian funerary inscriptions had already loosely used the term "formulae" ("Formel"), in addition to "prayers and acclamations", in order to refer to the various parts that make up the texts,[34] but it is Krause's merit to have recognized that his "Formulare" are the real essence of any Nubian epitaph. These formularies are best described as a set textual format, framing and presenting the core of each epitaph, the name of the deceased.[35] They permit the distinction of various textual types and the study of their development over time, the main source of variation being the absence/presence of their constitutive elements, the actual formulae (which, according to their form and contents, may be classed in turn as prayers, acclamations, etc.).

The importance of Krause's approach is that he chose the very feature that makes funerary inscriptions so unbearably dull to most scholars, their profoundly formulaic character, as his point of departure. For, paradoxically, it is precisely their predictability and repetitiveness that provide the key to their proper understanding. Formulae are still too often considered as meaningless, stereotypical elements that betray an ultimate lack of originality and creativity. From the point of view of western modernist literary criticism such a judgment may seem correct, but it is hopelessly inadequate as an approach to the literature that was and still is produced by the vast majority of pre-modern and non-western societies. In fact, ever since the "oral-formulaic theory" was developed by Milman Parry and Albert Lord, initially as a means for understanding epic poetry, a different view of formulae prevails. Instead of "empty" elements, formulae are the basic material for the oral production or reproduction of texts in all societies – including those of classical antiquity – where reciting and listening were the normal ways of processing a text, even a written text.[36]

It is by now a well-known fact that inscriptions were also meant to be read aloud. As has been said about the earliest Greek inscriptions, at the moment of reading, the reader lends his voice to the absent writer and becomes the vocal

"WHAT IS MAN?"

instrument that embodies the writing.[37] The writing remains, so to say, incomplete, if a voice is not attached to it.[38] This holds true for Byzantine inscriptions as well, which Amy Papalexandrou describes as "voiced texts" in the sense of John Miles Foley: texts that although written down were intended for oral delivery and, indeed, realize their full communicative function only when being performed.[39]

Also Coptic funerary inscriptions are written texts focused on oral performance and aural reception. This is most immediately apparent in the variety of interlocutory strategies deployed by the various classes of funerary inscriptions. The best known of these is undoubtedly the so-called "appel aux vivants", where the voice of the deceased addresses the reader, usually asking for a prayer or another ritual act, such as sacrifice.[40] The Coptic stela of a boy or young man Gregory, from Middle Egypt, clearly shows that epitaphs of this kind are typically engaging in a dialogue: "be so kind, everyone who will read (ⲱϣ) this epitaph (ἐπιτάφιον), pray (ϣⲗⲏⲗ) for me . . ." (ll. 9–11).[41] The reader, who "reads aloud" the text (Coptic ⲱϣ has as its first meaning "to cry, call"),[42] brings the voice of the deceased (the "me" who, in this epitaph, relates his early death) to life and is asked to answer with a prayer (ϣⲗⲏⲗ). The inscription, moreover, refers to itself ("this epitaph: ἐπιτάφιον") as the mediator of this dialogic event, indicating that the ancient authors of such texts had been consciously reflecting on their function.

The originally pre-Christian funerary acclamation "do not grieve", known in Greek and Coptic from both Egypt and Nubia, could at first sight also be taken to address the reader. In fact, it typically engages the reader to address the deceased: "do not grieve, N.N., that *you* have died".[43] Yet, whoever may be addressed formally, when read out aloud the text becomes an exhortation to reflection and acceptation, aimed at the mourning family. Once even brief and so-called "stereotypical" funerary inscriptions are studied as texts that were meant to be performed orally at a certain place and in certain circumstances, the attention automatically shifts from names and dates to form and meaning.

Reading as a communal rite

But there is more about reading than just sound. In "voiced texts", oral and written modes of communication merge – not, however, in the sense that they are oral sayings committed to writing afterwards.[44] Inscriptions are written texts, firmly rooted in scribal traditions, but at the same time embedded in oral practices. Funerary inscriptions, in particular, offer written scenarios for oral forms of commemoration through the intermediary of the formulae that determine their textual form. These are not merely rhetorical or mnemo-technical devices, but communicative tags that serve a variety of purposes. In the words of Viv Edwards and Thomas J. Sienkewicz, a formula is "a symbol of shared knowledge and experience",[45] and it therefore serves as a basis for understanding and interaction. Inscriptions in particular were – according to Amy Papalexandrou – "active . . . as a site of opportunity, one which fostered oral utterances and interactions demanded by all manner of social and ritual acts".[46]

395

In the case of funerary inscriptions these "social and ritual acts" are conditioned by the multiple event of the deceased's commemoration in which the various groups of social agents outlined above (family, class, church) participate. As is clearly shown by the few examples given above, the deceased is represented in this configuration by the funerary monument. The text of the epitaph may address the deceased (as in the "do not grieve" texts) or the deceased may speak through the epitaph (as in the "appel aux vivants"). Even very simple epitaphs contain the name of the deceased, and sometimes no more than this name. In fact, the tombstone can be said to replace the deceased materially.[47] In the ritual enactment of the deceased's commemoration, the tombstone and the tomb therefore occupy a central position, circumscribed topographically by their situation in a sacral space (cemetery, church or funerary chapel). And even when the epitaph is not actually recited anymore, it remains a written speech act that virtually perpetuates the ritual commemoration of the deceased on the spot and ensures his or her continuing presence within the community.[48]

Reciting texts, however, was but part of the complex of social practices that made up the commemoration of the dead. Even if not all of these practices can be reconstructed from the mere artifacts that we have at our disposal (tombs, lamps, inscribed tombstones), we can be quite sure that they must have been inspired primarily by Christian or Christianized ritual. This is duly reflected in the textual formulae themselves, which tend to be informed on a large scale by familiar phrases from the liturgy. A particularly obvious and well-known example is the prayer with the opening words "God of the spirits", which provided a favorite text for the longer Greek epitaphs of medieval Nubia and can still today be heard at Greek burials.[49] Reciting such texts is a way of sharing the creed, not in the narrow acceptation of a doctrinal statement, but in the broader sense of a set of commonly accepted views of death and afterlife. Since reading the texts was embedded in a wider ritual and monumental setting, which involved doing and saying things together, including praying and singing, the texts reproduce and reinforce social cohesion. Reciting the texts, listening to them, and joining in with prayers and hymns conveys and strengthens a sense of belonging to the same social and religious group.

It is important to emphasize that, although this ritual setting may be best known through the transmitted texts of the Christian funeral liturgy, it comprised much more than the sanctioned liturgical books that modern scholars are wont to see as its most authoritative expression. Also funerary inscriptions are never merely copies from liturgical handbooks, even if a part or the whole of their formulary may derive in some way from the official funeral liturgy. Liturgical books and epitaphs represent two distinct genres, each obeying its own rules and traditions. Variations and developments within the formulary of funerary inscriptions do not need to coincide, therefore, with changes within Christian liturgy in this limited understanding, as a set of prescriptive texts, but may reflect developments in a broader framework of conceptions and usages, or simply variations with a social or even regional background.

"WHAT IS MAN?"

As a manageable corpus of texts, yet one covering nearly the entire time span of Coptic funerary epigraphy in Nubia, from about the eighth till the very end of the eleventh century, the epitaphs from Qasr Ibrim are an ideal illustration of the preceding remarks.[50] As most of them bear no absolute dates, however, the development sketched below is likely to lack a firm chronological basis. Epitaphs of various types have clearly been used side by side, and the use of different texts may have been conditioned by differences in, for example, wealth, gender or social status. In other words, variation is not necessarily a correlate of chronology. Yet Nubian funerary texts do show a number of significant shifts and developments over time and these can be usefully studied with the help of the various formularies that are identified as the main distinctive element of the inscriptions. Martin Krause,[51] following Henri Munier,[52] used these formularies to postulate a chronological development from simple to more complicated, hoping thereby to obtain absolute dates for the usually undated Nubian tombstones. Even if it is doubtful whether this principle, which claims that formularies became longer and more composite over time, can be applied as an iron rule for dating purposes, there is sufficient reason to start our discussion with the simplest type of text that can be found on Coptic epitaphs from Qasr Ibrim. Apart from their plausible antiquity, these simple texts have the advantage of showing in a very explicit way what epitaphs are about, namely commemoration.

The "commemoration" type

The tombstones from Qasr Ibrim that exhibit the "commemoration" formulary are all rather unpretentious stone monuments, usually of modest size, typically about 30 cm high by 20–25 cm wide, but sometimes considerably smaller. Somewhat less than twenty examples are now known from Qasr Ibrim.[53] Curiously, but for two exceptions, they all commemorate women. Outside of Qasr Ibrim and its immediate surroundings, what might be called "Central Nobadia" (Ibrim, Toshka, Arminna), this type is rare and it has not been found in Nubia south of Faras.[54] By contrast, it was popular in Southern Egypt, in particular in the First Cataract area, where a majority of the stelae from Dayr Anba Hadra, the so-called Monastery of St. Simeon, near Aswan, shows a closely related formulary. The interrelations between the First Cataract area and the Qasr Ibrim region have been intensive ever since the days of the early Nobadian monarchy, as is shown for example by the Tantani letters, and the adoption of the "commemoration" type in Northern Nubia may have been a typical contact phenomenon.

From a textual point of view the Qasr Ibrim epitaphs of this type share a basically tripartite form, with an opening formula characterized by the phrase ⲡⲣⲡⲙⲉⲉⲩⲉ (alternatively, ⲅⲁ ⲡⲣⲡⲙⲉⲉⲩⲉ) ⲛ-/ⲙ-, "the memorial (or, for the commemoration) of", which introduces the name of the deceased, preceded by the epithet ⲧⲙⲁⲕⲁⲣⲓⲁ/ⲡⲙⲁⲕⲁⲣⲓⲟⲥ, "the blessed". Then, a second formula, characteristically beginning with ⲡⲉⲅⲟⲟⲩ, "the day (on which she/he died)", introduces the date of death. To describe the event of death itself, the verb ⲙⲧⲟⲛ ⲙⲙⲟⲝ, "to go to

397

rest, to find rest", corresponding to Greek ἀναπαύομαι, is used, which is standard in Nubian epitaphs.[55] A third element consists of a usually brief prayer, which may be reduced to a mere *pax* formula ("in the peace of God") and a concluding Amen.[56] Other than the first two constituents of this formulary, the prayer section is subject to great variation both in length and in substance. It usually consists of one or more formulae that ask for heavenly "rest" for the deceased (see below). The stela of a man Petros from Qasr Ibrim (*I. Qasr Ibrim* 31) may be given here as a representative example of this type of epitaph:

The memorial (ⲡⲣⲡⲙⲉⲉ̄ⲩⲉ) of the blessed Petros.
The day on which he went to rest is the thirteenth of Pakhon.
O Lord, Jesus Christ, may You grant rest to his soul in Your kingdom.
Amen, so be it.

Since Nubian epitaphs of the "commemoration" type usually do not bear an absolute date, for which there was no functional need (see below), dating them is not unproblematic. Actually, none of the examples from Qasr Ibrim can be dated precisely. From their relative frequency and the considerable variation shown in the actual wording of the texts, it may be concluded that this formulary was in use over a longer period. Even if one hesitates to follow a linear evolutionary model, simple epitaphs of this type could well represent a typological early stage of development.[57] In that case, the emergence of the type might date back to the seventh century when, as the epitaph of Bishop Joseph from Old Dongola would suggest, Coptic came into use in addition to Greek as a language for funerary inscriptions. On the other hand, epitaphs like that of Marianta (*I. Qasr Ibrim* 37, quoted *in extenso* below), with its opening *bismillah* that clearly reflects Islamic influence, cannot be earlier than the eighth century and are likely to be even younger. In fact, an identical opening *bismillah* is found in the Coptic stela of a woman Semne, presumably from Middle Egypt, which is dated to A.D. 863.[58] The hypothesis of a long time range, covering the entire period from the supposed incipience of Coptic funerary epigraphy in Nubia in the seventh century till somewhere well in the ninth or perhaps even the tenth century, appears to be confirmed by the material from Aswan. There, in the large series of tombstones from Dayr Anba Hadra, a monastery that maintained strong contacts with Nubia, absolute dates are far more frequent. These show that the "commemoration" formulary was in use over the full period of the eighth and ninth centuries, with dates ranging from 716 to 892.[59]

The characteristic opening formula with the noun ⲡⲣⲡⲙⲉⲉ̄ⲩⲉ is not always the actual beginning of the text, as it may be preceded by other elements. Thus, in the stela of a woman Eiktoume, it is preceded by the acclamation ⲟⲩⲁ ⲡⲉ ⲡⲛⲟⲩⲧⲉ ⲡⲃⲟⲏⲑⲟⲥ, "one is God, the helper", the Coptic rendering of εἷς θεὸς ὁ βοηθῶν, extremely popular in epitaphs from southernmost Egypt.[60] The stela of Marianta opens with a Coptic rendering of the Arabic *bismillah*, "in the name of God, the merciful and compassionate", already mentioned above.[61] These additional

"WHAT IS MAN?"

formulaic elements in no way affect the character of the following formulary, in which the main formative constituent remains the "commemoration" phrase.[62]

The distinctive noun πρπμεεγε, "the remembrance, commemoration", in the opening formula must correspond to μνήμη, μνημήσις, or even μνημεῖον in similar Greek formulae. In all likelihood, indeed, the Coptic formula translates Greek models. Greek formulae of this type occur frequently in Egypt and occasionally also in Nubia, for example in large numbers in Sakinya (Toshka West),[63] but have not (or, not yet) been found in Qasr Ibrim. Both Coptic forms of the opening formula, with and without the preposition ϩⲁ, are used without apparent distinction. Nevertheless, as with the various usages of Greek μνήμη, two different interpretations must be envisaged.[64] The nominal group πρπμεεγε ⲛ- N.N. (equivalent to Greek μνήμη τοῦ N.N.), without the preposition, appears to refer to the monument itself, and should accordingly be rendered as "the memorial of N.N.". This is confirmed by the variant ⲡⲁⲓ ⲡⲉ πρπμεγε ⲛⲧⲁ (*l.* πρπμεεγε ⲛⲧⲉ) ⲡⲙⲁⲕⲁⲣⲓⲟⲥ ⲁⲛⲁⲥⲧⲁⲥ[ⲓⲟⲥ], "*this is* the memorial of the blessed Anastasios".[65] Here, the "this" can only be the epitaph itself or, more likely, the entire funerary monument of which it was part. The phrase ϩⲁ πρπμεεγε ⲛ-/ⲙ- N.N. (equivalent to Greek ὑπὲρ μνήμης τοῦ N.N.), on the other hand, refers more specifically to the function of the monument: "(a monument erected) *for* the commemoration of N.N.".[66]

This distinction, imposed by the modern practice of translating a written text, is only apparent, however. Not only is – in both forms of the formula – the same word used, πρπμεεγε, which – as a noun derived from the light-verb compound ⲣ̄ πμεεγε, "to remember" – denotes an act of remembrance, but in the act itself both interpretations merge. The reader lends his voice to the monument that explains its *raison d'être* as being either erected for or visually representing the same ritual act of remembrance. In either case, moreover, the texts are focused on the name-date sequence. This is even more clearly visible in variants like (ϩⲁ) ⲡⲉϩⲟⲟⲩ ⲙⲡⲣⲡⲙⲉⲉⲅⲉ ⲛ-/ⲙ- N.N., "(for) *the day of* the commemoration of N.N.", that are frequently found in Aswan, Sakinya and elsewhere.[67] Also in each of the Qasr Ibrim representatives of this group, this date, introduced by the noun ⲡⲉϩⲟⲟⲩ, "the day", follows immediately. The act of remembrance consists therefore of remembering a precisely named individual, the deceased, on a precisely specified date, the day of his or her death.

In order to understand the nature of the act of remembrance implied in the "commemoration" formulary, it is necessary to bear in mind that remembering the dead in a Christian sense is not a matter of recalling the cherished image of an individual from the archive of memory, but an active form of intercession. Already in the *Euchologion* attributed to Serapion of Thmuis, from fourth-century Egypt,[68] intercessory prayer for the deceased is an integral part of the eucharistic anaphora (chap. 1, ed. Johnson, pp. 48–50). It comprises two elements: the prayer itself and the commemoration, which in turn consists of the recitation of the names of the dead: "We intercede (παρακαλοῦμεν) also on behalf of all who have fallen asleep and whose commemoration (ἀνάμνησις) we are making . . . (After the recitation [ὑποβολή] of the names): Sanctify these souls (etc.)". Commemorating the dead is essentially the liturgical act of naming the dead and praying for them.

In the very extensive *memento* of the later Egyptian *Anaphora of Saint Basil*, it is God himself who is asked, or rather ordered, to "remember" both the living Church and its deceased members. For example thus: "remember (μνήσθητι, ⲁⲣⲓ ⲡⲙⲉⲉⲩⲉ) likewise all those in priesthood who went to rest and all those from the ranks of laity, and grant them rest in the bosom of Abraham, Isaac, and Jacob (etc.)" (ed. Budde, pp. 190–1, here translated after the Sahidic). The priest asks God to "remember" the dead and specifies the sense of this "remembrance" in his following prayer. "To remember" here obviously means "to reward" the dead in afterlife. Here also names are read and the deacon asks the community formally to join the prayer of the priest, to which they answer with their *Kyrie eleisons*.[69]

Remembering the dead in a Christian sense is therefore in itself a beneficial act that, as far as the Church on earth is concerned, consists of reciting the names of the dead and collectively praying for them with a view to their eschatological salvation or, as the *Euchologion of Serapion* has it, their "sanctification". In the text of the Nubian funerary stelae, as in the liturgical handbooks, the name of the deceased occupies a central position in the commemoration process. The entire opening formula of the "commemoration" type supposes the recitation of the name of the deceased and would be pointless without it. In fact, the name is the minimal indispensable element in any Christian epitaph.[70] It marks the symbolic presence of the deceased in his funerary monument. Yet it is more than a mere owner's mark. In Nubia the name is as a rule preceded by ⲧⲙⲁⲕⲁⲣⲓⲁ/ⲡⲙⲁⲕⲁⲣⲓⲟⲥ, "the blessed".[71] In a Christian context, this epithet anticipates the eschatological fulfillment of the many *makarismoi* in the New Testament, perhaps in particular – in a funeral context – that of Rev. 14:13: "blessed are the dead who die in the Lord from now on". This may seem a vain pretense on behalf of the deceased's family, but makes excellent sense in the ritual experience of shared Christian hope conveyed by the recitation of the funerary formula.

A similar idea is articulated by reciting the name itself, which is the briefest possible form of commemorative prayer and the very core of the Christian act of remembrance, for the name of the Christian is fraught with eschatological significance.[72] Reciting it anticipates the acquittal of the dead in the individual judgment, when Christ will "confess" (ὁμολογέω) his or her name "before (His) Father and (His) angels" (Rev. 3:5). This "confession", significantly an originally juridical term, is in reality an act of reading aloud a written text, analogous to the reading of names from a tombstone or from the diptychs during the Eucharist. As indeed judgment was dispensed on the basis of books, the name of the deceased should be found written in the "Book of Life".[73] This explains the popularity of the Old Testament patriarch Enoch, who was particularly venerated in the Nile Valley as "the scribe of righteousness", the one who wields his "spiritual pen case" to register the sins and merits of mankind in book volumes (τομάριον).[74] Reading aloud the name of the "blessed" deceased from his tombstone in the

"WHAT IS MAN?"

course of his or her communal commemoration held a promise of future salvation for all present.

The fixed order "name – date – prayer" that underlies the "commemoration" formulary reflects a clear hierarchy. The name of the deceased, in Nubia normally preceded by the epithet "blessed", is the most crucial formulaic element in any epitaph, which the text immediately links to the date of his or her demise. This date, the hierarchically second element in the formulary, consists in all of the Qasr Ibrim stelae of a month plus day, while only very few add an indiction year, which is, strictly speaking, superfluous.[75] This practice has a precise liturgical background. The day of the month is the minimal information indispensable for the periodic commemoration of the dead, which already in late antiquity took place typically on the third, ninth and fortieth days after death.[76] This forty-day period is basically an extended marginal period, following the stage of *séparation* (separation of the soul from the body, of the dead from the society of the living) in A. van Gennep's famous triplet, during which the dead, newly dressed, assumes the journey towards his or her *agrégation* in the netherworld.[77] Following the burial rites proper, this period is marked by periodic commemorations that accompany the stepwise ascension of the soul, which leaves its mortal dwelling behind in order to meet its heavenly judge on the fortieth day. As a whole it mirrors the marginal period at the beginning of each human life, which spans the time from conception to baptism.[78] Just like every marginal period, it is a period of insecurity that demands prayers to be said for the soul in order to protect his frightening passage from this world to another. A considerable part of Christian funeral rites are geared precisely at affording this protection.

The Christian East still largely observes this periodical pattern of commemoration, although its actual rhythm appears to be subject to some variation over time and place. Thus the present-day Coptic (Bohairic) liturgy knows a "service of the lifting of the mat", performed in the house of the deceased on the third day and in the church on the fortieth day, a memorial service in the church that may be celebrated one month, forty days, six months or a year after the date of the death, and requiem services for the third day, the fortieth day and the anniversary of the death.[79] It can be safely assumed that the Nubian Church observed a more or less similar commemorative cycle. It would normally come to an end after forty days, with the final *agrégation* of the deceased's soul in the netherworld, which explains the nearly total absence of year dates in epitaphs of the present type. Adding a year date (either a year of the indiction or an era date) appears to become more common towards the end of the first millennium and, generally, may have served to enhance the status of wealthy families who could afford a prolonged series of anniversary celebrations.

The typical formulary of the "commemoration" type explicitly aims at making the sepulchral monument of the deceased a participant in this public cycle of commemorations. Whatever more strictly liturgical commemoration took place elsewhere, for example in a church or chapel, we can be sure that a visit to the tomb was part of the ritual. As was argued above, the tomb and its inscription take

401

the place of the deceased herself or himself. In as far as the ritual bore a social and public character, it must have also included the recitation of the epitaph, entirely or partly.

The prayers

The part of the epitaph that was most clearly meant to be recited aloud was its prayer section. Not only the "commemoration" type of epitaph, but most Nubian funerary inscriptions in Coptic contain such a prayer section as their third main constituent. Its invariable position, following the name of the deceased, echoes the order in the *Euchologion of Serapion* (quoted above), where the actual prayer also follows the recitation of the names of the departed. The prayer section is as a rule made up of formulae derived from the intercessory prayers of the eucharistic anaphora or the more specific services celebrated for the deceased. It usually bears an overtly interlocutory character, addressing the Godhead under one of its current designations, for example as "O Lord, Jesus Christ", in the epitaph of Petros, quoted above. Even when a third-person verbal form is occasionally used, as a rule a "directive" Third Future,[80] the divine addressee of the prayers is made explicit in a similar formulaic way. For instance in a bishop's epitaph from Faras: "May God, lover of mankind (ⲡⲙⲁⲓⲣⲱⲙⲉ, Gr. ὁ φιλάνθρωπος), grant him rest in the bosom of Abraham and Isaac and Jacob" (*I. Khartoum Copt* 4, ll. 10–13). The formulaic invocation of God signals the transition from the commemorative part of the epitaph, focused on the deceased and the death event, to the properly supplicatory part of the text.

All prayer formulae seek to obtain more or less specific benefits for the soul of the deceased in the netherworld.[81] Some of these formulae, such as the famous prayer for "rest in the bosom of Abraham, Isaac and Jacob", belong to the earliest Christian prayers for the dead, and can be found in respectable and ancient sources, such as the *Apostolic constitutions* (VIII, 41, 2) and the *Euchologion of Serapion of Thmuis* (chap. 18, ed. Johnson, p. 68), both from the fourth century, but also in the *memento* of the Bohairic *Anaphora of Saint Basil*, which is still currently used in the Coptic Church (ed. Budde, pp. 190–91, quoted above), or in the prayers of the Bohairic funerary ritual.[82]

Only rarely local particularities or, perhaps, formulae that have not survived in other sources come to light. Among these, a prayer that "the Archangel Michael may watch over the bones" of the deceased deserves to be mentioned. It occurs in tombstones from the Faras-Meinarti region, just north of the Second Cataract,[83] but not in those from Qasr Ibrim or elsewhere in Nubia. In time it is restricted, moreover, to the first half of the eleventh century and it has been found only in Coptic. The present-day Ethiopic funerary ritual includes a prayer for the protection of the body of the deceased, but this does not invoke Michael.[84] The Nubian evidence suggests that a similar prayer invoking St. Michael may have circulated locally in Nobadia, where the Archangel was an extremely popular saint.

"WHAT IS MAN?"

The familiarity and predictability of the prayer formulae are part of their function, for they are primarily a way of perpetuating the commemorative liturgy of the community on the burial spot, either virtually on the stone or by means of oral delivery, on the occasion of periodical commemorations. In a condensed and familiar form, they epitomize Christian expectations about afterlife. In spite of the formulaic character of its constituents, the composition of the prayer section as a whole does not obey iron rules. The prayers may vary greatly in length, organization and substance. In some of the early epitaphs of the "commemoration" type they are lacking altogether, as is shown by the material from Aswan and also a single instance from Qasr Ibrim (*I. Qasr Ibrim* 29). Instead, they may have been supplied from memory. Also very brief prayers were possibly supplemented with similar phrases in the course of their oral delivery, the text on the stone serving primarily as an *aide-mémoire*. Longer epitaphs tend to devote more space to the prayer section, which may grow into a fairly long and convoluted composition.[85]

In all these cases, the principle underlying the organization of the prayers is clearly not that of literally copying from liturgical books, but rather that of centonization: formulae are selected and combined in an order and form that are subject to a certain degree of variation. Since centonization underlies much literature that is orally produced or reproduced, in particular poetry, the procedure again shows these epitaphs to be "voiced texts". As a result, even otherwise very similar epitaphs may exhibit more or less subtle differences in the number, nature and form of the prayer formulae and the acclamations. In the Qasr Ibrim material, this is neatly illustrated by two probably ninth-century epitaphs for women that, judging from the strong similarities in text, script and execution, must not only be contemporaneous but even derive from the same workshop.[86] The following chart shows how in these two epitaphs, that of Marianta and Theognosta respectively, textual formulae are shifted, expanded and modified to obtain subtle variations on the pattern of a single formulary.[87]

stela of Marianta	*stela of Theognosta*
[opening acclamation]	[opening acclamation]
In the name of God,	–
the merciful and compassionate.	–
[naming lemma]	[naming lemma]
For the commemoration (ⲅⲁ ⲡⲣⲡⲙⲉⲉⲩⲉ)	The memorial (ⲡⲣⲡⲙⲉⲉⲩⲉ)
of the blessed Marianta.	of the blessed Theognosta.
[dating lemma]	[dating lemma]
The day on which she went to rest	The day on which she went to rest
is the 6th (?) of the month Toobe.	is Choiak 17.
[prayer: invocation]	[prayer: invocation]
O God	O God,
of Abraham and Isaac and Jacob,	the merciful and compassionate,
[prayer: request]	[prayer: request]

403

	may You grant rest to her soul
grant rest to her soul	in Your kingdom,
–	and feed her from the Tree of Life
–	and make her recline in the bosom of
in the bosom of	–
our Holy Fathers, whose names we gave,	
Abraham and Isaac and Jacob.	Abraham and Isaac and Jacob.
[closing formula]	[closing formula]
So be it, so be it. Amen.	Amen. So be it.

Several interesting observations can be made. First of all, the comparison of the two stelae confirms that both forms of the formulary based on the verbal noun ⲡⲣⲡⲙⲉⲉⲩⲉ, with and without the preposition ⲅⲁ, were used without apparent distinction by the redactors of the texts in spite of our different translations. Then, the stela of Marianta opens with the Coptic rendering of the Arabic *bismillah*, the main elements of which (the epithets "merciful and compassionate") recur in the invocation opening the prayer in the stela of Theognosta. In both cases, the phrasing clearly betrays the influence of Muslim scribal habits that is similarly found in contemporaneous Egyptian sources.[88] The *bismillah* had become acceptable as an opening formula also for Christian texts. Lastly, and perhaps most interestingly, the prayer for Marianta begins with the age-old invocation of "the God of Abraham, Isaac and Jacob".[89] In this position, it usually appears to oust the prayer for rest "in the bosom of Abraham, Isaac and Jacob" from the ensuing prayer.[90] Here both are found, perhaps by inadvertence, so that the prayer immediately follows the invocation. This obliged the redactor to the rather awkward repetition of the series of the names of the patriarchs in the prayer itself. In order to make the prayer sound somewhat less inelegant, he then made the *ad hoc* insertion of the unhabitual phrase "of our Holy Fathers, whose names we gave", before repeating the three names of the patriarchs. The prayer for Theognosta, on the other hand, begins with an invocation modelled after the *bismillah* that in Marianta's stela opens the entire text ("O God, the merciful and compassionate"). Thus the redactor saved space for inserting two other, equally formulaic prayers,[91] before introducing the prayer for rest in the bosom of the patriarchs, with which both epitaphs conclude. If more examples were available, further analysis could reveal significant patterns in the selection and seriation of the formulae that make up the prayer sections of Nubian epitaphs.

The "verdict" type

Much different in tone and substance and much less exclusively focused on the act of commemoration itself are the epitaphs characterized by the opening formula ⲕⲁⲧⲁ ⲧⲁⲡⲟⲫⲁⲥⲓⲥ, "according to the verdict (ἀπόφασις)", usually expanded by a relative clause explaining: "which God pronounced over Adam, saying . . .". Only three of them are known from Qasr Ibrim,[92] and also elsewhere in Nubia they are relatively rare.[93] These epitaphs, which generally tend to be somewhat longer than

"WHAT IS MAN?"

those of the commemoration type, can likewise be described as tripartite. The opening formula serves as an evocation of the "verdict" par excellence, the one pronounced by God over Adam after his fall, "earth you are, to the earth again you shall return", which is as a rule quoted verbatim after Gen. 3:19 (LXX). Normally, Adam is mentioned explicitly as the addressee of God's verdict and he is frequently styled as "our father" or "our first father". The epithet is significant, for it brings out that God's verdict did not concern just one historical person but his entire offspring. Only following the Genesis quote it is said that, by virtue of this same verdict, the blessed so-and-so, the owner of the stela, "went to rest" on that or that date. This name/date lemma is then followed by a prayer section of varying length. In the exceptionally brief epitaph of Kyriakos from Qasr Ibrim (*I. Qasr Ibrim* 38), this results in the following text:

> In accordance with the verdict (ἀπόφασις) which God gave over Adam, saying: "Adam, earth you are, to the earth again you shall return",
> the blessed Kyriakos went to rest in the month of Phaophi, (day) 16.
> O Lord God, may you grant rest to his soul.

Textually the epitaph of Kyriakos is practically identical to another Qasr Ibrim stela, that of a woman Reumata (?) from the North Cemetery.[94] Other examples show some variation in the introductory formula. Thus, two epitaphs from Sakinya (Toshka West) have ϩⲓⲧⲛ ⲧⲁⲡⲟⲫⲁⲥⲓⲥ, "through the verdict";[95] another one from Arminna opens with ⲕⲁⲧⲁ ⲡϣⲁϫⲉ ⲛⲧ{ⲁ}ⲁⲡⲟⲫⲁⲥⲓⲥ (etc.), "in accordance with the word of the verdict (etc.)".[96] A rare Greek example from Faras opens with a more elaborate phrase: κατὰ τὴν τοῦ παντοκράτορος θεοῦ ἀμετάθετον ἀπόφασιν, "in accordance with the immutable verdict of the almighty God", preceding the quote of Gen. 3:19.[97] For the "immutable" of the Faras epitaph, one may compare an Egyptian *Totenklage* stela, which exclaims: ⲱ ⲁⲡⲟⲫⲁⲥⲓⲥ ⲛⲁⲧⲡⲁⲣⲁⲓⲧⲉⲓ ⲙⲙⲟⲥ ⲁⲩⲱ ⲛⲁⲧⲃⲱⲗ ⲉⲃⲟⲗ, "o inexorable (+ παραιτέω) and unalterable verdict!"[98]

Also some variation can be observed in the verb used in the following relative clause. In the very brief epitaph of Kyriakos, it is a form of ϯ, "to give" (ⲧⲁⲁ⸗), as in the near-identical Qasr Ibrim stela of Reumata (?). The stela from Arminna, on the other hand, has ϫⲱ, "to say", and in Egyptian examples also ⲧⲁⲩⲟ, "to utter, proclaim", is found.[99] In the stela of a monk from Qasr Ibrim (*I. Qasr Ibrim* 39), the Greek loanword ὁρίζω, "to ordain", is used, which is a typical element in this context, well attested both in the Egyptian funerary stelae of the *Totenklage* type (about which see below) and in Coptic legal documents from the Theban area.[100] Thus, a passage from the elaborate testament of a woman Sousanne, from the first half of the eighth century, corresponds almost word for word to the opening lines of the epitaph of the monk from Qasr Ibrim: "when . . . I will leave this sojourn behind and go the way that every human being has to go, *in accordance with the verdict* (ἀποφάσις) *which God ordained* (ὁρίζω) *over Adam, saying: 'Earth you are, to the earth again you shall return'*, I desire and order . . .".[101] Likewise, in an undated stela of the *Totenklage* type from Aswan, it is stated that "all flesh"

405

must die ⲕⲁⲧⲁ ⲧⲁⲡⲟⲫⲁⲥⲓⲥ ⲛⲧⲁ ⲡⲛⲟⲩⲧⲉ ϩⲟⲣⲓⲍⲉ ⲙⲙⲟⲥ ⲉⲭⲉⲙ ⲡⲉⲛⲉⲓⲱⲧ ⲁⲇⲁⲙ ϫⲉ ⲛⲧⲕ <ⲟⲩ>ⲕⲁϩ ⲛⲕⲛⲁⲕⲟⲧⲕ (1. ⲉⲕⲛⲁⲕⲟⲧⲕ) ⲉⲡⲕⲁϩ, "in accordance with the verdict which God had ordained (ὁρίζω) over our father Adam, saying: 'Earth you are, to the earth you shall return'".[102] In Qasr Ibrim itself, the verb occurs again in the epitaph of a woman Eisousiko, dated to A.D. 1035: ⲕⲁⲧⲁ ⲑⲉ ⲛⲧⲁ ⲡⲉⲛϫⲟⲉⲓⲥ ⲛⲁⲅⲁⲑⲟⲥ ϩⲟⲣⲓⲍⲉ ⲙⲙⲟⲥ ⲙ̄ⲡⲉⲛⲉⲓⲱⲧ ⲛϣⲟⲣⲡ ⲁⲇⲁⲙ, "in the same way as our good Lord had ordained (ὁρίζω) to our first father Adam . . .".[103]

As is immediately apparent, these opening texts are not about commemoration, but about death and mortality. Two important aspects can be distinguished: first, an etiology of death is presented, of which the universal validity and ineluctability are underscored by referring to the fate of the first man, Adam, and, secondly, the death of not only Adam but every individual is said to depend on a divine verdict. In most cases, God's verdict (Gen. 3:19) is quoted literally. In the context of a public recitation, the full focus would be on this quote, God's own words, and reading the formula would become no less than a forceful *memento mori*. Yet the central term that characterizes the text is not a word denoting death or dying, but a Greek word for "decree, verdict".

In addition to the noun ἀπόφασις, also the related verb ἀποφαίνω, "to declare, proclaim; to render a verdict", may be used. The occurrence of both words is, furthermore, not strictly limited to epitaphs with the opening formula ⲕⲁⲧⲁ ⲧⲁⲡⲟⲫⲁⲥⲓⲥ, since they appear as well in texts of the *pronoia* type, discussed below.[104] In each case, the verdict or sentence to which they refer is the divine decree of Gen. 3:19, a full quote of which follows in most instances. Beyond its occurrence in epitaphs, the same terminological configuration is found in Coptic documentary texts, in particular testaments from the Theban area, an example of which was already quoted above.[105] Both ἀπόφασις and ἀποφαίνω are technical terms with clear juridical overtones.[106] Their use with reference to Gen. 3:19 appears to be, as far as epigraphy is concerned, limited to Egypt and Nubia. It is tempting to trace it to the work of Saint Athanasius, where this terminology occurs in various polemical contexts.

In a famous passage from his second *Discourse against the Arians*, the context is Christological. Athanasius seeks to refute the claim, imputed to the Arians, that the Son, Christ, did not share the full divinity of the Father. If the Logos, Athanasius argues, annulled the "verdict" pronounced by the Father in the beginning over all human beings, that is Gen. 3:19, it shows that He was God's own Word, as this could never have been achieved by someone who was a mere creature himself (cap. 67, 4–5; ed. Metzler and Savvidis, 244–45). More properly funeral is the context in *Festal letter* no. 41, of A.D. 369, where it occurs in a polemic against the pretended Meletian habit of exhibiting the bodies of the martyrs publicly instead of burying them. The quote of the "verdict" from Gen. 3:19 opens a passage that demonstrates the universal character and the biblical basis of the practice of interment.[107] Both passages underline the universal and ineluctable character of human death, which is operative by virtue of a law proclaimed by God the Father himself. They are equally "legalistic", but they differ in perspective. In the first case, a

"WHAT IS MAN?"

verdict passed by one divine lawgiver is annulled by another divine person. In the second case, the verdict declares interment (as opposed to other ways of disposing of the dead) a law.

How the ἀπόφασις terminology, as applied to Gen. 3:19, passed from the writings of Athanasius, if these really were its source, into the language of testaments and epitaphs, is not clear as yet. In any case, this passage is most likely to have occurred on Egyptian soil. Its use in Nubian epitaphs reflects the same juridical concept of death as found in Athanasius. It links the event of the individual's death directly to the prototypical events related in the early chapters of the biblical book of Genesis. Death is not an accident, but a punishment for human disobedience and the actualization of God's original verdict over Adam. It is also, by the same token, inevitable (see the "immutable" and related words used in the texts quoted above), universal (Adam being "our first father"), and a return to man's first lifeless element, earth, which in a sepulchral context can be taken to refer in a very precise way to the earth in which the deceased is interred (as in Athanasius' *Festal letter* no. 41).

Apart from having a strong juridical flavor, this concept of death may also strike us as fairly negative: every human individual falls victim to the same divine law, however hard he would like to avoid it. And if Christian logic demands that the original verdict over Adam is annulled by the obedience of the second Adam, Christ (as in Athanasius' second *Discourse against the Arians*), this is only rarely expressed in the epitaphs. It is true that the formulaic reference to the transgression and punishment of Adam may have automatically evoked its historical counterpart, human redemption in Christ. In fact, this happens in every celebration of the *Liturgy of Saint Basil*, where immediately after the Sanctus the eucharistic prayer begins by evoking the fateful events from Genesis and then proceeds to relating the advent of Christ (ed. Budde, 148–49, cf. 281–83). Yet, comparison with other funerary texts from Egypt and Nubia confirms that the verdict formulary is primarily concerned with death.

There can be little doubt that the particular vision of death, subsumed under the use of the legal terms ἀπόφασις/ἀποφαίνω and the reference to the etiological story from Genesis is, as far as funerary epigraphy is concerned, an innovation of late seventh-century Egypt. It can even be tentatively connected with the contemporaneous emergence of the *Totenklage* type of epitaphs in Upper Egypt, discussed below. Its etiological bias is articulated very explicitly in several lengthier epitaphs of the eighth-ninth centuries. Striking examples are the stelae of Kosmas, from Qau near Asyut in Egypt (A.D. 799 or 805),[108] and Mariankouda, tetrarch of Makuria, from Hambukol (north of Old Dongola, in Makuria; A.D. 887).[109] Both stelae recount the Genesis story *in extenso*, but only the Mariankouda stela refers, in the spirit of 1 Cor. 15:20–2 and clearly in imitation of the *Liturgy of Saint Basil*, to Christ's redemptory death as the positive counterpart of Adam's fall: "He came straightforwardly (and), having assumed the flesh and become man, by ascending onto the wood saved all those who had been enslaved by death" (ll. 10–12).[110] And if Gen. 3:19 is quoted verbatim to announce the death of Mariankouda (ll. 12–13), the word ἀπόφασις is not used here. The earlier stela of Kosmas, by contrast, omits any reference to redemption through Christ and is far more pessimistic in tone: "when God saw that man had become disobedient, he proclaimed (ἀποφαίνω) over him a bitter punishment, that is death, which He made lord over him and his offspring forever, saying: 'Adam, earth you are, to the earth you shall return'" (ll. 12–16). And again: Kosmas enjoyed life "in imitation of my first father Adam when he lived in Paradise, before the verdict (ἀπόφασις) of God came over him" (ll. 23–5). In both epitaphs, death is emphatically presented as inherent in the human condition ever since Adam's fall. In the Egyptian stela, where the redemptory moment embodied in Christ gives way to a complaint about death, also the ἀπόφασις terminology is prominently present.

The epitaph of Kosma, but not that of Mariankouda, belongs to the typical genre of the *Totenklage*, and in fact the same terminology can be found in various other representatives of that genre, which bear dates ranging from the very end of the seventh to the middle of the ninth century (see below). The opening phrases of the Leiden *Totenklage* stela of Eulogia, dated to A.D. 759 and probably from Antinoopolis, are unmistakably akin to the Nubian formula:

O fearful and terrifying verdict that God pronounced over Adam:
"Earth you are, to the earth again you shall return"![111]

Also the Coptic marble stela of a woman Trosis (Drosis), of unknown provenance and dated to A.D. 741, resembles the Nubian epitaphs of the "verdict" type very closely:

This is the first verdict that God pronounced over Adam, the first man, saying: "Adam, earth you are, to the earth you shall return".[112]

These dated Egyptian analogies suggest that probably from the first half of the eighth century onwards a new conception of the tombstone spread southward into Nubia. From a simple memorial, preserving a name and a date, the Nubian epitaph developed into a theological statement that reflects not merely on the fate of the deceased but also contains a forceful warning to the living who read the text, inviting them to consider the deeper meaning of their own mortality. Whereas the ἀπόφασις terminology seems to pass out of use after the middle of the tenth century,[113] the evocation of the Genesis story remained popular until the very end of the Coptic epigraphic tradition in Nubia (see below). In fact, the etiology of death which it proposes strongly shaped the entire Coptic funerary epigraphy of Nubia in the period from the eighth to the eleventh century. In Nubian Greek epitaphs, by contrast, it remains only marginally attested.[114]

The *Totenklage* type

The "verdict" type of funerary formulary and the radical change in outlook that it implies can – as was demonstrated above – be traced in all likelihood to the

"WHAT IS MAN?"

typical genre of the *Totenklage* stelae.[115] As far as we can judge now, the latter rose to prominence in Middle Egypt towards the end of the seventh century, and there had been little, so it seems, in the earlier Christian epigraphy of Egypt to prepare its sudden apparition. It revolutionized the concept of the epitaph in a way that even today can be considered as quite shocking. Instead of the usually brief pious formulae and prayers that ask for God's mercy for a deceased who had merely "fallen asleep" or "gone to rest", the *Totenklage* stelae are often lengthy poetic compositions that in sometimes harshly realistic images describe the terrors of dying and the feelings of bereavement and dismay caused by the sudden death of a beloved. Death is a brutal intrusion into man's life and both the moral and the physical effects of dying may be described in considerable detail. It leaves the deceased's family in sorrow and reduces the dead man or women to a corpse in a tomb, his or her true home. In the face of death, the futility of life itself becomes apparent: "today's man is tomorrow's earth and dust".[116] Life is not only, in biblical terms, as transient as grass, smoke or a shade, it is a burden as well: "I was urged by the cares of this life that weighed heavier on me than the waves of the sea".[117] A disenchanted view of both life and death predominates in these epitaphs. In spite of occasional claims such as "on the day of the resurrection both my soul and my body will go to the Lord",[118] the perspective of heavenly bliss is almost entirely absent from the epitaphs. Instead of μακάριος, "blessed", the deceased is typically qualified, with a deliberate inversion, as "wretched", ταλαίπωρος.[119]

The sudden emergence of the *Totenklage* genre is still something of a riddle, as is the reason for its apparent popularity in the region of Antinoopolis. In any case, this highly ambitious genre of epitaphs clearly addressed a sophisticated audience, able to appreciate the high-flown literary images and the scriptural and liturgical allusions of the texts. Its literary sources have been reasonably well studied. The earlier literature tended to overemphasize its indebtedness to pre-Christian, pharaonic tradition,[120] yet the texts themselves are clearly steeped in the language of the Bible, in particular the Old Testament, and dependent on the funeral liturgy for their attitudes towards death and dying as well as for much of their imagery. Different than the prayers discussed above, they draw their inspiration not from the intercessory prayers of the celebrant, however, but from the hymnody of the funeral liturgy, which accounts for a decisive difference in spirit and letter. The intercessory prayers and the hymnody of the Byzantine funeral rites represent, in the words of Elena Velkovska, "two different ways of seeing death": the first, a confident and consoling one, which she calls "theological"; the second, a realistic and outright macabre one, which she calls "anthropological" (although "anthropocentric" might be a better term).[121] It might seem as if two basically irreconcilable views are merely brought into juxtaposition here. Yet the paradox is resolved when we realize that in the hymns it is the people who sing of its grief, whereas in the presidential prayers the celebrant opens the perspective of redemption.[122] The *Totenklage* genre by its very nature adheres to the "anthropocentric" vision of death, the traces of which can still be found even in the "purged" Bohairic liturgy of present-day Egypt.[123]

409

The term *Totenklage*, "funeral lament, dirge", immortalized by Maria Cramer's monograph on the subject, is aptly chosen. Funeral laments are a traditional element in Mediterranean culture, even today, and the *Totenklage* poems must be connected with these traditional laments as much as with similar elements in the funeral liturgy.[124] Although there can be no doubt that the *Totenklage* epitaphs are deliberate poetic creations with undeniable literary ambitions, and not the unreflected echo of traditional folklore, they consciously and overtly adhere to the tradition of the sung, individualized lament that still accompanies many funerals in the Mediterranean world. They are not copied *en bloc* from standard models, however, since almost all are differently phrased. Several of the longer texts present themselves explicitly as "dirges" (ⲧⲟⲉⲓⲧ, corresponding to Greek θρῆνος), in which those who master the art are asked to join:

When there is a holy prophet who knows how to write dirges,
Let him stand by me in this role.
When there is one who knows how to weep with the weeping,
Let him join us as well.
When there is one who perceives the poverty (?) of his own head,[125]
Let him approach us.[126]

And the epitaph of a Deacon John invites the reader thus:

All who love weeping over those who passed away,
Let them sing a surpassing dirge about the unhappy fate of my youth.[127]

As this unhappy fate is narrated in the following lines, reading here apparently means joining in the singing of a dirge.

Other links with the traditional ritual lament are the strong emotional coloring of the texts and their outspoken interlocutory character. In a majority of cases, the texts either make the owner of the stela, as an "I", address the reader/listener of the epitaph, or they make the reader of the text address the deceased, as a "you" (in which case, however, the texts subsequently tend to slip into the third person). By these "oral" strategies they engage the reader in a dialogue in which he is either the confident of the deceased, who confesses his fate to him and his audience, or the latter's biographer (and sometimes also his panegyrist). Both strategies are geared at provoking a spoken or – rather – sung act of commemoration, but unlike the "objective" epitaphs of the "commemoration" type, the *Totenklage* stelae intend to implicate the reader and his audience emotionally and intellectually. Even when the epitaph is told in the objective third person, like the touching story of the early death of the girl Drosis, it switches in the end to the performative first-person plural: "Him (*scil.* God) we now pray and beseech that His mercy may befall her who went too early . . .".[128]

The interlocutory character of the texts was less of an innovation than their novel view of death, however. Grief is already thematized in those epitaphs

"WHAT IS MAN?"

where the reader asks the deceased "not to grieve" (see above). This originally pre-Christian acclamation was extremely popular in southernmost Egypt and in northern Nubia (Nobadia), where it occurs in both Greek and in Coptic forms, and a dated example from Sakinya (Toshka West), not far from Qasr Ibrim, suggests that it remained in use till around the turn of the millennium.[129] Yet the "do not grieve" formula is an appeal to serenity and acceptance, whereas the *Totenklage* stelae reveal a radically different attitude.

The revolutionary impact of the *Totenklage* genre was felt also in Nubia. Its influence can be found in the first place probably in the spread of the "verdict" type of text and the *apophasis* terminology, discussed above, but also in occasional phrases in other kinds of epitaphs.[130] More importantly, however, the genre itself is also attested in Nubia. The *Totenklage* type appears to have had its geographical center in Middle Egypt, in particular in the region of Antinoopolis, where in recent years several more examples have come to light (as yet unpublished). Yet one epitaph of this kind was already known to have been found in Aswan, on Egypt's southern frontier.[131] Now a much ruined example, composed for a woman, has been identified among the Coptic epitaphs from Qasr Ibrim.[132]

Regrettably, the stela is too fragmentary to reconstruct its opening lines, which apparently do not follow a standard formulary, as is frequently the case with this genre. Yet the occurrence of the verb ⲡⲁⲣⲁⲅⲉ, "to pass", in l. 1 and the word for "tomb", ⲙⲉⲁⲁⲩ, in ll. 2–3, link its text unmistakably to the *Totenklage* group with its macabre and disenchanted view of death. The opening section is furthermore concluded by a quote from Job 14, a chapter that also inspired several of the Egyptian *Totenklage* epitaphs: "for what is born of a woman, is short-lived" (Job 14:1, LXX).[133] This particular verse is not found in any of the known Egyptian *Totenklage* stelae, but it does occur in another Coptic epitaph from Qasr Ibrim itself,[134] and sums up neatly the attitude of these texts. Here it is followed by the well-known quote from Gen. 3:19 and a rather standard prayer section. The name of the deceased and the date of her death are regrettably lost.

However ruined, the Qasr Ibrim stela is an indisputable witness to the presence of the originally Middle Egyptian *Totenklage* tradition on Nubian soil and licenses us to look for more examples from elsewhere in Nubia. Thus, in spite of its laconic brevity, the epitaph of a girl or woman from Sakinya (Toshka West) that was already known for a long time can now be better understood.[135] Its text runs:

What is man?
He is earth (and) to the earth again he has returned.
Unique is the verdict (ἀπόφασις) of God.[136]
Thus went to rest the blessed [N.N.], the daughter of Chreste (?), on Phaoph 3.
O God, grant rest to her (soul in Your) kingdom.

The rather mutilated stela is dated in its top to the Diocletian year 575, which places the death of its owner on 30 September 858.

This epitaph addresses the reader with a poignant question about human nature, answered by a paraphrase of the classic verdict from Gen. 3:19, identified in the briefest possible way as the unique and hence universal source of death in the following line. The text is undoubtedly modeled on similar phrases in the hymnody of the funeral liturgy, such as the "What is our life?" that is still today sung in the Greek liturgy, where the answer is: "Really but bloom, vapor and morning dew!"[137] And it can easily be imagined that the readers of the brief and laconic phrases of the Sakinya epitaph were able to expand and supplement them from well-known liturgical hymns. But even in its present succinctness it is as fine a piece of funeral poetry as the lengthy *Totenklage* stelae from Egypt.

The *pronoia* type

Neither the *Totenklage* genre nor the "verdict" type of epitaph enjoyed great popularity in Nubia beyond the ninth century. The formulary that is best attested in the latest period of Coptic epigraphy in Nubia and that is generally considered to be the most characteristic Nubian funerary formulary in that language is the one with the opening phrase ϩⲓⲧⲛ ⲧⲉⲡⲣⲟⲛⲟⲓⲁ ⲙⲡⲛⲟⲩⲧⲉ, "through the providence of God". It attributes the individual's death not to a "verdict" of God, but to His "providence" (πρόνοια). Structurally but also intrinsically it closely resembles the formulary of the "verdict" type. A representative but incomplete example from Qasr Ibrim (*I. Qasr Ibrim* 47) reads:

> Through the providence (πρόνοια) of God,
> the creator (δημιουργός) of the universe,
> in accordance with what He had said to Adam:
> "Earth you are, to the earth again you shall return" (cf. Gen. 3:19),
> thus went to rest the blessed N.N. . . ..

Like the "verdict" type, the formulary displays a tripartite structure: a long opening phrase, which here includes the familiar quote from Genesis, introduces the lemma with the name of the deceased and the date of his or her demise (lost in the present case), which in turn is followed by a prayer section (again lost here). Within the present formulary, two basic textual types can be distinguished: with and without God's well-known words from Gen. 3:19.

The *pronoia* formulary characterizes the tenth-eleventh century Coptic bishops' stelae from Faras and many similar often elaborately sculptured high-status monuments from the same period. The formulary is not only found in northern Nubia, but as far south as al-Koro in the region between the Fourth and the Fifth Cataracts (*I. Khartoum Copt.* 123). In Qasr Ibrim it was apparently as popular as elsewhere in Nubia around the turn of the millennium. Presently thirteen tombstones opening with a wholly or partly preserved ϩⲓⲧⲛ ⲧⲉⲡⲣⲟⲛⲟⲓⲁ formula in Coptic are known from Qasr Ibrim.[138] Nine of them appear to include the quote from Gen. 3:19, whereas only two lack the quote (in two other cases its presence cannot be

"WHAT IS MAN?"

ascertained anymore). A further eight stelae from Qasr Ibrim bear the same opening formula in Greek.[139] Among the Qasr Ibrim material, the stela of Mariakyto, dated to A.D. 1059 (*I. Qasr Ibrim* 45), is a typical example of a high-status funerary monument from the latest phase of Coptic epigraphy in Nubia. Both the relief decoration, which places the text within a simplified *aedicula*, and the form of the text itself are characteristic and have close parallels from tenth-eleventh-century Faras (for example the stelae of bishops Aaron, A.D. 972, and George, A.D. 1097) and Meinarti (for example the stela of a woman Iesousyko, A.D. 1046).[140]

Both the Mariakyto stela and all four Coptic examples from Qasr Ibrim that bear a date belong to the eleventh century. Yet the use of the formula in Coptic epitaphs must go back to the early ninth century at the latest, since it appears in the funerary monument of Bishop Kerikos of Saï.[141] Remarkably, its Greek equivalent (opening with θεία προνοία) is less frequent in Nubian funerary epigraphy than its Coptic counterpart, but seems to have come into use earlier: a precisely dated example is from the middle of the eighth century.[142] The Greek formula, moreover, was widespread in Egypt, in particular in papyri, long before it found its way into Nubian funerary inscriptions.[143] The Coptic formula, on the other hand, while popular in Nubia, does not occur among the published Coptic epitaphs from Egypt, not even in the area of the First Cataract (Aswan).[144] Therefore, it would appear that its rise and diffusion in Nubia was a fairly late internal Nubian development, for which Greek-language examples were followed. In Coptic, then, it became far more popular than it had ever been in Greek, at least in Nubia.

In spite of its similarities to the "verdict" formulary, among which the frequent presence of the Genesis quote is certainly the most important, the *pronoia* formulary definitely strikes a different note. The originally Stoic concept of divine πρόνοια as a universal ruling principle was adopted and popularized in Christian thought very early. That it had retained its original meaning in the Nubian formulary is shown by the continuation of the text. It is a common characteristic of the formulary, proper to its Greek and Coptic forms alike, to expand the word for God with one or more standard epithets that underscore his omnipotence and his sole authority over life and death.[145] In the Greek epitaphs of this type from Qasr Ibrim, the epithet is normally the participle δεσπόζων, "who is lord".[146] In the Coptic stela of Mariakyto, God is called "almighty (παντοκράτωρ), the creator (δημιουργός) of the universe" (*I. Qasr Ibrim* 45, ll. 2–3).[147] A variant form of the Coptic formula, not attested in Qasr Ibrim but frequently found in the epitaphs from the important monastery of Wadi al-Ghazali, in Makuria, replaces the word πρόνοια, "providence", with κέλευσις, "order".[148] In its turn, this can be considered a variant of the popular Greek opening formula that attributes the death event to "the inclination and will" (νεύσει καὶ βουλήσει) of God and is also found at Qasr Ibrim.[149] As both the various divine epithets and the cited variants show, the texts of the *pronoia* type focus on the sovereign autonomy and omnipotence of God, from whom all human life and death depend, not as a judge (as in the "verdict" type), but as the creator and ruler of the universe.

413

One significant difference between the Greek and the Coptic versions of the formula remains to be observed. In the Greek version, the Genesis quote (Gen. 3:19) never intervenes. The lemma with the verb introducing the death event and the name of the deceased always follow the mention of the Godhead immediately. Even if the majority of the examples from Qasr Ibrim include the Genesis quote, it was fairly often omitted also in the Coptic version. In the earliest datable occurrence of the formula in Coptic, the stela of Bishop Kerikos of Saï (early ninth century; see above), the Genesis quote does occur already, although not yet in its standard place in the opening phrase but in a damaged context (presumably the opening invocation of the prayer), following the name and date lemmata (ll. 7–8). Assuming that the Coptic formula has been calqued on the earlier Greek one, it seems likely that the type with the Genesis quote is a typologically secondary development, that most likely took shape under the influence of the structurally very similar "verdict" formulary, with its clear etiological bias.

This supposition appears to be confirmed by those instances where either the noun ἀπόφασις or the verb ἀποφαίνω are inserted between the opening *pronoia* formula and the Genesis quote (some have been mentioned above, in our discussion of the "verdict" type). The significance of such "hybrid" textual types seems fairly obvious. They may be said to redress the balance between the negative view of death as punishment, articulated by the juridical *apophasis* terminology, and the positive statement of God's absolute sovereignty and providential care, found in the *pronoia* formula. God did not only pronounce a fatal verdict on entire mankind, He also cares for each individual. No less than the epitaphs of the "verdict" type, these texts invite the reader to reflect on death and mortality, but in a less grim way.

The interesting stela of Martha from Sakinya (Toshka West) permits us to be more precise, however.[150] Its text reads:

> Through the providence (πρόνοια) of God,
> the creator (δημιουργός) of the universe,
> He who gave verdict (ἀπόφασις) over our father Adam
> on account of his transgression (παράβασις) which he had committed,
> saying: "Adam, earth you are, to the earth you shall go again",
> thus then went to rest the blessed Martha on the twenty-fourth of Pachon, 24.
> May God make her recline in the bosom of Abraham and Isaac and Jacob.
> Amen. So be it, so be it.

The inserted phrase "on account of his transgression (παράβασις) which he (*scil.* Adam) had committed" is not normally found in either the *pronoia* or the "verdict" type of text. The word παράβασις is a juridical term again, just like ἀπόφασις, and denotes the violation of a contract.[151] It is also a technical term in the present context, for the related verb παραβαίνω is the very opening word of the eucharistic prayer in the *Liturgy of Saint Basil*: "when we had transgressed (παραβάντας) Your command through the ruse of the serpent and had fallen out of

"WHAT IS MAN?"

eternal life and had been banished from the paradise of delight . . .".[152] It shifts the responsibility for death from the Godhead to man. Death may be foreseen or even ordered by God, through His πρόνοια or κέλευσις, yet it was not wanted by Him but triggered by man's transgression: God's verdict was issued in response to the willful violation of His command. As a way of reflecting on death, the Coptic *pronoia* epitaphs that evoke man's punishment can be read as an attempt at resolving the eternal paradox of divine omnipotence and human *autexousia*.

The regular insertion of the Genesis quote and the quite frequent traces of the *apophasis* terminology in the Coptic version of the *pronoia* formula reveal the relative independence of the Coptic text tradition vis-à-vis its Greek-language model. Since its rise and development as a funeral formulary appears also to have been independent of Egyptian models, the *pronoia* text is undoubtedly the most Nubian of all Coptic formularies. In fact, already in 1925 Hermann Junker had described it as "nubisches Eigengut", a creation proper to Nubian Christianity.[153]

Conclusions

Within the Christian tradition, commemorating the dead is basically naming the dead and praying for them. The epitaphs that we have studied here were meant to assist the living in their prayers and offer them scenarios for commemoration. They may be characterized as liturgy by the square centimeter. Yet they are more. In an extremely condensed form they articulate ideas about death, mortality and afterlife, the omnipotence of God and the relation between God and mankind. Although epitaphs may be addressing God or the deceased formally, they are basically addressing the living, the reader and his audience, who are supposed to process the texts by oral recitation. Reading (or singing) an epitaph is no private prayer but part of a communal rite that reinforces shared religious views of death and afterlife, and forges social cohesion.

Following Martin Krause, the Coptic funerary inscriptions from Qasr Ibrim have here been grouped according to their distinctive formularies. Such a formulary has been defined as a set textual format, framing and presenting the core of each epitaph, the name of the deceased, and itself made up of a variable number of formulae. Four types are represented in the Coptic material from Qasr Ibrim. Each of these can be described as tripartite: one or more opening formulae introduce lemmata that contain the name of the deceased and the date of his or her demise. A third part is predominantly supplicatory in character and consists of a longer or shorter prayer section. It is made up of a selection of formulaic phrases, usually derived from the commemorative parts of the Sunday liturgy or the more properly funeral services. They normally address God in favor of the deceased.

The four types of formularies that we have been able to distinguish reflect a certain development. The presumably earliest type, which northernmost Nubia shared with the Egyptian frontier zone around Aswan, offers a text that is entirely focused on the act of commemoration. An even stronger influence of Egyptian scribal habits is visible in the adoption, probably from the early years of the eighth

415

century onwards, of the originally Middle Egyptian *Totenklage* genre and, probably in conjunction with the latter type, the juridical *apophasis* terminology that quotes God's "verdict" over Adam (Gen. 3:19) as an etiology of death and a virtual *memento mori*. With the discovery of a *Totenklage* stela in Qasr Ibrim, this textual genre that deploys various interlocutory strategies is now well attested on Nubian soil. Both the Egyptian *Totenklage* genre and the *apophasis* type of text bear witness to a remarkably negative and anthropocentric view of death. This seems to be partly corrected in an originally Nubian creation, the Coptic *pronoia* type that balances the quote of Gen. 3:19 with clear statements of God's omnipotence and providential care. This latter type, then, became the most favored Coptic funeral formulary in tenth-eleventh-century Nubia.

The development sketched above is not a clear-cut linear one, however. All types discussed, even the rare *Totenklage* type, appear to have remained in use over several centuries and there has certainly been much overlap chronologically. The adoption or abandonment of the various types over the years has, if our analysis can be accepted, primarily to do with variation in the societal perception of death and commemoration. In addition, other variables such as the wealth and social status of the deceased and his family must have determined both the length and the style of an epitaph at least as profoundly as this supposed conceptual evolution.

The written texts themselves are clearly neither born from the whims of individual scribes nor produced by mechanically copying immutable models. Centonization plays an important role in their redaction, accounting for considerable variety even among at first sight very similar pieces from one and the same workshop, as examples from Qasr Ibrim demonstrate neatly. Although commemorative liturgy provided a conceptual as well as a calendrical framework, the epitaphs studied above are no slavish copies from liturgical handbooks. In fact, in as far as the medium is the message, a considerable distance separates the inscribed stone immured in a tomb from the page of a liturgical codex. If individual prayers and formulae were borrowed freely and frequently from liturgical textbooks, it is because both textual genres each in its own way serve the same purpose: singing the commemoration of the dead.

Judged by their doctrinal contents, the Coptic formularies studied above seem hardly interested in the specifically Christian aspect of redemption through Christ.[154] The first Adam is nearly omnipresent, the second Adam hardly visible. This is all the more striking as it is in marked contrast with the Greek-language epitaphs from Nubia. The Greek epitaph of Tetrarch Mariankouda, already quoted above, takes the fall of primeval man as its point of departure, just like so many Coptic epitaphs do, but primarily to describe how the Son, "by ascending onto the wood saved all those who had been enslaved by death" (*I. Khartoum Greek* 18, ll. 11–12). Also the most popular Greek formulary from Nubia, that of the "God of the spirits" type, is focused on God's victory over death and hell and His love for mankind.[155] The Coptic-language formularies, on the other hand, focus on human mortality, endlessly repeating Gen. 3:19, a quote practically absent from the

"WHAT IS MAN?"

Greek material. Conceivably, a sociological explanation applies here. The Coptic focus on death as punishment for human sin might reflect the strong monastic bias of Coptic literary culture in general, which may have helped to shape the Coptic funerary formularies as well.

Several related questions deserve further investigation. The Greek and Coptic epitaphs from Nubia can justifiably be considered a single bilingual corpus of texts. Yet it is striking to observe how little textual overlap there is between the Greek and the Coptic funeral formularies. Apart from a significant number of exceptions, Greek and Coptic in Nubia may seem to represent two separate textual traditions, even when all other variables (social, geographical, chronological, etc.) are identical. Thus, the bishops of Faras in the eleventh century used alternately either the Greek "God of the spirits" formulary or the totally different Coptic *pronoia* formulary plus the Genesis quote, instead of, for example, at least remotely similar versions of one bilingual formulary.[156] A Coptic variant of the "God of the spirits" formulary was actually well known, but it is found only once or twice in Nubia, far less often than in Egypt.[157] Further study of the code-selecting strategies of the corpus may provide explanations along sociological or liturgical lines.

Finally, even though the Nubian tradition of funerary epigraphy has a specific character of its own and created textual types of its own, such as the Coptic *pronoia* formulary, Egyptian models have played a considerable role in shaping it. Yet these Egyptian models are only imperfectly known themselves. This applies in particular to the *Totenklage* genre. After many decades of undeserved neglect, this unique genre demands a thorough new study. Such a study should not merely collect the new material that has come to light since Maria Cramer's 1941 book, but map the genre's literary aims and strategies carefully and situate it within its proper social and historical context. One of its outcomes will certainly be that epigraphy is not such a dull subject as it is usually considered to be. Another one, predictably, will be that a funerary inscription is anything but an "Urkunde".

Notes

*   This essay is the outcome of work undertaken together with Adam Łajtar on the Greek and Coptic inscriptions from Qasr Ibrim and published in 2010, abbreviated in this volume as *I. Qasr Ibrim*. In its present form, it owes much to the critical acumen and wide reading of my wife, Ewa D. Zakrzewska.

1   Monastic community: C. Wietheger, *Das Jeremias-Kloster zu Saqqara unter besonderer Berücksichtigung der Inschriften*, Altenberge: Oros 1992; episcopal see: S. Jakobielski, *A history of the bishopric of Pachoras on the basis of Coptic inscriptions*, Warsaw: PWN–Éditions scientifiques de Pologne 1972.

2   E.g., in S.J. Davis, *The cult of Saint Thecla: a tradition of women's piety in late antiquity*, Oxford: Oxford University Press 2001.

3   Topographical: D. Zuntz, "Koptische Grabstelen: ihre zeitliche und örtliche Einordnung", *MDAIK* 2 (1932), 22–38; chronological: M. Krause, "Die Formulare der christlichen Grabsteine Nubiens", in: K. Michałowski (ed.), *Nubia: récentes recherches*, Warsaw: Musée national 1975, 76–82.

JACQUES VAN DER VLIET

4 See, for Egypt, Th.K. Thomas, *Late antique Egyptian funerary sculpture: images for this world and the next*, Princeton: Princeton University Press 2000; for a wider area, A. Petrucci, *Writing the dead: death and writing strategies in the western tradition*, Stanford: Stanford University Press 1998.

5 The literature about the site and its history is vast; an important recent title is: W.Y. Adams, *Qasr Ibrim: the earlier medieval period*, London: Egypt Exploration Society 2010; for an overview of much of past and current research, see J.L. Hagen and J. van der Vliet (eds.), *Qasr Ibrim, between Egypt and Africa: a case study in cultural exchange (NINO Symposium, Leiden, 11–12 December 2009)*, Leiden: NINO 2011.

6 See now F. Aldsworth, *Qasr Ibrim: the cathedral church*, London: Egypt Exploration Society 2010.

7 See W. Godlewski, "A short essay on the history of Nobadia from Roman to Mamluk times", in: Hagen and van der Vliet, *Qasr Ibrim*, 123–33.

8 Apart from *I. Qasr Ibrim* 29–40, 45–52, 66–72, 81, 87–90, the Coptic epitaphs from Qasr Ibrim comprise: a group acquired in 1900 by Georg Steindorff (see provisionally T.S. Richter, "Die neun Stelen Ägyptisches Museum der Universität Leipzig Inv.-Nr. 680–88 mit der Herkunftsangabe Qasr Ibrim", [in: S. Emmel et al. (eds.), *Ägypten und Nubien in spätantiker und christlicher Zeit: Akten des 6. Internationalen Koptologenkongresses, Münster, 20.-26. Juli 1996*, Wiesbaden: Reichert Verlag 1999, I, 295–304]); a group brought to light in the 1930s and for the greater part published by U. Monneret de Villard, *La Nubia medioevale* I, Cairo: Mission archéologique de Nubie 1935, 112–15; and a group excavated by W.B. Emery in the early 1960s and published by E.S. Meltzer, in: A.J. Mills, *The cemeteries of Qasr Ibrîm: a report of the excavations conducted by W.B. Emery in 1961*, London: Egypt Exploration Society 1982, 82–4. Some further pieces were published separately; for further details, see *I. Qasr Ibrim*, 3–4.

9 The position of Coptic in Christian Nubia is discussed more extensively in J. van der Vliet, "Coptic as a Nubian literary language: four theses for discussion", in: W. Godlewski and A. Łajtar (eds.), *Between the cataracts: proceedings of the 11th Conference for Nubian studies, Warsaw University, 27 August–2 September 2006*, II/2: *Session papers*, Warsaw: Warsaw University Press 2010, 765–71 [Study 21].

10 For the general picture, see now J.H.F. Dijkstra, *Philae and the end of ancient Egyptian religion: a regional study of religious transformation (298–642 CE)*, Louvain/Paris/Dudley, MA: Peeters 2008.

11 *FHN* III 320–22; a new edition is in preparation by J.L. Hagen, Leiden University.

12 Published by P. Hubai, *A Megváltó a keresztről: Kopt apokrifek Núbiából (A Kasr el-Wizz kódex)*, Budapest: Szent István Társulat 2006; idem, *Koptische Apokryphen aus Nubien: der Kasr el-Wizz Kodex*, Berlin: de Gruyter 2009.

13 Toshka: see next note; Arminna: H. Junker, *Ermenne: Bericht über die Grabungen ... auf den Friedhöfen von Ermenne (Nubien) im Winter 1911/12*, Vienna/Leipzig: Hölder-Pichler-Tempsky 1925, 150–52, and various isolated pieces; Faras-Meinarti region: Jakobielski, *Bishopric of Pachoras* III; *I. Khartoum Copt.* 1–21; Saï: A. Tsakos, "Medieval funerary inscriptions from Sai Island", *CRIPEL* [29 (2011/2012), 297–330]; Makurian sites: *I. Khartoum Copt.* 28–124; M. Cramer, *Koptische Inschriften im Kaiser-Friedrich-Museum zu Berlin*, Cairo: Société d'archéologie copte 1949, 9–13. For Coptic funerary texts from Dongola, see A. Łajtar, "New finds of Greek epitaphs at Dongola", in: *Nubian voices*, 89–94.

14 Edited by T. Mina, *Inscriptions coptes et grecques de Nubie*, Cairo: Société d'archéologie copte 1942 (replacing the incomplete publication by U. Monneret de Villard, *Le iscrizioni del cimitero di Sakinya (Nubia)*, Cairo: IFAO 1933); partly republished in S. Pernigotti, "Stele cristiane da Sakinya nel Museo di Torino", *Oriens antiquus* 14 (1975), 21–55; study of the formularies: Krause, "Die Formulare".

"WHAT IS MAN?"

15 Most recent edition: S.G. Richter, *Studien zur Christianisierung Nubiens*, Wiesbaden: Reichert Verlag 2002, 164–72; cf. Dijkstra, *Philae*, 299–302.

16 Recent efforts to obtain a more precise dating (see the literature cited in the preceding note) are flawed, since they are based on a non-existent indiction date; see G. Ochała, "The date of the foundation inscription from Dendur revisited", [*BASP* 48 (2011), 217–24].

17 See S. Jakobielski, and J. van der Vliet, "From Aswan to Dongola: the epitaph of Bishop Joseph (died AD 668)", in: A. Łajtar and J. van der Vliet (eds.), *Nubian voices: studies in Christian Nubian culture*, Warsaw: University of Warsaw/Raphael Taubenschlag Foundation 2011, 15–35 [Study 29].

18 Cf. J. Bingen, "L'épigraphie grecque de l'Égypte post-constantinienne", in: *Pages d'épigraphie grecque* II: *Égypte (1983–2002)*, Brussels: Epigraphica Bruxellensia 2005, 85–99 at 86.

19 For Nubian funerary architecture, see Monneret de Villard, *Nubia* III, Cairo 1957, 63–78; W.Y. Adams, "Towards a comparative study of Christian Nubian burial practice", *Archéologie du Nil moyen* 8 (1998), 13–41; D.A. Welsby, *The medieval kingdoms of Nubia: Pagans, Christians, and Muslims along the Middle Nile*, London: British Museum Press 2002, 57–61.

20 See F.L. Griffith, "Oxford excavations in Nubia", *LAAA* 14 (1927), 57–116, especially 62–81.

21 A. Łajtar, "New finds", 61–5, no. 6.

22 Examples from Qasr Ibrim include among others *I. Qasr Ibrim* 19–20, 55, 57 (Greek) and 45 (Coptic).

23 See Thomas, *Funerary sculpture*, 16–19.

24 See J-P. Sodini, "Les 'tombes privilégiées' dans l'Orient chrétien (à l'exception du diocèse d'Egypte)", in: Y. Duval and J.-Ch. Picard (eds.), *L'inhumation privilégiée du IVe au VIIIe siècle en Occident: actes du colloque tenu à Créteil les 16–18 mars 1984*, Paris: de Boccard 1986, 233–43, and other papers in the same volume.

25 For the tombs of the Faras bishops, see now W. Godlewski, *Pachoras: the cathedrals of Aetios, Paulos and Petros: the architecture*, Warsaw: Warsaw University Press 2006, 139–50.

26 E.g. the stelae of Bishop Joseph, see Jakobielski and van der Vliet, "From Aswan to Dongola" (presbytery of a church) [Study 29], and Archbishop Georgios, see A. Łajtar, "Georgios, Archbishop of Dongola (+ 1113) and his epitaph", in: *Εὐεργεσίας χάριν: studies presented to B. Bravo and E. Wipszycka*, Warsaw: University of Warsaw/Raphael Taubenschlag Foundation 2002, 159–92 at 161–63 (chapel).

27 See the brief description in Adams, *Qaṣr Ibrîm*, 82–3.

28 See U. Monneret de Villard, "Rapporto preliminare dei lavori della missione per lo studio dei monumenti cristiani della Nubia, 1930–1931", *ASAE* 31 (1931), 7–18 at 8–10; idem, *Nubia* I, 112–15; Mills, *Cemeteries*, 3 and pl. III.

29 Thus also M.A. Handley, *Death, society and culture: inscriptions and epitaphs in Gaul and Spain, AD 300–750*, Oxford: Archaeopress 2003, 172–75.

30 Petrucci, *Writing the dead*, xviii.

31 I owe this insight to Alexandros Tsakos.

32 E. Velkovska, "Funeral rites according to the Byzantine liturgical sources", *DOP* 55 (2001), 21–45 at 39–40. For further background, see U. Volp, *Tod und Ritual in den christlichen Gemeinden der Antike*, Leiden/Boston: Brill 2002, 214–39; E. Rebillard, *Religion et sépulture: l'Église, les vivants et les morts dans l'Antiquité tardive*, Paris: Éditions de l'École des hautes etudes en sciences socials 2003, 161–97.

33 Krause, "Die Formulare".

34 H. Junker, "Die christlichen Grabsteine Nubiens", *ZÄS* 60 (1925), 111–48.

35 Krause himself does not really define his "Formulare", but classifies them exclusively on the basis of their opening words (Krause, "Die Formulare", 77). Here a somewhat

less rigid procedure is followed, discussed below. The role of the name as the core of each epitaph is likewise discussed below.

36 For the ever-growing body of literature about orality and literacy in the ancient world, see the annotated bibliography by S. Werner, "Literacy studies in classics: the last twenty years", in: W.A. Johnson, and H.N. Parker (eds.), *Ancient literacies: the culture of reading in Greece and Rome*, Oxford: Oxford University Press 2009, 333–82. What follows owes much to the brilliant improvisations on the subject of early Greek inscriptions in J. Svenbro, *Phrasikleia: anthropologie de la lecture en Grèce ancienne*, Paris: Éditions La Découverte 1988, and the seminal essay of A. Papalexandrou, "Echoes of orality in the monumental inscriptions of Byzantium", in: L. James (ed.), *Art and text in Byzantine culture*, Cambridge: Cambridge University Press 2007, 161–87. For the role of textual formulae in Coptic literature, see E.D. Zakrzewska, "Masterplots and martyrs: narrative techniques in Bohairic hagiography", in: F. Hagen et al. (eds.), *Narratives of Egypt and the ancient near East: literary and linguistic approaches*, Louvain: Peeters 2010, 499–523.

37 Svenbro, *Phrasikleia*, 7, cf. 70–1.

38 Ibidem, 53; cf. M. Carroll, *Spirits of the dead: Roman funerary commemoration in Western Europe*, Oxford: Oxford University Press 2008, 54–5.

39 Papalexandrou, "Echoes"; J.M. Foley, *How to read an oral poem*, Urbana/Chicago: University of Illinois Press 2002, for "voiced texts" especially 43–5.

40 Cf. M. Cramer, *Die Totenklage bei den Kopten*, Vienna/Leipzig: Hölder-Pichler-Tempsky 1941, 56; for ancient Egypt, see J. Sainte Fare Garnot, *L'appel aux vivants dans les textes funéraires égyptiens des origines à la fin de l'Ancien Empire*, Cairo: IFAO 1938; cf. J. Assmann, *Stein und Zeit: Mensch und Gesellschaft im alten Ägypten*, Munich: Wilhelm Fink Verlag 1991, 167–78; for the Latin West, Carroll, *Spirits of the dead*, 53–5.

41 *SB Kopt.* III 1596, A.D. 775, collated with the original (Museum Liebieghaus, Frankfurt am Main; cf. J. van der Vliet, "*Parerga*: notes on Christian inscriptions from Egypt and Nubia", *ZPE* 164 (2008), 153–58 at 155–57 [Study 16]). Demands for prayer are not limited to funerary inscriptions and occur e.g. in the Dendur dedicatory inscription, ll. 12–14 (quoted above, n. 15), and in graffiti, such as the one from Derr in Richter, *Studien*, 158–59. In return, the reader may expect to receive "blessing", according to a fairly common Coptic formula, for which see G. Roquet, "Inscriptions bohaïriques de Dayr Abû Maqâr", *BIFAO* 77 (1977), 163–79 at 174–76.

42 For its etymology, see W. Vycichl, *Dictionnaire étymologique de la langue copte*, Louvain: Peeters 1983, 251, s.v.

43 See *I. Khartoum Copt.*, 42, n. 207, with further literature.

44 Cf. Svenbro, *Phrasikleia*, 35.

45 V. Edwards and Th.J. Sienkewicz, *Oral cultures past and present: rappin' and Homer*, Oxford/Cambridge, MA: B. Blackwell 1990, 199.

46 Papalexandrou, "Echoes of orality", 165–66.

47 Cf. Foley, *Oral poem*, 199, and the remarks on the "isonymie" of the tomb and the deceased in Svenbro, *Phrasikleia*, 92.

48 For this function of the funerary monument, see Y. Duval, *Auprès des saints corps et âme: l'inhumation «ad sanctos» dans la chrétienté d'Orient et d'Occident du IIIe au VIIe siècle*, Paris: Études augustiniennes 1988, 213–17.

49 As in Tassos Boulmetis' nostalgic film Πολίτικη κουζίνα ("A touch of spice", 2003); cf. Αγιασματάριον II, Rome 1955, 32 and 1–3.

50 See above note 8.

51 Krause, "Die Formulare".

52 H. Munier, "Les stèles coptes du Monastère de Saint-Siméon à Assouan", *Aegyptus* 11 (1930–31), 257–300 and 433–84.

53 *I. Qasr Ibrim* 29–37, plus eight or nine more that are listed in the commentary to *I. Qasr Ibrim* 29.

"WHAT IS MAN?"

54 Aswan: Munier, "Stèles coptes" (almost all of the inscriptions, which are reprinted in *SB Kopt.* I 498–675, as well as various isolated epitaphs with the same provenance); Toshka West (Sakinya): Mina, *Inscriptions, passim* (cf. Krause, "Die Formulare", 76–82, especially 78–9; Pernigotti, "Stele", 27); Arminna: Junker, *Ermenne,* 151–52 (cf. Junker, "Grabsteine", 131). Only a single example is known from Faras: *I. Khartoum Copt.* 10, quoted below (n. 67).

55 For the important concept of eschatological rest in early Christianity, see J. Helderman, *Die Anapausis im Evangelium Veritatis,* Leiden: Brill 1984, 47–70; for more references, *I. Khartoum Copt,* 27.

56 Thus *I. Qasr Ibrim* 29; for the *pax* formula, see Wietheger, *Jeremias-Kloster,* 153; to the references given there may be added: I. Kajanto, "The hereafter in ancient Christian epigraphy and poetry", *Arctos* 12 (1978), 27–53 at 39–41 (Latin sources only).

57 Following the analysis of Krause, "Die Formulare", 79–81.

58 S. Donadoni, "Ptahmose e Semne alla 'Sapienza'", in: C. Basile and A. Di Natale (eds.), *Atti del VII convegno nazionale di egittologia e papirologia, Siracusa, 29 novembre-2 dicembre 2001,* Siracusa: Istituto internazionale del papiro 2003, 15–22 at 18, ll. 1–2 (*SB Kopt.* III 1647).

59 See Munier, "Stèles coptes"; cf. Krause, "Die Formulare", 79.

60 The stela: Monneret de Villard, *Nubia* I, 115; for the formula ογλ πε πνογτε/εἷς θεός, see E. Peterson, *ΕΙΣ ΘΕΟΣ: Epigraphische, formgeschichtliche und religionsgeschichtliche Untersuchungen,* Göttingen: Vandenhoek & Ruprecht 1926; for its Nubian occurrences, Junker, "Grabsteine", 130, 138, 143; Pernigotti, "Stele cristiane da Sakinya", 28; cf. *I. Khartoum Copt.,* 47.

61 *I. Qasr Ibrim* 37, quoted *in extenso* below.

62 In this respect my concept of the formulary differs from that of Martin Krause, who in a somewhat mechanical manner distinguishes types of formularies strictly after the first words of the entire text (Krause, "Die Formulare", 77).

63 Over sixty occurrences of the formula ὑπερ μνήμης τοῦ N.N. (with variants) are published in Mina, *Inscriptions;* cf. Krause, "Die Formulare", 78 and 80; Pernigotti, "Stele", 26–7. Other provenances: Junker, "Grabsteine", 127.

64 On μνήμη in the Christian Greek epigraphy of Egypt, see G. Lefebvre, "Notes épigraphiques", *ASAE* 6 (1905), 188–91 at 189–90; A. Łajtar, in *I. Varsovie* 99, *ad* l. 1; cf. G. Bartelink, "Note sur μνήμη", *Eranos* 59 (1961), 84–5.

65 Mina, *Inscriptions,* no. 76, ll. 1–3.

66 Compare the Faras stela of the monk Dios, which bears the formula: ετβε πρπμεεγε μν ταναπαγcιc μπμακαριοc ncon διοc, "for the commemoration and the rest of the blessed brother Dios", the Coptic rendering of Greek ὑπερ μνήμης καὶ ἀναπαύσεως τοῦ μακαρίου N.N. (*I. Khartoum Copt.* 10, ll. 2–4).

67 See Junker, "Grabsteine", 131; Munier, "Stèles coptes", 259; M. Cramer, *Archäologische und epigraphische Klassifikation koptischer Denkmäler des Metropolitan Museum of Art, New York, und des Museum of Fine Arts, Boston, Mass.,* Wiesbaden: Harrassowitz 1957, 31–3; Krause, "Die Formulare", 78–9; Pernigotti, "Stele", 27, n. 35.

68 See M.E. Johnson, *The prayers of Sarapion of Thmuis: a literary, liturgical and theological analysis,* Rome: Pontifico Istituto Orientale 1995.

69 See A. Budde, *Die ägyptische Basilios-Anaphora: Text – Kommentar – Geschichte,* Münster: Aschendorff 2004, 504–06 (names), 437–38 and 514–15 (deacon's call and answer).

70 Epitaphs consisting of a name only abound in W.E. Crum, *Coptic monuments,* Cairo: IFAO 1902 (reprinted Osnabrück: Otto Zeller Verlag 1975), and G. Lefebvre, *Recueil des inscriptions grecques-chrétiennes d'Égypte,* Cairo: IFAO 1907 (reprinted Chicago: Ares 1978).

71 The translation "the late" obfuscates its Christian meaning; see, for further arguments, J. Pelsmaekers, "Een korte bemerking bij de vertaling van de term μακάριος", *Bulletin*

de l'Institut Historique Belge de Rome/Bulletin van het Belgisch Historisch Instituut te Rome 58 (1988), 5–9; cf. *I. Khartoum Copt*, 30, n. 157.

72 See e.g., for the Latin West, Carroll, *Spirits of the dead*, 59–62, and C.R. Galvao-Sobrinho, "Funerary epigraphy and the spread of Christianity in the West", *Athenaeum* 83 (1995), 431–62 at 454–55, who quotes Lactantius, *Divine institutes* IV, 26, 13: "iacentia mortuorum corpora erexit eosque *nominibus suis inclamatos* a morte revocavit" (ed. P. Monat, *SC* 377, 212, my emphasis), referring to *John* 11:43–4. Cf. the Roman rite of the *conclamatio*; see A.C. Rush, *Death and burial in Christian antiquity*, Washington, DC: The Catholic University of America Press 1941, 108–09; Volp, *Tod*, 71–4.

73 See e.g. Phil. 4:3, Rev. 20:15. Cf. W. Bauer, *Wörterbuch zu den Schriften des Neuen Testaments*, Berlin/New York: de Gruyter 1971 (fifth edition), col. 280, s.v. βίβλος. 2: "D(as) Gericht wird näml(ich) auf Grund von Büchern vollstreckt"; see also the often quoted epitaph *CIG* 9686, from Rome, *apud* C. Wessel, *Inscriptiones graecae christianae veteres Occidentis*, Bari: Edipugia 1988, no. 385 (= *ICUR* VIII 23378).

74 E.g. in the homily of Ps.-John Chrysostom, *On the four creatures* (*CPG* 5150 (9)), of which both Coptic and Old Nubian versions have been found in Nubia; see J.L. Hagen, "'The Great Cherub' and his brothers: Adam, Enoch and Michael and the names, deeds and faces of the Creatures in Ps.-Chrysostom, *On the Four Creatures*", in: N. Bosson and A. Boud'hors (eds.), *Actes du huitième Congrès international d'études coptes, Paris, 28 juin-3 juillet 2004*, Louvain/Paris/Dudley, MA: Peeters 2007, II, 467–80; Enoch: B.A. Pearson, "Enoch in Egypt", in: R.A. Argall, B. Bow, and R.A. Werline (eds.), *For a later generation: the transformation of tradition in Israel, early Judaism, and early Christianity (Festschrift G. Nickelsburg)*, Harrisburg 2000, 216–31 at 226–29.

75 For the occurrence of indiction dates in Nubian epigraphic sources, see now G. Ochała, *Chronological systems of Christian Nubia*, Warsaw: University of Warsaw/ Raphael Taubenschlag Foundation 2011.

76 Velkovska, "Funeral rites", 22 and 39–40; cf. Volp, *Tod*, 225–27. These dates are already mentioned in the *Apostolic constitutions* VIII, 42 (*CPG* 1730). The Coptic fragments, published by M. Hasitzka, "Koptische Totenklagen auf Papyrus", in: M. Capasso, R. Pintaudi, and G. Messeri Savorelli (eds.), *Miscellanea papyrologica in occasione del bicentenario dell'edizione della Charta Borgiana*, Florence: Gonnelli 1990, 297–303 at 300–03 (seventh-eighth century), may be connected with the calculation of such dates.

77 See A. van Gennep, *Les rites de passage*, Paris: Nourry 1909, especially 209–36, ch. VIII, "Les funérailles", where he underlines the predominantly marginal character of funerary rites. For death as a "journey", see Rush, *Death and burial*, 44–71.

78 See, in particular for the mirroring marginal periods, G. Dagron, "Troisième, neuvième et quarantième jours dans la tradition byzantine: temps chrétien et anthropologie", in: *Le temps chrétien de la fin de l'antiquité au moyen âge, IIIe-XIIIe siècles*, Paris: Éditions du CNRS 1984, 419–30; cf. A. van Lantschoot, "Révélations de Macaire et de Marc de Tarmaqâ sur le sort de l'âme après la mort", *Le Muséon* 63 (1950), 159–89; G. Giamberardini, *La sorte dei defunti nella tradizione copta*, Cairo: Edizioni del Centro Francescano di Studi Orientali Christiani 1965.

79 Thus according to O.H.E. Burmester, *The Egyptian or Coptic Church: a detailed description of her liturgical services and the rites and ceremonies observed in the administration of her sacraments*, Cairo: Société d'archéologie copte 1967, 216–19; but cf. L. Störck, in H. Becker and H. Ühlein (eds.), *Liturgie im Angesicht des Todes: Judentum und Ostkirchen* I, Sankt Ottilien: EOS Verlag 1997, 636–37. Both ancient and modern liturgical sources show that the exclusive focus on the annual celebration of the day of death as *dies natalis*, found in much secondary literature, see typically Rush, *Death and burial*, 72–87, is unwarranted, if not misleading.

"WHAT IS MAN?"

80 See Chr.H. Reintges, *Coptic Egyptian (Sahidic dialect): a learner's grammar*, Cologne: Köppe 2004, 268.

81 No review of the attested Nubian prayer formulae is given here; for fairly comprehensive lists, see the indices in *I. Khartoum Copt*, 210–11, and *I. Qasr Ibrim*, 324–25.

82 See e.g. B. Botte, "Les plus anciennes formules de prière pour les morts", in: *La maladie et la mort du chrétien dans la liturgie: Conférences Saint-Serge, XXIe Semaine d'études liturgiques*, Rome: Edizioni Liturgiche 1975, 83–99 at 93–5; Budde, *Basilios-Anaphora*, 509–11; Burmester, *The Egyptian or Coptic Church*, 203 and 213.

83 See *I. Khartoum Copt.*, 62.

84 See the German translation by F.E. Dobberahn, in Becker and Ühlein, *Liturgie im Angesicht des Todes* II, 879–81.

85 As e.g. in *I. Khartoum Copt.* 17 (Debeira West) and 18–19 (Meinarti), all from the eleventh century.

86 Stela of Marianta: *I. Qasr Ibrim* 37 (on its date, see above); stela of Theognosta: G. Schenke, "Ein koptischer Grabstein aus Nubien", *ZPE* 132 (2000), 176–78 (acquired in commerce in the 1960s; comparison with the stela of Marianta shows that it must likewise originate from Qasr Ibrim).

87 Between square brackets my analysis of the text.

88 An example quoted above, in note 58.

89 See M. Rist, "The God of Abraham, Isaac, and Jacob: a liturgical and magical formula", *Journal of Biblical Literature* 57 (1938), 289–303.

90 Cf. T. Derda and J. van der Vliet, "Four Christian funerary inscriptions from the Fayum (I. Deir al-'Azab 1–4)', *JJP* 36 (2006), 21–33 at 26, n. 16 [Study 9]; but see the Greek stela M.G. Tibiletti Bruno, *Iscrizioni nubiane*, Pavia: Fusi 1964, no. 43 (Northern Nubia, A.D. 699), where the two sets of names are not following each other immediately, however.

91 Prayer for rest "in Your kingdom", see e.g. *I. Qasr Ibrim* 30, with commentary; for being fed from "the Tree of Life", see *I. Qasr Ibrim* 21, with commentary to ll. 23–5.

92 *I. Qasr Ibrim* 38 (Kyriakos) and 39 (an unknown monk); a third one: Meltzer, in Mills, *Cemeteries*, 83–4, no. 5, pl. LXXVI, XCI.3 (a woman Reumata?).

93 E.g. only two real examples can be found among the several hundreds of stelae from Sakinya (Mina, *Inscriptions*, nos. 284 and 316).

94 See note 92 above.

95 Mina, *Inscriptions*, nos. 284 (republished in Pernigotti, "Stele", no. 18) and 316.

96 Stela of a woman Marianta, A.D. 921; *SB Kopt.* I 460 (reading of the owner's name corrected by me).

97 Junker, "Grabsteine", 118, ll. 1–3.

98 Cramer, *Totenklage*, no. 9, ll. 1–3 (*SB Kopt.* I 465).

99 E.g. in the stela of Trosis (Drosis) in the British Museum, quoted below (note 112).

100 See H. Förster, *Wörterbuch der griechischen Wörter in den koptischen dokumentarischen Texten*, Berlin/New York: de Gruyter 2002, 587–88, s.v.; T.S. Richter, *Rechtsemantik und forensische Rhetorik: Untersuchungen zu Wortschatz, Stil und Grammatik der Sprache koptischer Rechtsurkunden*, Leipzig: Wodtke und Stegbauer 2002, 143.

101 *P.KRU* 66, ll. 23–6, with my emphasis (*SB Kopt.* II 953).

102 Cramer, *Totenklage*, no. 14, ll. 5–7 (*SB Kopt.* I 675).

103 Richter, "Die neun Stelen", 302, ll. 10–13.

104 For example in *I. Khartoum Copt.* 23, ll. 3–4 (*SB Kopt.* III 1646; the noun); *I. Khartoum Copt.* 3, ll. 4–5, and 6, ll. 4–5 (the verb); at Qasr Ibrim: *I. Qasr Ibrim* 50, l. 5 (the verb).

105 See Cramer, *Totenklage*, 47–8; further examples: Förster, *Wörterbuch*, 91–2, s.v. ἀποφαίνω/ἀπόφασις; Richter, *Rechtsemantik*, 143.

106 For discussion, see Junker, "Grabsteine", 141; Pernigotti, "Stele", 28–9; *I. Khartoum Copt*, 27, n. 139; J. van der Vliet, "Judas and the stars: philological notes on the newly

JACQUES VAN DER VLIET

published Gospel of Judas (*GosJud*, Codex gnosticus Maghâgha 3)", *JJP* 36 (2006), 137–52 at 138–40.

107 Sahidic, ed. L-T. Lefort, *S. Athanase: lettres festales et pastorales en copte*, Louvain: L. Durbecq 1955, 23, ll. 22–7 (equals p. 63, ll. 6–10); translation: idem, *S. Athanase: lettres festales et pastorales en copte*, Louvain: L. Durbecq 1955, 42; for the transmission of this letter and commentary, see the annotated translation in A. Camplani, *Atanasio di Alessandria: lettre festali; Anonimo: indice delle lettere festali*, Milano: Paoline 2003, 526–37.

108 Cramer, *Totenklage*, no. 5 (*SB Kopt.* I 783).

109 *I. Khartoum Greek* 18, with commentary.

110 For the liturgical parallels, see *I. Khartoum Greek*, 86–7. The term that describes the Son's "ascension onto the wood" in ll. 11–12 is παράβασις, used here in the very rare sense of "passage", chosen deliberately to make it the positive counterpart of Adam's negative παράβασις, "transgression", the common acceptation of the word (in the latter sense it does not occur in this particular epitaph, but it can be presupposed in the given context; see our brief discussion of the term, below).

111 H.D. Schneider, "The lamentation of Eulogia: a Coptic dirge in the Leiden Museum of Antiquities" *Oudheidkundige mededelingen* 50 (1969), 1–7 at 4, ll. 1–4. Further close parallels are found in Cramer, *Totenklage*, nos. 11, ll. 17–19 (Antinoopolis; *SB Kopt.* I 785), 12, ll. 4–7 (Assiut, *SB Kopt.* I 698) and 14, ll. 5–7 (Aswan, *SB Kopt.* I 785).

112 British Museum, EA 1208, ed. H.R. Hall, *Coptic and Greek texts of the Christian period from ostraka, stelae, etc. in the British Museum*, London: British Museum 1905, 6, no. 1, pl. 7.

113 See for a late example: *I. Khartoum Copt.* 23 (*SB Kopt.* III1646), which strictly spoken already belongs to the *pronoia* type (see below).

114 Thus already A. Łajtar, "Two Greek funerary stelae from Polish excavations in Old Dongola", *Archéologie du Nil moyen* 5 (1991), 157–66 at 161, n. 4.

115 The principal studies are: Cramer, *Totenklage*, and Schneider, "Lamentation"; W.A. Ward, "The philosophy of death in Coptic epitaphs", *Journal of Bible and Religion* 24 (1957), 34–40, is disappointing. For the intertextual relations with contemporary legal documents from Thebes, see Richter, *Rechtsemantik*, 148–50 (cf. 143–44). Recent finds at Antinoopolis, about which I was kindly informed by Alain Delattre, show that the epigraphic genre arose already in the end of the seventh century and that, as for language, it does not only occur in Coptic but also in Greek.

116 Cramer, *Totenklage*, no. 2, ll. 3–4 (*SB Kopt.* I780).

117 Ibidem, no. 12, ll. 1–4 (*SB Kopt.* I 698).

118 Ibidem, no. 13, ll. 8–9 (*SB Kopt.* I 786).

119 Two Greek epitaphs from Ginari (Tafa/Taphis, in the Dodekaschoinos) call the deceased, in addition to "blessed (μακάριος)", also "miserable (οἰζυρός) and short-lived (μινυνθάδιος)", C.M. Firth, *The archaeological survey of Nubia: report for 1908–1909*, Cairo: Government press 1912, 45, Grave 37, and 49, Grave 842; for the readings, see *SEG* XLII 654 (1992).

120 Similarly, M. Krause, "Das Weiterleben ägyptischer Vorstellungen und Bräuche im koptischen Totenwesen", in: *Das römisch-byzantinische Ägypten: Akten des internationalen Symposions 26–30. September 1978 in Trier*, Mainz: Phillip von Zabern 1983, 85–92 at 88.

121 Velkovska, "Funeral rites", 43.

122 In my opinion, the fine study by Velkovska tends to overemphasize the irreconcilability of the two points of view, which disappears on the level of the performance and cannot, therefore, be an argument in support of her split-source theory.

123 See e.g. Burmester, *The Egyptian or Coptic Church*, 218; cf. Cramer, *Totenklage*, 41–7.

"WHAT IS MAN?"

124 Traditional laments, see M. Alexiou, *The ritual lament in Greek tradition*, Cambridge: Cambridge University Press 1974; cf. Foley, *Oral poem*, 195–99; funeral liturgy: Cramer, *Totenklage*, 41–7 and 57–72, and Velkovska (quoted above). For intriguing Coptic *Totenklage*-texts on papyrus, see M. Hasitzka, "Koptische Totenklagen", 297–300; L.S.B. MacCoull, "Further notes on P. Vindob. K8355v: a Coptic lament", *ZPE* 112 (1996), 287–88.

125 Uncertain translation.

126 Cramer, *Totenklage*, no. 3, ll. 7–12 (*SB Kopt.* I 781); for a similar invitation in the Bohairic funeral liturgy, see ibidem, 41–2.

127 Cramer, *Totenklage*, no. 7, ll. 9–14 (*SB Kopt.* I 464).

128 Ibidem, no. 6, ll. 13–15 (*SB Kopt.* I 784).

129 Mina, *Inscriptions*, no. 319, dated to A.D. 987.

130 See e.g. *I. Khartoum Copt*, 131–32 (about stelae from Ghazali), and the Qasr Ibrim stela of Patarmoute, quoted below (n. 135).

131 Cramer, *Totenklage*, no. 14 (*SB Kopt.* I 675).

132 *I. Qasr Ibrim* 40, found not on the Qasr Ibrim citadel, but on the mainland; the script suggests a date in the eighth or ninth century.

133 For Job 14 in Egyptian *Totenklage* epitaphs, see Cramer, *Totenklage*, 101; in a late ninth-century Ethiopic epitaph, M. Kropp, " 'Glücklich, wer vom Weib geboren, dessen Tage doch kurzbemessen, . . .!' Die altäthiopische Grabinschrift von Ham, datiert auf den 23. Dezember 873 n. Chr.", *Oriens christianus* 83 (1999), 162–76 at 170–73.

134 The lost stela of a priest Patarmoute, published in J. van der Vliet, "Two Coptic epitaphs from Qasr Ibrim", *JEA* 92 (2006), 217–23 at 219–23 [Study 25]; the epitaph belongs otherwise to the *pronoia* type.

135 Mina, *Inscriptions*, no. 88; the much reduced final prayer is in Greek.

136 This is Togo Mina's translation and renders the text as it stands; its slightly awkward use of masculine grammatical gender in ογⲁ and ⲡⲉ is hardly disturbing in a Nubian text. It looks as if the redactor of the text has expanded the well-known acclamation ογⲁ ⲡⲉ ⲡⲛογⲧⲉ (εἷς θεός), for which see above note 61.

137 *Αγιασματάριον* II, 30; also quoted in Cramer, *Totenklage*, 67–8.

138 See *I. Qasr Ibrim*, commentary to no. 45, for a list.

139 *I. Qasr Ibrim* 20–1, 23, 25 and 41–4.

140 These are *I. Khartoum Copt.* 3, 5 and 19; cf. *I. Qasr Ibrim*, commentary to no. 45.

141 Ed. J. van der Vliet, in G.Th. Martin, *Stelae from Egypt and Nubia in the Fitzwilliam Museum, Cambridge, c. 3000 BC–AD 1150*, Cambridge: Cambridge University Press 2005, no. 115.

142 The stela of Bishop Maththaios of Faras, see A. Łajtar, in *I. Varsovie* 105 (A.D. 766); cf. *I. Khartoum Greek*, 33–5, commentary to no. 5, ll. 2–4; *I. Qasr Ibrim* 20, commentary to ll. 3–4.

143 See the extensive commentary in *I. Khartoum Greek*, 33–5.

144 Nor was it common in other Egyptian contexts; see Förster, *Wörterbuch*, 684, s.v., where all epigraphic examples are from Nubia, apart from a very late Sahidic *dipinto* (of a commemorative nature) in Esna (R-G. Coquin, "Les inscriptions pariétales des monastères d'Esna: Dayr al-Šuhadâ' – Dayr al-Faḫûrî", *BIFAO* 75 (1975), 241–84 at 269–70, no. H, pl. XLVIII; *SB Kopt.* I, no. 342, probably *ca.* twelfth–fourteenth century); cf. *I. Khartoum Greek*, 34–5, for the very rare instances in Greek inscriptions beyond Nubia, among which a single Egyptian example from Philae (*I. Philae* II 216, building inscription, A.D. 577).

145 See Junker, "Grabsteine", 127 and 131; *I. Khartoum Copt*, 26–7; Van der Vliet, in Martin, *Stelae from Egypt*, 169–72.

146 See the commentary at *I. Qasr Ibrim* 20, ll. 3–4.

147 For the epithet παντοκράτωρ, see the study by O. Montevecchi, "Pantokrator", in: *Studi in onore di Aristide Calderini e Roberto Paribeni*, Milan: Ceschina 1957, II, 401–32, especially 418–20; cf. S. Pernigotti, "Una stele funeraria copta", *Oriens antiquus* 10 (1971), 53–6 at 54–5, who exaggerates the rarity of the term in Coptic epigraphy, however.

148 For a further variant, ϩⲙ ⲡⲟⲩⲱϣ ⲙⲡⲛⲟⲩⲧⲉ, "by the will of God", also from Ghazali, see *I. Khartoum Copt.* 72, with commentary.

149 *I. Qasr Ibrim* 53–4; cf. *I. Khartoum Greek* 1, commentary to ll. 1–2.

150 Mina, *Inscriptions*, no. 311, but preferably the more accurate publication by Pernigotti, "Stele", no. 19, with pl. XV, 2, should be used.

151 *WB* II, col. 234; Förster, *Wörterbuch*, 610–11.

152 Ed. Budde, *Basilios-Anaphora*, 148–49, after the Greek (the Bohairic phrases almost identically). In the liturgy, this introductory phrase is merely a prelude to the following evocation of God's solicitude for mankind that culminated in the Incarnation of the Son.

153 Junker, "Grabsteine", 140.

154 The same remark is made about the funeral hymnody of the Byzantine liturgy by Velkovska, "Funeral rites", 43.

155 Ibidem, 44. For several reasons, the designation "prayer of the Euchologion Mega type" for this formulary has been intentionally abandoned both here and in *I. Qasr Ibrim*. Methodologically, it obscures the distinction between the epigraphic formulary and its liturgical model (see my remarks above). Furthermore, it anachronistically claims a late and Byzantine (in the narrow sense) redaction as the model for an ancient prayer that was known far beyond the Greek-speaking world. For corrective remarks, see ibidem, 44; H. Brakmann, "*Defunctus adhuc loquitur*: Gottesdienst und Gebetsliteratur der untergegangenen Kirche in Nubien", *Archiv für Liturgiewissenschaft* 48 (2006), 283–333, especially 300–10.

156 See the complementary sets of stelae published in Jakobielski, *Bishopric of Pachoras* (Coptic), and J. Kubińska, *Inscriptions grecques chrétiennes*, Warsaw: PWN–Éditions scientifiques de Pologne 1974 (Greek).

157 The occurrences are listed in Roquet, "Inscriptions bohaïriques", 164–71.

INDEX

Indices

1   Biblical figures and saints
2   Personal names
3   Professions, offices and titles
4   Place names
5   Language
6   Epigraphic sources
7   Architectural, symbolic and iconographic features
8   Textual genres and features
9   Inscriptions
10 Modern researchers

1 Biblical figures and saints (selection)

Adam 138, 180, 320, 322, 349,
    404–408, 412, 414, 416; Eve
    (Zoe) 180
Apostles 69, 103, 309, 310, 311, 327,
    331–335; Andrew 35, 332; John, the
    Evangelist 31; Paul 30, 35; Peter 30,
    332, 352
Archangel: Gabriel 130, 131,
    134–138, 311, 312; Michael 31, 102,
    113, 130, 131, 134–138, 208, 263, 273,
    342, 349, 350, 360, 402; Raphael 134,
    353
Athanasius, archbishop of Alexandria 229,
    406, 407

Basil, Liturgy (Anaphora) of Saint 303,
    400, 402, 407, 414

Claudius of Antioch 195, 196

David, King 355, 356, 373

Four Living Creatures 87, 312, 313, 349,
    353, 355, 360

Isidore, martyr 19, 227, 332

James the Persian 90, 91
Jesus Christ (Jesus, Christ): 38, 40, 41, 66,
    70, 76, 87, 137, 138, 207, 247, 270, 400,
    406, 407, 408, 416; as an infant 130,
    134, 136, 137; *see* 8. Textual genres and
    features
John Kame 68, 69, 84, 89
John of Lykopolis, monk 193, 196, 197, 198
Joseph, bishop of Sbeht (Isfaht) 374
Joseph, patriarch 370, 372, 373

Mary, the Virgin 50, 87, 102, 130, 131,
    133–138, 247, 309, 310, 311, 328, 331,
    333, 354

Pachomius 36, 226, 230, 271
Phoibammon 30, 227

Severus, patriarch of Antioch 229, 263,
    358
Shenoute, abbot 205, 206, 207, 209, 215,
    218, 226

INDEX

Three Patriarchs 32, 89, 154; "Abraham, Isaac and Jacob" *see* 8. Textual genres and features

2 Personal names

Egypt:

Aaron, priest 88
Abraham, bishop of Aswan and Elephantine 261
Abu'l-Azz, amir 260
al-Mabrat 236, 237, 239, 240
Anatole, deacon, father of the painter Merkouri 239
Anni, (fem.) deceased 111, 112
Athanasius, bishop of Qus 240

Basile, (priest) 198, 199

Caesarius, Count 206, 207, 218
Chael, Abba, cleric 36
Chael, deacon, father of the deacon Severus 116
Chael, Lord 162, 163
Christodoros, deceased 173
Claudius (Klaute), priest 208
Constantine, bishop of Asyut 193, 196, 374

Damian, pastry baker 113, 114, 115, 123
Daniel, bishop of Philae 258
David, sponsor, from Hermonthis 263

Elias, administrator 136
Eulogia, (fem.) deceased 408

Gabriel (Kabri), archpriest 68, 103
Gabriel, deacon 87
George (Giorgios), deacon 37
George, deacon and steward 76
Gerontios, priest 151, 153, 154
Gregory, deceased 219, 220, 395

Hadra (Hatre), bishop of Aswan 243
Heliodoros, donor 207, 216, 217, 218

Ioannes, archimandrite 208, 209
Iokim, father of Lord Chael 163

James, bishop of Atfih 104
Joseph (Josephios) [I], bishop of Aswan 374

Victor, the general (son of Romanus) 194, 195, 196, 198

Joseph [II], former bishop of Aswan 259, 271, 365, 366, 370–375, 391, 398
Joseph III, bishop of Aswan 303
Joseph, (priest) from Talmarage 198
Joseph, deacon 69
Joseph, deceased 187

Kallirhoe, wife of Heliodoros, donor 207, 215, 216, 218
Kolthi, priest 37, 105, 174, 175, 176
Kosmas, deceased, from Qau 407, 408

Macarius (Makari), deacon 173
Mark III, archbishop of Alexandria 262
Marnitas, prefect 153
Mena Panau, ascetic 87
Menas, deceased 151, 153, 154
Merkouri, painter, monk 221n19, 239, 240
Moses of Nisibis, abbot 88, 91
Moses, bishop of Wasim (Letopolis) 77
Moses, priest, *hegoumenos* 88

Ôl, Apa, *meizoteros* 115

Paese, priest 68, 69
Pakire, deacon 198, 199
Pamin, monk 210
Paniskos, "the great archimandrite" 209
Papas, deceased 154, 155
Papas, son of Mercury 102
Papnoute, archimandrite 103
Paul of Tamma 193
Pesynthios, bishop of Koptos 228, 372, 374
Peter (Petro), monk, deceased 245–248, 262
Peter, deacon 29
Peter, *lashane* of Jeme 227
Phib, donor 123, 124, *131*, *132*, 133, 136, 138
Phibamon, priest, monk, sponsor 208
Phoibammon, donor 123, 124, *131*, *132*, 133, 136, 138

INDEX

Phoibammon, physician 157, 158
Pousi, bishop of Philae 188

Raphael, priest 33, 35

Samuel, son of Mark 32
Sarah, daughter of Paniskos 182
Sarah, daughter of Ysidoros 181
Semne, (fem.) deceased 398
Severus, bishop (Fayoum) 103
Severus, bishop of Aswan 263
Severus, deacon 115, 116, 117
Solomon, artist 86
Sousanne, (fem.) testator 405

Nubia:

Aaron, bishop of Faras 349, 413
Abraham, priest 257
Acutus, from Rome 384, 385
Angelosko, deceased 354, 360, 361
Aroumi, (fem.) *ara* from Taphis 279, 280,
  282

Basil, king 221

Chael, king 289

David, king 355, 356
Doulista a.k.a. Papsine, mother of a Faras
  bishop 342, 343, 344, 357
Drosis, (fem.) deceased 283, 284, 408,
  410

Edra, (fem.) deceased 298, 299
Eikkir, (fem.) deceased 358, 359
Eiñitta, owner of the Jesus-Church, Qasr
  Ibrim 270, 353
Eirpanome, Nobadian king 257, 270, 271,
  391
Eisousiko, (fem.) deceased 406
Eli, owner of a church at Faras 356

George (Georgios), archbishop of Dongola
  372
George (Georgios, Georgiou), bishop of
  Faras (A.D. 1097) 273, 349, 354, 391,
  413
George [II], king (A.D. 969) 247, 248
George [IV], king (A.D. 1157) 84
Goassi, son of Sentiko(l), eparch of
  Nobadia and Choiakeikšil 348, 362n9

Stephanos, Apa, deacon, architect 209
Stephanos, spiritual son of John Kame 68

Takathon, wife of the deacon Severus 116
Theodore, Abba, *proedros* 76, 77
Theodore, Armenian painter 83, 207
Theodore, bishop of Aswan 262
Theodore, bishop of Misr (Old Cairo) 77
Theodore, bishop of Philae 256, 257, 258,
  259, 391
Theodosius, archbishop of Alexandria 66

Zacharias, archbishop of Alexandria 68,
  103

Iesou, bishop of Saï 221
Iesou, eparch 329, 330, 331, 332, 333, 334
Iesousinkouda, eparch of Nobadia 301,
  357
Iesousyko, (fem.), church owner 348, 349,
  350, 351, 360, 413
Ioannes (Ioannou), bishop of Faras 272
Ioannes, priest of (the Church of St.)
  Gabriel 312
Iohannou, priest 348, 350, 360

Joseph, exarch of Talmis 257

Kartolaos, courtier 263, 264
Kolothos, deacon 302, 303
Kudanbes (Koudanpes) 263, 264
Kyriakos (Kerikos), bishop of Sai 413, 414
Kyriakos, deceased, from Qasr Ibrim 405
Kyros, bishop of Faras 334

Manna, (fem.) from Kalabsha 280, 281,
  282, 283, 284
Maraña George, the Builder 355
Mariakyto, deceased 413
Mariankouda, tetrarch of Makuria 407,
  408, 416
Marianos (Marianou), bishop of Faras,
  archimandrite of Pouko 349, 352, 353
Marianos (Marianou), eparch 304
Marianos (Marianou), priest of (the
  Church of) the Four Living Creatures
  312, 353
Marianta, (fem.) deceased 398, 403, 404
Martha, deceased 317, 318, 323, 414
Maššuda, buys land 353

429

INDEX

Mena, priest 320, 321
Merke, priest of the Cross-Church of Faras 360
Merkourios, king of Makuria 260, 279, 374, 375
Michaeliko (Michaelikol), (fem.) owner of (the Church of St) Michael, Argine 348, 349, 350, 351, 359, 360
Moses George, king of Dowato 270, 273, 355, 356

Nonnen, daughter of Mena 353

Patarmoute, priest of the Catholic Church, Ibrim 319, 320, 321, 322, 323
Paul (Paulos), bishop of Faras 77, 327, 328, 329, 330, 331, 332, 333, 335, 371
Paul (Paulos), *meizoteros*, from Kalabsha 284, *285*, 286, 287, 303, 304
Paul, priest 256
Persi, daughter of Poñitta 353
Peter (Petrou), deacon, son of the bishop of Kourte 272, 349
Peter, deceased, from Qasr Ibrim 398, 402
Peter, priest of the Christ-Church of Pachoras 352
Petronia, (fem.) deceased 349
Petros [I], bishop of Faras 271
Philotheis, deacon, son of [M]arianta 353

Poñitta, sells land 353
Protakia, (fem.) deceased 296, 297

Reumata, (fem.) deceased, from Qasr Ibrim 405

Severus, priest 299, 300, 301
Staurophoros, Nobadian official 272, 349, 358
Stephanos, deceased, from Qasr Ibrim 319
Symeon, Abba, bishop of Saï 221

Tantani, Nobadian chieftain 269, 270, 271, 273, 274, 390, 397
Taria, (fem.) deceased 297
Theognosta, (fem.) deceased 403, 404
Theophilus, deacon 332
Thomas, bishop of Faras 304, 339n61, 376n10
Timotheos, bishop of Faras and Ibrim 240; "scrolls of" 275n5, 309, 310, 338n42
Tirsakouni 270
Titta, family of bishops of Faras 342, 343, 344, 357
Tôkiltôeton, Nobadian king 288, 289

Zacharias (III, Zachari), king (A.D. 962) 243–248, 262, 263

3 Professions, offices and titles

administrator (*actuarius*) of the *kastron* of Philae 261
ara 279, 280
archbishop, of Alexandria 31, 66, 68, 69, 103, 129, 262; of Dongola 372
archimandrite 68, 103, 208, 209, 226, 349, 353
archpriest 68, 103

bishop 31, 32, 34, 63, 65, 68, 69, 76, 77, 78, 103, 104, 151, 188, 193, 199, 220, 221, 240, 243, 256, 257, 258, 261, 262, 263, 271, 273, 286, 304, 309, 310, 323, 327–335, 341–344,, 349, 351–357, 365, 391, 392, 412, 413, 414, 417; former bishop 259, 271, 365, 370–375, 390, 391, 392, 398, 402

deacon 29, 37, 68, 69, 76, 77, 87, 103,115, 116, 117, 158, 163, 173, 198, 209, 226,

239, 272, 280, 302, 332, 349, 351, 353, 400, 410
domestikos (*domesticus*) of Faras 301, 357; *protodomestikos* 67

envoy to Babylon 349, 353
eparch, of Nobadia 260, 287, 301, 304, 329, 330, 332, 333, 334, 390; of Nobadia and Choiakeikšil 348
exarch, of Talmis 257, 283

"having the church of so-and-so" (ἔχων, ⲡⲉⲧⲟⲩⲁⲛⲧⲧⲥ̄, ⲕⲟⲛ) 347, 350–360
hegoumenos 88, 104

King, Nubian: of Dotawo (Nobadia, Alwa and Makuria) 355; Meroitic 255 257; Nobadian 256, 257, 288, 289, 375, 391; Makurian 221, 243, 246, 247, 248, 260–264, 270, 273, 279, 289, 351, 355, 374

INDEX

Lady Queen (Kandake of Meroe) 384, 385
Lord 65, 162, 163

meizoteros 115, 280, 284, *285*, 286, 287, 303; *protomeizoteros* 67
metropolitan (*mutranus*) of Misr 77

nauarches of the Nobadae 357, 358
naukrates of the Seven Lands 358

pastry baker 113, 114
physician 157, 158

priest 33, 35, 37, 65, 68, 69, 88, 103, 104, 105, 153, 175, 208, 256, 257, 263, 280, 289, 299, 300, 301, 309, 312, 319, 320, 321, 322, 323, 348, 350, 351, 352, 353, 359, 360, 400; *papa* 37, 88, 176, 198; pagan priest 217, 255
proedros 76, 77

steward (*oikonomos*) 76, 77, 88, 136

tetrarch of Makuria 407, 416

4 Place names

Egypt:

Abydos 19, 283, 322
Akhmim (Panopolis) 29, 34, 188, 203, 204, 205, 207, 216, 218, 248; Panopolite (district) 18, 203, 215
Alexandria 15, 78, 135, 255, 259; Gabbari 15, 70
Antinoe (Antinoopolis) 6, 17–18, 127, 166, 186, 188, 189, 190, 195, 219, 408, 409, 411; Dayr Abu Hennis 18, 127, 128
Aswan (Syene) 19–20, 188, 244, 246, 253, 254, 259, 260, 262, 263, 264, 284, 303, 304, 318, 365, 371, 373, 374, 399, 403, 405, 411, 413, 415; Dayr al-Shaykha (al-Qubbaniya) 134; Elephantine 19, 253, 255, 261, 374; Monastery of Anba Hadra (St. Hatre; Monastery of St. Simeon) 5, 6, 84, 239, 243, 245, 246, 247, 261, 262, 263, 264, 397, 398; Qubbat al-Hawa (Church of St. Severus) 262, 263
Asyut (Lykopolis) 193, 194, 195, 196, 198, 199; Dayr al-Gabrawi 193, 194, 195, 198; Dayr al-'Izam (Monastery of Apa John of the Desert) 44, 193, 194, 196, 197, 198

Bawit 5, 6, 134, 135, 163, 167

Cairo 36, 37, 165, 167, 221, 248, 263, 264, 309, 310; Babylon 65, 77, 349, 353; Old Cairo (Misr, Fustat) 16, 75, 77, 78, 135, 166, 320; Church of the Virgin Mary, al-Mu'allaqa 75

Edfu 15, 230, 271
Esna 5, 7, 180, 225, 226, 230, 231, 235, 240, 271, 286; hermitages 225, 230;

Dayr al-Shuhada (Monastery of the Martyrs) 231; Dayr al-Fakhuri (Monastery of the Potter) 231, 237, 328

Fayoum 4, 5, 8, 9, 13, 16, 17, 33, 37, 44, 45, 47, 67, 68, 88, 99, 100, 101, 104, 105, 112, 113, 114, 116, 117, 123, 133, 135, 151, 153, 154, 155, 157, 158, 161, 163–167, 173–176, 231, 274; Abu Hamid (Hamed), Church of the Virgin 101, 102, 102, 117, 164; al-Nazlah, Church of St. Menas 101; Bulgusuq (Peljisok), Church of the Virgin 8, 101, 104, 105, 117; Dayr Abu Lifa 102; Dayr al-'Azab (Monastery of Anba Abraam) 151, *152, 155, 156, 157*, 158; Dayr al-Hammam 173; Great Church (cathedral of Arsinoe) 153; Hamuli 175; Medinet al-Fayoum (Arsinoe, Fayoum town) 123, 151, 153; Tutun (Tebtynis) 17, 32, 33, 47, 68, 101, 102, 104, 105, 117, 135, 164, 167, 175; *see* Naqlun, Sinnuris

Hermonthis 181, 182, 205, 230, 263

Kellia 4, 7, 15, 64, 66, 64, 84, 89, 228

Luxor (Thebes) 226, 282; Karnak 226

Middle Egypt 5, 17, 68, 112, 137, 165, 180, 186, 188, 189, 190, 197, 219, 230, 261, 274, 284, 286, 322, 395, 398, 409, 411, 416; Pemje (Oxyrhynchus) 29, 261; Spania 229

431

INDEX

Monastery of St. Antony, Red Sea 32, 83, 135, 141, 231

Naqada 235, 240; Dayr Mari Girgis 235, 236; Dayr Mari Buqtur 235, 240
Naqlun 16, 32, 33, 38, 43, 102, 103, 105, 157, 173, 174, 176; Monastery of Naqlun (Dayr al-Malak, Monastery of the Archangel Gabriel, monastery church) 33, 37, 43, 68, 69, 88, 102, 104, 113, 117, 135, 138, 151, 158, 164, 167, 173, 174, 175, 231

Philae 7, 19, 20, 21, 77, 195, 255, 257, 258, 259, 260, 295; castle of 260, 261

Saqqara 6, 17, 43, 135, 179, 286; Monastery of Apa Jeremiah 5, 67, 84, 180
Sinai, Monastery of St. Catherine 126
Sinnuris 16, 113, 123, 139: church of St. Michael (church of Sinnuris) 113, 123,

124, *124, 125*, 126, 127, 128, 129, 130, 136, 139; Dayr Sinnuris *114*, 115, 123; South Church 126, 127, 141
Sohag 18, 19; White Monastery 196, 203, 204, 205, 207, 209, 210, 215, 216, 218, 226, 239, 270; Red Monastery 203, 239

Upper Egypt, Church of St. Abraham the Anchorite 65, 67

Wadi al-Natrun (Scetis) 6, 9, 15, 68, 69, 83, 84, 85, 86, 88, 89, 90, 91, 92, 128, 134; Dayr Abu Maqar (Monastery of St. Macarius), 68, 85, 86, 88, 90; Dayr al-Suryan, Church of the Virgin 4, 15–16, 79, 84, 86, 87, 88, 89, 90, 91, 167, 207, 228
Western Thebes 19, 65, 158, 227, 229, 230; Jeme (Medinet Habu) 227; Monastery of Epiphanius 228, 229, 231; Monastery of St. Phoibammon 90, 227, 228, 231

Nubia:

al-Koro 391, 412
Alwa (Alodia), kingdom of 269, 355, 390
Argini (Argine), Church of Michael 350, 359, 360
Arminna 311, 391, 397, 405
Ashkeit 352, 358, 359

Banganarti 22–23, 274

Dakka 282, 385
Debeira (West, Komangana) 272, 304, 349; Church of Severus 358
Dendur, temple of 256, 257, 270, 273, 390, 391
Dodekaschoinos 66, 254, 255, 279, 282, 287, 288, 289, 295, 297, 304, 322, 391
Dongola (Old Dongola), capital of Makuria 22, 259, 263, 264, 269, 271, 273, 274, 322, 347, 351, 360, 361, 365, 371, 373, 374, 375, 376, 390, 391, 392; overview of churches 375

Faras (Pachoras), capital of Nobadia 7, 21, 77, 240, 247, 260, 269, 271, 273, 279, 280, 286, 304, 309, 310, 327, 329, 332, 341, 342, 343, 347, 349, 353, 357, 371, 374, 391, 402, 405, 412, 413, 417;

"Bishop's Palace" 328, 330, 331, 333, 335; Cathedral (Church of the Virgin Mary) 310, 312, 327, 329, 330, 331, 333, 334, 335, 353, 356; Christ Church 352; (Church of St.) Michael 342, 357; (Church of) the Four Living Creatures 312, 313; "Church on the Mastabafield" 333; "Church on the south slope of the Kom" (Church of the Twelve Apostles, former cathedral) 69, 327, 330, 332, 333, 334, 335; Cross Church 352, 360; Rivergate Church 356
Furgundi (Purgundi) 353, 361

Ghazali 23, 269, 270, 271, 273, 286, 304, 381, 382, 383, 391, 413

Ikhmindi (Mehendi) 287, 288, 289

Kalabsha (Talmis) 21, 254–257, 279, 280–287, 295–304
Komati 354, 360

Maharraqa (Hiera Sykaminos) 282, 287, 295, 301
Makuria, kingdom of 269, 273, 286, 375, 376, 390

INDEX

Meinarti 22, 273, 304, 347–351, 360, 391, 402, 413; "Island of Michael" 348
Musawwarat al-Sufra, Meroitic temple site 382, 383, 384, 385, 386
Mushu (Mesho) 269, 391

Nobadia, kingdom of 7, 257, 259, 260, 269, 271, 279, 286, 304, 317, 356, 375, 389, 390, 391, 402, 411

Pouko 349, 353

Qasr Ibrim (Phrim, Ibrim) 31, 240, 260, 262, 263, 269, 270, 273, 274, 279, 282, 287, 304, 309, 310, 311, 317–321, 323, 344, 349, 352, 353, 358, 389, 390, 391, 392, 397, 398, 399, 401, 403–406, 411–416; Cathedral (Church of the apostles, Catholic Church) 309, 310, 311, 317, 319, 320, 322, 323, 390;

Church of the Archangel Gabriel 312, 357, 358; Church of the Virgin Mary 309, 310; Jesus-Church 353, 358; bishop's tombs 317, 319, 323

Saï 22, 220, 273, 295, 391, 413, 414
Sakinya (Toshka West) 6, 270, 286, 297, 303, 309, 312, 319, 321, 322, 349, 391, 397, 399, 405, 411, 412, 414; Church of St. Gabriel 311, 312
Sonqi Tino 246, 247, 312, 353, 360

Tafa (Taphis) 21, 254, 255, 260, 279, 283, 295, 296, 297, 299, 303; Tafa-Ginari 279, 280, 282, 297, 298, 300, 301
Tamit 361; Church of the Archangel Raphael 353; Church of the Angels 354
Teme, Church of the Four Living Creatures on the island of 312, 349–350, 353
[. . .]touri 354, 355

5 Language

Arabic 4, 5, 13, 27, 28, 32–36, 41, 83, 85, 88, 90, 99, 101, 103, 104, 105, 118, 124, 126, 137, 139, 167, 173, 209, 237, 245, 248, 254, 260, 261, 269, 270, 274, 310, 311, 348, 391; in Coptic characters 208; Kufic 33, 104; Kufesque (pseudo-Arabic) 33, 45, 104
Armenian 83, 85, 207

bilingual(ism) 4, 8, 22, 33, 34, 35, 84, 179, 197, 207, 238, 239, 259, 263, 321, 323, 365, 371, 376, 417

Coptic: 3–8, 32, 33, 34, 83, 85, 88, 90, 99, 100, 101, 103, 104, 105, 111, 112, 123, 137, 139, 173, 174, 175, 179, 180, 185, 197, 205, 207, 209, 219, 227–230, 243, 254, 261, 262, 263, 269, 270–273, 286, 303, 311, 312, 317, 323, 327, 329, 331, 350, 351, 353, 360, 365, 370, 371, 381, 389, 390, 391, 393, 398, 413, 414, 415; Bohairic 4, 7, 15, 16, 20, 46, 64, 66, 68, 69, 84, 86–90, 139, 161, 173, 228, 270, 274, 300, 310, 391, 401, 402, 409; Bohairicism 237; Copto-Greek, Greek-Coptic 180, 185; Fayoumic proper 4, 33, 100, 101, 105, 112, 115, 116, 164; Fayoumi-Sahidic 16, 17, 33, 37, 42–47, 48, 68, 69, 100, 101, 104, 115, 161,

164, 174; Middle Egyptian 46, 173; Sahidic 4, 14, 18–22, 33–38, 42, 43, 44, 46, 65–69, 90, 100, 105, 116, 137, 164, 174, 185, 190, 195, 196, 198, 220, 226, 229, 236, 237, 238, 244, 245, 254, 256–259, 269, 270, 271, 273, 274, 284, 286, 302, 303, 304, 317, 318, 321, 322, 350, 367, 372, 376, 390, 391, 400

Demotic 78, 257, 272

Ge'ez (Ethiopic) 83, 137
Greek 3, 4, 5, 8, 13–23, 27, 28, 32, 33, 34, 36, 37, 41–44, 64, 75, 78, 83, 84, 88, 91, 97, 99, 100, 101, 104, 105, 112, 113, 114, 123, 131, 136, 151, 154, 156, 157, 173, 179, 189, 190, 197, 203, 205, 206, 209, 215, 220, 221, 227, 229, 230, 238, 239, 247, 254, 255, 257–261, 263, 269–274, 279, 280, 281, 284, 286–289, 296, 298, 299, 301–304, 311, 312, 317, 318, 320–323, 327, 328, 334, 335, 341, 342, 347–350, 356, 359, 360, 367, 370, 371, 372, 376, 381, 382, 390–394, 398, 399, 411, 413, 414, 415; mixture of Greek and Old Nubian 351–355, 357, 361

hieroglyphic 4, 83, 254

INDEX

Latin 3, 51, 83, 194, 195, 196, 257, 263, 381, 382, 384, 385

Meroitic 254, 255, 256, 257, 390
multilingual(ism) 4, 8, 229, 269, 288

Old Nubian 20, 22, 83, 84, 221, 269, 274, 288, 311, 342, 343, 344, 349–355, 357, 358, 359, 361, 382, 393

Syriac 4, 15, 16, 83, 85, 86, 88, 90, 97, 98, 229

6 Epigraphic sources

altar screen 124, 139, *140*, 141, *141*, 365
amphora 20, 22, 196, 198

building or foundation inscription 77, 103, 206, 327, 328, 329, 331, 371; dedicatory inscription 4, 69, 101, 102, 117, 131, 162, 206, 261; founder's memento 85, 206

chalice 102, 104

dipinto (plur. *dipinti*) 4, 66, 85, 89, 90, 103, 206, 208, 209, 228, 230, 237, 239, 243, 244, 247, 248, 261, 262, 349, 353, 354, 356

flabellum 86, 97
funerary inscription (epitaph): 4, 67, 78, 99, 101, 112, 113, 151, 259, 261, 271, 272, 273, 283, 297, 300, 303, 305, 311, 312, 321, 333, 354, 357–360, 376, 389, 390, 391, 392, 396, 411; stela (tombstone) 4, 6, 84, 99, 100, 101, 111, 112, 114, 115, 123, 151, 154, 155, 157, 166, 173, 179, 180, 185, 190, 203, 204, 205, 210, 219, 220, 230, 243, 244, 245, 250, 261, 270–273, 279–284, 286, 287, 295, 296, 298–302, 304, 305, 310, 312, 317, 318, 341, 347, 348, 349, 353, 356, 365, 366, 371, 372, 381, 383, 392, 397, 400, 407,

411; mummy-label 17, 100, 158; on a wooden tablet 100, 151, 157, 158

graffito (plur. graffiti) 4, 85, 90, 102, 103, 243, 261, 349, 352, 353, 354, 359, 360, 361, 384, 385

icon 85, 104, 135, 136, 137, 138, 139

key, ceremonial 209, 210
keystone 328, 329

legend 4, 13, 16, 18, 20, 28, 30, 85, 86, 87, 89, 90, 91, 96, 97, 98, 102, 104, 132, 134, 135, 207, 231, 248, 355
lintel: 101, 102, 115, 116, 117; Caesarius lintel 206, 207, 218; Mu'allaqa lintel 75–81

painter's memento 85, 86, 87, 103

reliquary 91

tirâz (texts on textiles): in general 5, 27, 31–42, 104, 166, 167; hood 43; shawl 38, 39, 42, 43, 44–49; shroud 36, 175; tunic 33, 42, 43, 174, 175

votive (offering, *ex-voto*) 124, 133, 136, 138, 207, 218, 301

7 Architectural, decorative and iconographic features

acroteria 111, 115
aedicula 111, 113, 115, 157, 173, 348, 392, 413
Alpha, Omega 162, 165, 167, 205

cross: 23, 32, 34, 45, 48, 86, 91, 100, 101, 104, 112, 113, 134, 139, 155, 161, 166, 173, 205, 209, 281, 296, 298, 299, 301, 302, 305, 317, 318, 327, 334, 366, 367; *ankh*-shaped 29, 204; *croix pattée* 284, 286; Greek cross 111, 114, 165, 244;

Latin cross 207, 215; Maltese cross 111, 154; sculptured (in raised relief) 116, 157, 161, 185, 207, 217, 218, 329, 371

griffin 165, 166, 167

haykal 91, 113, 126, 127, 130, 139, 354

khurus 87, 88, 91, 92, 126, 127, 129, 130, 136, 139

INDEX

laurel wreath 154, 155

medallion 138, 165, 166, 167

palmette 165, 167
pastophorium 127, 130, 208

8 Textual genres and features

"appel aux vivants" 395, 396; "do not grieve"
(var. "be sorrowful") 205, 395, 396, 411

biblical quotes and reminiscences: Genesis
(Gen. 3:19) 322, 349, 405, 406, 407,
411, 412, 414, 416; (Gen. 37) 374; (Gen.
39) 374; Numbers (Num. 6) 35; (Num.
16:22) 230; (Num. 27:16) 230; Job 14
(14:1) 322, 323, 411; Psalms, in general
27, 34, 37, 38, 39, 40, 41, 42; (Ps. 2)
38, 42, 44; (Ps. 8) 14, 34, 37, 42; (Ps.
17:33–34) 38, 42, 46; (Ps. 19:2–4) 34;
(Ps. 22) 42, 43; (Ps. 26:9b) 38, 43; (Ps.
32:1) 42; (Ps. 33:20) 373; (Ps. 44:9–11)
37, 38, 39, 40, 41, 43, 51, 55n106; (Ps.
46:2–3) 38, 43, 105; (Ps. 71) 34, 37, 38,
43, 44; (Ps. 90) 39; (Ps. 109) 37–38, 44,
51, 55n97; (Ps. 120:7) 51; (Ps. 133) 39,
44, 49; Sir. 42:16 382; Matthew (Mt.
16:18) 332; (Mt. 25:21/23, 25:34) 287,
304; (1Cor. 15:20–21) 407; (1Peter 2:9)
41; Revelations (Rev. 3:5) 400; (Rev. 4)
312; (Rev. 14:13) 400

closing formula: "Amen" 297, 299, 300,
302, 318, 320, 371, 398, 404, 414; "So
be it" 162, 398, 404, 414

deceased: "blessed" 282, 286, 318, 320,
322, 360, 370, 397, 398, 399, 400, 403,
411; "the body of" 322; age 205, 349,
350, 357, 359; "the mortal dwelling"
322; "the mortal remains of" 322, 370;
"saintly" (ὁ ἐν ἁγίοις) 368–369, 370,
371, 373; "wretched" 409
dating formula: "Christ being king" 41, 66,
67, 68, 69–71; Diocletian (year) 64, 68,
75, 76, 77, 78, 100, 115, 116, 133, 162,
190, 219, 220, 245, 286, 327, 329, 330,
348, 368, 371, 411; "during the feast of"
353; Hijra ("Saracene" era) 64, 67, 78;
indiction (year) 64, 65, 66, 67, 76, 111,
114, 153, 154, 157, 187–190, 205, 219,

polychrome (polychromy) 131, 161, 209,
392

tabula (*ansata*) 38,155, 157, 158, 209

220, 281, 288, 289, 296, 297, 298, 300,
371, 401; Martyrs, Era of the (A.M.) 64,
78, 199, 219, 220, 239, 262, 341, 349,
359; Moon (day) 236, 237, 262; "written
in" 302; "year 1 of King Zachari" 245,
262
death formula: "to come to perfection"
(ἐτελεώθη) 279, 281, 282, 283, 298,
299, 370; "to end his/her life" (τέλει
τοῦ βίου) 205, 282, 300; "to fall asleep"
(ἐκοιμήθη) 114, 153, 279, 282, 296,
297; "for the falling asleep" 312; "to lay
down the body" 111, 112, 205; "to reach
perfection" 239, 370, 371; "to go to
rest" (ⲙⲧⲟⲛ ⲙⲙⲟ=, ἀναπαύομαι) 66, 154,
187, 189, 205, 220, 245, 302, 318, 320,
398, 400, 403, 405, 409, 411, 412, 414

invocation: Jesus, or Jesus Christ 34, 35,
64, 102, 116, 153, 162, 174, 138, 239,
299, 300, 301, 320, 367, 370, 398,
402; "Christ our God" 64, 70, 138;
"the Father and the Son and the Holy
Spirit, one single Godhead" 199; "God
of (Saint) so-and-so. . . " 90, 91, 332;
"God of the Apostles" 332; "God of
the spirits (and Lord of all flesh)" 100,
230, 286, 300, 341, 396, 416, 417; God
of (our fathers) Abraham and Isaac and
Jacob 153, 283, 403, 404; "Hail, temple
of God!" 139, 141; "Holy Trinity" 137,
180, 188, 190, 297, 298; "one is God,
the helper" 181, 205, 398; Savior 32,
35, 64; "treasury of all mercy" 302, 304;
"True God" 35, 111, 112, 116

opening formula: "In the name of the
Father, the Son, and the Holy Spirit"
187; "In the name of God" 34, 174;
"In the name of God, the merciful and
compassionate" (*bismillah*) 34, 398,
403, 404; "Here lie the mortal remains
of" 320, 322, 370, 371; "Stela of the

435

INDEX

deceased N.N." 205; "This good work was accomplished (under/by)" 328, 329

pax formula: "in peace" 153, 154; "in the peace of God" 318, 398
prayer: "bless" 35, 116; "have mercy upon (the soul of)" 36, 112, 187, 188, 205, 245, 286, 320; "grant rest" (to the soul of)/ "rest the soul of" 114, 153, 157, 220, 221, 286, 297, 298, 299, 300, 303, 320, 342, 349, 398, 402, 404, 405, 411; "guard" 35; "help" 35, 138, 163, 174; "in a place of rest" 284; "in a place of verdure" 302, 303; "in the bosom of (our holy fathers) Abraham, Isaac and Jacob" 100, 281, 283, 286, 297, 299, 300, 302, 303, 342, 400, 402, 404, 414; "in the Kingdom of Heavens" (var. "Your Kingdom")

36, 398, 404; "protect" 35, 116, 261; God's blessing (*baraka*) 40, 85, 90; "with all His saints" 302, 303; "you are the rest of your servant" 358, 359

type of epitaph: commemoration type 318, 397–403, 405, 410; ἔνθα κατάκειται-type ("here lies. . . ") 279–284, 296–299, 322; Euchologion Mega type 352, 356, 357, 358, 360; litany type 137, 179, 230; *pronoia* type ("through the providence of God") 302, 303, 320, 321, 349, 412, 414, 415, 416, 417; Totenklage type 6, 219, 371, 406, 412–417; verdict type (*apophasis*) 404, 405, 406, 407, 408, 411–416

wish for prosperity: "flourish!" 30, 35

9 Inscriptions (selection)

CIL III 22 194
CIL III 83 381, 384

I. Khartoum Copt. 1 69–70, 79n18, 251n39, 260, 331, 333, 335n3, 337n26, 371, 379n33
I. Khartoum Copt. 2 250n18, n30, 251n39, 329, 332, 337n27, n30, 345n12
I. Khartoum Copt. 4 402, 324n18, 402
I. Khartoum Copt. 5 273, 301, 391, 425n140
I. Khartoum Copt. 12 298
I. Khartoum Copt. 17 272, 273, 304, 363n24, 423
I. Khartoum Copt. 18 324n23, 360, 362n13, 416, 423n85, 424n109
I. Khartoum Copt. 19 275n14, 324n23, 360, 362n15, 363n26, 423n85, 425n140
I. Khartoum Copt. 123 412
I. Khartoum Greek 5 357, 425n142
I. Khartoum Greek 7 358, 359, 362n9
I. Khartoum Greek 9 298
I. Khartoum Greek 18 416
I. Khartoum Greek 14 22, 275n19
I. Qasr Ibrim 20 297, 419n22, 425n146
I. Qasr Ibrim 22 316n51, 352
I. Qasr Ibrim 29 298, 317, 403, 420n53, 421n56
I. Qasr Ibrim 31 398
I. Qasr Ibrim 37 398, 421n61, 423n86

I. Qasr Ibrim 38 405, 423n92
I. Qasr Ibrim 39 405
I. Qasr Ibrim 45 413, 419n22, 425n138, n140
I. Qasr Ibrim 47 412
I. Varsovie 101 251n39, 260, 329, 334, 335n4
I. Varsovie 118 179
I. Varsovie, appendix I, A8 180

SB I 1597 215
SB III 6250 78
SB III 6089 280, 281, 290n16
SB III 6311 218
SB IV 7429 360
SB V 8737 280
SB V 8728 341
SB V 7944 385
SB VIII 10074 287, 294n67
SB X 10516 300
SB Kopt. I 44 69
SB Kopt. I 299 198
SB Kopt. I 302 20, 21, 168n15
SB Kopt. I 348 238
SB Kopt. I 432 284, *285*, 292n42
SB Kopt. I 498–675 249n4, 261, 303, 421n54
SB Kopt. I 719 304, 399n61, 376n10
SB Kopt. III 1538 227
SB Kopt. III 1596 191n5, 219–220, 222n18
SB Kopt. III 1645 304

INDEX

10 Modern researchers (selection)

Bagnall, Roger S. 78, 219
Bernand, Étienne 7, 217
Bilabel, F. 280, 281, 341
Bingen, Jean 3, 288, 289
Boud'hors, Anne xxi, 100, 101

Calament, Florence xxi, 100, 101
Coquin, René-Georges 7, 216, 217, 226,
230, 237, 238, 239
Cramer, Maria 6, 116, 381, 383, 410, 417
Criscuolo, Lucia 203, 217
Crum, Walter E. 173, 194, 196, 198, 206,
208, 228, 244, 245, 319, 320

Dekker, Renate xv, xx, 263
Derda, Tomasz xvi, xx–xxi, 124, 130, 139
de Ricci, Seymour 180, 181, 182, 190
Dijkstra, Jitse H.F. xvi, xx, 258
Donadoni, Sergio 288, 289, 350, 353, 360
Durand, Maximilien 30, 174, 175

Evelyn White, Hugh G. 6, 84, 91, 92

Firth, Cecil M. 279, 280, 282, 298, 300
Fluck (Wietheger), Cäcilia xxi, 7, 44, 49,
204

Griffith, Francis L. 221, 280, 343, 352
Grossmann, Peter xvi, xxi, 124, 130, 136,
207, 218

Innemée, Karel xxi, 88, 90, 91, 203

Jakobielski, Stefan xvi, xxi, 7, 327–331,
352, 354, 355, 357, 365

Junker, Hermann 6, 282, 283, 297, 300,
394, 415

Khurshid, Fathy 113, 123
Krause, Martin 3, 284, 318, 394, 397, 415

Łajtar, Adam xvi, xx, xxi, 8, 173,
179–182, 220, 227, 261, 279, 282, 288,
299, 317, 342, 354, 359, 371
Lefebvre, Gustave 5, 6, 8, 163, 190, 216,
217, 218, 280, 281, 282, 341

MacCoull, Leslie S.B. 75, 76, 77, 78, 219
Mina, Togo 6, 311, 312
Monneret de Villard, Ugo 216, 243, 247,
321, 322, 348
Munier, Henri 6, 226, 243, 244, 261, 284,
397

Roquet, Gérard 116, 117, 161, 165

Sauneron, Serge 7, 225, 230
Schaten, Sofia xvi, xxi, 99, 100, 161, 163,
164, 165, 203

Tibiletti Bruno, Maria Grazia 279–283,
288, 300, 301, 341, 360
Török, László 75, 76, 77, 78, 385

Van der Vliet, Jacques xv, xix, xx, 124,
130, 139, 161, 203, 305, 357, 365

Worp, Klaas A. xx, xxi, 78, 219, 305

Young, Thomas 282, 283, 284, 287, 296